Conference on Neutrino and Nuclear Physics (CNNP2017)

Journal of Physics: Conference Series
Volume 1056

Catania, Italy
17 – 21 October 2017

ISBN: 978-1-5108-6793-2
ISSN: 1742-6588

Printed from e-media with permission by:

Curran Associates, Inc.
57 Morehouse Lane
Red Hook, NY 12571

Some format issues inherent in the e-media version may also appear in this print version.

Printed by Curran Associates, Inc. (2018)

For permission requests, please contact the Institute of Physics
at the address below.

Institute of Physics
Dirac House, Temple Back
Bristol BS1 6BE UK

Phone: 44 1 17 929 7481
Fax: 44 1 17 920 0979

techtracking@iop.org

Additional copies of this publication are available from:

Curran Associates, Inc.
57 Morehouse Lane
Red Hook, NY 12571 USA
Phone: 845-758-0400
Fax: 845-758-2633
Email: curran@proceedings.com
Web: www.proceedings.com

Table of contents

Volume 1056

Conference on Neutrino and Nuclear Physics (CNNP2017)

17–21 October 2017, Catania, Italy

Accepted papers received: 2 July 2018
Published online: 19 July 2018

Preface

Papers

IOP Conf. Series: Journal of Physics: Conf. Series **1056** (2018) 011001 doi:10.1088/1742-6596/1056/1/011001

International Advisor Committee

Agodi Clementina (INFN-LNS, Italy)
Auerbach Naftali (Tel Aviv University, Israel)
Azaiez Faical (iThemba LABS, South Africa)
Balantekin Akif Baha (University of Wisconsin-Madison, USA)
Beacom John F. (Ohio State University, USA)
Brofferio Chiara (University of Milano Bicocca and INFN, Italy)
Broglia Ricardo A. (Niels Bohr Institute, Denmark)
Brown Alex B. (MSU, USA)
Cappuzzello Francesco (University of Catania and INFN-LNS, Italy)
De Geronimo Gianluigi (Stony Brook University, New York, USA)
Ejiri Hiro (RCNP, Japan)
Faessler Amand (University of Tuebingen, Germany)
Frekers Dieter (University of Muenster, Germany)
Galindo-Uribarri A. (Oak Ridge National Laboratory, USA)
Harakeh Muhsin N. (KVI-CART, Netherlands)
Hix William R. (Oak Ridge National Laboratory, USA)
Horowitz Charles J. (Indiana University, USA)
Iachello Francesco (Yale University, USA)
Kajino Toshitaka (University of Tokyo, Japan)
Lenske Horst (University of Giessen, Germany)
Lindner Manfred (Max Planck Institute for Nuclear Physics, Heidelberg, Germany)
Lubian Jesus R. (University Federal Fluminense, Brazil)
Masiero Antonio (University of Padova and INFN, Italy)
Mittig Wolfgang (MSU, USA)
Otsuka Takaharu (University of Tokyo, Japan)
Pallavicini Marco (University of Genova and INFN, Italy)
Sato Akira (Osaka University, Japan)
Scholberg Kate (Duke University, USA)
Suhonen Jouni (University of Jyvaskyla, Finland)
Terranova Francesco (University of Milano Bicocca and INFN, Italy)
Tretyak Vladimir (Kiev Institute of Nuclear Research, Ukraine)
Uesaka Tomohiro (RIKEN, Japan)
Vissani Francesco (GSSI and INFN-LNGS, Italy)
Zuber Kai (TU Dresden, Germany)

Organizing Committee

Agodi Clementina (INFN-LNS, Italy) *chair*
Cappuzzello Francesco (University of Catania and INFN-LNS, Italy) *chair*
Carbone Diana (INFN-LNS, Italy) scientific secretary
Calvo Daniela (INFN-Torino, Italy)
Cavallaro Manuela (INFN-LNS, Italy)
Colonna Maria (INFN-LNS, Italy)
Lo Presti Domenico (University of Catania and INFN-Catania, Italy)
Pandola Luciano (INFN-LNS, Italy)
Santopinto Elena (INFN-Genova, Italy)
Sapienza Piera (INFN-LNS, Italy)

IOP Conf. Series: Journal of Physics: Conf. Series **1056** (2018) 011001 doi:10.1088/1742-6596/1056/1/011001

The present volume contains the lectures and poster presentations given at the Conference on Neutrino and Nuclear Physics (CNNP2017), that was hosted, from 15-21 October 2017, jointly by the Istituto Nazionale di Fisica Nucleare - Laboratori Nazionali del Sud (INFN – LNS), and the Department of Physics and Astronomy of Catania University (DFA – Uni Ct), in the venue of the baroque Benedictine Monastery in Catania (Italy).
The main objective of Conference on Neutrino and Nuclear Physics is to promote a collaborative framework of researchers from the fields of nuclear, neutrino, astro and dark-matter physics to discuss experiments and theories in which nuclear physics aspects are particularly relevant. The overlap between different communities is an unavoidable, despite demanding, feature in order to face several challenges of modern fundamental physics. The sharing of different experimental and theoretical techniques, the exchange of technical experiences and know-how is perhaps the best resource to build a unified view of the afore mentioned scientific problems.
The list of the topics developed during its first edition (CNNP2017) is below:

- Nuclear double beta decays
- Nuclear structure in connection with neutrino physics
- Nuclear reactions as a probe for weak decays
- Neutrino-nucleus interaction at low and high energy
- Supernova models and detection of supernova neutrinos
- Solar models and detection of solar neutrinos
- Direct and indirect dark-matter searches
- Rare beta decays of nuclei for neutrino-mass measurements
- Neutrino oscillations and matter effects
- Anomalies in reactor neutrinos
- New related detection technologies

During the Conference it emerged that nuclear structure plays an ever growing role in the mentioned fields of fundamental physics, but still a lot needs to be done. In this context, it is of utmost importance to allow people working at the intersections of these fields to meet regularly to exchange ideas and results. For this reason, the International Advisory Committee has decided to accedt the candidacy of South Africa iThemba LABS to host the second edition to be organized in Cape Town in february 2020.

Additional information about the CNNP2017, including program details and presentation materials, are available on : http://agenda.infn.it/event/CNNP2017.

Clementina Agodi
Istituto Nazionale di Fisica Nucleare
Laboratori Nazionali del Sud – Catania, Italy

Francesco Cappuzzello
Department of Physics and Astronomy of Catania University, Italy

IOP Conf. Series: Journal of Physics: Conf. Series **1056** (2018) 011001 doi:10.1088/1742-6596/1056/1/011001

IOP Conf. Series: Journal of Physics: Conf. Series **1056** (2018) 011002 doi:10.1088/1742-6596/1056/1/011002

Peer review statement

All papers published in this volume of *Journal of Physics: Conference Series* have been peer reviewed through processes administered by the proceedings Editors. Reviews were conducted by expert referees to the professional and scientific standards expected of a proceedings journal published by IOP Publishing.

Neutrino Interactions in the few-GeV region and the MiniBooNE anomaly

L Alvarez-Ruso[1], J Nieves[1], and E Saúl Sala[1] and E Wang[2]

[1]Departamento de Física Teórica and Instituto de Física Corpuscular
Centro Mixto Universidad de Valencia-CSIC, 46071 Valencia, Spain
[2]Department of Physics, Zhengzhou University, Zhengzhou, Henan 450001, China

E-mail: Luis.Alvarez@ific.uv.es

Abstract. The MiniBooNE experiment reported results from the analysis of ν_e and $\overline{\nu}_e$ appearance searches, which showed an excess of signal-like events at low reconstructed neutrino energies with respect to the expected background. The origin of this anomaly could reside in poorly understood neutrino interactions in the detector, leading to unaccounted backgrounds. We discuss photon emission in neutral current interactions with nucleons and nuclei, finding it insufficient to explain the observed event excess. A proposed alternative scenario in which photons arise from the radiative decay of heavy ($\sim$ 50 MeV) sterile neutrinos produced by ν_μ electromagnetic and weak interactions in the detector is also considered. We have calculated the expected signal at MicroBooNE. The distinctive shape and total number of photon events from this mechanism makes its experimental investigation appealing.

1. The MiniBooNE anomaly

The paradigm of three mixing neutrino flavors emerges from oscillation experiments with solar, atmospheric, reactor and accelerator neutrinos in which the square-mass differences and mixing angles have been determined with ever growing precision. Nevertheless, a number of anomalies that challenge this picture has been observed. One of them, reported by MiniBooNE, has found an excess of electron-like events over the predicted background in both ν and $\bar{\nu}$ modes [1, 2]. The excess is concentrated at $200 < E_\nu^{\rm QE} < 475$ MeV, where $E_\nu^{\rm QE}$ is the neutrino energy reconstructed assuming a charged-current quasielastic (QE) nature of the events.

Existing analyses struggle to accommodate this result together with world oscillation data, even in presence of one or more families of sterile neutrinos [3, 4]. The effect of multinucleon interactions in E_ν reconstruction is insufficient to remove the tension in global analyses [5], pointing at an explanation that does not invoke oscillations.

2. Photon-emission backgrounds in the Standard Model

One of the possible origins of the MiniBooNE anomaly resides in poorly understood backgrounds. At low $E_\nu^{\rm QE}$ the background is dominated by photon emission because Cherenkov detectors like MiniBooNE cannot distinguish electrons from single photons. The largest source of single photons is NCπ^0 production where one of the photons from the $\pi^0 \to \gamma\gamma$ decay is not identified. This background has been constrained by the MiniBooNE's NCπ^0 measurement [6]. The second most important process is photon emission induced by NC interactions (NCγ), which can take place on single nucleons

$$\nu(\bar{\nu})\, N \to \nu(\bar{\nu})\, \gamma\, N\,; \tag{1}$$

and on nuclear targets

$$\nu(\bar{\nu})\, A \;\to\; \nu(\bar{\nu})\, \gamma\, X \tag{2}$$
$$\nu(\bar{\nu})\, A \;\to\; \nu(\bar{\nu})\, \gamma\, A \tag{3}$$

via incoherent [Eq. (2)] or coherent [Eq. (3)] scattering. This background could not be directly measured in the experiment. Instead, it was estimated using the NCπ^0 measurement, assuming that NCγ events come from the radiative decay of weakly produced resonances, mainly $\Delta \to N\gamma$ [1, 2]. This procedure neither takes into account the existence of non-resonant terms (first and second diagrams in figure 1 with virtual nucleons in the intermediate state), nor the coherent channel. If the NCγ emission estimate were not sufficiently accurate, this would be relevant to track the origin of the observed excess.

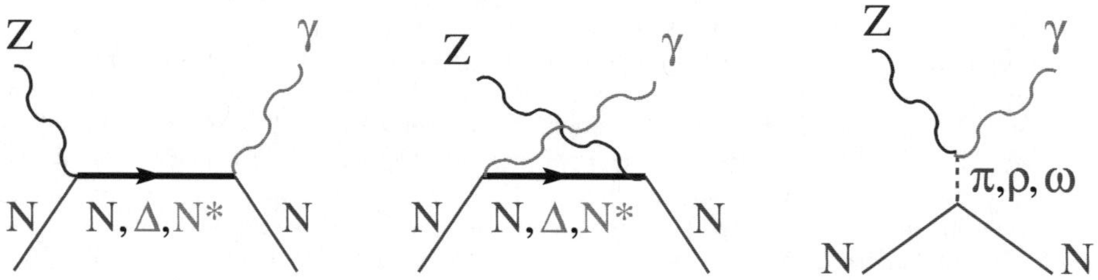

Figure 1. Feynman diagrams for NC photon emission considered in the literature. The first two diagrams stand for direct and crossed baryon pole terms with nucleons and baryon resonances $\Delta(1232)$, $N^*(1440)$, $N^*(1520)$, $N^*(1535)$ in the intermediate state. The third diagram represents t-channel meson (π, ρ, ω) exchange contributions.

Theoretical models for the reaction of Eq. (1) in the few-GeV region have been developed in Refs. [7, 8, 9]. These calculations incorporate s- and u-channel amplitudes with nucleons and $\Delta(1232)$ in the intermediate state (see figure 1). The structure of nucleon pole terms at threshold is fully determined by the symmetries of the Standard Model. The extension towards higher energy transfers required to make predictions for neutrino cross sections is performed by the introduction of phenomenological parametrizations of the weak and electromagnetic form factors. The same strategy has been followed for the resonance terms. The $\Delta(1232)$ excitation followed by its radiative decay is the dominant mechanism. The contribution from the $D_{13}(1520)$ on proton targets is sizable above $E_\nu \sim 1.5$ GeV while $P_{11}(1440)$ and $S_{11}(1535)$ are negligible (see figure 5 of Ref. [9]). The pion-pole mechanism, which originates from the $Z^0\gamma\pi$ vertex, fixed by the axial anomaly of QCD, is nominally of higher order [10] and, indeed, gives a very small contribution to the cross section. Among the terms with heavier meson t-channel exchange, the ω one was proposed as a solution for the MiniBooNE anomaly [11] because of the rather large (although uncertain) couplings and the ω isoscalar nature, which enhances its impact on the coherent reaction of Eq. (3). However, actual calculations found this contribution small compared to $\Delta(1232)$ excitation [7, 12].

The incoherent NCγ reaction on nuclear targets, Eq. (2), has been studied in Refs. [13, 9] using the relativistic local Fermi gas approximation to take into account Fermi motion and Pauli blocking. The broadening of the Δ resonance in the medium has also been incorporated using a spreading potential in Refs. [13, 8], while Ref. [9] uses the parametrization of the imaginary part of the in-medium Δ selfenergy as a function of the local nuclear density derived in Ref. [14]. The neglect of nuclear medium corrections is a poor approximation: by taking into account Fermi motion and Pauli blocking, the cross section already goes down by more than 10%. With the full model the reduction is of the order of 30% (see figure 9 of Ref. [9]). Coherent photon emission [Eq. (2)] is computed in Refs. [15, 9] using the same microscopic input as for the incoherent process. The sum of amplitudes over all nucleons leads to the Fourier transform of different operators proportional to the nuclear density. The resulting integrated cross sections are about a factor 10-15 smaller than the incoherent ones.

The NCγ model outlined above has been applied to calculate the number and distributions of single photon events at MiniBooNE [16], using the available information about the detector mass and its composition (CH_2), the total number of protons on target (POT), [2], the flux prediction [17] and photon detection efficiency [18]. The yields from the incoherent channel are the largest ones. Those from the coherent channel and the reaction on protons, which are comparable, are smaller but significant. The coherent contribution is particularly important for antineutrinos and in the forward direction (see the plots in figure 3 of Ref. [16]).

Our results for the $E_\nu^{\rm QE}$ distributions are shown in figure 2. The error bands correspond to a 68 % confidence level according to the error budget in Table 1 of Ref. [16]. The comparison with the MiniBooNE in situ estimate [2, 18] shows a good agreement: the shapes are similar and the peak positions coincide. The largest discrepancy is observed in the lowest energy bin. In the two bins with the largest number of events, the two calculations are consistent within our errorbars. The overall agreement is also good in comparison to the result of Zhang and Serot [8]. Therefore,

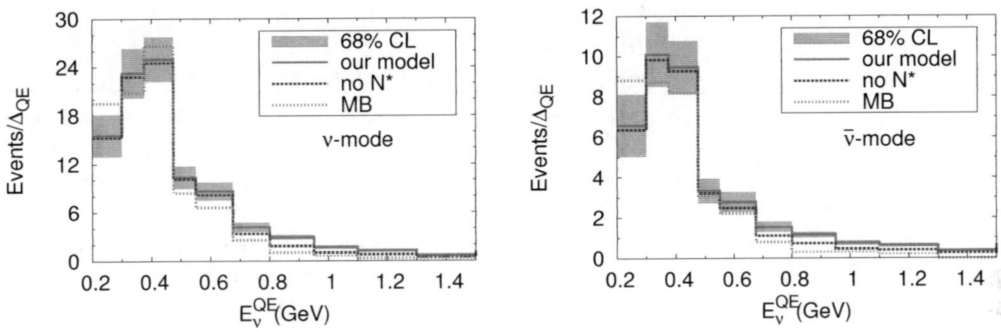

Figure 2. $E_\nu^{\rm QE}$ distributions of total NCγ events for the ν (left) and $\bar{\nu}$ (right) modes. The "MB" histograms display the MiniBooNE estimates [18].

we conclude that neutral current photon emission from single-nucleon currents is insufficient to explain the event excess observed by MiniBooNE in both neutrino and antineutrino modes.

3. Photons from radiative decay of heavy neutrinos

Gninenko proposed that additional photons could originate in the weak production of a heavy ($m_h \approx 50$ MeV) sterile neutrino slightly mixed with muon neutrinos, followed by its radiative decay [19]. In Ref. [20] it was pointed out that the ν_h could also be electromagnetically produced, alleviating tensions in the original proposal with other data such as those from radiative muon capture measured at TRIUMF.

We have revisited the scenario presented in Ref. [20] using present understanding of electromagnetic (EM) and weak interactions on nucleons and nuclei [21, 22]. For a more detailed analysis, we compare to the MiniBooNE excess of events in the originally measured electron-like energy and angle [18] (being the photon ones in our case) rather than in $E_\nu^{\rm QE}$. We have studied

ν_h EM and weak production in the following processes

$$\nu_\mu\,,\bar{\nu}_\mu(k)\,+\,N(p)\;\rightarrow\;\nu_h\,,\bar{\nu}_h(k')\,+\,N(p')\,, \tag{4}$$

$$\nu_\mu\,,\bar{\nu}_\mu(k)\,+\,A(p)\;\rightarrow\;\nu_h,\bar{\nu}_h(k')\,+\,A(p')\,, \tag{5}$$

$$\nu_\mu\,,\bar{\nu}_\mu(k)\,+\,A(p)\;\rightarrow\;\nu_h,\bar{\nu}_h(k')\,+\,X(p')\,. \tag{6}$$

Reaction (5) is coherent while (6) is incoherent; excited states X include any number of knocked out nucleons but no meson production.

In the EM case, following Ref. [20], we have adopted the effective interaction

$$\mathcal{L}_{eff} = \frac{1}{2}\mu^i_{tr}\left[\bar{\nu}_h\sigma_{\mu\nu}\left(1-\gamma_5\right)\nu_i + \bar{\nu}_i\sigma_{\mu\nu}\left(1+\gamma_5\right)\nu_h\right]\partial^\mu A^\nu\,, \tag{7}$$

in terms of a real transition coupling μ^i_{tr}; ν_h is assumed to be a Dirac fermion of mass m_h. In the weak case, the neutrino vertex has the same structure as in the Standard Model and is proportional to the mixing $U_{\mu h}$. The EM (NC) hadronic tensors are the same probed in the corresponding electron (neutrino) scattering processes. For the nucleon, it is given in terms of electromagnetic and axial form factors, for which we have adopted dipole parametrizations. For coherent scattering (5), the tensor is proportional to the square of the nuclear form factor, obtained as the Fourier transform of the empirical charge-density distribution. For the incoherent reaction, as in the previous section, we take into account particle-hole excitations in infinite nuclear matter, adapted to finite nuclei using the local density approximation.

With the parameters of Ref. [20], $m_h = 50$ MeV, $\mu^\mu_{tr} = 2.4 \times 10^{-9}\mu_B$. and $|U_{\mu h}|^2 = 0.003$, the EM cross section on nuclei is dominated by the coherent mechanism, while the incoherent one is suppressed by Pauli blocking at low four-momentum transfers, where the amplitude is enhanced by the photon propagator. On the contrary, the incoherent reaction is the largest contribution to the weak NC part (figure 1 of Ref. [21]). Interference terms between the EM and NC amplitudes are allowed but negligible.

We have investigated the ν_h propagation and radiative decay inside the MiniBooNE detector, obtaining the photon energy and angular distributions. We have taken advantage of the fact that, as pointed out in Ref. [20], the beam energies are large compared to m_h and only an insignificant amount of the electromagnetically (weakly) produced heavy neutrinos have the spin against (aligned with) its momentum. Radiative decay photons are emitted predominantly in the direction opposite to the ν_h spin. The ν_h lifetime in its rest frame $\tau = 5\times10^{-9}$ seconds [20]. To compare to the measured excess of events, the detection efficiency [18] has to be taken into account. The number of low energy events is underestimated in ν-mode, while the agreement is good in $\bar{\nu}$-mode. The predominantly EM coherent contribution is strongly forward peaked. This leads to a very narrow angular distribution not observed in the experiment as can be seen in figure 2 of Ref. [21]. This result is in line with the findings of Ref. [23].

The agreement can be improved by fitting the parameters in the allowed range established in Ref. [24]. With these values, $m_h = 68.6$ MeV, $\mu^\mu_{tr} = 6.4 \times 10^{-10}\mu_B$., $|U_{\mu h}|^2 = 0.01$ and $\tau = 2.5\times10^{-9}$ seconds, the resulting event energy and angular distributions are given in figure 3. The MiniBooNE excess of events is now better described, particularly the angular distributions. This better agreement is obtained at the price of reducing the EM strength, while increasing the NC one by setting $|U_{\mu h}|$ to its maximal allowed value; this upper limit in $|U_{\mu h}|$ prevents from obtaining a more satisfactory description of the data.

4. Single-photon events at MicroBooNE

The radiative decay hypothesis can be further tested at the Short Baseline Neutrino (SBN) detectors at Fermilab: SBND, MicroBooNE (currently taking data) and ICARUS, with liquid

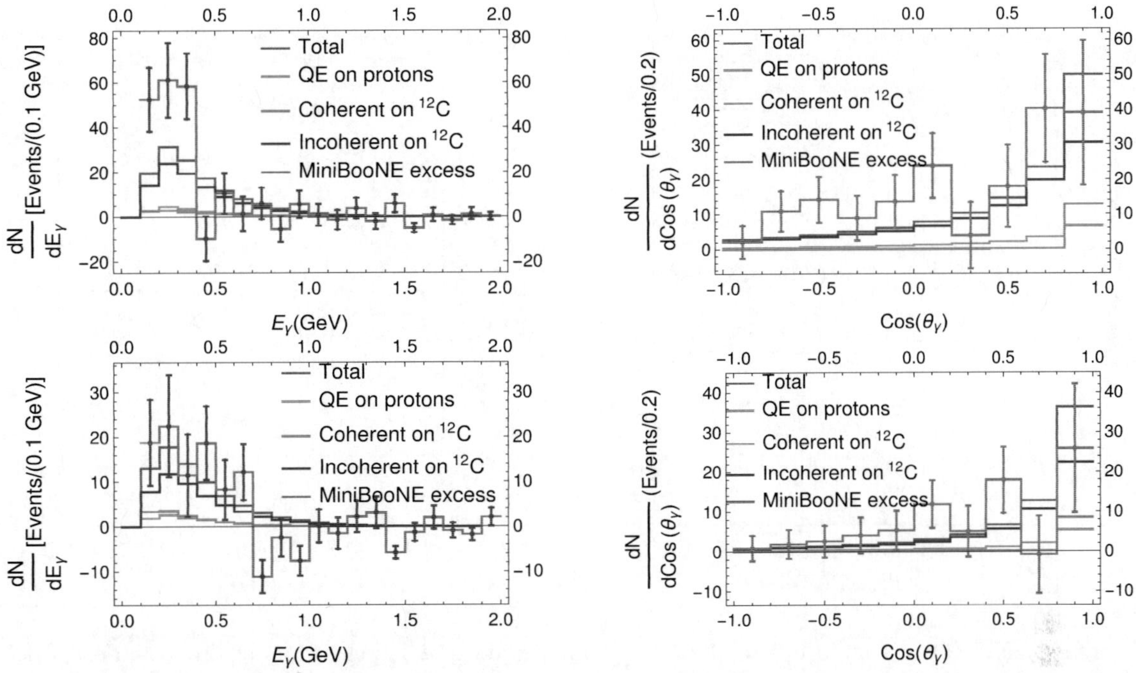

Figure 3. Photon events from radiative decay of ν_h, $\bar{\nu}_h$ at the MiniBooNE detector in ν-mode (top) and $\bar{\nu}$-mode (bottom) compared to the MiniBooNE excess [18].

Argon targets. Our predictions for the photon distributions at MicroBooNE, with the flux in ν-mode [25] and assuming $N_{POT} = 6.6 \times 10^{20}$, are displayed in figure 4. The shape and number of events appears distinctive from those from conventional photon emission mechanisms shown in figure 5. In particular, events from ν_h decay are clearly more forward-peaked.

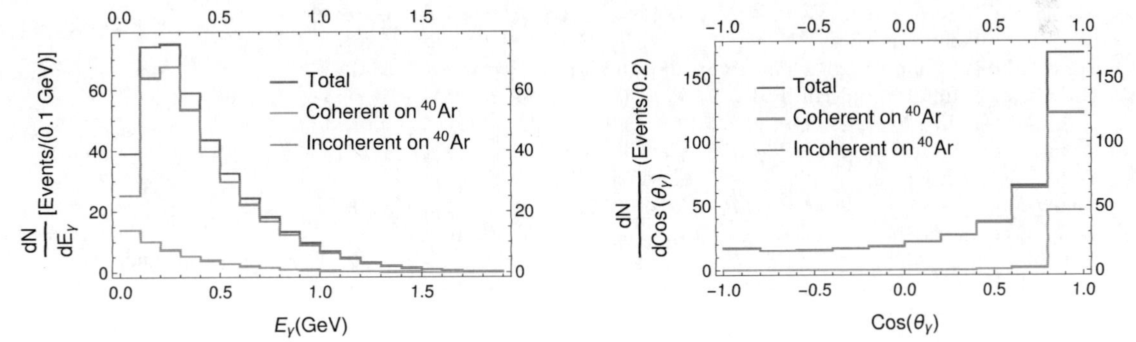

Figure 4. Photon events from radiative decay of ν_h at MicroBooNE in ν-mode.

Acknowledgments

Work supported by the Spanish Ministerio de Economía y Competitividad and European Regional Development Fund under Contracts FIS2014-51948-C2-1-P, FIS2014-51948-C2-2-P, FIS2017-84038-C2-1-P, FIS2017-84038-C2-2-P and SEV-2014-0398, and by Generalitat Valenciana under contract PROMETEOII/2014/0068.

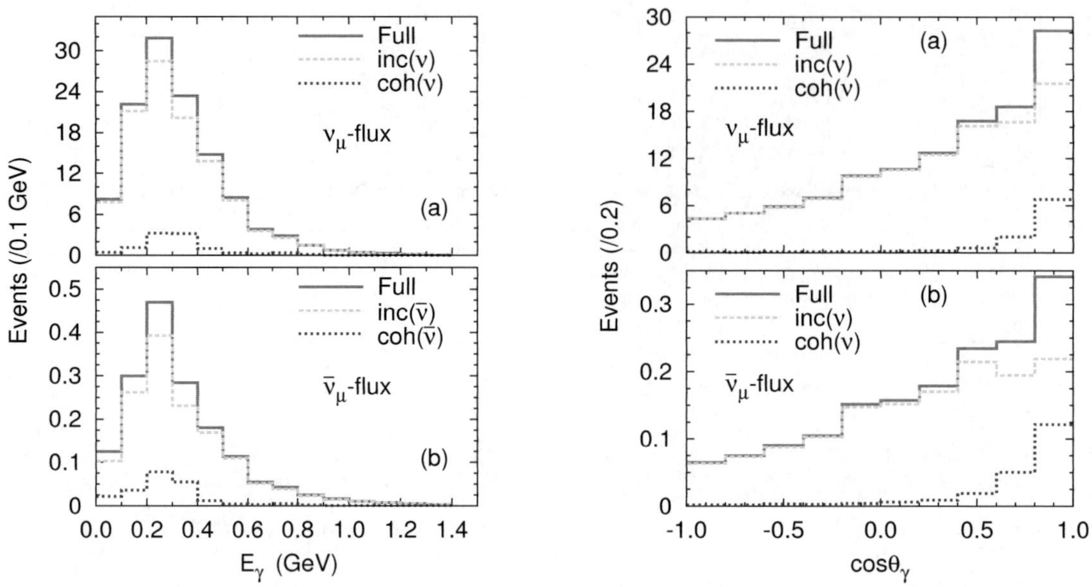

Figure 5. NCγ events at MicroBooNE in ν-mode predicted with the model of Ref. [9] described in Sec. 2. The (negligible) contribution from $\bar{\nu}$ in the flux is also shown in the bottom panels.

References

[1] Aguilar-Arevalo A *et al.* (MiniBooNE Collaboration) 2009 *Phys.Rev.Lett.* **102** 101802 (*Preprint* 0812.2243)
[2] Aguilar-Arevalo A *et al.* (MiniBooNE Collaboration) 2013 *Phys.Rev.Lett.* **110** 161801 (*Preprint* 1207.4809)
[3] Giunti C, Laveder M, Li Y F and Long H W 2013 *Phys. Rev.* **D88** 073008
[4] Gariazzo S, Giunti C, Laveder M and Li Y F 2017 *JHEP* **06** 135 (*Preprint* 1703.00860)
[5] Ericson M, Garzelli M V, Giunti C and Martini M 2016 *Phys. Rev.* **D93** 073008
[6] Aguilar-Arevalo A A *et al.* (MiniBooNE Collaboration) 2010 *Phys.Rev.* **D81** 013005 (*Preprint* 0911.2063)
[7] Hill R J 2010 *Phys.Rev.* **D81** 013008 (*Preprint* 0905.0291)
[8] Zhang X and Serot B D 2013 *Phys.Lett.* **B719** 409–414 (*Preprint* 1210.3610)
[9] Wang E, Alvarez-Ruso L and Nieves J 2014 *Phys.Rev.* **C89** 015503 (*Preprint* 1311.2151)
[10] Serot B D and Zhang X 2012 *Phys.Rev.* **C86** 015501 (*Preprint* 1206.3812)
[11] Harvey J A, Hill C T and Hill R J 2007 *Phys.Rev.Lett.* **99** 261601 (*Preprint* 0708.1281)
[12] Rosner J L 2015 *Phys. Rev.* **D91** 093001 (*Preprint* 1502.01704)
[13] Zhang X and Serot B D 2012 *Phys.Rev.* **C86** 035502 (*Preprint* 1206.6324)
[14] Oset E and Salcedo L 1987 *Nucl.Phys.* **A468** 631–652
[15] Zhang X and Serot B D 2012 *Phys.Rev.* **C86** 035504 (*Preprint* 1208.1553)
[16] Wang E, Alvarez-Ruso L and Nieves J 2015 *Phys.Lett.* **B740** 16–22
[17] Aguilar-Arevalo A *et al.* (MiniBooNE Collaboration) 2009 *Phys.Rev.* **D79** 072002 (*Preprint* 0806.1449)
[18] Miniboone collaboration: https://www.boone.fnal.gov/for_physicists/data_release/
[19] Gninenko S N 2009 *Phys. Rev. Lett.* **103** 241802
[20] Masip M, Masjuan P and Meloni D 2013 *JHEP* **1301** 106
[21] Alvarez-Ruso L and Saul-Sala E 2017 *Proceedings, Prospects in Neutrino Physics (NuPhys2016): London, UK, December 12-14, 2016* (*Preprint* 1705.00353)
[22] Alvarez-Ruso L and Saul-Sala E *in preparation*
[23] Radionov A 2013 *Phys. Rev.* **D88** 015016
[24] Gninenko S N 2011 *Phys. Rev.* **D83** 015015 (*Preprint* 1009.5536)
[25] Pavlovic Z and Palamara O *private communication*

IOP Conf. Series: Journal of Physics: Conf. Series **1056** (2018) 012002 doi:10.1088/1742-6596/1056/1/012002

Bound-state double-beta decay

A Babič[1,2,3], D Štefánik[4], M I Krivoruchenko[3,5] and F Šimkovic[2,3,4]

[1] Faculty of Nuclear Sciences and Physical Engineering, Czech Technical University in Prague, 115 19 Prague, Czech Republic
[2] Institute of Experimental and Applied Physics, Czech Technical University in Prague, 128 00 Prague, Czech Republic
[3] Bogoliubov Laboratory of Theoretical Physics, Joint Institute for Nuclear Research, 141980 Dubna, Russia
[4] Faculty of Mathematics, Physics and Informatics, Comenius University in Bratislava, 842 48 Bratislava, Slovakia
[5] Institute for Theoretical and Experimental Physics, 117218 Moscow, Russia

E-mail: `andrej.babic@utef.cvut.cz`

Abstract. We study a new mode of the neutrinoless and two-neutrino double-beta decays in which a single electron is emitted from the atom. The other electron is assumed to occupy one of the vacant $s_{1/2}$ or $p_{1/2}$ subshells of the daughter ion. Such process could manifest itself through an additional background signal in the single-electron spectra, which will be accessible in the next-generation experiment SuperNEMO. We calculate the phase-space factors in terms of relativistic electron wave functions obtained as the solutions to the Dirac equation and evaluated at the nuclear radius, while taking into account the shielding effect of nuclear charge via the multiconfiguration Dirac–Hartree–Fock package GRASP2K. Half-lives are estimated for the most relevant double-beta-decay isotopes and experimental significance is discussed.

1. Introduction

Observation of the neutrinoless double-beta decay ($0\nu\beta\beta$) would mark a revolution in the theory of neutrino masses and a huge step towards new physics beyond the Standard Model. It would bring us direct evidence that the massive neutrinos ν_i ($i = 1, 2, 3$) are in fact Majorana fermions, implying the identity of the flavor neutrinos ν_α ($\alpha = e, \mu, \tau$) and their respective antineutrinos $\overline{\nu}_\alpha$, and at the same time demonstrate that the total lepton number L is not strictly conserved [1]. Measurement of the $0\nu\beta\beta$ half-life (for which only lower bounds currently exist) would provide us with key to the absolute scale of neutrino masses m_i, as well as with the possibility to infer the mechanism of leptonic CP violation required in order to explain the observed baryon asymmetry of the Universe [2]. Given the potential to answer so many fundamental questions in particle and astroparticle physics, it is understandable that in the recent decades the search for neutrinoless double-beta decay has drawn great attention of both theorists and experimentalists.

The double-beta decay is a transmutation of an even-even parent nucleus ${}^{A}_{Z}\mathrm{X}$ into a daughter nucleus ${}_{Z+2}^{\ \ A}\mathrm{Y}$, two electrons e^- (and a pair of antineutrinos $\overline{\nu}_e$), denoted $0\nu\beta^-\beta^-$ ($2\nu\beta^-\beta^-$):

$$ {}^{A}_{Z}\mathrm{X} \longrightarrow {}_{Z+2}^{\ \ A}\mathrm{Y} + e \quad + e \quad + (\nu_e + \nu_e). \tag{1}$$

The neutrinoless mode $0\nu\beta^-\beta^-$ increases L by 2 units and could be discovered by revealing a monoenergetic peak at the spectrum endpoint in calorimetric measurements of the sum of electron energies. The two-neutrino mode $2\nu\beta^-\beta^-$ has so far been observed in 11 out of 35 isotopes for which the ordinary β^- decay into the odd-odd intermediate nucleus is either energetically forbidden or substantially suppressed by spin selection rules. In what follows, we restrict ourselves to the dominant ground-state $0^+ \longrightarrow 0^+$ nuclear transitions.

By analogy with the bound-state β^- decay first observed about 25 years ago [3], we study the bound-state double-beta-decay modes, denoted $0\nu\mathrm{EP}\beta^-$ ($2\nu\mathrm{EP}\beta^-$):

$$ {}^{\mathrm{A}}_{\mathrm{Z}}\mathrm{X} \longrightarrow {}_{\mathrm{Z+2}}^{\mathrm{A}}\mathrm{Y} + e_{\mathrm{b}}^- + e^- + (\overline{\nu}_e + \overline{\nu}_e). \tag{2} $$

In these new modes, the emission of a single free electron e^- is accompanied by an electron production (EP) of a bound electron e_{b}^- in one of the available $\mathrm{s}_{1/2}$ or $\mathrm{p}_{1/2}$ subshells above the valence shell of the daughter 2+ ion. Contribution of bound states with higher angular momenta can be safely disregarded since their wave functions vanish at the origin, and hence exhibit only a minimal overlap with the nucleus; already the $\mathrm{p}_{1/2}$ states yield only a negligible input to the decay rates. Since the free $0\nu\mathrm{EP}\beta^-$ electron carries away the entire released kinetic energy Q, this mode could be in principle recognized as a peak near the endpoint of the single-electron spectrum which will be studied in the next-generation experiment SuperNEMO [4].

2. Calculation of phase-space factors

The double-beta decay is a 2nd-order process governed by the effective beta-decay Hamiltonian of the following V − A structure [5]:

$$ \mathcal{H}_\beta(x) = \frac{G_\beta}{\sqrt{2}}\, \overline{e}(x)\, \gamma^\mu \left(1 - \gamma^5\right) \nu_e(x)\, j_\mu(x) + \mathrm{H.c.} \tag{3} $$

Here, $G_\beta \equiv G_{\mathrm{F}} \cos\theta_{\mathrm{C}}$ includes the Fermi constant G_{F} together with the Cabibbo angle $\theta_{\mathrm{C}} = 13°$ due to quark mixing [6], $e(x)$ and $\nu_e(x)$ are the electron and electron-neutrino fields, respectively, and the hadronic charged current $j_\mu(x) = \overline{p}(x)\, \gamma_\mu \left(g_V - g_A \gamma^5\right) n(x)$ couples the proton $p(x)$ and neutron $n(x)$ fields via the vector $g_V = 1$ and (unquenched) axial-vector $g_A = 1.27$ coupling constants. Due to neutrino mixing, the left-handed components of the flavor- and massive-neutrino fields are related by a unitary transformation: $\nu_{\alpha L}(x) = \sum_i U_{\alpha i}\, \nu_{iL}(x)$, where $U_{\alpha i}$ are the elements of the PMNS matrix. The most plausible seesaw mechanism incorporates the massive neutrinos into the theory via Majorana mass terms, in which case the L-violation can be realized by a light Majorana-neutrino exchange between the two weak-interaction vertices.

The formulae for the inverse $0\nu(2\nu)\beta^-\beta^-$ half-lives [7]:

$$ \begin{aligned} \left(T_{1/2}^{0\nu\beta\beta}\right)^{-1} &= g_A^4\, G^{0\nu\beta\beta}(Z,\, Q) \left|M^{0\nu\beta\beta}\right|^2 \left|\frac{m_{\beta\beta}}{m_e}\right|^2, \\ \left(T_{1/2}^{2\nu\beta\beta}\right)^{-1} &= g_A^4\, G^{2\nu\beta\beta}(Z,\, Q) \left|m_e\, M^{2\nu\beta\beta}\right|^2 \end{aligned} \tag{4} $$

in general factorize in terms of a kinematical phase-space factor $G^{0\nu(2\nu)\beta\beta}(Z,\, Q)$, a nuclear matrix element $M^{0\nu(2\nu)\beta\beta}$, and the effective Majorana neutrino mass $m_{\beta\beta} = \sum_i U_{ei}^2\, m_i$ (normalized to the electron mass m_e). The value of $|m_{\beta\beta}|$ is currently unknown due to undetermined Majorana phases contained in U_{ei}; nevertheless, experimental lower bounds on $T_{1/2}^{0\nu\beta\beta}$ imply an upper bound on $|m_{\beta\beta}|$, provided that the nuclear matrix elements $M^{0\nu\beta\beta}$ have been evaluated within the theory of nuclear structure. The most stringent bound has so far been obtained in the ^{136}Xe experiment KamLAND-Zen [8]: $|m_{\beta\beta}| < 61$–$165\,\mathrm{meV}$ at 90% C.L. In case of the inverted hierarchy of neutrino masses, the effective mass lies within the interval: $|m_{\beta\beta}| = 20$–$50\,\mathrm{meV}$; we have estimated the $0\nu\beta^-\beta^-$ and $0\nu\mathrm{EP}\beta^-$ half-lives assuming the optimistic value: $|m_{\beta\beta}| = 50\,\mathrm{meV}$.

The relativistic electron wave functions $\psi(\vec{r})$ are obtained as the solutions to the Dirac equation with Coulomb potential $V(r) = -\alpha Z_{\text{eff}}/r$, where $\alpha = 1/137$ is the fine-structure constant and Z_{eff} is the effective atomic number of the daughter nucleus experienced by the final-state electrons. The bispinor $\psi(\vec{r})$ with separated radial $r \equiv |\vec{r}|$ and angular $\hat{r} \equiv \vec{r}/|\vec{r}|$ variables in general takes the form [9]:

$$\psi_{\kappa\mu}(\vec{r}) = \begin{pmatrix} f_\kappa(r)\,\Omega_{\kappa\mu}(\hat{r}) \\ i g_\kappa(r)\,\Omega_{-\kappa\mu}(\hat{r}) \end{pmatrix}, \tag{5}$$

where $\kappa = (l-j)(2j+1) = \pm 1, \pm 2, \ldots$ collectively labels all possible combinations of the orbital $l = 0, 1, \ldots$ and spin $s = \pm 1/2$ angular momenta, $\mu = -j, \ldots, +j$ denotes the projection of the total angular momentum $j = |l+s|$ onto the z-axis, $f_\kappa(r)$ and $g_\kappa(r)$ are the relativistic radial wave functions which further depend on the electron energy, and the two-component angular functions $\Omega_{\kappa\mu}(\hat{r})$ are known as the spinor spherical harmonics.

In the continuous spectrum, it is sufficient to consider the dominant $s_{1/2}$ term from the partial-wave expansion, in which case the free-electron wave functions are evaluated at the nuclear radius $R = 1.2\,\text{fm}\,A^{1/3}$ and reduced to the Fermi function [10]:

$$F(Z, E) = f_{-1}^2(E, R) + g_{+1}^2(E, R) \approx 4\left[\frac{|\Gamma(\gamma + i\nu)|}{\Gamma(2\gamma+1)}\right]^2 (2pR)^{2\gamma-2}\, e^{\pi\nu}, \tag{6}$$

where $E = \sqrt{\vec{p}^2 + m_e^2}$ is the energy of a free electron with momentum $\vec{p}$, and we have defined the parameters $\gamma = \sqrt{\kappa^2 - (\alpha Z)^2}$ and $\nu = \alpha Z E/p$ $(p \equiv |\vec{p}|)$. For the free electron, the shielding effect was shown to be rather insignificant, and hence we retain the full charge of the daughter nucleus [11]: $Z_{\text{eff}} = Z + 2$.

In the discrete spectrum, the bound-electron wave functions can be approximated by the bound-state Fermi function:

$$B_n(Z, A) = f_{n,-1}^2(R) + g_{n,+1}^2(R), \tag{7}$$

where $n = 1, 2, \ldots$ is the principal quantum number and the two terms emerge from the inclusion of $s_{1/2}$ and $p_{1/2}$ states, respectively. For the bound electron, the shielding effect of nuclear charge has been taken into account by means of relativistic many-electron wave functions via the multiconfiguration Dirac–Hartree–Fock package GRASP2K [12].

Under the standard approximations, the bound-state double-beta-decay modes $0\nu(2\nu)\text{EP}\beta^-$ do not alter the nuclear matrix elements: $M^{0\nu(2\nu)\text{EP}\beta} \approx M^{0\nu(2\nu)\beta\beta}$ and only affect the kinematical phase-space factors:

$$G^{0\nu\text{EP}\beta}(Z, Q) = \frac{G_\beta^4\, m_e^2}{32\pi^4 R^2 \ln 2} \sum_{n=n_{\min}}^{\infty} B_n(Z, A)\, F(Z+2, E)\, E\, p,$$

$$G^{2\nu\text{EP}\beta}(Z, Q) = \frac{G_\beta^4}{8\pi^6 m_e^2 \ln 2} \sum_{n=n_{\min}}^{\infty} B_n(Z, A) \int_{m_e}^{m_e+Q} \mathrm{d}E\, F(Z+2, E)\, E\, p \int_0^{m_e+Q-E} \mathrm{d}\omega_1\, \omega_1^2\, \omega_2^2. \tag{8}$$

In both scenarios, the total released kinetic energy equals: $Q = M_i - M_f - 2m_e$, where $M_{i,f}$ denote the masses of the initial- and final-state nuclei and we have neglected the small neutrino masses m_i. The summation runs over all available electron shells n above the valence shell of the resulting 2+ ion, while E and $\omega_{1,2}$ represent the total energies of the free electron and the two antineutrinos, respectively. Neglecting the electron binding energies and nuclear recoil, from energy conservation it follows that in $G^{0\nu\text{EP}\beta}$: $E = m_e + Q$ (i.e., the single free electron carries away the entire kinetic energy released in the decay), while in $G^{2\nu\text{EP}\beta}$: $\omega_2 = m_e + Q - E - \omega_1$ and the infinite sum of integrals separates into a product of the Fermi sum $\sum_{n=n_{\min}}^{\infty} B_n(Z, A)$ and just one double integral independent of n.

Table 1. $\beta^-\beta^-$-decay isotopes ${}^A_Z$X with corresponding Q values [13], decay-rate ratios $\Gamma^{0\nu\mathrm{EP}\beta}/\Gamma^{0\nu\beta\beta} \approx G^{0\nu\mathrm{EP}\beta}/G^{0\nu\beta\beta}$ and $\Gamma^{2\nu\mathrm{EP}\beta}/\Gamma^{2\nu\beta\beta} \approx G^{2\nu\mathrm{EP}\beta}/G^{2\nu\beta\beta}$, half-lives $T^{0\nu\mathrm{EP}\beta}_{1/2}$ and $T^{0\nu\beta\beta}_{1/2}$ estimated assuming $g_A = 1.27$, $|m_{\beta\beta}| = 50\,\mathrm{meV}$ and nuclear matrix elements $M^{0\nu\beta\beta}$ calculated in [14, 15], and half-lives $T^{2\nu\mathrm{EP}\beta}_{1/2}$ extracted from measured values $T^{2\nu\beta\beta}_{1/2}$ [16].

${}^A_Z$X	Q [MeV]	$\Gamma^{0\nu\mathrm{EP}\beta}/\Gamma^{0\nu\beta\beta}$	$\Gamma^{2\nu\mathrm{EP}\beta}/\Gamma^{2\nu\beta\beta}$	$T^{0\nu\mathrm{EP}\beta}_{1/2}$ [y]	$T^{0\nu\beta\beta}_{1/2}$ [y]	$T^{2\nu\mathrm{EP}\beta}_{1/2}$ [y]	$T^{2\nu\beta\beta}_{1/2}$ [y]
${}^{48}_{20}$Ca	4.268	3.51×10^{-7}	3.72×10^{-6}	1.23×10^{34}	4.32×10^{27}	1.18×10^{25}	4.40×10^{19}
${}^{76}_{32}$Ge	2.039	3.63×10^{-6}	3.07×10^{-5}	1.36×10^{32}	4.95×10^{26}	5.38×10^{25}	1.65×10^{21}
${}^{82}_{34}$Se	2.998	1.97×10^{-6}	1.83×10^{-5}	7.05×10^{31}	1.38×10^{26}	5.04×10^{24}	9.20×10^{19}
${}^{96}_{40}$Zr	3.356	7.65×10^{-7}	7.23×10^{-6}	2.46×10^{32}	1.88×10^{26}	3.18×10^{24}	2.30×10^{19}
${}^{100}_{42}$Mo	3.034	3.59×10^{-6}	3.29×10^{-5}	1.73×10^{31}	6.21×10^{25}	2.16×10^{23}	7.10×10^{18}
${}^{110}_{46}$Pd	2.017	9.69×10^{-6}	7.91×10^{-5}	1.83×10^{31}	1.78×10^{26}		
${}^{116}_{48}$Cd	2.813	1.45×10^{-6}	1.28×10^{-5}	7.13×10^{31}	1.03×10^{26}	2.24×10^{24}	2.87×10^{19}
${}^{124}_{50}$Sn	2.291	2.77×10^{-6}	2.31×10^{-5}	1.51×10^{32}	4.18×10^{26}		
${}^{128}_{52}$Te	0.867	1.56×10^{-5}	1.09×10^{-4}	1.36×10^{32}	2.13×10^{27}	1.84×10^{28}	2.00×10^{24}
${}^{130}_{52}$Te	2.528	2.68×10^{-6}	2.29×10^{-5}	4.33×10^{31}	1.16×10^{26}	3.02×10^{25}	6.90×10^{20}
${}^{134}_{54}$Xe	0.824	1.67×10^{-5}	1.15×10^{-4}	1.89×10^{32}	3.16×10^{27}		
${}^{136}_{54}$Xe	2.458	2.84×10^{-6}	2.40×10^{-5}	1.24×10^{32}	3.52×10^{26}	9.12×10^{25}	2.19×10^{21}
${}^{150}_{60}$Nd	3.371	6.16×10^{-7}	5.61×10^{-6}	6.51×10^{31}	4.01×10^{25}	1.46×10^{24}	8.20×10^{18}
${}^{238}_{92}$U	1.145	2.95×10^{-6}	2.10×10^{-5}			9.52×10^{25}	2.00×10^{21}

3. Results and discussion

In table 1, we present the results for the ground-state $0^+ \longrightarrow 0^+$ nuclear transitions of the most relevant $\beta^-\beta^-$-decay isotopes ${}^A_Z$X listed together with their respective Q values [13]. The ratios $\Gamma^{0\nu\mathrm{EP}\beta}/\Gamma^{0\nu\beta\beta} \approx G^{0\nu\mathrm{EP}\beta}/G^{0\nu\beta\beta}$ and $\Gamma^{2\nu\mathrm{EP}\beta}/\Gamma^{2\nu\beta\beta} \approx G^{2\nu\mathrm{EP}\beta}/G^{2\nu\beta\beta}$ between the corresponding decay rates are independent of the nuclear matrix elements $M^{0\nu\beta\beta}$ and $M^{2\nu\beta\beta}$ as well as the effective Majorana neutrino mass $m_{\beta\beta}$, and hence free of the peculiarities of the nuclear and neutrino physics. The half-lives $T^{0\nu\mathrm{EP}\beta}_{1/2}$ and $T^{0\nu\beta\beta}_{1/2}$ are estimated assuming the unquenched value of the axial-vector weak coupling constant: $g_A = 1.27$, the effective Majorana neutrino mass corresponding to the top of the allowed inverted-hierarchy region: $|m_{\beta\beta}| = 50\,\mathrm{meV}$, and the nuclear matrix elements $M^{0\nu\beta\beta}$ calculated via the spherical pn-QRPA approach including the realistic CD-Bonn nucleon-nucleon potential with short-range correlations and partial isospin-symmetry restoration [14], except for the isotope ^{150}Nd which was treated separately within the deformed pn-QRPA model [15]. On the other hand, the half-lives $T^{2\nu\mathrm{EP}\beta}_{1/2}$ are extracted regardless of any theoretical assumptions from the experimentally measured values $T^{2\nu\beta\beta}_{1/2}$ [16]. We observe that the $0\nu\mathrm{EP}\beta^-$ mode is strongly suppressed and out of reach of the running and near-future experiments. Conversely, its $2\nu\mathrm{EP}\beta^-$ counterpart is relatively more significant by one order of magnitude; indeed, the $2\nu\mathrm{EP}\beta^-$ half-lives are already at the level of the present $0\nu\beta^-\beta^-$-decay sensitivity. Finally, the overall suppression is mainly attributed to the presence of other electrons in the atom: the lowest-lying orbitals are already occupied, while the shielding effect of nuclear charge substantially reduces the bound-electron wave functions on the surface of the nucleus.

In figure 1, we show the calculated single-electron spectra $1/\Gamma\, \mathrm{d}\Gamma/\mathrm{d}\varepsilon$ as functions of the electron kinetic energy $\varepsilon = (E - m_e)/Q$ for the isotope ^{82}Se, which will serve as the primary source isotope in the next-generation tracking-and-calorimetry double-beta-decay experiment SuperNEMO located at LSM, France [4]. The discrete $0\nu\mathrm{EP}\beta^-$ peak in principle consists of

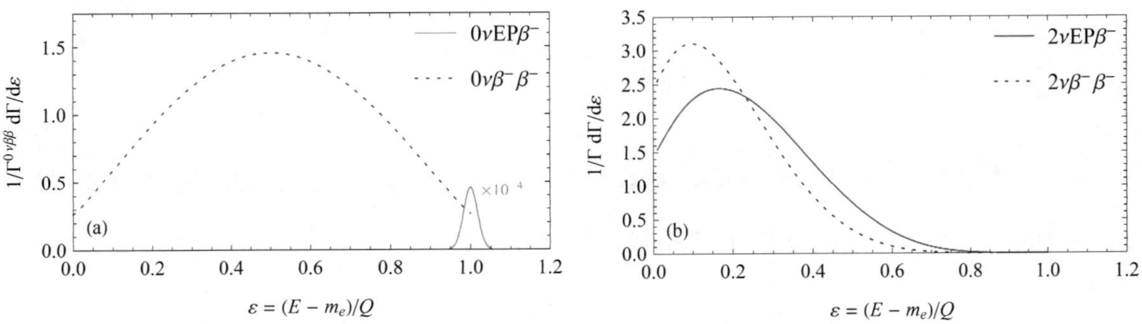

Figure 1. Single-electron spectra $1/\Gamma\, d\Gamma/d\varepsilon$ (normalized to unity) as functions of the dimensionless portion of electron kinetic energy $\varepsilon = (E - m_e)/Q$ for the source isotope ^{82}Se ($Q = 2.998\,\text{MeV}$) which will be studied in SuperNEMO [4], in case of the decay modes: (a) $0\nu\text{EP}\beta^-$ and $0\nu\beta^-\beta^-$, (b) $2\nu\text{EP}\beta^-$ and $2\nu\beta^-\beta^-$. For illustration, the $0\nu\text{EP}\beta^-$ peak is represented by a Gaussian with the desired energy resolution of SuperNEMO calorimeters: $\text{FWHM}/Q = 7\%/\sqrt{Q/\text{MeV}}$ and scaled by a factor of 10^4.

an indefinite number of contributions, each shifted beyond the Q value by the electron binding energy; however, these would be indistinguishable under any achievable energy resolution. In conclusion, the $2\nu\text{EP}\beta^-$ mode exhibits a different shape of the spectrum, which points to more optimistic prospects towards its observation as a background process in SuperNEMO.

Acknowledgments

This work was supported by the VEGA Grant Agency of the Slovak Republic under Contract No. 1/0922/16, the Slovak Research and Development Agency under Contract No. APVV-14-0524, the Ministry of Education, Youth and Sports of the Czech Republic under the ERDF Grant No. CZ.02.1.01/0.0/0.0/16_013/0001733, the Grant of the Plenipotentiary Representative of the Czech Republic in JINR under Contract No. 203 from 31/03/2017, the RFBR Grant under Contract No. 16-02-01104, and Grant of the Heisenberg–Landau Program under Contract No. HLP-2015-18. The authors express their sincere gratitude towards R. Dvornický for enlightening ideas regarding the atomic-structure calculations.

References

[1] Vergados J D, Ejiri H and Šimkovic F 2016 *Int. J. Mod. Phys.* E **25**(11) 1630007
[2] Babič A and Šimkovic F 2013 *AIP Conf. Proc.* **1572** 7–10
[3] Jung M *et al.* 1992 *Phys. Rev. Lett.* **69**(15) 2164–2167
[4] Arnold R *et al.* (SuperNEMO Collaboration) 2010 *Eur. Phys. J.* C **70** 927–943
[5] Bilenky S M and Petcov S T 1987 *Rev. Mod. Phys.* **59** 671
[6] Patrignani C *et al.* (Particle Data Group) 2016 *Chin. Phys.* C **40**(10) 100001
[7] Kotila J and Iachello F 2012 *Phys. Rev.* C **85** 034316
[8] Gando A *et al.* (KamLAND-Zen Collaboration) 2016 *Phys. Rev. Lett.* **117**(8) 082503
[9] Rose M E 1961 *Relativistic Electron Theory* (Wiley)
[10] Doi M, Kotani T and Takasugi E 1985 *Prog. Theor. Phys. Suppl.* **83** 1–175
[11] Štefánik D, Dvornický R, Šimkovic F and Vogel P 2015 *Phys. Rev.* C **92**(5) 055502
[12] Jönsson P, Gaigalas G, Bieroń J, Fischer C F and Grant I P 2013 *Comput. Phys. Commun.* **184**(9) 2197–2203
[13] Wang M, Audi G, Kondev F G, Huang W J, Naimi S and Xu X 2017 *Chin. Phys.* C **41**(3) 030003 URL `http://stacks.iop.org/1674-1137/41/i=3/a=030003`
[14] Šimkovic F, Rodin V, Faessler A and Vogel P 2013 *Phys. Rev.* C **87**(4) 045501
[15] Fang D L, Faessler A and Šimkovic F 2015 *Phys. Rev.* C **92**(4) 044301
[16] Barabash A S 2015 *Nucl. Phys.* A **935** 52–64

IOP Conf. Series: Journal of Physics: Conf. Series **1056** (2018) 012003 doi:10.1088/1742-6596/1056/1/012003

Single Particle Levels and $\beta\beta$-Decay Matrix Elements in The Interacting Boson Model

J Barea[1] and J Kotila[2]

[1] Departamento de Física, Universidad de Concepción, Casilla 160-C, Concepción 4070386, Chile

[2] University of Jyvaskyla, Department of Physics, B.O. Box 35, FI-40014, University of Jyvaskyla, Finlan

E-mail: jbarea@udec.cl,jenni.kotila@jyu.fi

Abstract. Recently a new method to calculate the occupancies of single particle levels in atomic nuclei was developed in the context of the microscopic interacting boson model, in which neutron and proton degrees of freedom are treated explicitly (IBM-2). The energies of the single particle levels constitute a very important input for the calculation of the occupancies in this method, and further they play important role in the calculation of double beta decay nuclear matrix elements. Here we discuss how the $0\nu\beta\beta$, $0\nu_h\beta\beta$, and $2\nu\beta\beta$-decay nuclear matrix elements (NMEs) are affected when the energies of single particle levels are changed.

1. INTRODUCTION

The question of whether neutrinos are Majorana or Dirac particles, and of what are their masses and phases in the mixing matrix, remains one of the most important in physics today. A direct measurement of the average mass can be obtained from the observation of the neutrinoless double beta decay, $0\nu\beta\beta$. Several experiments are underway to detect this decay, and others are in the planning stage (for review see e.g. [1]). The half-life for this decay can be written as

$$[\tau_{1/2}^{0\nu}]^{-1} = G_{0\nu} \left| M_{0\nu} \right|^2 \left| f(m_i, U_{ei}) \right|^2 , \tag{1}$$

where $G_{0\nu}$ is a phase space factor, $M_{0\nu}$ the nuclear matrix element and $f(m_i, U_{ei})$ contains physics beyond the standard model through the masses m_i and mixing matrix elements U_{ei} of neutrino species. Concomitant with the neutrinoless modes, there is also the process allowed by the standard model, $2\nu\beta\beta$, which has been observed in several nuclei. For this process, the half-life can be, to a good approximation, factorized in the form

$$\left[\tau_{1/2}^{2\nu}\right]^{-1} = G_{2\nu} \left| M_{2\nu} \right|^2 . \tag{2}$$

In order to extract physics beyond the standard model, contained in the function f in Eq. (1), an accurate calculation of both $G_{0\nu}$ and $M_{0\nu}$ is needed. These calculations serve the purpose of extracting the neutrino mass $\langle m_\nu \rangle$ if $0\nu\beta\beta$ is observed, and of guiding searches if $0\nu\beta\beta$ is not observed.

The fact that $0\nu\beta\beta$-decay is a unique process, and there is no direct probe which connects the initial and final states other than the process itself makes the prediction challenging for

IOP Conf. Series: Journal of Physics: Conf. Series **1056** (2018) 012003 doi:10.1088/1742-6596/1056/1/012003

theoretical models. Recently we have calculated the occupancies of the single particle levels in order to satisfy a twofold goal: to asses the goodness of the single particle energies and check the reliability of the used wave functions. Both tests are particularly important in the case of nuclei involved in double beta decay, as they affect the evaluation of the NMEs and then their reliability [2]. As part of the calculation single particle energies were updated as shown in Fig. 1 and further discussed in Ref. [3]. The previous and updated sets of single particle levels are obtained from the experimental excitation energies of one-particle states, but in the second case they correspond to nuclei with mass number closer to the values of the nuclei studied in this work, since we are assuming that the energies of the single particle levels change with the value of A. Due to the lack of experimental data in some cases we used interpolated values or we extract the values from systematics in several nuclei. More technical details can be found in [3].

In what follows we report the $0\nu\beta\beta$, $0_h\nu\beta\beta$, and $2\nu\beta\beta$-decay nuclear matrix elements calculated using these updated single particle energies.

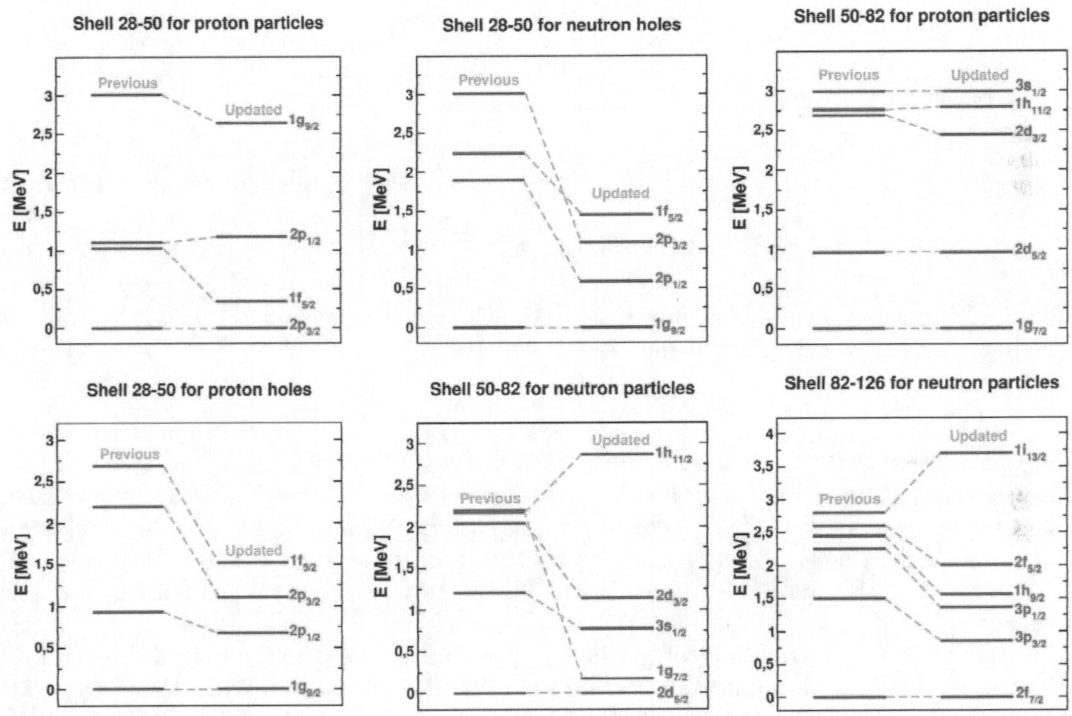

Figure 1. Comparison of previous and updated single particle energies in different shells 28-50 and 50-82 for protons and 28-50 50-82, and 82-126 for neutrons.

2. RESULTS

The method of calculating $\beta\beta$-decay NMEs in IBM-2 is described in detail in Refs. [4, 5] and the method for isospin restoration in Ref. [6]. The used parameters apart from single particle energies are listed in Refs. [5, 7].

2.1. Double beta decay with light neutrino exchange, $0\nu\beta\beta$

Comparison of previous and updated $0\nu\beta\beta$-decay NMEs calculated using IBM-2 is presented in Fig. 2. In general, the use of the updated single particle energies increase the NMEs. The increment is especially notable in the $A = 76 - 100$ region. These nuclei correspond to the proton shell 28-50 and neutron shells 28-50, 50-82 and Fig. 1 reveals that, indeed, the change in single particle energies is prominent in these shells.

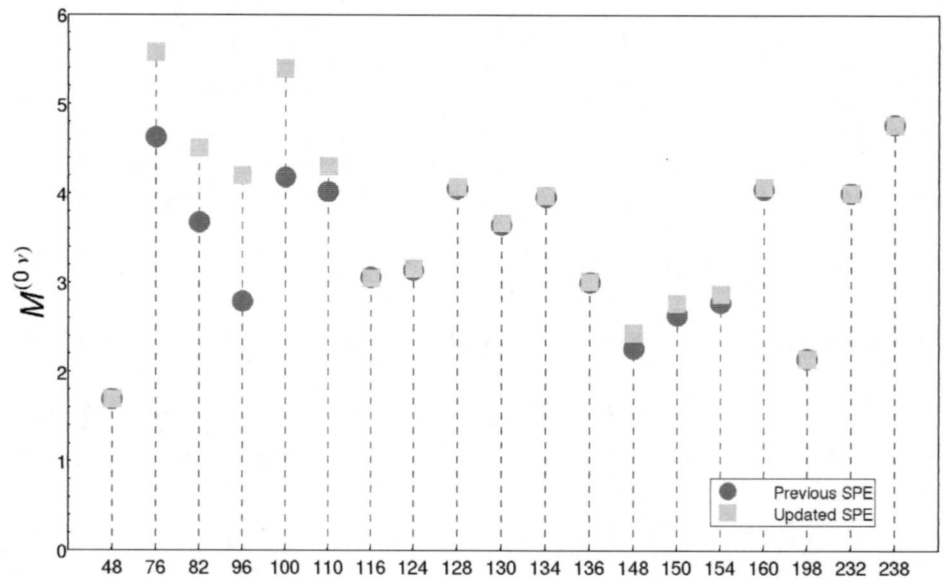

Figure 2. Comparison of previous and updated nuclear matrix elements for $0\nu\beta\beta$-decay with light neutrino exchange calculated using IBM-2.

2.2. Double beta decay with heavy neutrino exchange, $0\nu_h\beta\beta$

Fig. 3 shows the comparison of previous and updated $0\nu_h\beta\beta$-decay NMEs calculated using IBM-2. In case of heavy neutrino exchange the increase is also notable in $A = 76 - 100$ region as in the case of light neutrino exchange, but there is a decrease in $A = 150 - 160$ region. For $A < 76$, $100 < A < 150$, and $A > 160$ nuclei the NMEs stay intact, even though single particle energies are updated.

Note that in Ref. [8] a method of predicting possible contributions of sterile neutrinos to neutrinoless double beta decay using $0\nu\beta\beta$ and $0\nu_h\beta\beta$ NMEs was discussed. Using the NMEs obtained using updated single particle energies, also these predictions can be updated.

2.3. $2\nu\beta\beta$-decay

For comparison we also show $2\nu\beta\beta$-decay NMEs in Fig. 4. Like in the case of light neutrino exchange, the NMEs calculated with the updated single particle energies are generally larger than the previous ones, with the notable exception of $A = 116$. In $A = 76-100$ and $A = 148-160$ regions the increment is notable and when combined with phase space factors, [9], will affect the predicted half-lives. Thus the extracted effective values of g_A will also be affected, if procedure of [5] is followed.

3. DISCUSSION AND SUMMARY

The observed increase in the matrix elemens shown in Figures 2, 3 and 4 can be explained in our calculation scheme by considering what is affected when the single particle energies are changed. In the cases where the increase is observed the single particle energies decrease in general. This

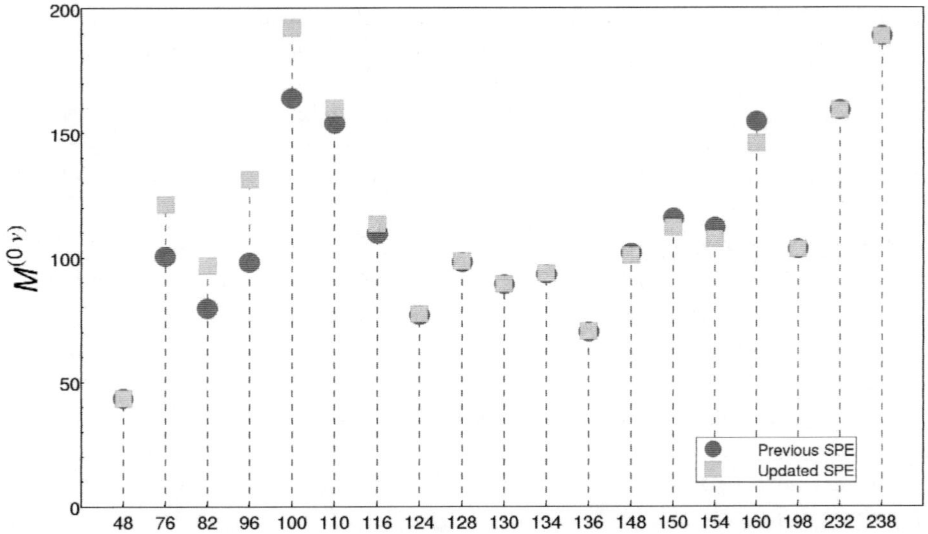

Figure 3. Comparison of previous and updated nuclear matrix elements for $0\nu\beta\beta$-decay with heavy neutrino exchange calculated using IBM-2.

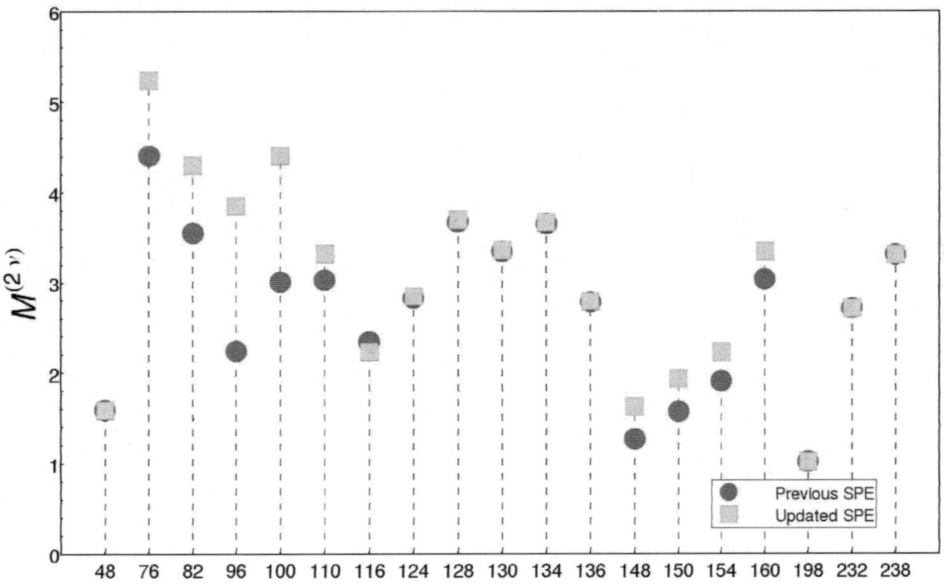

Figure 4. Comparison of previous and updated nuclear matrix elements for $2\nu\beta\beta$-decay calculated using IBM-2.

decrease or compressing of the levels produces higher values of the structure coefficients of the S and D pairs, which appear in products and raised to exponents which depends on the number of pairs or equivalently the number of bosons (see reference [4] for details). This effect is combined in protons and neutrons for $A = 76 - 100$ and only in neutrons for $A = 148 - 160$, which explains why the increasing values are at lesser extend in the last cases. Also the case $A = 110$ is less affected because of the low number of protons outside closed shell.

In summary, we have calculated nuclear matrix elements relevant for double beta decay (with two and zero neutrinos for light and heavy neutrino exchange) using an updated set of single particle energies which describe their measured occupancies fairly well. In some cases the new values of the matrix elements are higher than those of previous calculations, showing that the single particle level energies have an impact in our calculations, especially when both protons and neutrons are affected and the number of valence particles outside closed shells is higher.

4. ACKNOWLEDGMENTS

This work was supported by Chilean Ministry of Education (Fondo Nacional de Desarrollo Científico y Tecnológico) (Grant No. 1150564) and the Academy of Finland (Suomen Akatemia) under the Finnish Center of Excellence Program 2012-2017 (Nuclear and Accelerator Based Program at JYFL).

References

[1] Cremonesi O and Pavan M 2014 *Adv. High. Energy Phys.* **2014** 951432
[2] Engel J 2015 *J. Phys. G: Nucl. Part. Phys.* **42** 034017
[3] Kotila J and Barea J 2016 *Phys. Rev. C* **94** 034320
[4] Barea J and Iachello F 2009 *Phys. Rev. C* **79** 044301
[5] Barea J, Kotila J and Iachello F 2013 *Phys. Rev. C* **87** 014315
[6] Barea J, Kotila J and Iachello F 2015 *Phys. Rev. C* **91** 034304
[7] Beller *et al* 2013 *Phys. Rev. Lett.* **111** 172501
[8] Barea J, Kotila J and Iachello F 2015 2015 *Phys. Rev. D* **92** 093001
[9] Kotila J and Iachello F 2012 *Phys. Rev. C* **85** 034316

IOP Conf. Series: Journal of Physics: Conf. Series **1056** (2018) 012004 doi:10.1088/1742-6596/1056/1/012004

Probing beta decay matrix elements through heavy ion charge exchange reactions

J I Bellone[1,2 †], M Colonna[2], H Lenske[3] and J A Lay[4], within the NUMEN project

[1] Dipartimento di Fisica e Astronomia, Universitá degli studi di Catania, Italy
[2] INFN/LNS , via S. Sofia, 62, CT 95123, Catania, Italy
[3] Institut für Theoretische Physik, Justus-Liebig-Universität Giessen, D-35392 Giessen, Germany
[4] Departamento de FAMN, Universidad de Sevilla, Apartado 1065, E-41080 Sevilla, Spain

E-mail: † `bellone@lns.infn.it`

Abstract. To access information on neutrinoless double beta decay ($0\nu\beta\beta$) nuclear matrix elements, it has been proposed by the NUMEN collaboration to exploit the analogies between double beta decay processes and double charge exchange (DCE) nuclear reactions, looking in particular at the conditions where the corresponding cross section can be factorized into nuclear reaction and nuclear structure terms. DCE reactions can be treated as a convolution of two correlated or uncorrelated single charge exchange (SCE) processes, resembling $0\nu\beta\beta$ and $2\nu\beta\beta$, respectively. Thus it is important to model first SCE processes, to get a deeper insight into the possibility to factorize the corresponding cross section, so one can gain a better understanding of DCE cross section factorization. In this contribution, DCE reactions are discussed in terms of the convolution of two uncorrelated SCE processes, which should allow one to extract information on $2\nu\beta\beta$ nuclear matrix elements. These theoretical investigations are performed in close synergy with the experimental activity running at INFN-LNS within the NUMEN project.

1. Introduction

Nowadays neutrinoless double beta decay ($0\nu\beta\beta$) represents one of the main probes of Physics beyond the Standard Model, because of its sensitivity to Majorana phases, which appears within the neutrino effective mass, $m_{2\beta}$. Indeed, it can be shown that the half - life of a nucleus $\mathcal{N}$, which may undergo ($0\nu\beta\beta$), can be given by the expression $\left[T^{0\nu}_{1/2}(\mathcal{N})\right]^{-1} = G^{\mathcal{N}}_{0\nu}|\mathcal{M}^{\mathcal{N}}_{0\nu}|^2 f(m_{2\beta})$, which implies that, if such decay were observed, then from half - life measurements, it would be possible to extract the neutrino effective mass, which enters through the function $f(m_{2\beta})$, once determined the phase - space factor $G^{\mathcal{N}}_{0\nu}$ and the beta decay strength, $|\mathcal{M}^{\mathcal{N}}_{0\nu}|^2$. Actually, NME values, obtained through the various nuclear structure models, differ by about a factor of 3 [1]. For this reason, is has been proposed by the NUMEN collaboration the study of Heavy Ion Double Charge Exchange (DCE) nuclear reactions, due to a lot of analogies between these processes and $0\nu\beta\beta$ [2–5]. The main goal of this kind of study is to determine the conditions under which it is possible to factorize Heavy Ion DCE reaction Cross Section into the product of a nuclar reaction and a nuclear structure term, in order to extract information on NMEs from the latter term, once measured DCE Cross Section. In first approximation, the DCE process can be interpreted like a correlated or uncorrelated sequence of two Single Charge Exchange (SCE) nuclear reactions; in the first case it is possible to establish an analogy between $0\nu\beta\beta$ decay and DCE reaction only on a diagrammatic level, while in the second case there is an analogy between the latter process and the weak $2\nu\beta\beta$ decay, allowed within the Standard Model, which goes far beyond the mere diagrammatic structure, residing in the kind of spin - isospin operators involved in both weak and strong transitions. Such descriptions of DCE processes show the importance of the study of SCE processes as well .

2. Heavy Ion Charge Exchange Cross Section

The Cross Section for a generic process, in the Center of Mass frame, is given by

$$\frac{d^2\sigma}{dEd\Omega}^{(CE)} = \frac{E_\alpha E_\beta}{4\pi^2(\hbar c)^4}\frac{k_\beta}{k_\alpha}\frac{1}{(2J_A+1)}\frac{1}{(2J_a+1)}\sum_{\substack{m_a,m_A\\m_b,m_B}}|\sum_{\substack{\tau=C,T\\SO}}\mathcal{M}^{(\tau)}_{\alpha\beta}(\mathbf{k}_\alpha,\mathbf{k}_\beta)|^2 \tag{1}$$

where J_A and J_a are target and projectile total angular momenta; $k_{\alpha/\beta}$ are the initial/final - channel relative momenta in projectile - target center of mass frame, respectively; $E_{\alpha/\beta}$ is the reduced relative energy of the system and $\mathcal{M}^{(\tau)}_{\alpha\beta}(\mathbf{k}_\alpha,\mathbf{k}_\beta)$ is the transition matrix element for central ($\tau = C$), tensor ($\tau = T$) and spin−orbit ($\tau = SO$) components, respectively. Calculations are performed within Distorted Wave Born Approximation (DWBA) framework, in zero range approximation and in momentum space, because in the latter representation projectile and target coordinates can be separated and $\mathcal{M}^{(\tau)}_{\alpha\beta}(\mathbf{k}_\alpha,\mathbf{k}_\beta)$ becomes the momentum integral of the product of two factors:

$$\mathcal{M}^{(\tau)}_{\alpha\beta}(\mathbf{k}_\alpha,\mathbf{k}_\beta) = \sum_{ST}\int d^3p\, K^{(\tau,ST)}_{\alpha\beta}(\mathbf{p})N^D(\mathbf{k}_\alpha,\mathbf{k}_\beta,\mathbf{p}) \tag{2}$$

the distortion factor, N^D, which accounts for the relative projectile - target motion, i.e. the nuclear reaction term, being defined in terms of initial/final - channel distorted waves, $\chi_{\alpha/\beta}$, as $N^D = \frac{1}{(2\pi)^3}\int d^3r\,\chi^*_\beta(\mathbf{r})\chi_\alpha(\mathbf{r})e^{-i\mathbf{p}\cdot\mathbf{r}}$; the reaction kernel, given by

$$K^{(\tau,ST)}_{\alpha\beta}(\mathbf{p}=\mathbf{q}_{\alpha\beta}-\mathbf{q}) = V^{(C)}_{ST}(p)F^{(ST)\dagger}_{ab}F^{(ST)}_{AB} + \delta_{S1}\sqrt{\frac{24\pi}{5}}V^{(T)}_{ST}(p)Y^*_2(\hat{p})\left[F^{(ST)\dagger}_{ab}\otimes F^{(ST)}_{AB}\right]_2 \tag{3}$$

$\mathbf{q}_{\alpha\beta} = \mathbf{k}_\alpha - \mathbf{k}_\beta$ is the momentum transfer and $\mathbf{q}$ represents the displacement of the integration variable $\mathbf{p}$ from $\mathbf{q}_{\alpha\beta}$. The reaction kernel contains the nuclear structure information, because of its dependence on the Fourier transform of the effective nuclear interaction potential, $V^{(\tau)}_{ST}(p)$, parameterized according to [9], and on both projectile and target form factors $F^{(ST)}_{ab/AB}(\mathbf{p}) =< J_{b/B}m_{b/B}|\frac{1}{4\pi}e^{i\mathbf{p}\cdot\mathbf{r}_{a/A}}\mathcal{O}^{(ST)}|J_{a/A}m_{a/A} >= \sum_{\substack{L,M_L,\\J_1,M_1}}\left(J_{a/A},m_{a/A},J_{b/B},m_{b/B}|J_1,M_1\right)(L,M_L,S,M_S|J_1,M_1)\,\rho^{(L,S,J_1)}_{ab/AB}(p)i^LY_{L,M_L}(\hat{p})$. The latter ones are in turn proportional to the Fourier transform of the transition densities, $\rho(p)$, corresponding to the (S,T) transition operators,$\mathcal{O}^{(ST)}$, which are the same for both SCE and single beta decay and for both uncorrelated DCE process and $2\nu\beta\beta$. Hence, cross section factorization implies to take out of the momentum integral at least one of the two terms, N^D or $K_{\alpha\beta}$.

3. Cross Section factorization: Gaussian reaction kernel

In this work, calculations are performed only for central nuclear interaction and only for $L = 0$, for $^{40}Ca_{g.s.}\left(^{18}O_{g.s.},^{18}F_{g.s.}\right)^{40}K_{1^+}$ system, studied within the NUMEN collaboration [5]. From the direct, i.e. surface, nature of Charge Exchange reactions, follows that both projectile and target form factors can be parameterized with gaussians peaked at the nuclear surface, which fit well the transition densities deduced from microscopic calculations. Once provided that the interaction potential is nearly constant as a function of momentum p, also the reaction kernel can be described through a gaussian, centered at $R = \sqrt{R_a^2 + R_A^2}$ and with a spread given by $\sigma = \sqrt{\sigma_a^2 + \sigma_A^2}$, where $R_{a/A}$and $\sigma_{a/A}$ are projectile and target nuclear radius and surface thickness, respectively. Hence, in momentum space, the reaction kernel is factorized into a fuction depending only on momentum transfer $\mathbf{q}_{\alpha\beta}$ and a "residual function", $h_{\alpha\beta}(\mathbf{q}, \boldsymbol{\rho}) = e^{-\frac{1}{2}\sigma^2 q^2} e^{-i\mathbf{q}\cdot\boldsymbol{\rho}}$, with $\boldsymbol{\rho} = \mathbf{R} + i\sigma^2 \mathbf{q}_{\alpha\beta}$, whose dependence on $\mathbf{q}_{\alpha\beta}$ makes such factorization not exact, but for $\mathbf{q}_{\alpha\beta} = \mathbf{0}$. A check has been made in order to determine the $q_{\alpha\beta}$ values for which this factorization works. Such test has been exploited within a quite realistic approximation for low energy Heavy Ion reactions: the Black Disk Approximation (BDA), that assumes total suppression of the distorted waves inside the reaction region (Black Disk) and plane waves outside that region, allowing to find an analytical expression of the distortion factor N^D, which is found to be inversely proportional to the target nucleus mass. This (partial) factorization thus obtained, in turn, implies transition matrix element factorization

$$\mathcal{M}_{\alpha\beta} = \sum_{ST} K_{\alpha\beta}^{(S,T)}(\mathbf{q}_{\alpha\beta}) \int d^3q\, h_{\alpha\beta}(\mathbf{q}, \boldsymbol{\rho}) N^D(\mathbf{k}_\alpha, \mathbf{k}_\beta, \mathbf{q}) \tag{4}$$

4. Double Charge Exchange reaction kernel

In this work, DCE process is studied as a sequence of two uncorrelated SCE reactions, i. e. DCE nuclear interaction potential can be expressed in terms of SCE potential according to the following relation $\mathbb{V}^{(DCE)} = \mathbb{V}^{(SCE)} G_0 \mathbb{V}^{(SCE)}$, where $G_0 = \frac{1}{\omega_\alpha - \omega_{\gamma\Gamma} + i\epsilon}$ is the free propagator, expressed in terms of the total on - shell (ω_α) and off - shell (ω_γ) energies, so that DCE reaction kernel can be expressed as $\sum_\Gamma \int \frac{d^3k_\gamma}{(2\pi)^3} < \psi_B\varphi_b|\mathbb{V}^{(SCE)}|\Gamma, \mathbf{k}_\gamma > \mathcal{G}_{0,\gamma\Gamma} < \Gamma, \mathbf{k}_\gamma|\mathbb{V}^{(SCE)}|\psi_{aA}\varphi_a >= \sum_\Gamma \int \frac{d^3k_\gamma}{(2\pi)^3} K^{(SCE)}(\mathbf{q}_{\alpha\gamma}, \mathbf{R})\mathcal{G}_{0,\gamma\Gamma} K^{(SCE)}(\mathbf{q}_{\gamma\beta}, \mathbf{R}')$. Assuming Gaussian SCE reaction kernels and using the pole approximation, $\mathcal{G}_{\gamma\Gamma} \simeq \delta(\omega_{\gamma\Gamma} - \omega_\alpha)$, and Single State Dominance (SSD), DCE reaction kernel can be further factorized into the product of the two SCE reaction kernels . Such a proportionality relation, in turn, implies a proportionality between $2\nu\beta\beta$ strength and the product of the single beta decay strengths; indeed, factorized DCE Cross Section becomes

$$\frac{d^2\sigma}{dEd\Omega} = \frac{E_\alpha E_\beta}{(2\pi\hbar c)^4} \frac{k_\beta}{k_\alpha} \frac{1}{(2J_A+1)} \frac{1}{(2J_a+1)} \sum_{\substack{m_a,m_A \\ m_b,m_B}} |j_0(k_\omega|\mathbf{R} - \mathbf{R}'|)|^2 k_\omega^2 \mu_\omega^2 |K^{(SCE)}(q_{\alpha\omega})|^2 |K^{(SCE)}(q_{\omega\beta})|^2 |N^D|^2 \tag{5}$$

where k_ω is the on - shell intermediate relative momentum for one fixed intermediate nuclear state, "ω"; $q_{\alpha\omega}$ and $q_{\omega\beta}$ are the on - shell momenta transfer moduli, averaged over Ω_ω.

5. Results for Single and Double Charge Exchange Cross Sections

Calculations are performed by employing the gaussian form factors described in the previous sections, within different approximations for the separation function $h_{\alpha\beta}(\mathbf{q}, \boldsymbol{\rho})$, and BDA, assuming spherical symmetric distortion factor. The results thus obtained have been compared to that determined relaxing the factorization hypothesis, as shown in figure 1, where Plane Wave Born Approximation (PWBA) calculations have been included in order just to display distortion factor magnitude and to show the changes in diffraction pattern, strongly influenced by the optical potential, both in full DWBA (solid curve) and in the DWBA - factorized cases

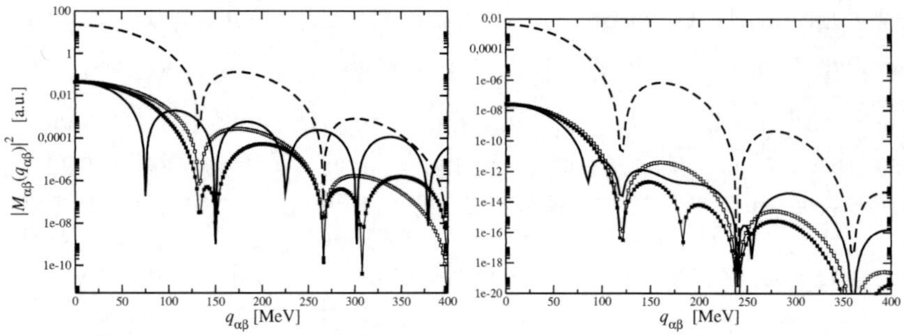

Figure 1. Comparison of transition matrix element for SCE (left panel) and DCE (right panel) reactions obtained within (curves with full and empty squares) and without (solid curve) factorized reaction kernel and in PWBA (dashed curve). For details see the text.

(curves with full and empty squares for two choises of $h_{\alpha\beta}$).In both SCE and DCE cases, the three DWBA curves coincide up to $q_{\alpha\beta} \simeq 25 - 30\,MeV$, meaning that even though factorization is exact only for zero momentum transfer, it works pretty well in a small range around $q_{\alpha\beta} = 0$, that is the $q_{\alpha\beta}$ range of interest if one wants to gain information on single beta and $2\nu\beta\beta$ decays, but probably not for $0\nu\beta\beta$, which is governed by a mechanism different from $2\nu\beta\beta$ one. The other important result, that can be extracted from the latter figure, is that DCE distortion factor is smaller (greater absorption) than that for SCE process.

6. Conclusions and Outlooks

This work allows to provide Heavy Ion Charge Exchange Cross Section factorization by assuming a Gaussian Reaction Kernel, just from considerations on the direct nature of such reactions. The Cross Section factorization obtained within these calculations is "exact" for momentum transfer $q_{\alpha\beta} = 0$, but works well up to $q_{\alpha\beta} \simeq 25 - 30\,MeV$. BDA allows to determine an analytical expression for the distortion factor, which is smaller for DCE than for SCE processes. DCE formalism code implementation in DWBA, without separation ansatz is a work in progress. Once developed the latter calculations, the main focus will be the comparison with data on DCE cross section, in order to check the possible influence of the coherent DCE mechanism [10], which could carry information on $0\nu\beta\beta$ decay process.

References

[1] Barea J, Kotila J, Iachello F, 2012 *Phys. Rev. Lett.* **109** 042501 URL `https://journals.aps.org/prl/abstract/10.1103/PhysRevLett.109.042501`.
[2] Cappuzzello F *et al.*, 2015 *J. Phys.: Conf. Ser.* **630** 12018 URL `http://iopscience.iop.org/article/10.1088/1742-6596/630/1/012018/pdf`
[3] Agodi C *et al.*, 2015 *Nuclear and Particle Physics Proceedings* **265-266** 28-30 URL `https://www.sciencedirect.com/science/article/pii/S2405601415003491`
[4] Cappuzzello F *et al.*, 2018 *Eur. Phys. J. A* **54** 72.
[5] Cappuzzello F, Cavallaro M, Agodi C, Bondí M, Carbone D, Cunsolo A, Foti A, 2015 *Eur. Phys. J. A* **51** URL `https://link.springer.com/article/10.1140/epja/i2015-15145-5`.
[6] Lenske H, Bellone J I, Colonna M, Lay J A 2018 *Phys. Rev. C* (submitted).
[7] Cappuzzello F, Lenske H, Cunsolo A, Beaumel D, Fortier S, Foti A, Lazzaro A, Nociforo C, Orrigo S E A, Winfield J S, 2004 *Nucl. Phys. A* **739** 30-56 URL `http://www.sciencedirect.com/science/article/pii/S0375947404005366`.
[8] Franey M A, Love W G, 1981 *Phys. Rev. C* **24** 1073 URL `https://journals.aps.org/prc/abstract/10.1103/PhysRevC.24.1073`.
[9] Franey M A, Love W G, 1985 *Phys. Rev. C* **31** 488 URL `https://journals.aps.org/prc/abstract/10.1103/PhysRevC.31.488`.
[10] Lenske H 2018 *see this volume* **ibid**.

DAMA/LIBRA results and perspectives

R Bernabei[1,2], P Belli[1], F Cappella[3], V Caracciolo[4], R Cerulli[1], A d'Angelo[3], A Di Marco[1], A Incicchitti[3], V Merlo[1,2], F Montecchia[1,5], C J Dai[6], H L He[6], H H Kuang[6], X H Ma[6], X D Sheng[6], R G Wang[6] and Z P Ye[6,7]

[1] INFN sez. Roma "Tor Vergata", Rome, Italy
[2] Dip. di Fisica, Università di Roma "Tor Vergata", Rome, Italy
[3] INFN sez. Roma, Rome, Italy
[4] Laboratori Nazionali del Gran Sasso, Assergi (AQ), Italy
[5] Dip. di Ingegneria Civile e Ingegneria Informatica, Università di Roma "Tor Vergata", Rome, Italy
[6] Key Laboratory of Particle Astrophysics, Institute of High Energy Physics, Chinese Academy of Sciences, Beijing, China
[7] University of Jinggangshan, Ji'an, Jiangxi, P. R. China

E-mail: rita.bernabei@roma2.infn.it

Abstract. The DAMA/LIBRA-phase2 experiment ($\simeq$ 250 kg highly radiopure NaI(Tl) equipped with new high quantum efficiency photomultipliers) is in progress in the Gran Sasso National Laboratory of the I.N.F.N.. When considering the data collected by DAMA/LIBRA–phase1 and by the former DAMA/NaI experiment ($\sim$ 100 kg of highly radio-pure NaI(Tl)) the data over 14 independent annual cycles, for a total exposure of 1.33 ton $\times$ yr, have already been released by exploiting the model-independent Dark Matter (DM) annual modulation signature. Cumulatively a DM annual modulation effect has been observed at 9.3 σ C.L., supporting the presence of DM particles in the galactic halo. No systematic effect or side reaction able to mimic the observed DM annual modulation has been found. Recent corollary analyses in frameworks of Mirror DM candidates and R&D efforts for a possible future DAMA/LIBRA-phase3 are mentioned.

1. DAMA/LIBRA results and a few comments

The DAMA project develops and uses low background scintillators and set-ups to investigate rare processes. DAMA/LIBRA is the main experiment [1, 2, 3, 4, 5, 6, 7, 8, 9, 10, 11, 12, 13, 14, 15, 16]; it is the second generation highly radiopure NaI(Tl) set-up after the pioneering DAMA/NaI [17, 18]. DAMA/LIBRA is further investigating the presence of DM particles in the galactic halo by exploiting the model independent DM annual modulation signature [19, 20].

The DM annual modulation signature is due to the Earth's revolution around the Sun, which is moving in the Galaxy; as a consequence of the velocities composition, the flux of DM particles impinging a terrestrial detector is expected to follow a cosinusoidal behaviour with maximum around $\simeq$ June 2nd (when the projection of the Earth orbital velocity on the Sun velocity with respect to the Galactic frame is maximum) and minimum around $\simeq$ December 2nd (when the two velocities are opposite). This signature is very effective because the signal induced by DM particles must simultaneously satisfy many requirements: the rate must contain a component modulated according to a cosine function (1) with one year period (2) and a phase peaked roughly at $\simeq$ June 2nd (3); the modulation must only be present in a well-defined low energy range (4); it must apply only to those events in which just one detector among many actually "fires" (*single-hit* events), since the DM particle multi-interaction probability is negligible (5); the modulation amplitude in the region of maximal sensitivity must be $\simeq 7\%$ for usually adopted halo distributions (6), but it can be larger (even up to $\simeq 30\%$) in case of some possible scenarios. This signature is model independent and no systematic effect or side reaction able to mimic the effect (i.e. able to account for the observed modulation amplitude and to simultaneously satisfy all the requirements of the signature) is available [1, 2, 3, 4, 7, 8, 12, 13, 17, 18, 21].

The full description of the DAMA/LIBRA set-up in all its phases is described in details in Refs. [1, 2, 3, 4, 6, 7, 8, 13].

Many independent analyses have been performed on the data of the 14 independent annual cycles; all the analyses confirm the presence of a peculiar annual modulation satisfying all the features of the investigated DM signature [2, 3, 4, 8]. In Figure 1, as an example, the time behaviour of the experimental residual rate of the *single-hit* scintillation events for DAMA/LIBRA–phase1 in the (2–6) keV energy interval is plotted.

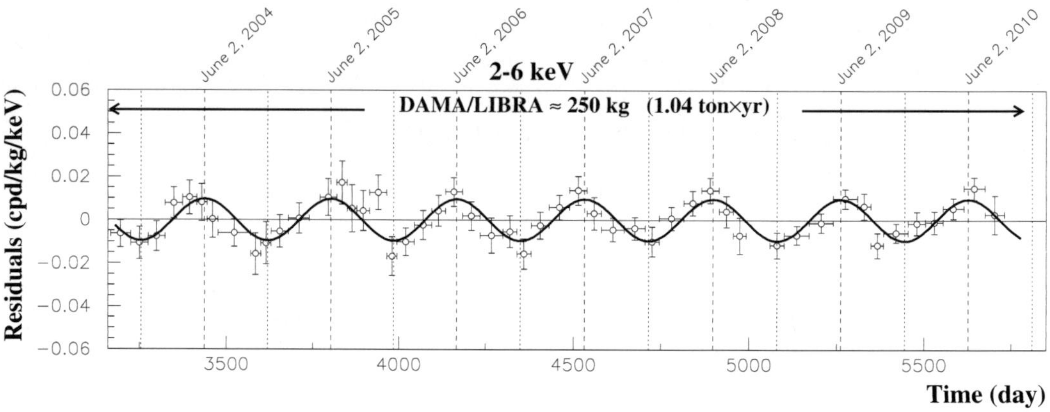

Figure 1. Experimental residual rate of the *single-hit* scintillation events measured by DAMA/LIBRA–phase1 in the (2–6) keV energy interval as a function of the time. The superimposed curve is the cosinusoidal function behaviour $A\cos\omega(t-t_0)$ with a period $T = \frac{2\pi}{\omega} = 1$ yr, a phase $t_0 = 152.5$ day (June 2nd) and modulation amplitude, A, equal to the central values obtained by best fit on the data points of the entire DAMA/LIBRA–phase1. The dashed vertical lines correspond to the maximum expected for the DM signal (June 2nd), while the dotted vertical lines correspond to the minimum.

When fitting the *single-hit* residual rate of DAMA/LIBRA–phase1 together with the DAMA/NaI ones, with the function: $A\cos\omega(t-t_0)$, considering a period $T = \frac{2\pi}{\omega} = 1$ yr and a phase $t_0 = 152.5$ day (June 2nd) as expected by the DM annual modulation signature, the following modulation amplitude in NaI(Tl) is obtained: $A = (0.0110 \pm 0.0012)$ cpd/kg/keV, corresponding to 9.2 σ C.L.. When the period, and the phase are kept free in the fitting procedure the modulation amplitude is (0.0112 ± 0.0012) cpd/kg/keV (9.3 σ C.L.), the period $T = (0.998 \pm 0.002)$ year and the phase $t_0 = (144 \pm 7)$ day, values well in agreement with expectations for a DM annual modulation signal. In particular, the phase is consistent with about June 2nd and is fully consistent with the value independently determined by Maximum Likelihood analysis [4]. The run test and the χ^2 test on the data have shown that the modulation

Table 1. Summary of the contributions to the total neutron flux at LNGS; the value, $\Phi^{(n)}_{0,k}$, the relative modulation amplitude, η_k, and the phase, t_k, of each component is reported. It is also reported the counting rate, $R_{0,k}$, in DAMA/LIBRA for *single-hit* events, in the $(2-6)$ keV energy region induced by neutrons, muons and solar neutrinos, detailed for each component. The modulation amplitudes, A_k, are reported as well, while the last column shows the relative contribution to the annual modulation amplitude observed by DAMA/LIBRA, $S_m^{exp} \simeq 0.0112$ cpd/kg/keV [4]. For details see Ref. [13] and references therein.

	Source	$\Phi^{(n)}_{0,k}$ (neutrons cm^{-2} s^{-1})	η_k	t_k	$R_{0,k}$ (cpd/kg/keV)	$A_k = R_{0,k}\eta_k$ (cpd/kg/keV)	A_k/S_m^{exp}
SLOW neutrons	thermal n ($10^{-2} - 10^{-1}$ eV)	1.08×10^{-6}	$\simeq 0$ however $\ll 0.1$	-	$< 8 \times 10^{-6}$	$\ll 8 \times 10^{-7}$	$\ll 7 \times 10^{-5}$
	epithermal n (eV-keV)	2×10^{-6}	$\simeq 0$ however $\ll 0.1$	-	$< 3 \times 10^{-3}$	$\ll 3 \times 10^{-4}$	$\ll 0.03$
FAST neutrons	fission, $(\alpha, n) \to$ n (1-10 MeV)	$\simeq 0.9 \times 10^{-7}$	$\simeq 0$ however $\ll 0.1$	-	$< 6 \times 10^{-4}$	$\ll 6 \times 10^{-5}$	$\ll 5 \times 10^{-3}$
	$\mu \to$ n from rock ($>$ 10 MeV)	$\simeq 3 \times 10^{-9}$	0.0129	end of June	$\ll 7 \times 10^{-4}$	$\ll 9 \times 10^{-6}$	$\ll 8 \times 10^{-4}$
	$\mu \to$ n from Pb shield ($>$ 10 MeV)	$\simeq 6 \times 10^{-9}$	0.0129	end of June	$\ll 1.4 \times 10^{-3}$	$\ll 2 \times 10^{-5}$	$\ll 1.6 \times 10^{-3}$
	$\nu \to$ n (few MeV)	$\simeq 3 \times 10^{-10}$	0.03342*	Jan. 4th*	$\ll 7 \times 10^{-5}$	$\ll 2 \times 10^{-6}$	$\ll 2 \times 10^{-4}$
	direct μ	$\Phi^{(\mu)}_0 \simeq 20\ \mu$ m^{-2}d^{-1}	0.0129	end of June	$\simeq 10^{-7}$	$\simeq 10^{-9}$	$\simeq 10^{-7}$
	direct ν	$\Phi^{(\nu)}_0 \simeq 6 \times 10^{10}\ \nu$ cm^{-2}s^{-1}	0.03342*	Jan. 4th*	$\simeq 10^{-5}$	3×10^{-7}	3×10^{-5}

* The annual modulation of solar neutrino is due to the different Sun-Earth distance along the year; so the relative modulation amplitude is twice the eccentricity of the Earth orbit and the phase is given by the perihelion.

amplitudes singularly calculated for each annual cycle of DAMA/NaI and DAMA/LIBRA–phase1 are normally fluctuating around their best fit values [2, 3, 4].

No modulation was found in any possible source of systematics or side reactions; thus, cautious upper limits on possible contributions to the DAMA/LIBRA–phase1 measured modulation amplitude were obtained (see Refs. [2, 3, 4]). It is worth noting that they do not quantitatively account for the measured modulation amplitudes, and are even not able to simultaneously satisfy all the many requirements of the signature. Similar analyses were also performed for the DAMA/NaI data [17, 18]. In particular, the case of neutrons, muons and solar neutrinos has been discussed in details in Refs. [7, 13], where it has been demonstrated that they cannot give any significant contribution to the DAMA annual modulation result. Table 1 summarizes the safety upper limits on the contributions to the observed modulation amplitude due to the total neutron flux at LNGS, either from (α, n) reactions, from fissions and from muons and solar-neutrinos interactions in the rocks and in the lead around the experimental set-up; the direct contributions of muons and solar neutrinos are reported there too. Other arguments can be found in Refs. [1, 2, 3, 4, 7, 8, 12, 13, 17, 18, 21].

In conclusion, DAMA gives model-independent evidence (at 9.3 σ C.L. over 14 independent annual cycles) for the presence of DM particles in the galactic halo.

As regards comparisons, we recall that no direct model independent comparison is possible in the field when different target materials and/or approaches are used; the same is for the strongly model dependent indirect searches. In particular, the DAMA model independent evidence is compatible with a wide set of scenarios regarding the nature of the DM candidate and related astrophysical, nuclear and particle Physics scenarios; some given scenarios and parameters are discussed e.g. in Refs. [2, 5, 8, 12, 14, 15, 16, 17, 18, 22, 23, 24, 25, 26, 27, 28, 29, 30, 31, 32, 33, 34, 35, 36] and references therein. Further large literature is available on the topics.

In conclusion, both negative results and possible positive hints reported in literature can be compatible with the DAMA model-independent DM annual modulation results in various scenarios considering also the existing experimental and theoretical uncertainties. Let us also recall for the case of liquid noble gases the comments on fundamental aspects pointed out e.g. in Ref. [37, 38]. Finally it is worth noting that experiments applying statistical multiple rejections of their counting rate and/or use neutron source (and even more high intensity neutron generator as e.g. XENON experiment) cannot reliably investigate the DM annual modulation signature in their target materials.

Finally, for completeness, let us remind the recent investigation on other approaches and second order effects carried out with the DAMA/LIBRA-phase1 data [12, 14] as well as investigation on rare processes other than DM [9, 10, 11]

2. A recent example of corollary analyses

Among the many possible corollary analyses we just mention here the recent one performed on the DAMA/LIBRA–phase1 data in frameworks of mirror dark matter candidates [15, 16], which originate from hidden gauge sectors. Two scenarios have been considered: the asymmetric Mirror DM [15] and the symmetric one [16]. In both cases, the values obtained for the $\sqrt{f}\epsilon$ (where f is the fraction of DM in the Galaxy in form of mirror atoms and ϵ is the coupling constant) parameter are well compatible with cosmological bounds [15, 16]. In these analyses several uncertainties on the astrophysical, particle physics and nuclear physics models have been taken into account in the calculation. In particular, in the symmetric mirror DM scenario the

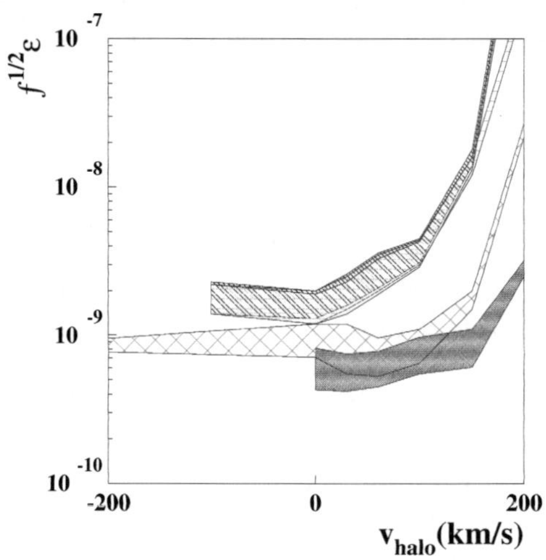

Figure 2. Example of allowed regions for the $\sqrt{f}\epsilon$ parameter as a function of v_{halo} (halo temperature $T = 5 \times 10^5$ K). The regions have been obtained by considering a composite dark halo H′(20%), He′(74%), C′(0.9%), O′(5%), Fe′(0.1%), with $v_0 = 220$ km/s and parameters as given in the set A of ref. [16]. The five contours correspond to different quenching factor modeling (see [16]).

DM particles are expected to form bubbles in the Galaxy with diameter which could be even as large as the size of the solar system. The dark halo is composed by different species of mirror DM particles (different mirror atoms) and, at the present epoch, is crossing a region close to the Sun with a constant velocity, v_{halo} in the Galactic frame. The velocity distribution of the particles in the bubble can be considered maxwellian; it is assumed that the halo has its own local equilibrium temperature, T. In the analysis halo temperature in the range $10^4 - 10^8$ K has been considered. As an example, in Figure 2, the allowed regions for the $\sqrt{f}\epsilon$ parameter as a function of v_{halo} in different scenarios are reported; in particular, the regions have been obtained by considering a dark halo made of H′(20%), He′(74%), C′(0.9%), O′(5%), Fe′(0.1%) with a temperature $T = 5 \times 10^5$ K. The depicted contours correspond to different possible quenching factors values (see Ref. [16]). Many other scenarios are discussed in the literature.

3. Perspectives

After DAMA/LIBRA-phase1, an important upgrade has been performed by replacing all the PMTs with new ones having higher Quantum Efficiency (QE). In this new configuration a software energy threshold at 1 keV has been reached [6]. Thus, DAMA/LIBRA-phase2 is continuously running in order: 1) to increase the experimental sensitivity thanks to the lower software energy threshold of the experiment; 2) to improve the corollary investigation on the nature of the DM particle and related astrophysical, nuclear and particle physics arguments; 3) to investigate second order effects; 4) to investigate other signal features; 5) to investigate rare processes other than DM with high sensitivity.

Future improvements to further increase the sensitivity of the set-up (possible DAMA/LIBRA-phase3) can be considered by using new metallic high QE and ultra-low background PMTs to be directly coupled to the NaI(Tl) crystals. In this way a further large improvement in the light collection and a further lowering of the software energy threshold would be obtained. The developments with Hamamatsu Co. are successful and 4 PMT prototypes are at hand. New protocols, and also alternative ideas, are ongoing that regard the possibility to dismount the light guides of the detectors which also act as optical windows. Finally, the possibility of a pioneer experiment with anisotropic $ZnWO_4$ detectors to further investigate, with the directionality approach, those DM candidates inducing just nuclear recoils is in progress [39]; as e.g. the use of the present detectors untouched but with the new metallic PMTs and new ultra low background crystal scintillators (e.g. $ZnWO_4$) placed among the DAMA/LIBRA detectors in order to reach a high sensitivity and contemporarily to perform directionality measurements is ongoing.

References

[1] R. Bernabei et al. 2008 *Nucl. Instr. and Meth.* **A 592** 297
[2] R. Bernabei et al. 2008 *Eur. Phys. J.* **C 56** 333
[3] R. Bernabei et al. 2010 *Eur. Phys. J.* **C 67** 39
[4] R. Bernabei et al. 2013 *Eur. Phys. J.* **C 73** 2648
[5] P. Belli et al. 2011 *Phys. Rev.* **D 84** 055014
[6] R. Bernabei et al. 2012 *J. of Instr.* **7** P03009
[7] R. Bernabei et al. 2012 *Eur. Phys. J.* **C 72** 2064
[8] R. Bernabei et al. 2013 *Int. J. of Mod. Phys.* **A 28** 1330022
[9] R. Bernabei et al. 2009 *Eur. Phys. J.* **C 62** 327
[10] R. Bernabei et al. 2012 *Eur. Phys. J.* **C 72** 1920
[11] R. Bernabei et al. 2013 *Eur. Phys. J.* **A 49** 64
[12] R. Bernabei et al. 2014 *Eur. Phys. J.* **C 74** 2827
[13] R. Bernabei et al. 2014 *Eur. Phys. J.* **C 74** 3196
[14] R. Bernabei et al. 2015 *Eur. Phys. J.* **C 75** 239
[15] A. Addazi et al. 2015 *Eur. Phys. J.* **C 75** 400
[16] R. Cerulli et al. 2017 *Eur. Phys. J.* **C 77** 83
[17] R. Bernabei el al. 2003 *La Rivista del Nuovo Cimento* **26 n.1** 1-73
[18] R. Bernabei et al. 2004 Int. J. Mod. Phys. **D 13** 2127
[19] A.K. Drukier, K. Freese, D.N. Spergel 1986 *Phys. Rev.* **D 33** 3495
[20] K. Freese, J.A. Frieman, A. Gould 1988 *Phys. Rev.* **D 37** 3388
[21] R. Bernabei et al. 2000 *Eur. Phys. J.* **C 18** 283
[22] R. Bernabei et al. 2016 *Int. J. of Mod. Phys.* **A 31** 1642009

[23] R. Bernabei et al. 1996 *Phys. Lett.* **B 389** 757
[24] R. Bernabei et al. 1998 *Phys. Lett.***B 424** 195
[25] R. Bernabei et al. 1999 *Phys. Lett.***B 450** 448
[26] P. Belli et al. 2000 *Phys. Rev.* **D 61** 023512
[27] R. Bernabei et al. 2000 *Phys. Lett.* **B 480** 23
[28] R. Bernabei et al. 2001 *Phys. Lett.* **B 509** 197
[29] R. Bernabei et al. 2002 *Eur. Phys. J.* **C 23** 61
[30] P. Belli et al. 2002 *Phys. Rev.* **D 66** 043503
[31] R. Bernabei et al. 2006 *Int. J. Mod. Phys.* **A 21** 1445
[32] R. Bernabei et al. 2006 *Eur. Phys. J.* **C 47** 263
[33] R. Bernabei et al. 2007 *Int. J. Mod. Phys.* **A 22** 3155
[34] R. Bernabei et al. 2008 *Phys. Rev.***D 77** 023506
[35] R. Bernabei et al. 2008 *Mod. Phys. Lett.* **A 23** 2125
[36] R. Bernabei et al. 2008 *Eur. Phys. J.* **C 53** 205
[37] R. Bernabei et al. 2009 *ISBN 978-88-95688-12-1, Exorma Ed.* 1-53 [arXiv:0806.0011v2]
[38] R. Bernabei et al. 2015 *Int. J. of Mod. Phys.* **A 30** 1530053-1
[39] R. Bernabei et al. 2013 *Eur. Phys. J.* **C 73** 2276

The read-out and data transmission for the MAGNEX focal plane detector for the NUMEN project.

D Bonanno[1], D Lo Presti[1,2], D Bongiovanni[3], G Gallo[2,3], F Longhitano[1] and S Reito[1] for the NUMEN collaboration

[1] National Institute of Nuclear Physics (INFN-CT), Via Santa Sofia 64, 95123 Catania, I
[2] Department of Physics and Astronomy, University of Catania, Via Santa Sofia 64, 95123 Catania, I
[3] Laboratori Nazionali del Sud (INFN-LNS), Via Santa Sofia 62, 95123 Catania, I

E-mail: danilo.bonanno@ct.infn.it

Abstract. The main task of the read-out electronics for the new NUMEN focal plane detector (FPD) is the real-time data collection from the front-end ASICs and the high bandwidth transmission to data acquisition, in addition to the slow control of the front-end electronics and the synchronization of the whole detector. The read-out electronics architecture, designed to be modular and expandable to the final size of the detectors, is based on System On Module (SOM). This very versatile device couples high performance FPGA to powerful processor architecture and allows a graphical approach to the programming and interfacing. A detailed description of the architecture and the specifications of the read-out electronics are presented.

1. Introduction

With the ongoing upgrades of the cyclotron accelerator [1] at INFN Laboratori Nazionali del Sud (LNS) foreseen for the NUMEN project [2, 3, 4], an upgrade of the FPD [5] of the MAGNEX spectrometer [6, 7, 8] will be necessary. It includes the design of new front-end (FE) and read-out (RO) electronics, able to couple with the increased event rate (up to 10MHz in the NUMEN experiments). The Segmented anode and the FE electronics were totally redesigned with a solution that includes the VMM3 chip [9]. This solution led to the development of a totally digital RO based on System On Module (SOM) by National Instruments [10].

2. FE-RO Architecture

The main idea for the design of the electronics of the new FPD is the use of the same architecture of FE and RO (Figure 1) both for the tracker and the Particles Identification Detector (PID) wall. This choice allows an easier syncronization of the detectors, the uniformity of the data format, a contained software development as well as the reduction of the costs. The requirements for such system are: readout of about 1400 channels for the tracker and 4000 channels for the PID-wall, capability to handle a foreseen event rate of 100 kHz/cm, modularity, scalability, and radiation hardness. The architecture is based on a replica of the same module shown in Figure 2. Each module is equipped with a variable number of VMM3 chips according to the number of the corresponding channels to be read-out. These chips are read-out by an FPGA based system manufactured by National Instruments (SOM). A single SOM has 160 General Purpose Input Output pin (GPIO) that could be used, appropriately configured, for data read-out, slow control and communication ports. Taking into account the GPIOs required for slow control and for an additional Gigabit Ethernet connection, a SOM would be able to manage and read-out the data of 10 VMM chips. In particular for the tracker, if we keep in mind that the pitch of the strip of the segmented anode is 0.75 mm, and that a VMM reads 64 strip, each SOM should transfer data with a bandwidth of about 3 Gb/s, at the highest event-rate foreseen for the experiment. So the final number of VMM chips coupled to a SOM will be a compromise between the best bandwidth used by two Gigabit Ethernet connections, available in the SOM, and the possibility to design an unique kind of board for the front-end of each detector in the experiment.

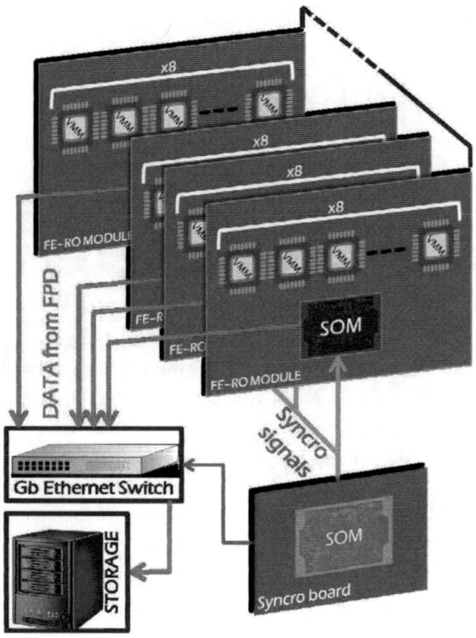

Figure 1. FE-RO architecture.

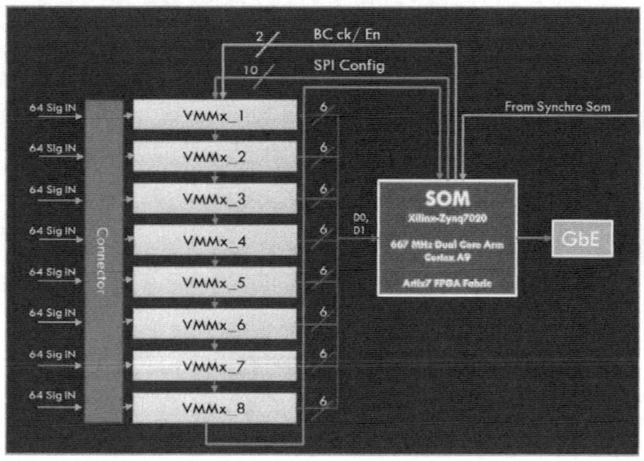

Figure 2. FE-RO module.

3. System on Module (SOM)

The System on Module (SOM) is a very compact, flexible, reliable, high-performance, industrial-grade, deployment-ready embedded computer. It combines a 667 MHz Dual-Core ARM Cortex-A9 processor, NI Linux Real-Time OS and an Artix-7 programmable Xilinx FPGA. A functional block diagram of the SOM is presented in figure 3. Both the real-time processor, whose main functions are communication and signal processing, and the FPGA, usefull to implement high-speed controls, custom timings and triggering, can be programmed with LabVIEW, allowing a very short development time, and a high debugging capacity [11]. In addition, the processor allows to run any application developed for linux-based operative system. Through a single high-density connector, it is possible to access lines dedicated to the main communication standards managed by the processor, as well as 160 digital I / O lines managed directly at the FPGA level: 16 Single-Ended 3.3 V and 144 configurable also as 72 differentials. This powerful combination of resources in a single system allows the management of the slow control of front-end ASICs, the read-out of data and their transfer, through a standard ethernet protocol, towards the system of storage. The SOM is also able to generate and distribute, with a very low skew, all the clock signals necessary for the correct functioning of the apparatus. The computational resources of this system would also allow to implement triggering and filtering strategies in the case it is necessary to reduce the data bandwidth at larger event rate. Radiation tolerance tests were also conducted at the University of Sao Paulo, in the Nuclear and Energy Research Institute (IPEN) reactor. Although data analysis is still in progress, the tests have demostrated that the SOM, exposed to a neutron flux (from 10^4 to 10^6 n/cm^2*s) well above that estimated by the simulations for the NUMEN experiment, at maximum beam intensity, is subject to some effects (single event upset) recoverable by a system reset operation. Due to the flexibility of the system and its described features, its application as read-out electronics for all the spectrometer detectors is promising and reliable.

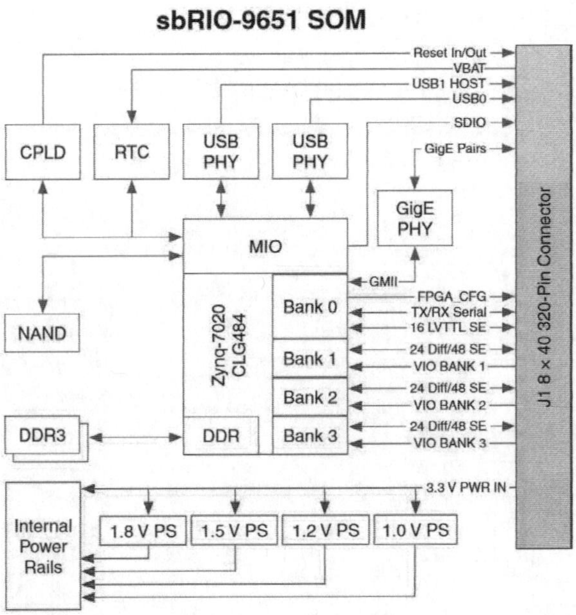

Figure 3. Functional block diagram of the SOM (Courtesy by National Instruments).

4. Status

The developers of the chip sent us a complete system to testing and studying the second version of the chip (VMM2). When all the features that could be of interest for the experiment were investigated and verified, was developed a new read-out board equipped with a SOM that allowed, coupled with the original front-end board, to test the control and the communication between the SOM and the chip. Then was developed a front-end board with VMM2 useful to study the coupling between the whole acquisition system (composed by front-end and read-out custom boards) and some detector prototypes intended to use in the experiment. Meanwhile a new version of the chip (VMM3) was available and the front-end board was upgraded with a new design that is now under construction.

References

[1] Calabretta L *et al.* 2017 *Modern Physics Letters A* **32** 1740009
[2] Cappuzzello F *et al.* 2015 *Journal of Physics: Conference Series* **630** 012018
[3] Agodi C *et al.* 2015 *Nuclear and Particle Physics Proceedings* **265-266** 28 – 30 ISSN 2405-6014 proceedings of the Neutrino Oscillation Workshop
[4] Cappuzzello F *et al.* 2018 *The European Physical Journal A* **54** 72

[5] Cavallaro M *et al.* 2012 *The European Physical Journal A* **48** 59 ISSN 1434-601X
[6] Cappuzzello F *et al.* 2016 *The European Physical Journal A* **52** 167 ISSN 1434-601X
[7] Cappuzzello F *et al.* 2014 *Nuclear Instruments and Methods in Physics Research Section A: Accelerators, Spectrometers, Detectors and Associated Equipment* **763** 314 – 319 ISSN 0168-9002
[8] Cunsolo A *et al.* 2007 *The European Physical Journal Special Topics* **150** 343–346
[9] De Geronimo *et al.* 2013 *IEEE Trans.Nucl.Sci.* **60** 2314–2321
[10] URL `http://www.ni.com/pdf/manuals/376962c.pdf`
[11] URL `http://www.ni.com/white-paper/6983/en/`

IOP Conf. Series: Journal of Physics: Conf. Series **1056** (2018) 012007 doi:10.1088/1742-6596/1056/1/012007

The Front-end for the new focal plane detector for the NUMEN project

D G Bongiovanni[1], C Agodi[1], D Bonanno[2], D Carbone[1], F Cappuzzello[1,3], M Cavallaro[1], G De Geronimo[4], G Gallo[1,3], F Longhitano[2], D Lo Presti[1,3] and S Reito[2] for the NUMEN collaboration

[1] INFN Laboratori Nazionali del Sud, Catania, Italy
[2] INFN Sezione di Catania, Catania, Italy
[3] Dipartimento di Fisica e Astronomia, Università di Catania, Italy
[4] Stony Brook University,100 Nicolls Rd, Stony Brook, NY 11794, USA

E-mail: daniele.bongiovanni@ct.infn.it

Abstract. The design of the front-end electronics for the new tracker of the MAGNEX Focal Plane Detector (FPD) for the NUMEN project is presented. The front-end is based on the VMM chip, developed for ATLAS experiment at CERN. The architecture of the front-end (FE) electronics is designed to be modular and scalable to the final detector. The segmented anode board is designed in order to take advantage of the unique performances of the VMM chip, allowing a digital reconstruction of the track at high event rate. This anode board is connected to front-end by mean of Micro Coax Cable Assembly and does not make use of vacuum connectors. The front-end boards will be placed in air, facilitating in this way the heat dissipation and the connection to the read-out (RO) electronics. An innovative anode read-out strategy allows the reduction of the total number of channels to about 1400 and the measurement of the track at different depth in the detector with 750 μm spatial resolution.

1. Introduction

The foreseen upgrade of the Superconducting Cyclotron of INFN-LNS to reach the high beam intensity necessary for the NUMEN [1, 2, 3, 4] experimental program demands substantial changes in the technologies used for the MAGNEX [5, 6, 7] Focal Plane Detector (FPD) [8] together with new front-end electronics.

The design of front-end (FE) and read-out (RO) electronics was conducted in parallel with the design of the new FPD. In particular, one of the main objectives was to design a modular, scalable, radiation hard architecture, which, in addition, fulfilled the strong requirements in terms of high event rate, easy maintenance and precise synchronization.

The FPD tracker was radically redesigned taking advantage of the possibility to design a brand new full-custom FE and RO electronics.

A new concept for the tracker was developed. The starting point was to extract, for each track, the position of the ions in the volume of the tracker in a fully digital way.

IOP Conf. Series: Journal of Physics: Conf. Series **1056** (2018) 012007 doi:10.1088/1742-6596/1056/1/012007

2. The New Focal Plane Detector

The ionization charge is driven by an electric field in the gas towards a multiplying stage, typically a triple GEM or THGEM. Then the electrons are directed towards a first layer of 750 μm pitch strips.

Each strip of this layer is capacitive coupled to a twin strip in a second layer. The charge pulse inducted in the twin strip is then integrated by the FE and shaped.

The shaped signal is compared to a suitable threshold and the logic output of the comparator identifies the hit strip.

In this scenario the position is extracted by only one strip without the need for the calculation of the center of mass.

The value of the capacitance for each channel of the tracker is optimized to ensure maximum chip resolution in terms of charge and time. The drift time is also measured by the FE at sub-ns resolution.

An innovative scheme for the connection of FE electronics to the anode board was developed. The main objective was to place the FE in air to simplify the heat dissipation, the maintenance and, above all, the interconnections to RO. An advantage of this strategy is also the possibility to adopt countermeasure against to high level of radiation during the experiment.

The working principle of the segmented readout board of the new FPD tracker is represented in figure 1.

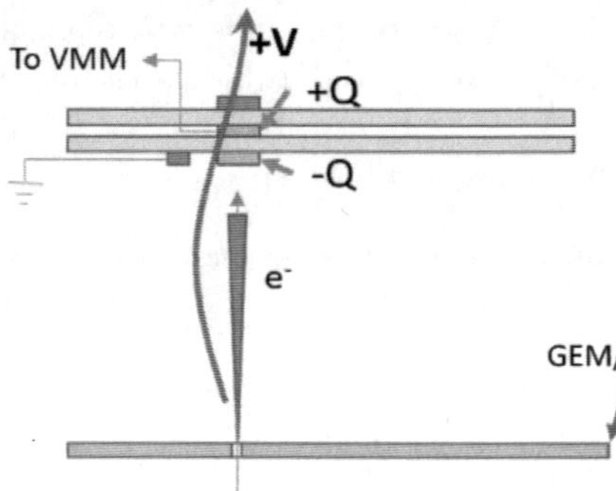

Figure 1. Working principle of the segmented readout board of the new FPD Tracker.

2.1. *Design of the FPD*

The FPD board has been designed in such a way as to cover the active area of the GEM foil.

This board is made up of 4 layers the first consists of a power supply plane in such a way as to be able to apply a potential difference to align the lines of the electric field, the second is the routing plane of the coupled strips, the third is the layer in which are present the strips exposed to the gas and finally the fourth layer is the ground plane in which there are trenches.

In figure 2 shows the presence of tranches that are transversal to the strips and identify the area in which the charge is collected so as to reduce the number of collisions on the individual strips.

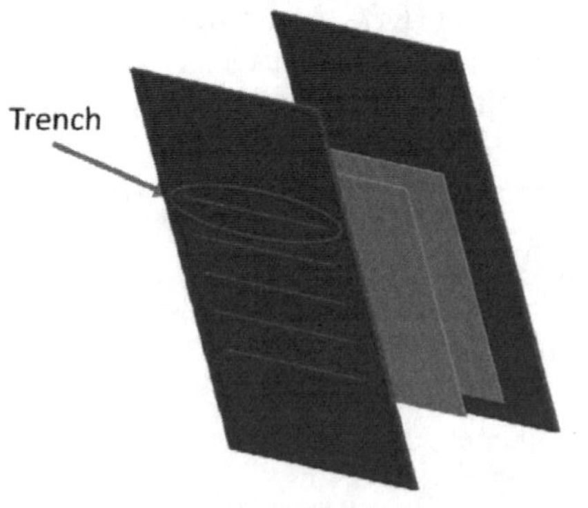

Figure 2. FPD segmented readout board in 3D view

IOP Conf. Series: Journal of Physics: Conf. Series **1056** (2018) 012007 doi:10.1088/1742-6596/1056/1/012007

3. The Front-End

The design of the front-end electronics for the new tracker of the NUMEN Focal Plane Detector is presented. The front-end is based on the VMM chip [9], developed for ATLAS experiment at CERN.

The architecture of the front-end electronics is thought to be modular and scalable to the final dimensions of the detector. The segmented anode board was designed in order to take advantage of the unique performances of the VMM chip [10], allowing a digital reconstruction of the track at high event rate.

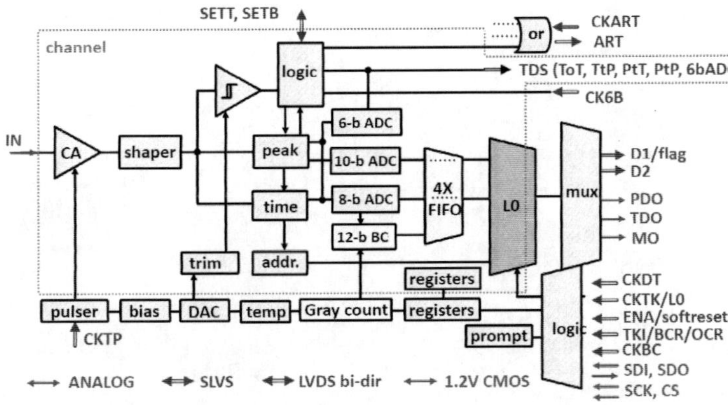

Figure 3. Architecture of the chip VMM3.

Fig. 3 shows the architecture of the VMM3, which is the actual evolution of VMM1, but with a much higher complexity and functionality. Each of the 64 channels provides the peak amplitude and time with respect to the bunch crossing clock signal or other trigger signal in a data driven mode. This is accomplished as follows. Each channel is equipped with a fast comparator with an individually adjustable threshold. When a signal crosses a set threshold a peak detection circuit is enabled.

Neighbour-enable logic allows setting the threshold relatively high and yet recording very small amplitudes. At the peak a time-to-amplitude converter is started and stopped by the trigger signal. The two amplitudes are digitized and stored in a de-randomizing buffer and readout serially with a smart token passing scheme that only reads out the amplitude, timing, and addresses of the channels with information, thus dramatically reducing the data bandwidth required and resulting in a very simple readout architecture.

The VMM3 has 64 channels easing the handling of a large number of channels. It also provides prompt information that can be used to provide pre-trigger signals.

3.1. Mode continuous

In the Numen experiment, the chip will be used using continuous mode.

The ASIC can operate in three modes: direct-outputs, two-phase or analog and continuous or digital.

In continuous (digital) mode a total of 38 bits are generated for each event in the VMM3. The first bit is used as a readout flag, the second is the threshold crossing indicator (it allows discrimination between above-threshold and neighbour events). Next is a 6 bits word for the channel address, followed by 10 bits associated with the peak amplitude, and 20 bits associated with the timing.

The 38-bit word is stored in a 4-events deep de-randomizing FIFO (there is one FIFO per channel) and it is read out using a token-passing scheme where the token is passed first-come first-serve only among those FIFOs that contain valid events. The first token is internally generated as needed and advanced with the token clock signal. The data in the FIFOs is thus sequentially multiplexed to the two digital outputs data0 and data1. The first output data0 is also used as a flag, indicating that events need to be read out from the chip. The external electronics releases a sync signal using the token clock as well (i.e. the token clock provides both advancement and data output synchronization), after which the 38-

IOP Conf. Series: Journal of Physics: Conf. Series **1056** (2018) 012007 doi:10.1088/1742-6596/1056/1/012007

bit data is shifted out in parallel to the data0 and data1 outputs using 19 clock edges of the external data clock.

4. Scheme of connections

The connection between the FPD readout board and FE is made via a vacuum seal board. In this board SAMTEC (SEARAY™ High-Speed/High-Density Array Micro Coax Cable Assembly) connectors have been used.

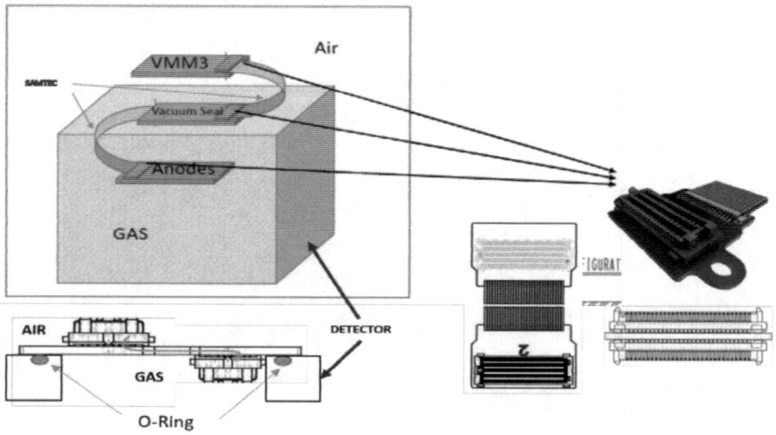

Figure 4. Scheme of connections

This choice allows for easy mounting and no need for vacuum connectors and it was possible to put the electronics in the air, avoiding the problems of heat dissipation in gas.

5. Conclusions

The research and development work on the design of FE-RO electronics for the NUMEN experiment was presented.

In the near future, the FE and RO board produced for a small-scale FPD will be tested, which will adopt all the strategies presented.

During the same period, a dedicated measurement campaign is planned to characterise the radiation tolerance of the FE-RO architecture presented.

References

[1] Cappuzzello F *et al.* 2015 *Journal of Physics: Conference Series* **630** 012018
[2] Agodi C *et al.* 2015 *Nuclear and Particle Physics Proceedings* **265-266** 28-30 ISSN 2405-6014
[3] Cappuzzello F *et al.* 2018 *Eur. Phys. J.* A **54** 72
[4] Cappuzzello F *et al.* 2015 *Eur. Phys. J.* A **51** 145
[5] Cappuzzello F *et al.* 2016 *Eur. Phys. J.* A **52** 167
[6] Cappuzzello F *et al.* 2014 *Nucl. Instrum. Methods Phys. Res.* A **763** 314-319 ISSN 0168-9002
[7] Cunsolo A *et al.* 2007 *Eur. Phys. J.* A ST **150** 343
[8] Cavallaro M *et al.* 2012 *The European Physical Journal* A **48** 59 ISSN 1434-601X
[9] De Geronimo G *et al.* 2007 *IEEE Transactions on Nuclear Science* **54** 541
[10] De Geronimo G *et al.* 2013 *IEEE Transactions on Nuclear Science* **60** 3

IOP Conf. Series: Journal of Physics: Conf. Series **1056** (2018) 012008 doi:10.1088/1742-6596/1056/1/012008

Light yield studies for SoLid phase I.

Delphine Boursette for the SoLid collaboration
LAL, Univ Paris-Sud, CNRS/IN2P3, Université Paris-Saclay, Orsay, France
E-mail: delphine.boursette@lal.in2p3.fr

Abstract. The SoLid experiment is searching for sterile neutrinos at the BR2 nuclear research reactor. Anti-neutrinos are detected through inverse beta decays (producing a positron and a neutron in delayed coincidence) with a very segmented detector made of thousands of scintillating cubes. SoLid has a very innovative hybrid technology with two different scintillators which have different temporal signature: polyvynil-toluene cubes (PVT) to detect the positrons and ^{6}LiF:ZnS sheets to detect the neutrons. The 288 kg detector prototype (SM1) took data in 2015. It demonstrated the detection principle and background rejection efficiency. The construction of SoLid phase I ($\sim$ 1.6 t) is now finished and data taking has started. To improve the energy resolution of SoLid phase I, we have tried to increase the light yield studying the PVT scintillator design. To study the positron light yield on the PVT, we have built a test bench with a ^{207}Bi source. We have improved the design of the cubes, their wrapping or the type and the configuration of the fibers. After its construction, the SoLid detector has been calibrated with a ^{22}Na source and preliminary results seem to show that the light yield has increased by a factor 3 compared to SM1 so it should improve the energy resolution of the positron from 21% for SM1 down to 12% for SoLid.

Introduction

Different experiments have seen anomalies in the neutrino flux which could be explained by the oscillation into a new sterile neutrino [1]. To test this hypothesis, the SoLid experiment will look for neutrino oscillations at very short baseline (~6-9 m) [2]. It is a reactor neutrino experiment at SCK•CEN (Mol, Belgium) which will search for oscillations to sterile neutrinos, measure precisely the ^{235}U flux and spectrum, and demonstrate the ability of a neutrino detector to monitor reactors. The oscillations to sterile neutrinos should be visible both as a change in the anti-neutrino flux with the distance and as a distortion of their energy spectrum. For SoLid phase I, we wanted to optimize the light yield compared to SM1 to get a better energy resolution. It would also allow us to lower the detection threshold to get a more efficient neutron trigger and improve the background rejection. Furthermore we wanted to have a better detector uniformity.

1. The SoLid detector

The SoLid experiment detects neutrinos through inverse beta decay: $\overline{\nu}_e + p \rightarrow e^+ + n$. The detector is highly segmented with cubes of 5×5×5 cm^3. The phase I detector will consist of 50 frames of 16 × 16 cubes. It has an innovative technology as it is composed of two scintillators with different time response. Positrons (fast scintillation signal) are detected in cubes made of PVT. They also give us a measurement of the anti-neutrino energy. Neutrons, after thermalization, can capture on the ^{6}LiF:ZnS sheets (slow scintillation signal) put on each PVT cube. The cubes are wrapped in Tyvek reflective coating and wavelength shifting fibers bring the scintillating light to silicon photomultiplier (SiPM) whose pulses are digitised (40 MHz). Mirrors were placed at the end of the fibers which do not have a SiPM. An efficient pulse shape analysis allows us to distinguish the signals from neutrons in ^{6}LiF:ZnS and positrons in PVT [3]. These delayed coincidences sign a $\overline{\nu}_e$ interaction. To reduce backgrounds, SoLid can take advantage of its time, spatial topology and energy measurements.

2. The PVT test bench

In order to improve the energy resolution, we have built a test bench to study the positron light yield in the PVT (Fig. 1). We use a ^{207}Bi source which produces conversion electrons with a peak energy of 1 MeV by decays through electron capture to excited states of ^{207}Pb. An external trigger with two photomultiplier tubes (PMTs) and a 110 μm scintillator triggers in coincidence on the 1 MeV conversion electrons only and reject the source gamma background. Above this external trigger, we put a SoLid type detector unit composed of one or several cubes with optical fibers and SiPM. We can change the wrapping of the cube, the number of ZnS sheets, the number and type of fibers and their mirrors or the cube position along the fiber. Using the single photo-avalanche (PA) peaks in the spectrum, we can convert the charge of the SiPM pulses in PA (Fig. 2) and compare the position on the 1 MeV peak for different configurations to measure the improvements. We have measured a systematic error of 5 % and a statistical error of 0.2 %.

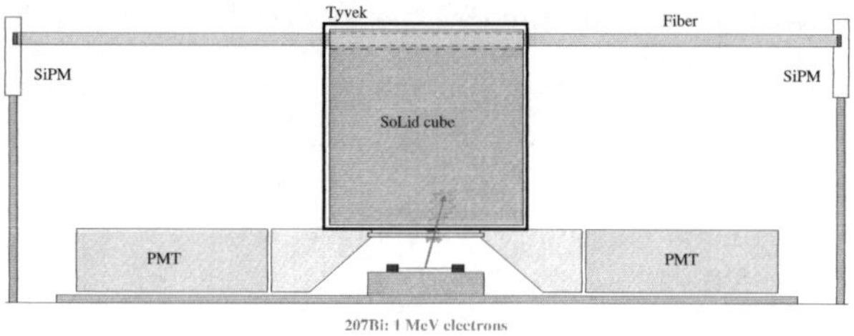

Figure 1. The PVT test bench with the exteral trigger composed of two PMTs and the SoLid type detector unit above it.

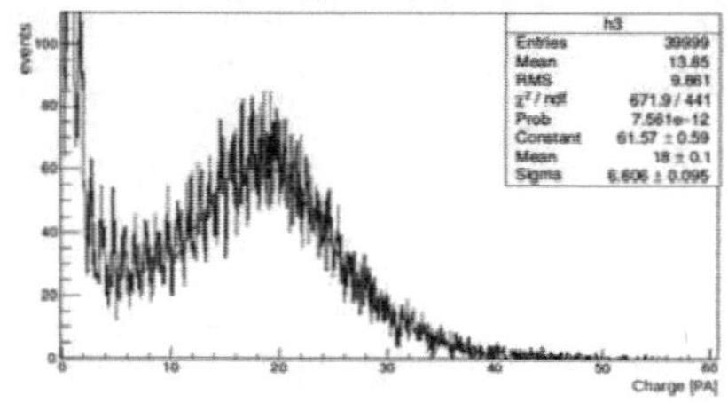

Figure 2. Distribution of the SiPM charge signal in PA on the test bench. We fit in red the 1 MeV peak from ^{207}Bi.

3. Results of PVT light yield studies

After testing several materials and configurations for the SoLid detector, we have found the improvements shown in Table 1. Adding a second ZnS layer reduces the light yield but it also increases the neutron detection efficiency and reduces its capture time (to reduce background).

Table 1. Examples of light yield improvements found with the PVT test bench for a cube located at the center of the fibers.

Detector component	**SM1**	**SoLid phase I**	**Gain**
Wrapping of the cubes	Tyvek of 75 g·m^{-3}	Tyvek of 105 g·m^{-3}	+10% per cube
Number of ZnS sheets	1	2	-4 per cube %
Optical fibers	single-clad fibers	double-clad fibers	+15 % per fiber
Mirrors	aluminium mirror	aluminised mylar mirror	+10 % per fiber
Number of fibers	2 fibers	4 fibers	+40% per cube

To get an idea of the increase in light yield expected for SoLid phase I compared to SM1, we have done measurements with one cube in the closest setup as possible to the 2 detectors. We have also measured the attenuation length of the different fibers used to extrapolate this one cube measurement to a whole plane (16 × 16 cubes).

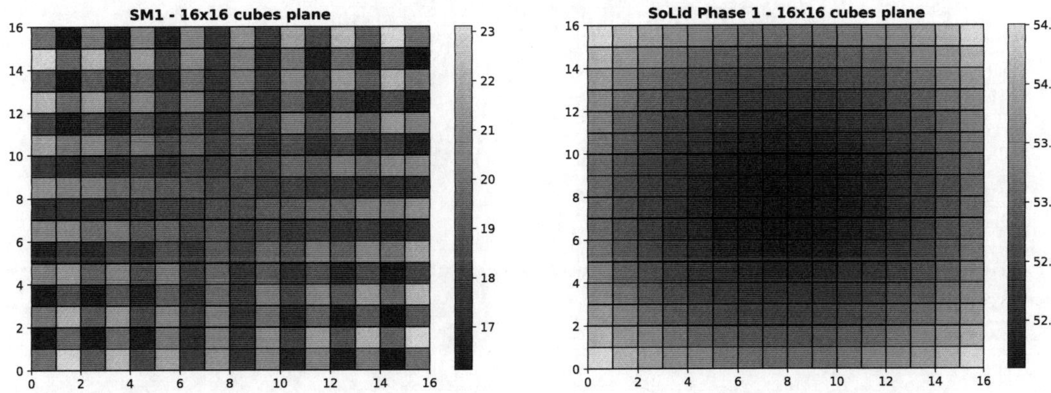

Figure 3. Detector plane light-yield maps for SM1 and SoLid Phase 1 extrapolated from our light-yield and attenuation measurements. The average light-yield are 18.9 and 52.3 PA/MeV for SM1 and SoLid Phase 1 respectively. The maximal cube difference is only 4 % for the Phase 1 compared to 43 % for SM1.

We expect thus an increase in light yield of 180% for SoLid phase I compared to SM1.

4. Calibration of the SoLid detector

After their construction, each frame of the SoLid detector (16 × 16 cubes) has been calibrated with ^{22}Na, ^{252}Cf and AmBe sources. An automated system has been used to move the sources and qualify all the cubes in a frame. The ^{22}Na source emits gammas at 511 keV and 1270 keV in coincidence. While one of the two 511 keV gammas was used to start the external trigger, the Compton edge of the 1270 keV gamma interaction in the cubes of the detector could be fitted to measure the light yield (Fig. 4).

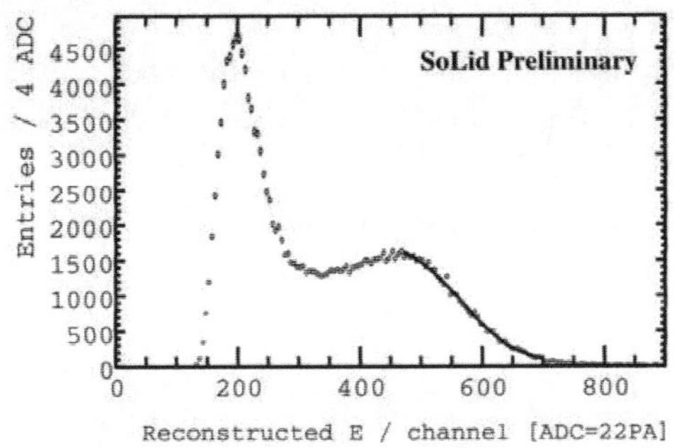

Figure 4. The ^{22}Na energy spectrum measured in a cube of the SoLid detector. The first peak around 200 ADC is due to the 511 keV gammas and the compton edge around 500 ADC corresponds to the 1270 keV gammas.

IOP Conf. Series: Journal of Physics: Conf. Series **1056** (2018) 012008 doi:10.1088/1742-6596/1056/1/012008

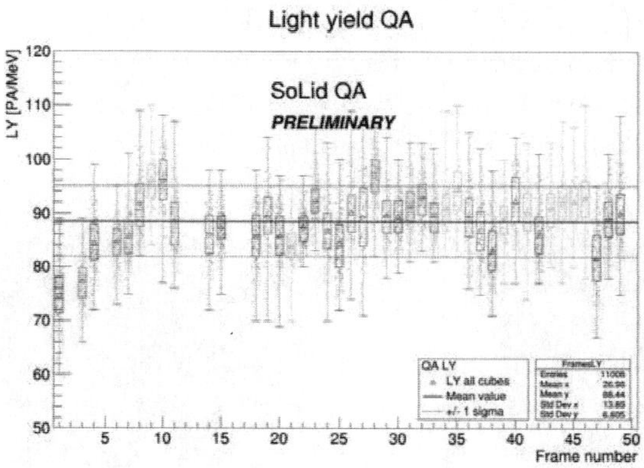

Figure 5. Average number of PA/MeV/cube (not corrected from cross-talk) for each frame. A frame represents 16 $\times$ 16 cubes.

All the frames are very uniform in light yield (Fig. 5). A preliminary light yield of 73 PA/MeV/cube was detected in average after cross-talk substraction which should lead to an energy resolution of 12%. This is an increase of the light yield by a factor three compared to SM1 which is better than expected. This bigger increase compared to test bench measurements may be due to the different electronics used for the test bench or to light leaks to the neighbouring cubes which can be collected in the detector while we use only one cube on the test bench.

Conclusion

With the test-bench measurements, we have found improvements to increase the light yield by changing the wrapping, the type and number of optical fibers, mirrors and cube design. These changes have improved the uniformity of the SoLid detector and according to preliminary results it seems that the light yield has been multiplied by 3 from SM1 to SoLid phase I. This should improve the energy resolution for SoLid phase I down to 12 % at 1 MeV which is very promising for its sensitivity to the reactor anomaly.

References

[1] C. Giunti et al. "Pragmatic View of Short-Baseline Neutrino Oscillations". In: *Phys. Rev.* D88 (2013), p. 073008. DOI: `10.1103/PhysRevD.88.073008`. arXiv: `1308.5288 [hep-ph]`.

[2] Y. Abreu et al. "A novel segmented-scintillator antineutrino detector". In: *JINST* 12.04 (2017), P04024. DOI: `10.1088/1748-0221/12/04/P04024`. arXiv: `1703.01683 [physics.ins-det]`.

[3] S. Vercaemer. "Neutron identification in the SoLid experiment ". In: *The European Physical Society Conference on High Energy Physics (EPS-HEP2015)* 234 (2016).

IOP Conf. Series: Journal of Physics: Conf. Series **1056** (2018) 012009 doi:10.1088/1742-6596/1056/1/012009

The CUORE and CUORE-0 experiments at LNGS

C Alduino[1], K Alfonso[2], F T Avignone III[1], O Azzolini[3], G Bari[4], F Bellini[5,6], G Benato[7], A Bersani[8], M Biassoni[9], A Branca[10,11], C Brofferio[12,9], C Bucci[13], A Camacho[3], A Caminata[8], L Canonica[14,13], X G Cao[15], S Capelli[12,9], L Cappelli[7,16,13], L Cardani[6], P Carniti[12,9], N Casali[6], L Cassina[12,9], D Chiesa[12,9], N Chott[1], M Clemenza[12,9], S Copello[17,8], C Cosmelli[5,6], O Cremonesi[9], R J Creswick[1], J S Cushman[18], A D'Addabbo[13], D D'Aguanno[13,19], I Dafinei[6], C J Davis[18], S Dell'Oro[20,13,21], M M Deninno[4], S Di Domizio[17,8], M L Di Vacri[13,22], V Dompè[5,6], A Drobizhev[7,16], D Q Fang[15], M Faverzani[12,9], E Ferri[9], F Ferroni[5,6], E Fiorini[9,12], M A Franceschi[23], S J Freedman[16,7,a], B K Fujikawa[16], A Giachero[12,9], L Gironi[12,9], A Giuliani[24], L Gladstone[14], P Gorla[13], C Gotti[12,9], T D Gutierrez[25], K Han[26], K M Heeger[18], R Hennings-Yeomans[7,16], H Z Huang[2], G Keppel[3], Yu G Kolomensky[7,16], A Leder[14], C Ligi[23], K E Lim[18], Y G Ma[15], L Marini[17,8], M Martinez[5,6,27], R H Maruyama[18], Y Mei[16], N Moggi[28,4], S Morganti[6], S S Nagorny[13,21], T Napolitano[23], M Nastasi[12,9], C Nones[29], E B Norman[30,31], V Novati[24], A Nucciotti[12,9], I Nutini[13,21], T O'Donnell[20], J L Ouellet[14], C E Pagliarone[13,19], M Pallavicini[17,8], V Palmieri[3], L Pattavina[13], M Pavan[12,9], G Pessina[9], C Pira[3], S Pirro[13], S Pozzi[12,9], E Previtali[9], F Reindl[6], C Rosenfeld[1], C Rusconi[1,13], M Sakai[2], S Sangiorgio[30], D Santone[13,22], B Schmidt[16], J Schmidt[2], N D Scielzo[30], V Singh[7], M Sisti[12,9], L Taffarello[10], F Terranova[12,9], C Tomei[6], M Vignati[6], S L Wagaarachchi[7,16], B S Wang[30,31], H W Wang[15], B Welliver[16], J Wilson[1], K Wilson[1], L A Winslow[14], T Wise[18,32], L Zanotti[12,9], G Q Zhang[15], S Zimmermann[33] and S Zucchelli[28,4]

[1] Dept of Physics and Astronomy, Univ of South Carolina, Columbia, SC 29208, USA
[2] Dept of Physics and Astronomy, Univ of California, Los Angeles, CA 90095, USA
[3] INFN – Laboratori Nazionali di Legnaro, Legnaro (Padova) I-35020, Italy
[4] INFN – Sezione di Bologna, Bologna I-40127, Italy
[5] Dipartimento di Fisica, Sapienza Università di Roma, Roma I-00185, Italy
[6] INFN – Sezione di Roma, Roma I-00185, Italy
[7] Department of Physics, University of California, Berkeley, CA 94720, USA
[8] INFN – Sezione di Genova, Genova I-16146, Italy
[9] INFN – Sezione di Milano Bicocca, Milano I-20126, Italy
[10] INFN – Sezione di Padova, Padova I-35131, Italy
[11] Dipartimento di Fisica e Astronomia, Università di Padova, I-35131 Padova, Italy
[12] Dipartimento di Fisica, Università di Milano-Bicocca, Milano I-20126, Italy
[13] INFN – Laboratori Nazionali del Gran Sasso, Assergi (L'Aquila) I-67100, Italy
[14] Massachusetts Institute of Technology, Cambridge, MA 02139, USA
[15] Shanghai Institute of Applied Physics, Chinese Academy of Sciences, Shanghai 201800, China
[16] Nuclear Science Division, Lawrence Berkeley Nat Lab, Berkeley, CA 94720, USA
[17] Dipartimento di Fisica, Università di Genova, Genova I-16146, Italy

IOP Conf. Series: Journal of Physics: Conf. Series **1056** (2018) 012009 doi:10.1088/1742-6596/1056/1/012009

[18] Wright Laboratory, Department of Physics, Yale University, New Haven, CT 06520, USA
[19] Dipartimento di Ingegneria Civile e Meccanica, Università degli Studi di Cassino e del Lazio Meridionale, Cassino I-03043, Italy
[20] Center for Neutrino Physics, Virginia Polytechnic Institute and State University, Blacksburg, Virginia 24061, USA
[21] INFN – Gran Sasso Science Institute, L'Aquila I-67100, Italy
[22] Dip di Scienze Fisiche e Chimiche, Università dell'Aquila, L'Aquila I-67100, Italy
[23] INFN – Laboratori Nazionali di Frascati, Frascati (Roma) I-00044, Italy
[24] CSNSM, Univ. Paris-Sud, CNRS/IN2P3, Univ. Paris-Saclay, 91405 Orsay, France
[25] Physics Dept, California Polytechnic State Univ, San Luis Obispo, CA 93407, USA
[26] INPAC and School of Physics and Astronomy, Shanghai Jiao Tong University; Shanghai Laboratory for Particle Physics and Cosmology, Shanghai 200240, China
[27] Lab de Fisica Nuclear y Astroparticulas, Univ de Zaragoza, Zaragoza 50009, Spain
[28] Dipartimento di Fisica e Astronomia, Università di Bologna, Bologna I-40127, Italy
[29] Service de Physique des Particules, CEA / Saclay, 91191 Gif-sur-Yvette, France
[30] Lawrence Livermore National Laboratory, Livermore, CA 94550, USA
[31] Dept of Nuclear Engineering, University of California, Berkeley, CA 94720, USA
[32] Department of Physics, University of Wisconsin, Madison, WI 53706, USA
[33] Engineering Division, Lawrence Berkeley National Lab, Berkeley, CA 94720, USA

[a] Deceased

E-mail: chiara.brofferio@unimib.it

Abstract. The Cryogenic Underground Observatory for Rare Events (CUORE) is the first bolometric experiment searching for neutrinoless double beta decay that has been able to reach the 1-ton scale. The detector consists of an array of 988 TeO_2 crystals arranged in a cylindrical compact structure of 19 towers. The construction of the experiment and, in particular, the installation of all towers in the cryostat was completed in August 2016 and commissioning started in fall 2016. The experiment has completed the pre-operation phase and is currently in data taking. We present here the achievements of CUORE during the commissioning phase and the limit on the ^{130}Te half-life for the neutrinoless double beta decay that has been released after the first 3 weeks of collected data. Physics results from CUORE-0 will also be updated.

1. Introduction

Neutrinoless Double Beta Decay (0νDBD) searches [1][2][3] represent a unique tool to assess the Dirac/Majorana nature of neutrino and to check the Lepton Number Conservation law. The experimental approach is usually based on the observation of a large number of nuclei, stable in normal beta decay but for which a double weak interaction process, changing the nuclear charge by two units, is energetically favorable. The decay involving the simultaneous emission of two electrons and two neutrinos (2νDBD) is allowed by the Standard Model of Electroweak Interactions and has been observed for several nuclides [4]. However, if neutrinos have a Majorana nature (and mass), a process without the emission of any neutrino is possible. A 0νDBD is an unambiguous signal of a Majorana mass and of a violation of the lepton number conservation. Moreover, the study of such decay can put a constraint on the absolute neutrino mass scale, which is at the moment unknown [5].

Most sensitive experiments are presently based on the homogeneous approach in which the nuclei under observation are part of the detector itself. The sensitivity of such experiments scales with the square root of the exposure (the live time times the detector mass) and the inverse of the observed background rate, since it is supposed that this also scales with the detector mass.

Low temperature detectors (LTD) are naturally suitable for this approach. These detectors are simply based on the possibility to register a temperature increase in a crystal, acting as absorber, when a particle interacts with it. In fact, the heat capacity of any cold diamagnetic and dielectric crystal is proportional to the cube of the ratio between the operating and Debye temperatures and can therefore become so small that even the tiny energy released by a particle in form of heat can be revealed by the temperature increase of the crystal by means of a suitable thermal sensor. Many interesting compounds have already been studied as candidates for this approach [6] with excellent results both on intrinsic radio-purity and on energy resolution, which is fundamental to reduce the number of counts due to 2νDBD at the Q-value of the decay in the 2-electrons energy sum spectrum, where the events from 0νDBD are expected. This background is in fact the only unavoidable one that will always be present.

2. The TeO_2 choice and CUORE-0

^{130}Te is an optimal candidate to search for 0νDBD [7] due to its high transition energy (2527.5 keV), and large isotopic abundance (34.2 %), which allows performing a sensitive experiment with natural tellurium. TeO_2 has demonstrated to be an optimal choice for LTD crystals to be used in 0νDBD searches, thanks to its high enough Debye temperature [8], good crystal properties, very low intrinsic radioactivity [9] and favorable Te mass percentage in the compound (27%). TeO_2 LTDs have been pioneered by the Milano group in a series of constantly increasing mass (from 6 g to 6.8 kg) experiments carried out at Laboratori Nazionali del Gran Sasso (LNGS) [10], which opened the way to CUORE [11], a 1-ton LTD experiment for ^{130}Te 0νDBD.

Suggested by Ettore Fiorini in 1998, CUORE physics potential attracted many international collaborators, that convinced the community (and the funding agencies) of the feasibility of the project with a smaller (41 kg) experiment, CUORICINO [12]. This experiment was in fact able to show not only the technical possibility of scaling in mass by using arrays of LTD towers, but also that the technique was mature to compete in background reduction (0.169 c/keV kg y) and 0νDBD half-life limits (2.8×10^{24} y at 90% C.L.) [13]. Once funded, CUORE detector was designed and optimized under all the possible aspects [14][15], to push the sensitivity as far as possible [16]. The demonstrator of all these efforts has been the single tower CUORE prototype, CUORE-0.

CUORE-0 consisted of 52 $5\times5\times5$ cm^3 TeO_2 crystals assembled in a tower of 13 floors for a total TeO_2 mass of 39 kg. All the technical details of the detector and its performance can be found in the literature [17]. CUORE-0 was not only the first prototype of a CUORE tower, but a small scale competing experiment, that after a 9.8 kg y exposure was able to improve the background (0.058 c/keV kg y) and 0νDBD half-life limits (2.7×10^{24} y at 90% C.L.) with respect to CUORICINO [18] and to measure with the highest precision ever obtained the half-life of ^{130}Te 2νDBD ($T_{2\nu} = (8.2 \pm 0.2_{stat} \pm 0.6_{syst}) \times 10^{20}$ y) [19]. Many other searches are still under study, like the decays of ^{120}Te and of ^{130}Te on the first 0^+ excited state, and will be published soon.

A detailed Monte Carlo simulation that exploits the information about the contaminations of materials obtained through radio-assay screening campaigns and bolometric measurements was built and fine-tuned on CUORE-0 data with excellent results [19]. The information extracted from CUORE-0 background model were then used to tune the general CUORE Monte Carlo simulation [20], that confirmed that the goal of a Background Index (BI) in the Region Of Interest (ROI) of 10^{-2} c/keV kg y was within reach. With CUORE-0 the 5 keV FWHM energy resolution goal was also achieved, on the full 9.8 kg y exposure.

3. CUORE construction and commissioning

The CUORE detector is composed by a closely packed array of 19 towers CUORE-0-like, for a total of 988 TeO_2 bolometers operated at a temperature of ~ 10 mK in a custom-made cryostat installed at Laboratori Nazionali del Gran Sasso (LNGS) in Italy. Each bolometer is equipped with a neutron transmutation doped (NTD) thermal sensor and a silicon heater, for gain stabilization [21][22]. With a total detector mass around 740 kg of TeO_2 (206 kg of ^{130}Te), CUORE is the first ton-scale cryogenic detector ever put into operation.

To cope with the strong requirements for surface (re)contamination, the detector components were produced, handled and cleaned according to specific protocols, the towers were assembled by use of specific glove-boxes always flushed with nitrogen to avoid ^{222}Rn or ^{210}Pb [23], and their installation in the cryostat was performed using a radon free modular clean room [24]. To shield the detector from gamma and neutron environmental background different layers of copper, lead (70 ton aprox, of which 8 ton inside the cryostat, at low temperature), borated polyethylene and boric acid are used.

The installation of the 19 towers was successfully completed in Summer 2016, obtaining 984 functioning bolometers out of 988. The cryostat interfaces and radiation shields were assembled in the following months. Cool down started in Dec. 2016. The custom-made dilution refrigerator system works without LHe and the cooling at intermediate temperatures (~ 40 K and ~ 4 K) is based on five pulse tube cryo-coolers. The dimensions, experimental volume (~ 1 m^3), mass (~17 t), and cooling power (3 μW at 10 mK) make it the largest and most powerful cryogen-free dilution refrigerator system in operation. Nonetheless, it took almost 1 month to cool to base temperature. A stable temperature of ~ 7 mK was reached on January 27, 2017. We then started the detector pre-operation phase, to optimize the signal readout, the mechanical and electrical noise reduction and the working points of the bolometers. In April 2017 the optimization was not yet completed but the Collaboration decided to start a preliminary science run to extract the very first hints on energy resolution and BI in the ROI.

4. First data-set results

In May 2017, we collected three weeks of physics data bracketed by two calibration periods. During calibration 12 "sausage-like" strings containing ^{232}Th sources are temporarily deployed inside the detector region in order to guarantee an approximately uniform γ-ray illumination of the detectors [25]. Six γ-lines (from 239 keV to 2615 keV) are then used to perform the energy calibration of the 984 bolometers. During the physics run we acquired an exposure of 38.1 kg y of TeO_2 (10.6 kg y of ^{130}Te), already greater than the total exposure collected by CUORE-0.

For this initial analysis 889 detectors (~90 %) were used, discarding those that would have required more efforts on stability and noise reduction. In our analysis approach, once the calibration is successfully applied, the physics data are blinded. This is obtained by introducing an artificial peak at the Q-value. The model and fitting strategy are then optimized and fixed before unblinding. A detailed description of the full procedure, that was developed for CUORE-0 and then used on this first dataset, is available in the literature [26]. The physics spectrum, see figure 1, is built after applying a series of selection criteria aimed at improving the experimental sensitivity. We therefore remove periods of low quality data (caused by noisy laboratory conditions) and reject pile-up events. Then we select only signals consistent with a proper template waveform (pulse shape analysis) in order to identify real particle events and finally we exclude events that simultaneously trigger more than one crystal, to reduce the background due to events depositing energy in multiple crystals.

For this first run we evaluated an overall detection efficiency of (55.3 ± 3.0)%, which includes a (88.35 ± 0.09)% probability that a 0νDBD is fully contained in a single crystal and a (62.6 ± 3.4)% probability that a physics event is not discarded when the selection criteria are applied.

We evaluate the detector energy resolution near the ROI by fitting the 2615 keV line in the physics spectrum. The harmonic mean of the detector FWHM resolutions for this preliminary run has been found to be 7.9 ± 0.6 keV.

To estimate the background in the ROI and the 0νDBD rate ($\Gamma_{0\nu}$) we perform an Unbinned Extended Maximum Likelihood fit in the [2465-2575] keV range (the ROI) with the same procedure used for CUORE-0 [26]. The best-fit values are $0.98^{+0.17}_{-0.15}\times10^{-2}$ counts/(keV kg y) for the background rate, and $(-0.03^{+0.07}_{-0.04}\,\text{(stat)} \pm 0.01\text{(syst)})\times10^{-24}\,\text{y}^{-1}$ for $\Gamma_{0\nu}$. We find no evidence for the 0νDBD of ^{130}Te and we can only calculate an upper limit of $\Gamma_{0\nu}$ by integrating the profile likelihood in the physical region ($\Gamma_{0\nu} \geq 0$). This corresponds to a half-life lower limit of $T_{1/2}^{0\nu} > 4.5 \times 10^{24}$ y (90% C.L.).

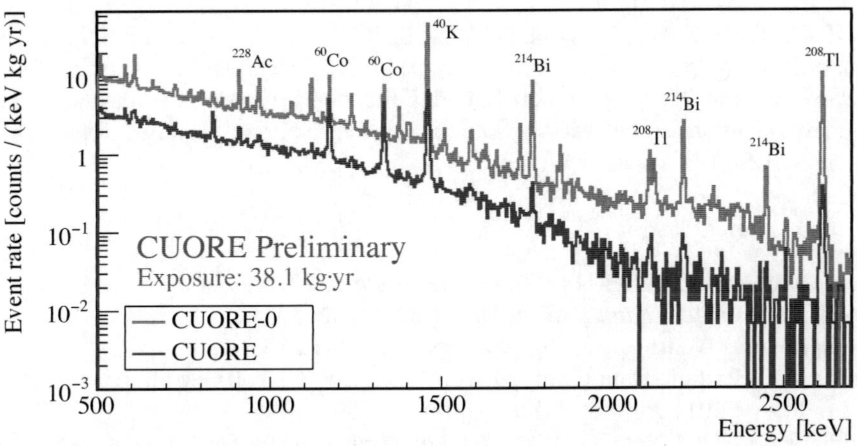

Figure 1: Comparison of physics spectra in the gamma region measured with CUORE and CUORE-0, with prominent γ-lines labeled.

Finally, we combine the first results from CUORE with those obtained from CUORE-0 [18] and Cuoricino [13] with 9.8 kg y and 19.8 kg y exposure of ^{130}Te, respectively. The half-life lower limit obtained by combining the profile negative-log-likelihood curves of the three experiments is $T_{1/2}^{0\nu} > 6.6 \times 10^{24}$ y (90% C.L.).

The combined half-life limit can be then interpreted as a limit on the effective Majorana neutrino mass ($m_{\beta\beta}$) in the framework of models of 0νDBD mediated by light Majorana neutrino exchange. When using the phase-space factors from [27] and nuclear matrix elements from a broad range of recent calculation models [28][29][30][31][32], with the nucleon axial coupling constant $g_A = 1.27$ we get $m_{\beta\beta} <$ (210-590) meV at 90% C.L., depending on the nuclear matrix element estimates employed. We do not consider other g_A values since there is still quite a lack of certainties on what to expect to be the correct number to be used for 0νDBD. See for instance the discussions in [33] and [34].

5. Another optimization campaign and a new science run

After this preliminary science run another optimization campaign of the detector operating conditions was started. We implemented an active noise cancellation system on the pulse tube cryo-coolers using micro-step motor linear drives to control the relative phases of the pulse tube pressure oscillations and fixing the configuration that maximizes the noise cancellation, leveraging the interference between the noise sources [35]. We improved the electrical grounding of the experiment and we performed a temperature scan of the base temperature of the detectors, choosing to work at 15 mK. We optimized the software bandwidth for the pulse amplitude analysis and extended the software trigger window from 5 s to 10 s.

A new science run was then carried out during August 2017. The corresponding dataset as well as the one described in this paper were completely re-processed with a slightly improved analysis procedure and results were presented after the CNNP conference, during CUORE inauguration at LNGS on October, 23rd and will be published on PRL [36].

Acknowledgments
The CUORE Collaboration thanks the directors and staff of the Laboratori Nazionali del Gran Sasso and the technical staff of our laboratories. This work was supported by the Istituto Nazionale di Fisica Nucleare (INFN); the Director, Office of Science, of the U.S. Department of Energy under Contract Nos. DE-AC02-05CH11231 and DE-AC52-07NA27344; the DOE Office of Nuclear Physics under Contract Nos. DE-FG02-08ER41551 and DEFG03-00ER41138; the National Science Foundation under Grant Nos. NSF-PHY-0605119, NSF-PHY-0500337, NSF-PHY-0855314, NSF-PHY-0902171, and NSF-PHY-0969852; the Alfred P. Sloan Foundation; the University of Wisconsin Foundation; and Yale University. This research used resources of the National Energy Research Scientific Computing Center (NERSC).

References

[1] Cremonesi O and Pavan M 2014 *Adv. High Energy Phys.* **2014** 951432
[2] Dell'Oro S *et al.* 2016 *Adv. High Energy Phys.* **2016** 2162659
[3] Giuliani A and Poves A 2012 *Adv. High Energy Phys.* **2012** 857016
[4] Patrignani C *et al.* (Particle Data Group) 2016 *Chin. Phys.* C 40:100001
[5] Ge S and Lindner M 2017 *Phys Rev* D 95 033003
[6] Kim Y H 2017 *Proc. 17th Int. Workshop on Low Temperature Detectors* Kurume City, Japan, July 17-21. To appear on a special issue of *J Low Temp Phys*
[7] Barabash A S 2012 *J. Phys. G: Nucl. Part. Phys.* **39** 085103
Artusa D R *et al.* 2014 *Eur. Phys. J.* C **74** 3096
[8] Barucci M 2001 *J. Low Temp. Phys.* **123** 303
[9] Arnaboldi C *et al.* 2010 *J. Cryst. Growth* **312** 2999
[10] Brofferio C and Dell'Oro S 2018 *arXiv:* 1801.03580 [hep-ex]
[11] Artusa D R *et al.* 2015 *Adv. High Energy Phys* **2015** 879871
[12] Arnaboldi C *et al.* 2008 *Phys. Rev.* C **78** 035502
[13] Andreotti E *et al.* 2011 *Astropart. Phys.* **34** 822
[14] Alessandria F *et al.* 2012 *Astropart. Phys.* **35** 839
[15] Alessandria F *et al.* 2013 *Astropart. Phys.* **45** 13
[16] Alduino C *et al.* 2017 *Eur. Phys. J.* C **77** 532
[17] Alduino C *et al.* 2016 *J. Instrum.* **11** P07009
[18] Alfonso K *et al.* 2015 *Phys. Rev. Lett.* **115** 102502
[19] Alduino C *et al.* 2017 *Eur. Phys. J.* C **77** 13
[20] Alduino C *et al.* 2017 *Eur. Phys. J.* C **77** 543
[21] Andreotti E *et al.* 2012 *Nucl. Instrum. Meth.* A **664** 161–170
[22] Carniti P *et al.* 2017 *arXiv*:1710.05565 [physics.ins-det]
[23] Buccheri A *et al.* 2014 *Nucl. Instrum. Meth.* A **768** 130
[24] Benato G *et al* 2018 *J. Instrum.* **13** P01010
[25] Cushman J S *et al.* 2017 *Nucl. Instrum. Meth.* A **844** 32–44
[26] Alduino C *et al.* 2016 *Phys. Rev.* C **93** 045503
[27] Kotila J and Iachello F 2012 *Phys. Rev.* C **85** 034316
[28] Barea J, Kotila J and Iachello F 2015 *Phys. Rev.* C **91** 034304
[29] Simkovic F, Rodin V, Faessler A and Vogel P 2013 *Phys. Rev.* C **87** 045501
[30] Hyvarinen J and Suhonen J 2015 *Phys. Rev.* C **91** 024613
[31] Menendez J, Poves A, Caurier E and Nowacki F 2009 *Nucl. Phys.* A **818** 139
[32] Rodriguez T R and Martinez-Pinedo G 2010 *Phys. Rev. Lett.* **105** 252503
[33] Suhonen J 2017 *Phys. Rev.* C **96** 055501
[34] Suhonen J 2017 *Frontiers in Physics* **5** 55
[35] D'Addabbo A *et al.* 2017 *arXiv*:1712.02753 [physics.ins-det]
[36] Alduino C *et al.* 2018 *Phys. Rev. Lett.* in the press (arXiv:1710.07988 [nucl-ex])

Data reduction for experimental measurements within the NUMEN project

S Calabrese[a,b], F Cappuzzello[a,b], C Agodi[a], M Cavallaro[a], D Carbone[a], L Acosta[c], D Bonanno[d], D Bongiovanni[a], I Boztosun[e], D Calvo[f], E R Chàvez Lomeli[c], F Delaunay[f,g], N Deshmukh[a], P Finocchiaro[a], M Fisichella[f], A Foti[d], G Gallo[b,d], A Hacisalihoglu[a,h], F Iazzi[f,i], R Introzzi[f], G Lanzalone[a,j], R Linares[k], F Longhitano[d], D Lo Presti[b,d], N Medina[l], A Muoio[a], J R B Oliveira[l], A Pakou[m], L Pandola[a], F Pinna[f,i], S Reito[d], G Russo[b,d], G Santagati[a], O Sgouros[m], S O Solakci[e], V Soukeras[m], G Souliotis[m], A Spatafora[a,b], D Torresi[a], S Tudisco[a], A Yildirim[e] and V A B Zagatto[k]

for the NUMEN collaboration

[a] INFN, Laboratori Nazionali del Sud, Catania, Italy
[b] Dipartimento di Fisica e Astronomia, Università di Catania, Catania, Italy
[c] Instituto de Fisica, Universidad Nacional de Mèxico, Mexico City, Mexico
[d] INFN, Sezione di Catania, Catania, Italy
[e] Departament of Physics, Akdeniz Üniversitesi, Antalya, Turkey
[f] INFN, Sezione di Torino, Turin, Italy
[g] LPC Caen, Normandie Universit, ENSICAEN, CNRS/IN2P3, Caen, France
[h] Institute of Natural Sciences, Karadeniz Teknik Üniversitesi, Trabzon, Turkey
[i] Dipartimento di Scienza Applicata e Tecnologia, Politecnico di Torino, Turin, Italy
[j] Facoltà di Ingegneria e Architettura, Università di Enna "Kore", Enna, Italy
[k] Instituto de Física, Universidade Fluminense, Niteròi, Rio de Janeiro, Brazil
[l] Departamento de Física Nuclear, Universidade de São Paulo, São Paulo, Brazil
[m] Physics Department, University of Ioannina, Ioannina, Greece

E-mail: calabrese@lns.infn.it

Abstract. Within the NUMEN project, we present experimental issues and data reduction for the ^{116}Cd(^{20}Ne,^{20}F)^{116}In single charge exchange, ^{116}Cd(^{20}Ne,^{18}O)^{118}In two-proton transfer and ^{116}Cd(^{20}Ne,^{19}F)^{117}In one-proton transfer reactions at 15 AMeV incident energy.

1. Introduction

The NUMEN (NUclear Matrix Elements of Neutrinoless Double Beta Decay) project proposes an original experimental method to get information on the nuclear matrix elements involved in neutrinoless double beta decay ($0\nu\beta\beta$), exploring its connection with double charge exchange nuclear reactions between heavy-ions. $0\nu\beta\beta$ Nuclear Matrix Elements (NMEs) are at the moment evaluable only through theoretical calculations [1, 2, 3] with not satisfactory agreement between the different models. In this view, the project [4, 5, 6] aims to use Double Charge Exchange (DCE) reactions as a tool to get experimentally driven information on $0\nu\beta\beta$ NMEs. Moreover, the absolute cross sections of competing multi-nucleon transfer channels on nuclei candidate for $0\nu\beta\beta$ decay are required. Indeed, to interpret the experimental cross sections and properly isolate the DCE contribution, the description of competing processes leading to the same final channel is mandatory. These are essentially one- and two-proton and neutron transfers. The physics case and the details of the project are discussed in Refs. [6, 7]. Experimentally, it is based on the use of the Superconducting Cyclotron high resolution beams and MAGNEX magnetic spectrometer [8, 9, 10], which is a relevant instrument in the research of heavy-ion physics [11, 12, 13, 14]. High performance are indeed required, since DCE as well as multi-nucleon transfer reactions are challenging from the experimental side due to the very low cross section expected and the large experimental background of competing processes; moreover, high mass, angular and energy resolution are needed to isolate specific transitions in such heavy-ions collisions. In addition, these experiments are more sensitive at very forward angles, so an important requirement is to access energy spectra and absolute cross sections angular distributions down to zero degree. In this contribution the experimental measurement of one-proton (1-p) and two-proton (2-p) transfer as well as single-charge exchange (SCE) reaction on ^{116}Cd target through the ^{20}Ne beam at 15 AMeV is presented. The experimental issues, the solutions adopted and the data reduction procedure are discussed.

2. Experimental Setup

The experiment was performed at the INFN-LNS laboratory in Catania. A $^{20}Ne^{10+}$ beam was extracted by the K800 Superconducting Cyclotron, with an energy of 15 AMeV and impinged on a 1360±140 μg/cm^2 thick ^{116}Cd target coupled to 1000±100 μg/cm^2 carbon post-stripper. The latter is added in order to decrease the contribution due to lower charge states (8^+ and 9^+) of the beam generated by the charge redistribution in the target. These ions, expecially the $^{20}Ne^{9+}$ elastically scattered at forward angles, have a magnetic rigidity close to that of some ejectiles examined, e.g. $^{20}F^{9+}$ for the SCE channel. Thus, they enter in the focal plane detector (FPD) acceptance, introducing a large background which damages the detector, requiring a limitation of the beam intensity. In order to further reduce these contributions, the spectrometer momentum acceptance was limited: two aluminium screens were mounted before the entrance of the focal plane to stop these unwanted ions.

The beam charge was collected by a Faraday cup located 15 cm downstream of the target. An integrated charge of 3.4±0.4 mC was measured. The outgoing ejectiles were momentum analyzed by the MAGNEX spectrometer [8] and detected by its FPD [15]. Thanks to the large momentum acceptance all the channels – 1-p, 2-p transfer and SCE – were simultaneously detected within the same magnetic fields. A representation of the measured transitions at the target is shown in Figure 1.

The measurement was performed with the spectrometer optical axis centered at $\theta_{lab} = 8°$. Thanks to the large angular acceptance of MAGNEX, an angular range of $3° < \theta_{lab} < 14°$ in the laboratory frame was covered.

3. Data Reduction

The focal plane data analysis begins with the calibration of the horizontal X_i and vertical position Y_i measured by the FPD, as described in [16]. The next step is the identification of the different ejectiles, which are first identified in atomic number (Z) and then in mass number (A) and charge state (q) exploiting a characteristic magnetic technique [17] that provides a mass resolution sufficient to clearly separate, in these conditions, the different isotopes of interest [18].

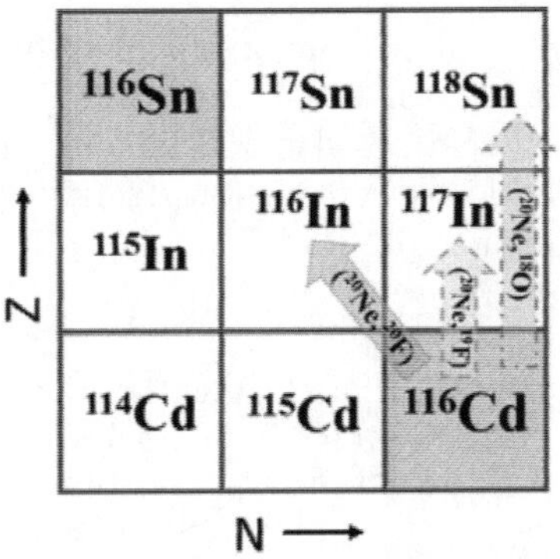

Figure 1. Reaction channels simultaneously measured using the same magnetic fields: (^{20}Ne,^{20}F) SCE in dotted ↖; (^{20}Ne,^{19}F) 1-p transfer in dashed ↑; (^{20}Ne,^{18}O) 2-p transfer in dashed-pointed ↑.

After the particle identification, the positions and angles of the selected ions measured at the focal plane are used to extract, event by event, the phase space parameters back to the collision point. To perform this back reconstruction, the global transport operator inside the magnetic elements of MAGNEX is required; setting the magnetic fields values and the geometry used during the experiment, the COSY INFINITY code [19] exploits an algebraic method to determine the spectrometer transport map (M) up to the 10th order [20, 21, 22, 23]. Then, Monte Carlo simulated events are generated for each analyzed reaction channel and transported by M to the FPD, to be compared to the experimental data in order to test the accuracy of M in reproducing the high-order aberrations clearly present in the experimental scatter plots. As an example, some characteristic plots of the spectrometer for the ^{116}Cd(^{20}Ne,^{19}F)^{117}In data are shown in Figure 2 along with their simulated events.

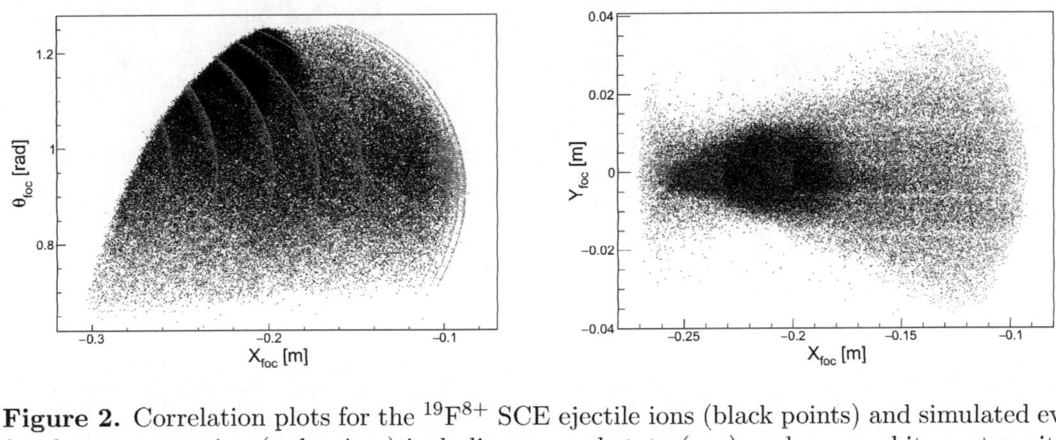

Figure 2. Correlation plots for the $^{19}F^{8+}$ SCE ejectile ions (black points) and simulated events for the same reaction (red points) including ground state (g.s.) and some arbitrary transitions. *Left panel:* horizontal angles θ_{foc} versus horizontal positions X_{foc} at the FPD. The first simulated profile – which corresponds to the g.s. transition – overlaps well with the experimental data. *Right panel:* vertical Y_{foc} versus horizontal X_{foc} positions at the FPD. The "half-butterfly" shape is due to the vertical non-linear focusing effect of the quadrupole field which is well reproduced by the simulated events.

Once a reliable direct matrix M is obtained, the inverse map (M^{-1}) is generated using COSY INFINITY and applied to the measured FPD data, thus returning the initial momentum vector at the target point, concluding the whole reconstruction procedure. The applicability of the described procedure to the presented data represents a relevant achievement in view of the development of the NUMEN experimental program.

References

[1] Barea J, Kotila J and Iachello F 2015 *Phys. Rev.* C **91** 034304.
[2] Vergados J D, Ejiri H and Simkovic F 2012 Rep. Prog. Phys. **75** 106301.
[3] Engel J and Menèndez J 2017 *Rep. Prog. Phys.* **80** 046301.
[4] Cappuzzello F et al. 2015 *J. Phys. Conf.: Conf. Ser.* **630** 12018.
[5] Agodi C et al. 2015 *Nucl. Part. Phys. Proc.* **265** 28.
[6] Cappuzzello F et al. 2018 *Eur. Phys. J.* A **54** 72.
[7] Cappuzzello F et al. 2015 *Eur. Phys. J.* A **51** 145.
[8] Cappuzzello F et al. 2016 *Eur. Phys. J.* A **52** 167.
[9] Cappuzzello F et al. 2014 *Nucl. Instr. and Meth.* A **763** 314.
[10] Cunsolo A et al. 2007 *Eur. Phys. J.: Spec. Top.* **150** 343.
[11] Carbone D et al. 2011 *J. Phys. Conf.: Conf. Ser.* **312** 082016.
[12] Cavallaro M et al. 2016 *Phys. Rev.* C **93** 064323.
[13] Carbone D et al. 2014 *Phys. Rev.* C **90** 064621.
[14] Soukeras V et al. 2015 *Phys. Rev.* C **91** 057601.
[15] Cavallaro M et al. 2012 *Eur. Phys. J.* A **48** 59.
[16] Carbone D et al. 2015 *Eur. Phys. J. Plus* **130** 143.
[17] Cappuzzello F et al. 2010 *Nucl. Instr. and Meth.* A **621** 419.
[18] Calabrese S et al. 2018 *Acta Phys. Pol.* B **49** 275.
[19] Makino K and Berz M 2006 *Nucl. Instr. And Meth.* A **558** 346.
[20] Lazzaro A et al. 2008 *Nucl. Instr. and Meth.* A **591** 394.
[21] Lazzaro A et al. 2008 *Nucl. Instr. and Meth.* A **585** 136.
[22] Lazzaro A et al. 2009 *Nucl. Instr. and Meth.* A **602** 494.
[23] Lazzaro A et al. 2007 *Nucl. Instr. and Meth.* A **570** 192.

IOP Conf. Series: Journal of Physics: Conf. Series **1056** (2018) 012011 doi:10.1088/1742-6596/1056/1/012011

Experimental issues for the measurement of the double charge exchange reactions within the NUMEN project

D Carbone[1], F Cappuzzello[1,2], C Agodi[1], M Cavallaro[1], L Acosta[3], D Bonanno[4], D Bongiovanni[1], I Boztosun[5], S Calabrese[1,2], D Calvo[6], E R Chávez Lomelí[3], F Delaunay[6,7], N Deshmukh[1], P Finocchiaro[1], M Fisichella[6], A Foti[4], G Gallo[1,2], A Hacisalihoglu[1,8], F Iazzi[6,9], R Introzzi[6,9], G Lanzalone[1,10], R Linares[11], F Longhitano[4], D Lo Presti[2,4], N Medina[12], A Muoio[1], J R B Oliveira[12], A Pakou[13], L Pandola[1], F Pinna[6,9], S Reito[4], G Russo[2,4], G Santagati[1], O Sgouros[1,13], S O Solakcı[5], V Soukeras[1,13], G Souliotis[14], A Spatafora[1,2], D Torresi[1], S Tudisco[1], A Yildirim[5] and V A B Zagatto[11] for the NUMEN collaboration

[1] INFN-Laboratori Nazionali del Sud, Catania, Italy
[2] Università degli Studi di Catania, Catania, Italy
[3] Universidad Nacional Autònoma de México, Ciudad de México, Mexico
[4] INFN-Sezione di Catania, Catania, Italy
[5] Akdeniz University, Antalya, Turkey
[6] INFN-Sezione di Torino, Turin, Italy
[7] LPC Caen, Normandie Université, ENSICAEN, UNICAEN, CNRS/IN2P3, Caen, France
[8] Istitute of Natural Science, Karadeniz Teknik Universitesi, Trabzon, Turkey
[9] DISAT, Politecnico di Torino, Turin, Italy
[10] Università di Enna "Kore", Enna, Italy
[11] Universidade Federal Fluminense, Niteroi, Brazil
[12] Universidade de Sao Paulo, Sao Paulo, Brazil
[13] University of Ioannina, Ioannina, Greece
[14] Laboratory of Physical Chemistry, Department of Chemistry, National and Kapodistrian University of Athens, Athens, Greece

Abstract. The NUMEN project proposes to study heavy-ion induced Double Charge Exchange (DCE) reactions with the final goal to get information on the nuclear matrix elements for neutrinoless double beta (0νββ) decay. The knowledge of the nuclear matrix elements is crucial to infer the neutrino average masses from the possible measurement of the half-life of 0νββ decay. DCE reactions and 0νββ decay present some similarities, the initial and final-state wave functions are the same and the transition operators are similar. The experimental measurements of DCE reactions induced by heavy ions present a number of challenging aspects, since they are characterized by very low cross sections.

1. Introduction
The physics of neutrinoless double beta (0νββ) decay has fundamental implications on particle physics, cosmology and fundamental physics. The observation of this rare process would signal that the total lepton number is not conserved and also it is considered one of the most promising ways to establish the Majorana or Dirac nature of neutrino and to have access to its effective mass. Therefore, this physics case is presently one of the most important researches Beyond the Standard Model and

might guide the way towards a Grand Unified Theory of fundamental interactions. The 0νββ decay basically involves nuclei, thus its analysis necessarily implies nuclear structure items. Indeed, the 0νββ decay rate can be expressed as a product of three independent factors: the phase-space factor, the Nuclear Matrix Element (NME) and a term containing the effective neutrino masses. The precise knowledge of NMEs is thus mandatory to extract information on the neutrino masses, when the decay rate will be possibly measured.
The evaluation of the NMEs is presently limited to state-of-the-art model calculations based on different approaches (QRPA, shell-model, IBM etc.) [1-4]. However, the presence of ambiguities in the models and the lack of strong constraints correspond to significant differences in the obtained values. Moreover, possible common approximations can correspond to systematic uncertainties.
In order to obtain experimentally driven information on the NMEs, the NUMEN [5,6,7] and the NURE [8-9] projects are exploring the possibility of using heavy-ion induced Double Charge Exchange (DCE) reactions. Even if DCE reactions and 0νββ decay are mediated by different interactions, there are some important similarities between them: i) the initial and final state wave functions in the two processes are the same, ii) the transition operators are similar, in both cases Fermi, Gamow-Teller and rank-two tensor components are present, iii) a large linear momentum (~100 MeV/c) is available in the virtual intermediate channel, iv) the two processes are non-local and are characterized by two vertices localized in a pair of valence nucleons, v) they take place in the same nuclear medium, vi) a relevant off-shell propagation through virtual intermediate channels is present.
The advantage for DCE reaction is to be “accessible” in laboratory, but a simple relation between DCE cross sections and ββ-decay half-lives is not trivial and needs to be explored.
In this context, an experimental campaign has started at the INFN-Laboratori Nazionali del Sud in Catania using the MAGNEX large acceptance magnetic spectrometer [10] focused on DCE reactions involving the nuclei of interest for 0νββ decay. Important experimental challenges must be addressed to measure DCE reactions since they are characterized by very low cross-sections and require a high energy resolution to distinguish the transitions in the region of the ground state. Both constraints are guaranteed by the use of the MAGNEX spectrometer, a tool with high performance and flexibility. In particular, the (^{20}Ne,^{20}O) DCE reactions at 15 AMeV on ^{116}Cd, ^{76}Ge and ^{130}Te, which are nuclei candidates for the 0νββ decay, were recently measured for the first time. Some details about the experimental issues of these measurement are discussed in this paper.

2. Heavy-ion double charge exchange reactions

The experimental study of nuclear transitions where the nuclear charge is changed by two units leaving the mass number unvaried was explored in the past with the aim to obtain information on the NMEs of ββ processes. The first studies were concentrated on pion double charge exchange reactions (π^+,π^-) [11-12], which were then abandoned, due to the large differences in the momentum transfers and in the nature of the operators between them and the 0νββ decay, as reported in [13]. Some early studies of heavy-ion induced DCE reactions are also reported. They were inconclusive due to the lack of zero-degree data and the poor yields in the measured energy spectra and angular distributions. The limitation was the very low cross sections involved, ranging from about 5-40 nb/sr [14] to 10 μb/sr [15]. Actually, this wide range of observed cross sections has never been deeply discussed. An additional complication in the interpretation of the data arose from possible contributions of multi-nucleon transfer reactions leading to the same final states [16].
Nowadays these experimental limitations are almost overcome, as we demonstrated in the pilot experiment performed at the INFN-LNS laboratory devoted to ^{40}Ca(^{18}O,^{18}Ne)^{40}Ar DCE reaction at 15 AMeV with the aim to measure the cross section at zero degrees [17]. The key elements in the experiment were the high resolution Superconducting Cyclotron (CS) beams and the use of MAGNEX, a modern large acceptance magnetic spectrometer characterized by high resolution in energy, mass and angle [18-19]. The high-order solution of the equation of motion is the key feature of MAGNEX, which guarantees the above mentioned performances and its relevance in the research of heavy-ion physics [20-25]. The pilot experiment demonstrated that high resolution and statistically

significant experimental data can be measured for DCE processes and that precious information towards NME determination could be at our reach [17].

3. The NUMEN and NURE projects

The aim of the NUMEN project is to measure the absolute cross section for DCE reactions on target nuclei candidates for the $0\nu\beta\beta$ decay and find a connection between the NMEs of the two processes. To move from the pilot experiment, performed on ^{40}Ca target, towards such target nuclei, important experimental limits need to be overcome. The challenge is to measure a rare nuclear transition under a very high rate of heavy ions produced by the beam-target interaction. In the exploration of nuclei of interests for $0\nu\beta\beta$ we consider that:
a) The Q-value for DCE reactions on such nuclei is typically more negative than in the case of ^{40}Ca explored in ref. [17]. This could strongly reduce the cross section.
b) The (^{18}O,^{18}Ne), which emulate the $\beta^+\beta^+$ decays, is particularly advantageous, due to the large value of the B(GT) strengths. To explore reactions of the $\beta^-\beta^-$ kind, none of the reactions looks like as favourable as the (^{18}O,^{18}Ne). For example, the (^{18}Ne,^{18}O) requires a radioactive beam, which cannot be available with comparable intensity. The proposed (^{20}Ne,^{20}O) has smaller B(GT), so a sensible reduction of the yield could be expected;
d) In some cases, gas or implanted targets are necessary, e.g. ^{136}Xe or ^{130}Xe, which are normally much thinner than solid state ones, with a consequent reduction of the collected yield;
e) In some cases the energy resolution is not enough to separate the ground from the excited states in the final nucleus. Thus, the coincident detection of γ-rays from the de-excitation of the populated states is mandatory, but at the price of the collected yield.
As a consequence, the present limits of beam power (~100 W) for the CS accelerator and acceptable rate for the MAGNEX focal plane detector (few kHz) allow us to concentrate on only few cases, which are planned in the NURE project [7-8] (e.g. ^{116}Cd, ^{130}Te, ^{76}Ge). In order to start a systematic exploration of all the nuclei of interest for $0\nu\beta\beta$ decay, an upgraded set-up, able to work with at least two orders of magnitude more luminosity than the present, is necessary. This goal can be achieved by a substantial change in the technologies implemented in the beam extraction [26], in the control of the beam induced radioactivity, in the detection of the ejectiles [27-31] and in the power dissipation of the thin targets [32]. In addition, in order to explore a wider range of incident energy, an increase of the maximum accepted magnetic rigidity is foreseen. This will be done preserving the geometry and field uniformity of the magnetic field [33-36], in order to keep the high-precision of the present trajectory reconstruction.
Finally, the development of a specific theory program to allow an accurate extraction of nuclear structure information from the measured cross sections is an important pillar of the NUMEN project. Relying on the use of the DWBA approximation for the cross section, the theory is focused on the development of microscopic models for heavy-ion DCE reactions, employing several approaches (QRPA, shell model, IBM) for inputs connected to nuclear structure quantities. We are also investigating the possible link between the theoretical description of the $0\nu\beta\beta$ decay and DCE reactions.

4. First experimental measurements on the (^{20}Ne,^{20}O) reaction

We performed first experimental investigations of the (^{20}Ne,^{20}O) DCE reaction on ^{116}Cd,^{76}Ge and ^{130}Te targets, which are candidates for the $0\nu\beta\beta$ decay. These are the first measurements of such a reaction, there are no data available in literature. A ^{20}Ne^{10+} cyclotron beam at 15 AMeV was delivered by the CS of INFN-LNS and impinged on ^{116}Cd rolled target of 1370 μg/cm^2 thickness and ^{76}Ge (386 μg/cm^2 thickness) and ^{130}Te (247 μg/cm^2 thickness) both evaporated on a C backing of ~ 50 μg/cm^2. The thickness of the various targets was carefully chosen in order to obtain an energy resolution which allows to distinguish the transition to the residual nucleus ground state from its first excited state. Indeed, the selected thickness of ^{116}Cd is much higher than that of ^{76}Ge and ^{130}Te, because the first excited state in ^{116}Sn case is at 1.293 MeV, to be compared to 0.559 MeV in ^{76}Se and 0.536 MeV in

^{130}Xe. The MAGNEX spectrometer was placed at forward angles including zero degree in the full acceptance mode (~50 msr). The total covered angular range was $0° \leq \theta_{lab} \leq 8°$.
Usually when a measurement is performed at zero degree, the beam enters in the spectrometer acceptance and the magnetic fields guide it to a place in the focal plane region away from the detectors. Finally it is collected by a faraday cup [17]. In the present case, the fields are set in order to transport the $^{20}Ne^{10+}$ ions towards the faraday cup position. However, when the beam passes through the targets a charge state distribution is originated. The maximum amount corresponds to the fully stripped $^{20}Ne^{10+}$ (~ 99%) but a sizeable amount of beam in the 9^+ and 8^+ charge states is also produced. These lower charge state components have a magnetic rigidity similar to that of the ejectiles of interest: $^{20}F^{9+}$ for the Single Charge Exchange (SCE) and $^{20}O^{8+}$ for DCE. Consequently, they enter in the FPD acceptance causing a limitation in beam intensity tolerable by the detector. In order to minimize the amount of $^{20}Ne^{9+}$ and $^{20}Ne^{8+}$ beams, a second target was placed downstream of the primary one to be used as a post-stripper material [37]. Different materials where tested and the final choice was a thick C foil of ~ 800 μg/cm^2. With this configuration the charge state distribution is ~ 99.1 % of 10^+, ~ $9.0 \cdot 10^{-3}$ % of 9^+ and ~ $2.0 \cdot 10^{-5}$ % of 8^+ [38]. This solution allowed only partially to reduce the background and thus a system of shields before the FPD entrance was also equipped to stop such ejectiles.
Despite these experimental limitations, we were able to measure energy spectra and absolute cross sections for the DCE reaction channel. Moreover, we measured also other reaction channels (one- and two-proton transfer, one- and two-neutron transfer and SCE), in order to estimate the role of the sequential multi-nucleon transfer routes on the diagonal DCE process. The data reduction [39][40] and analysis are almost completed and the results will be published soon.

Acknowledgments
This project has received funding from the European Research Council (ERC) under the European Union's Horizon 2020 research and innovation programme (grant agreement No 714625).

References
[1] Vergados J D, Ejiri H, and Simkovic F 2012 *Rep. Prog. Phys.* **75** 106301
[2] Vogel P 2012 *Jour. of Phys. G: Nucl. and Part. Phys.* **39** 124002
[3] Engel J and Menéndez J (2017) *Rep. Prog. Phys.* **80** 046301
[4] Dell'Oro S, Marcocci S, Viel M and Vissani F 2016 *Adv. High Energy Phys.* **2016** 2162659
[5] Cappuzzello F et al. 2015 *J. Phys.: Conf. Ser.* **630** 12018
[6] Agodi C et al. 2015 *Nucl. Part. Phys. Proc.* **265** 28
[7] Cappuzzello F et al. 2018 *Eur. Phys. J.* A **54** 72
[8] Cavallaro M et al. 2017 *PoS*(BORMIO2017)
[9] Cavallaro M 2017 *PoS*(NEUTEL2017) 031
[10] Cappuzzello F et al. 2016 *Eur. Phys.* J. A **52** 167
[11] Fazely A, Liu L C 1986 *Phys. Rev. Lett.* **57** 968
[12] Wu H, Wang R and Zhao E 1992 *Phys. Rev.* C **45** 1644
[13] Report to the Nuclear Science Advisory Committee, Neutrinoless Double Beta Decay, 2014.
[14] Blomgren J et al. 1995 *Phys. Lett.* B **362** 34
[15] Drake D M et al. 1980 *Phys. Rev. Lett.* **45** 1765
[16] Dasso C H and Vitturi A 1986 *Phys. Rev.* C **34** 743
[17] Cappuzzello F et al. 2015 *Eur. Phys. J.* A **51** 145
[18] Cunsolo A et al. 2007 *Eur. Phys. Jour. ST* **150** 343
[19] Cappuzzello F et al. 2014 *Nucl. Instr. and Meth.* A **763** 314
[20] Cappuzzello F et al. 2015 *Nature Communications* **6** 6743
[21] Cavallaro M et al. 2016 *Phys. Rev.* C **93** 064323
[22] Carbone D et al. 2017 *Phys. Rev.* C **95** 034603
[23] Carbone D 2015 *Eur. Phys. J Plus* **130** 143
[24] Oliveira J R B 2013 *J. Phys.* G **40** 105101

[25] Carbone D et al. 2011 *J. Phys.: Conf. Ser.* **312** 082016
[26] Calabretta L et al. 2017 *Modern Physics Letters* A **32** 17
[27] Cortesi M et al. 2017 *Review of Scientific Instruments* **88** 013303
[28] Muoio A et al. 2016 *EPJ Web of Conferences* **117** 10006
[29] Carbone D et al. 2016 *Results in Physics* **6** 863
[30] Carbone D et al. *Nucl. Instr. and Meth.* (in press). DOI:10.1016/j.nima.2017.10.095
[31] De Geronimo G et al. 2013 *IEEE Transactions on Nuclear Science* **60** 2314
[32] Iazzi F et al. 2017 *WIT Transactions on Engineering Sciences* DOI: 10.2495/MC170071
[33] Lazzaro A et al. 2008 *Nucl. Instr. and Methods* A **591** 394
[34] Lazzaro A et al. 2008 *Nucl. Instr. and Methods* A **585** 136
[35] Lazzaro A et al. 2007 *Nucl. Instr. and Methods* A **570** 192
[36] Lazzaro A et al. 2009 *Nucl. Instr. and Methods* A **602** 494
[37] Santagati G et al. 2018 *see this volume* ibid
[38] Shima K et al. 1992 *Atom. Data and Nucl. Data Tables* **51** 2
[39] Deshmukh N et al. 2018 *see this volume* ibid
[40] Calabrese S et al. 2018 *see this volume* ibid

IOP Conf. Series: Journal of Physics: Conf. Series **1056** (2018) 012012 doi:10.1088/1742-6596/1056/1/012012

Two-neutrino Double-Beta Decay within the Realistic Shell Model

L Coraggio[1], L De Angelis[1], T Fukui[1], A Gargano[1], and N Itaco[1,2]
[1]Istituto Nazionale di Fisica Nucleare,
Complesso Universitario di Monte S. Angelo, Via Cintia - I-80126 Napoli, Italy
[2]Dipartimento di Matematica e Fisica, Università degli Studi della Campania "Luigi Vanvitelli", viale Abramo Lincoln 5 - I-81100 Caserta, Italy

Abstract. We report on the calculation of double-β decay properties of ^{130}Te and ^{136}Xe within the framework of the realistic shell model. The effective shell-model Hamiltonian and Gamow-Teller transition operator are derived by way of many-body perturbation theory, in order to not to resort to an empirical quenching of the axial coupling constant g_A. The results compare well with experimental data, paving the way to the calculation of the neutrinoless double-β decay nuclear matrix element for the nuclei that are currently the experimental target for the detection of this process. In this work we investigate about the perturbative properties of our approach, and the theoretical reliability of the realistic shell model.

1. Introduction

One of the crossovers of current interest between nuclear structure and the physics "beyond the Standard Model" is the calculation of the nuclear matrix element (NME) of the neutrinoless double-β ($0\nu\beta\beta$) decay, that is directly related to the half life of such a process in the following form:

$$\left[T^{0\nu}_{1/2}\right]^{-1} = G^{0\nu} \left|M^{0\nu}\right|^2 \langle m_\nu \rangle^2 \quad , \tag{1}$$

where $G^{0\nu}$ is the so-called phase-space factor (or kinematic factor), $\langle m_\nu \rangle$ is the effective neutrino mass that takes into account the neutrino parameters associated with the mechanisms of light- and heavy-neutrino exchange, and $M^{0\nu}$ is the NME of the decay connecting the wave functions of the parent and grand-daughter nuclei.

The detection of $0\nu\beta\beta$ decay is nowadays the target of experiments in many laboratories, since its observation would correspond to a lepton number violation and provide direct information on the nature and properties of the neutrino [1, 2].

From Eq. 1, it is very clear that a crucial role is played by a reliable estimate of the NME in order to assess the necessary sensitivity of a detecting device and link the experimental results to a measurement of $|\langle m_\nu \rangle|$.

Currently, calculations of the NME are carried out by resorting to different nuclear structure models, such as the Interacting Boson Model (IBM) [3–5], the Quasiparticle Random-Phase Approximation (QRPA) [6–9], Energy Density Functional methods [10], and the Shell Model (SM). [11–17]. In the present work we report about calculations performed within the realistic shell model, where all the parameters appearing in the SM Hamiltonian, namely the single-particle (SP) energies and the two-body matrix elements of the residual interaction (TBME), and in the transition operators are derived from a realistic free nucleon-nucleon (NN) potential V_{NN}.

As a matter of fact, effective operators derived by way of perturbation many-body theory take into account the truncation of the full Hilbert space into the reduced SM model space, and calculations may be performed without resorting to any empirical parameter. In other words, in our approach we do not employ, for example, effective charges to calculate electromagnetic transition strengths, and we do not quench empirically the axial and vector current coupling constants. As a first step, we check this approach to the calculation of the $0\nu\beta\beta$ NME by calculating the two-neutrino double-beta ($2\nu\beta\beta$) decay NME and Gamow-Teller (GT) strengths of nuclei involved in possible $0\nu\beta\beta$ decays, and comparing the obtained results with the experimental counterparts. In Ref. [18] we have reported the results of our realistic SM calculation for the GT strengths and $2\nu\beta\beta$-decay NME of ^{130}Te and ^{136}Xe, that compare well with the available data.

In this paper, we will present a study of the convergence properties of the perturbative expansion of the effective SM Hamiltonian and operators we have employed in Ref. [18], derived within a model space spanned by the five proton and neutron orbitals $0g_{7/2}, 1d_{5/2}, 1d_{3/2}, 2s_{1/2}, 0h_{11/2}$ outside the doubly-closed ^{100}Sn, starting from the high-precision NN potential CD-Bonn [19].

In the next section we sketch out the derivation of our SM Hamiltonian and of the effective transition and decay operators. In Section 3, a detailed discussion about the perturbative expansion of the products of the effective SM Hamiltonian $H_{\rm eff}$, namely SP energies and TBME, and of the effective Gamow-Teller operator GT$_{\rm eff}$. In the last section some considerations and conclusions about the outcome of our study are are drawn.

2. Outline of calculations

We derive the SM effective Hamiltonian within the framework of the time-dependent degenerate linked-diagram expansion [20].

To this end, an auxiliary one-body potential U is introduced to write down the many-body Hamiltonian as the sum of an unperturbed term H_0, which describes the independent motion of the nucleons, and a residual interaction H_1:

$$H = \sum_{i=1}^{A} \frac{p_i^2}{2m} + \sum_{i<j} V_{NN}^{ij} = T + V_{NN} = (T + U) + (V_{NN} - U) = H_0 + H_1 \ ,$$

where V_{NN} represents the input NN potential.

The effective Hamiltonian $H_{\rm eff}$ is obtained by way of the Kuo-Lee-Ratcliff folded-diagram expansion in terms of the vertex function $\hat{Q}$-box, which in the perturbative approach is built by a collection of irreducible valence-linked diagrams [21]. We calculate the $\hat{Q}$-box including one- and two-body Goldstone diagrams through third order in H_1 [22]. Calculations beyond the third order in perturbation theory are computationally prohibitive, so we have calculated also the Padé approximant [2|1] [23] of the $\hat{Q}$-box in order to obtain the convergence value of the perturbation series, as suggested in [24]. The folded-diagram series is then summed up to all orders using the Lee-Suzuki iteration method [25].

The $\hat{Q}$-box contains one-body contributions, the collection of which is the so-called $\hat{S}$-box [26]. The folded-diagram expansion of the $\hat{S}$-box represents the theoretical SP energies, that will be employed in the SM calculations. The TBME will be then obtained by way of a subtraction procedure of these SP energies from $H_{\rm eff}$.

We employ as V_{NN} a low-momentum potential $V_{\rm low-k}$ defined within a cutoff momentum Λ by way of a similarity transformation [27]. This is a smooth potential which preserves exactly the on-shell properties of the original V_{NN} and is suitable for being used directly in nuclear structure calculations [28]. This has been done starting from the high-precision CD-Bonn NN potential [19], and then deriving a $V_{\rm low-k}$ corresponding to $\Lambda = 2.6$ fm^{-1}.

Employing this $V_{\rm low-k}$, we have calculated effective decay operators, namely the matrix elements of the effective GT operator, consistently with SM $H_{\rm eff}$ up to third-order in perturbation theory, within an approach that is strictly based on the one presented by Suzuki and Okamoto in Ref. [29].

In the following Section, we will present the results of our study about the convergence properties of theoretical SP energy spectra, TBME, and $2\nu\beta\beta$ NMEs as a function both of the dimension of the intermediate-state space and of the order of the perturbative expansion in order to assess the reliability of our realistic SM calculation.

3. Results

In Fig. 1, we report the single-neutron energy spectrum relative to the $0g_{7/2}$ orbit, as a function of the maximum allowed excitation energy of the intermediate states expressed in terms of the oscillator quanta $N_{\rm max}$. From the inspection of Fig. 1, it is clear that our results have practically achieved convergence at $N_{\rm max} = 14$.

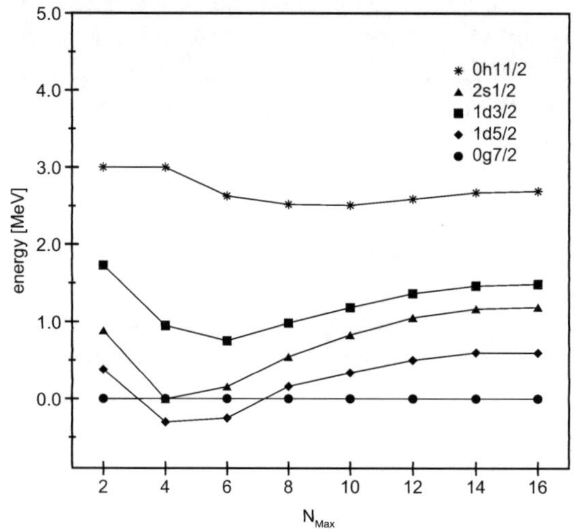

Figure 1. Neutron SP energies as a function of $N_{\rm max}$.

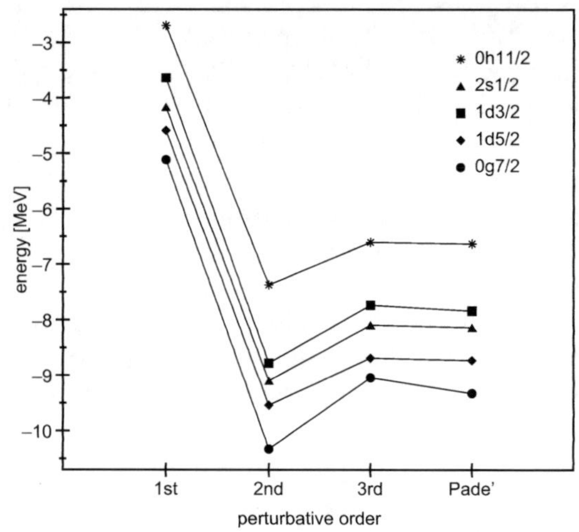

Figure 2. Same as in Fig. 1 as a function of the perturbative order.

In Fig. 2, the calculated neutron SP energies, using a number of intermediate states corresponding to $N_{\rm max} = 16$, are reported as a function of the order of the perturbative expansion up to the perturbative order $n = 3$, and compared with the Padé approximant [2|1] of the $\hat{Q}$-box. It can be seen that the results at third order are very close to those obtained with the Padé approximant, indicating that the truncation at third order may provide a good estimate of the sum of the series.

As regards the TBME, we report in Figs. 3 and 4 the neutron-neutron diagonal $J^{\pi} = 0^{+}$ TBME as a function both of $N_{\rm max}$ and the perturbative order n. We have chosen these TBMEs - that contain the pairing properties of our effective Hamiltonian - to be shown, since they are the largest in size of our calculated matrix elements, and the most sensitive to the behavior of the perturbative expansion.

From the inspection of Fig. 3 we see that the convergence with respect to $N_{\rm max}$ is very fast for diagonal matrix elements $(1d_{5/2})^2$, $(1d_{3/2})^2$, and $(2s_{1/2})^2$, while those corresponding to orbitals lacking of their own spin-orbit partner, $(0g_{7/2})^2$ and $(0h_{11/2})^2$, show a slower convergence.

The order-by-order convergence presented in Fig. 4 is satisfactory and as for the SP energies the results at third order are very close to those obtained with the Padé approximant, we may therefore conclude that our $H_{\rm eff}$, obtained from a $V_{\rm low-k}$ with a cutoff equal to 2.6 fm^{-1}, is a good estimate of the sum of its perturbative expansion.

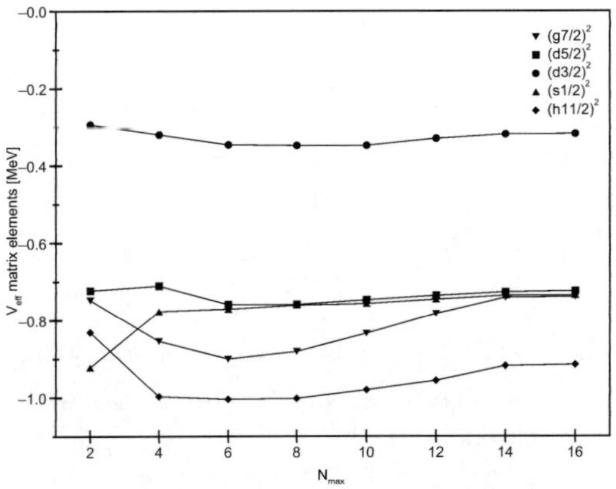

Figure 3. Neutron-neutron diagonal $J^\pi = 0^+$ TBME as a function of $N_{\rm max}$.

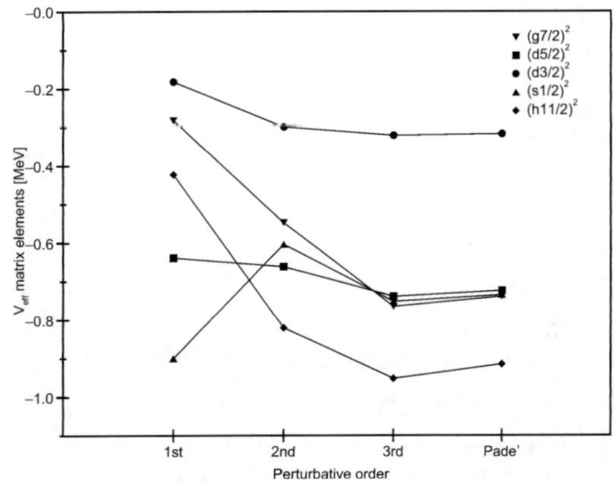

Figure 4. Same as in Fig. 3 as a function of the perturbative order.

We now turn our attention on the perturbative expansion of the GT effective operator $\rm GT_{eff}$. Because of the selection rules of the GT operator, the convergence of its matrix elements with respect to $N_{\rm max}$ is very fast. As a matter of fact, their values do not change from $N_{\rm max} = 2$ on at second order in the perturbative expansion [30], while their third decimal digit values do not change from $N_{\rm max} = 12$ on at third order.

The order-by-order convergence can be inferred from Table 1, where the results of the calculated NME of the $2\nu\beta\beta$ decays $^{130}{\rm Te}_{\rm g.s.} \rightarrow {}^{130}{\rm Xe}_{\rm g.s.}$ and $^{136}{\rm Xe}_{\rm g.s.} \rightarrow {}^{136}{\rm Ba}_{\rm g.s.}$ are reported and compared with the experimental results [31, 32]. As can be seen, also the convergence of the $2\nu\beta\beta$ NMEs is very satisfactory.

Table 1. Order-by-order $2\nu\beta\beta$ NME (in MeV^{-1}) for ^{130}Te and ^{136}Xe.

Decay	1st ord $M_{2\nu}^{\rm GT}$	2nd ord $M_{2\nu}^{\rm GT}$	3rd ord $M_{2\nu}^{\rm GT}$	Expt.
$^{130}{\rm Te} \rightarrow {}^{130}{\rm Xe}$	0.085	0.034	0.044	0.034 ± 0.003
$^{136}{\rm Xe} \rightarrow {}^{136}{\rm Ba}$	0.0564	0.0239	0.0285	0.0218 ± 0.0003

In order to focus on the convergence of the $\rm GT_{eff}$ matrix elements, in Table 2 we have reported also the results obtained with first-, second-, and third-order matrix elements, but employing always the same $H_{\rm eff}$, namely the one calculated at third order in the perturbative expansion. As can be seen, the results change for both transitions about 260% from the first- to second-order calculations, while the change is 9% and 5% from the second- to third-order results for ^{130}Te and ^{136}Xe decays, respectively.

Table 2. Same as in Table 1, but all calculated with the same third-order $H_{\rm eff}$.

Decay	1st ord $M^{\rm GT}_{2\nu}$	2nd ord $M^{\rm GT}_{2\nu}$	3rd ord $M^{\rm GT}_{2\nu}$	Expt.
^{130}Te $\rightarrow$ ^{130}Xe	0.142	0.040	0.044	0.034 ± 0.003
^{136}Xe $\rightarrow$ ^{136}Ba	0.0975	0.0272	0.0285	0.0218 ± 0.0003

4. Conclusions

In this paper we have presented the outcome of a study of the convergence properties of the effective Hamiltonian and GT operator, derived starting from a $V_{\rm low-k}$, that aims to assess the reliability of the calculation of the $2\nu\beta\beta$ NME for ^{130}Te and ^{136}Xe decays within the framework of the realistic shell model. We have calculated the dependence of SP energies, TBME, and $2\nu\beta\beta$ NME with respect to both the size of the intermediate-state space and the order by order convergence, up to the third order of the perturbative expansion. Our results show a rather good behavior of our perturbative expansion, besides a good agreement with the available data, providing support for our nearby goal, that will be the calculation of $0\nu\beta\beta$ NME for ^{130}Te and ^{136}Xe by way of the realistic shell model.

References

[1] Avignone F T, Elliott S R and Engel J 2008 *Rev. Mod. Phys.* **80**(2) 481–516
[2] Vergados J D, Ejiri H and Šimkovic F 2012 *Rep. Prog. Phys.* **75** 106301
[3] Barea J and Iachello F 2009 *Phys. Rev. C* **79**(4) 044301
[4] Barea J, Kotila J and Iachello F 2012 *Phys. Rev. Lett.* **109**(4) 042501
[5] Barea J, Kotila J and Iachello F 2013 *Phys. Rev. C* **87**(1) 014315
[6] Šimkovic F, Faessler A, Rodin V, Vogel P and Engel J 2008 *Phys. Rev. C* **77**(4) 045503
[7] Šimkovic F, Faessler A, Müther H, Rodin V and Stauf M 2009 *Phys. Rev. C* **79**(5) 055501
[8] Fang D L, Faessler A, Rodin V and Šimkovic F 2011 *Phys. Rev. C* **83**(3) 034320
[9] Faessler A, Rodin V and Simkovic F 2012 *J. Phys. G* **39** 124006
[10] Rodríguez T R and Martínez-Pinedo G 2010 *Phys. Rev. Lett.* **105**(25) 252503
[11] Caurier E, Nowacki F and Poves A 2008 *Eur. Phys. J. A* **36** 195
[12] Menéndez J, Poves A, Caurier E and Nowacki F 2009 *Phys. Rev. C* **80**(4) 048501
[13] Menéndez J, Poves A, Caurier E and Nowacki F 2009 *Nucl. Phys. A* **818** 139
[14] Horoi M 2013 *Phys. Rev. C* **87**(1) 014320
[15] Horoi M and Brown B A 2013 *Phys. Rev. Lett.* **110**(22) 222502
[16] Neacsu A and Horoi M 2015 *Phys. Rev. C* **91**(2) 024309
[17] Brown B A, Fang D L and Horoi M 2015 *Phys. Rev. C* **92**(4) 041301
[18] Coraggio L, De Angelis L, Fukui T, Gargano A and Itaco N 2017 *Phys. Rev. C* **95**(6) 064324
[19] Machleidt R 2001 *Phys. Rev. C* **63** 024001
[20] Kuo T T S and Osnes E 1990 *Lecture Notes in Physics, vol. 364* (Springer-Verlag, Berlin)
[21] Kuo T T S, Lee S Y and Ratcliff K F 1971 *Nucl. Phys. A* **176** 65
[22] Coraggio L, Covello A, Gargano A, NItaco and Kuo T T S 2012 *Ann. Phys.* **327**(9) 2125
[23] Ayoub N and Mavromatis H A 1979 *Nucl. Phys. A* **323** 125
[24] Hoffmann H M, Starkand Y and Kirson M W 1976 *Nucl. Phys. A* **266** 138
[25] Suzuki K and Lee S Y 1980 *Prog. Theor. Phys.* **64** 2091
[26] Shurpin J, Kuo T T S and Strottman D 1983 *Nucl. Phys. A* **408** 310
[27] Bogner S, Kuo T T S, Coraggio L, Covello A and Itaco N 2002 *Phys. Rev. C* **65** 051301(R)
[28] Coraggio L, Covello A, Gargano A, Itaco N and Kuo T T S 2009 *Prog. Part. Nucl. Phys.* **62** 135
[29] Suzuki K and Okamoto R 1995 *Prog. Theor. Phys.* **93** 905
[30] Towner I S and C K F 1983 *Nucl. Phys. A* **399** 334

IOP Conf. Series: Journal of Physics: Conf. Series **1056** (2018) 012012 doi:10.1088/1742-6596/1056/1/012012

[31] Barabash A S 2010 *Phys. Rev. C* **81**(3) 035501
[32] Albert J B *et al* (EXO Collaboration) 2014 *Phys. Rev. C* **89**(1) 015502

IOP Conf. Series: Journal of Physics: Conf. Series **1056** (2018) 012013 doi:10.1088/1742-6596/1056/1/012013

Experimental search of neutrino-less double beta decay in ^{130}Te

Oliviero Cremonesi
INFN – Sezione di Milano Bicocca, Milano I-20126, Italy
E-mail: oliviero.cremonesi@mib.infn.it

Abstract. ^{130}Te is one of the best candidates for the experimental search of neutrinoless double-beta decay. Thanks to its exceptionally large natural isotopic abundance ^{130}Te represents the only isotope for which the use of isotopic enrichment is practically not indispensable. Furthermore pure tellurium metal can be procured in large amounts and radio-pure crystals of tellurium compounds are commercially available. Different experimental techniques has been envisaged and a number of tellurium based experiments have been proposed and are starting taking data or are in advanced phase of construction.

The status of current experiments is reviewed, outlining the factors characterizing their sensitivity and thier role in the international competition. Future prospects and requirements are discussed.

1. Introduction

Experiments searching for neutrinoless double beta decay (NDBD) are attracting a lot of interest since they offer a unique opportunity to determine the Dirac/Majorana nature of neutrinos [1]. Indeed a direct observation of the decay would unambiguously demonstrate that neutrinos are massive Majorana particles providing at the same time a measurement of their absolute mass scale.

Based on the trend of the nuclear masses, only a limited number of nuclei are viable candidates for NDBD searches. Among them, only a few isotopes have the right features to guarantee the best sensitivities. Indeed, a simple argument shows that the experimental sensitivity can be parametrized as

$$T^{0\nu}_{exp} \propto \frac{\eta \cdot \epsilon}{A} \sqrt{\frac{M \cdot T}{\Delta \cdot B}} \tag{1}$$

where ϵ is the the detection efficiency, η the isotopic abundance, A the molecular mass, M the detector mass, T the observation time, Δ the energy resolution, and B the background index (rate per unit energy and mass of the detector). Furthermore, ultimate goal of all the experiments is the measurement of the effective neutrino mass $\langle m_{ee} \rangle = \sum_k U^2_{ek} m_k$ which is related to the oserved NDBD lifetime by the relation $T^{-1}_{0\nu} = G^{(0\nu)} |M^{(0\nu)}|^2 (\frac{|\langle m_{ee} \rangle|}{m_e})^2 \equiv F^{(0\nu)} |^2 |\langle m_{ee} \rangle|^2$, where $G^{(0\nu)}$ is the phase space term and $M^{(0\nu)}$ is the nuclear matrix element (NME) of the transition, both of which can only be calculated. Although the nuclear factor of merit $F^{(0\nu)}$ shows only a weak dependence on the nuclear species [2], the strong dependence of the experimental sensitivity on the choice of the candidate nucleus is evident from eq. (1). Large masses, high efficiencies, and extremely low bakground levels are the big challenges common to all the experiments.

IOP Conf. Series: Journal of Physics: Conf. Series **1056** (2018) 012013 doi:10.1088/1742-6596/1056/1/012013

^{130}Te is presently investigated with different experimental techniques. Some of them have already reached the maturity for a competitive experiment while others are still at the R&D level. Merits and limitations of the different approaches are reviewed together with a summary of the status of the art results.

2. Tellurium

An exceptionally large natural abundance (34.2%) makes ^{130}Te the only practical case for which the expensive and critical isotopic enrichment can be avoided. When coupled to a relatively high transition energy (2527.5 keV), in the middle of the gap between the Compton edge and the full energy peak of ^{208}Tl at 2615 keV, and to the possibility to get large samples of high radio-purity, the experimental interest for this isotope is evident. In the past, ^{130}Te has been the subject of a number of experimental searches with geochemical techniques. Since then, the need to pursue a direct measurement in order to get higher detection efficiencies has favored other isotopes (e.g. germanium and xenon) for which the homogeneous approach (where the detector acts at the same time as source of the decay) was possible. The development of the low temperature detectors and of the large mass scintillators has eliminated the restrictions on the choice of the detector materials and revived the interest on ^{130}Te. In the meanwhile also the progress of the tellurium based semiconductors has broken down some of the initial limitations (e.g. crystal dimensions) and presently a number of experiments on NDBD of ^{130}Te has been devised in the homogeneous approach. Apart from the already competitive results of CUORE and the forthcoming start of SNO+, the promising results from a number of R&D projects increase the interest on tellurium for future NDBD searches.

3. Bolometers

The use of single particle bolometers for rare event searches has been introduced in the 80's [3], mainly because of their excellent intrinsic energy resolution and the practical absence of any limitation on the choice of the detector materials.

The concept of these detectors is quite simple: when cooled down at very low temperatures (of the order of 10 mK), the heat capacity of a dielectric and diamagnetic crystal can be so low that even very small energy depositions (of the order of 10–1000 keV) can give rise to a measurable temparture increase. The structure of the detector is also very simple: a single crystal (absorber) in thermal contact with a heat sink (a cryostat) and a proper temperature sensor. The energy resolution of these devices is only limited by the statistical fluctuations of the phonons maintaining the thermal equilibrium with the heat sink ($\Delta = \sqrt{k_B C T^2}$). It does not depend on the deposited energy and can reach extremely low values. Actual detectors are still far from reaching this condition, and are still limited by a number "extrinsic" noise sources arising from instabilities the cryogenic system, electronics noise from the signal read-out chain, electromagnetic interferences and mechanical vibrations. Eventually, these contributions result in energy resolutions which for O(1 kg) bolometers can be as low as few keV in the MeV energy range. A value comparable to germanium diodes and still exceptionally better than mmany ionization or scintillation based detectors.

Tellurium dioxide (TeO_2) has been soon recognized as an ideal material for a bolometric experiment on NDBD given its good thermal performance and the existence of a well established industry for the production of large mass radio-pure crystals. After a number of initial experiments on single TeO_2 crystals of increasing mass, and the first small arrays of detectors developed in the 90's by the Milano group [4], the tonne-sized experiment CUORE (Cryogenic Underground Observatory for Rare Events) was eventually proposed in 1998 [5]. Since then, two large scale prototypes (Cuoricino and CUORE-0) have been constructed and operated to demonstrate the feasibility of CUORE.

3.1. CUORE

CUORE is hosted in the Laboratori Nazionali del Gran Sasso (LNGS) of INFN, in central Italy. The CUORE detector consists of 988 cubic TeO_2 crystals with natural ^{130}Te isotopic abundance, 5 cm side and a 750 g average mass, operated as bolometers at very low temperature around 10 mK [6] .

The 988 TeO_2 bolometers are arranged into 19 towers, each consisting of 13 floors of 4 detectors. The towers are arranged in a close-packed array and thermally connected to the mixing chamber of a ^{3}He/^{4}He dilution refrigerator (Leiden Cryogenics DRS-CF3000 continuous-cycle). The cooling of the system at intermediate temperatures ($\sim$40 K and $\sim$4 K) is based on five pulse tube cryocoolers (Cryomech PT415-RM) with a Joule-Thomson expansion valve. The dimensions, experimental volume ($\sim$1 m^3), mass ($\sim$17 t), and cooling power (3 μW at 10 mK) make this the largest and most powerful cryogen-free dilution cryostat in operation [7]. To minimize transmission of vibrations from the cryostat to the bolometers, the detector towers are independently supported by a Y-shaped beam that is vibrationally isolated from the cryostat support structure [8]. The cryostat and detector supporting systems, as well as the front-end electronics are located together in a Faraday Room [9] at the second floor of the CUORE underground building.

The commissioning of the cryogenic system was completed in 2015, followed by the installation of the detector in the summer 2016. The CUORE detector commissioning phase eventually started at the beginning of 2017. The initial two months were dedicated to the system optimization and setting of the optimal working points. In April 2017, although the optimization phase was not yet complete, CUORE was eventually ready for a preliminary science run (Dataset 1). A strategy based on alternated science and technical runs (aiming at optimizing the detector performance) was chosen for the first year of CUORE data taking. A second science run was therefore started on August 2017 (Dataset 2) after a summer optimization campaign.

The data from the first two datasets were released on October 2017 [10]. Main goal of these initial physics runs was to provide preliminary information on the parameters that most strongly affect the NDBD sensitivity: the energy resolution and the background index (BI) in the NDBD region of interest (ROI). Physics data were preceded and followed by calibration periods during which 12 Kevlar strings populated with low-intensity ^{232}Th sources were temporarily deployed (from room temperature across the cryogenic volume) inside the detector region in order to guarantee an approximately uniform γ-ray illumination of the detectors [11].

A total of 984 over 988 detectors are functioning, but a smaller subset of channels was selected for analysis based on the overall quality of the event reconstruction.

A number of improvements to detector performance were obtained during the 2017 summer and fall optimization campaigns, mainly based on the synchronization of the pulse-tube cryocoolers and the grounding scheme of the readout electronics and the working point selection.

The data collected in the first two datasets amount to a total TeO_2 exposure of 86.3 kg$\cdot$yr, characterized by an effective energy resolution of (7.7$\pm$ 0.5) keV FWHM and a background in the region of interest of (0.014 $\pm$ 0.002) counts/(keV$\cdot$kg$\cdot$yr). No evidence for NDBD of ^{130}Te was obtained from the analysis of the collected data, corresponding to a lower limit on the decay half-life of T $>$ 1.3 $\cdot 10^{25}$ yr (90% C.L.) including systematic uncertainties. Combining this result with those of Cuoricino and CUORE-0, a limit of T $>$ 1.5 $\cdot 10^{25}$ yr (90% C.L.) is obtained, which is the most stringent limit to date. In terms of the effective Majorana neutrino mass, this corresponds to an upper limit of $\langle m_{ee} \rangle <$ (140–400) meV, where the range reflects the uncertainty on the nuclear matrix element estimates employed [10].

CUORE is the first tonne-scale cryogenic detector array in operation, more than an order of magnitude larger than its predecessors. The successful commissioning and operation of this large-mass, low-background, cryogenic bolometer array represents a major advancement in the application of this technique to NDBD searches and demonstrates the feasibility of future large-mass bolometer arrays for rare-event searches.

3.2. CUPID

Common features of any next generation experiment are isotopic enrichment and active background rejection. Since the dominant background component in bolometric detectors is due to degraded alphas from surface radioactive contaminations [12], the discrimination of this component can imply a sensitive reduction of the background index.

This is the main goal of the CUPID (CUORE Upgrade with Paticle Identification) program [13]. Indeed, as realized already in the 90's, hybrid bolometric detectors based on the simultaneous detection of scintillation or Cerenkov light can accomplish this goal. A number of small R&D projects mainly aimimg at the development of very sensitive (bolometric) light detectors have beed proposed in recent years in the framework of CUPID [14] and the first kg-scale demonstrator, CUPID-0 based on isotopically enriched scintilllating ZnSe bolometers, has been recently put in operation at LNGS [15]. Demonstrators for other promising scintillating crystals (e.g. $LiMoO_4$) are under preparation [16]. Since no viable tellurium compound has scintillating properties, the observation of the Cerenkov signal has been suggested in order to implement hybrid bolometric detectors. Here the challenge is the oservation of the faint Cerenkov signal, which requires very sensitive light detectors with very low eergy thresholds. In these framework, KID's [17] and thermistors with Neganov-Luke effect [18, 19, 20] are the most developed light detectors with very promising results which show that the concept is proven and actual detectors are already at hand. A TeO_2 array demonstrator with Cerenkov light observation is anticipated in 2018.

4. Scintillators: SNO+

The dispersion of a NDBD candidate nucleus in a high purity scintillator has been proposed since the end of 90's [21] as an effective way of exploiting the high purity environments of large scale neutrino detectors after the completion of the neutrino program of measurements. It has been recently implemented with success for ^{136}Xe in the framework of the KamLAND infrastructure.

The transformation the SNO infrastructure in a large liquid scintillator-based experiment is the basic idea of SNO+ [22]. Initially devised for ^{150}Nd , it has been later converted in a ^{130}Te based NDBD experiment. The 12 m diameter acrylic vessel of SNO will be filled with about 780 tonnes of ultra-pure liquid scintillator (LABPPO) which will be loaded with 0.5% natural tellurium, corresponding to nearly 1300 kg of ^{130}Te . The expected effective Majorana neutrino mass sensitivity is in the region of 55-133 meV, just above the inverted mass hierarchy lower bound. The possibility of deploying up to ten times more natural tellurium is under investigation. The physics program of SNO+ is however more rich and includes the study of reactor antineutrino oscillations, low energy solar neutrinos, and geoneutrinos, to be sensitive to supernova neutrinos, and to search for exotic physics.

SNO+ is located in Sudbury, Ontario, Canada. 2 km (6000 m.w.e) below ground where the total muon flux is less than 10^{-9} cm^{-2}s^{-1}. Besides using the same acrylic vessel (AV), SNO+ is also observed by the same 9300, 8 inch Hamamatsu R1408 photomultiplier tubes of SNO. The detector is housed in a cavity filled with ultra pure water.

The transformation of the SNO detector from a heavy water based detector to a liquid scintillator detector required a substantial amount of work and is now almost complete. As the LABPPO is less dense than water, a hold down rope system is needed during operation. The system was installed in 2012 and has been tested at various levels throughout commissioning. A new processing plant had been constructed to achieve the purity of the scintillator needed to ensure the low backgrounds required ($< 10^{-17}$ g/g). The plant is currently being commissioned with 40 tonnes of LABPP.

Telluric Acid (TeA) is stored underground since 2015. A full scale purification plant is now in construction underground. It will produce the final LAB+PPO+Te-ButaneDiol cocktail which will guarantee the expected initial 0.5% Te concentration. The nitrogen cover gas already existing on SNO and aimimg to protect the detector from high levels of Radon gas in the mine air, has been upgraded and commissioned in 2014. As the light output of liquid scintillator is larger, the SNO+ electronics have also been upgraded. New XL3 readout cards as well as new MTC/A+ trigger cards will handle the increase in PMT hits. Also, the data acquisition system (DAQ) has undergone a complete renovation, with a complete decoupling of the data flow from detector controls. The original in situ LED/Laser optical calibration system has been completely redesigned in order to monitor the scintillator quality. It is expected to come online for scintillator phase. A substantial effort has been devoted to the update of the deployed calibration sources. Also this system is expected to be ready for the SNO+ scintillator phase.

SNO+ is currently filled with water. The detector electronics were switched on in early 2017 and are working properly. Both in situ and deployed source calibration is on going.

When detector calibration will be complete, SNO+ will collect data to search for nucleon decay as well as characterising the state of the detector. In 6 months of running, SNO+ will set world leading limits on invisible nucleon decay.

Scintillator fill will follow. The Scintillator phase will build upon the work done in the previous phase and assess the background levels present in the detector. Reactor and Geo anti-neutrino studies will be undertaken in this phase, as well as low energy solar neutrino observations. With the background studies complete, ^{130}Te loading is expected to begin in late 2018 [23].

5. Semiconductors: COBRA

Tellurium based semiconductors have well known features, with a number of applications in radiation detecttion [24]. With respect to germanium diodes, CdTe and CdZnTe devices suffer from a rather poor collection efficiency of holes which strongly limits the dimensions of the available detectors. With a proper readout electronics, large arrays of (very) small CdZnTe pixels can be a valid altenative with tracking capabilities which can help in strongly reducing the background contributions.

COBRA [25] is the only NDBD experiment based on CdZnTe detectors, and is planning to use a large amount of them, operated at room temperature, to search for NDBD. CdZnTe contains 9 NDBD candidate isotopes, including electron capture and positron decay modes. Two of them are tellurium isotopes (^{130}Te and ^{128}Te), although the isotope with the highest Q-value (2813.50 keV) is ^{116}Cd , which represents therefore the main target of the search. Indeed, the sensitivities for the other NDBD candidate isotopes are lower, given the lower Q-value and the unavoidable background coming from the ^{116}Cd two neutrino double beta decays.

After a number of performance studies on small scale detectors, a demonstrator setup is operated at LNGS to investigate the experimental challenges of operating CdZnTe detectors in low background mode and to identify potential background components. The experiment consists of an array of 16 monolithic, calorimetric detectors of coplanar grid design (CPG detectors) of 1×1×1 cm^3, arranged in four layers of 4×4 detectors. This setup has already demonstrated the high potential of pulse-shape analyses to reduce the background. Further improvements are expected by the use of the pulse-shapes and the coincidence analysis between multiple CPGs.

Furthermore, larger detetors (2×2×1.5) cm^3 are commercially available with a better surface-to-volume ratio. The final goal is to operate 11,000 of such detectors. With an enrichment of 90% in ^{116}Cd , an energy resolution of better than 1.5% (FWHM) and a background rate of less than $0.5{\cdot}10^{-3}$ counts keV^{-1} kg^{-1} y^{-1}, COBRA should be able to reach a sensitivity of less than 50 meV on $\langle m_{ee} \rangle$.

IOP Conf. Series: Journal of Physics: Conf. Series **1056** (2018) 012013 doi:10.1088/1742-6596/1056/1/012013

6. Conclusions

^{130}Te is stil one of the most interesting NDBD isotopes. It currently shares with ^{76}Ge and ^{130}Xe the most sensitive experimental NDBD seraches.

Two next generation tellurium based experiments have been constructed: the bolometric CUORE has started operation at LNGS early in 2017 while the SNO+ scintillator is expected to start the NDBD program late in 2018. Both of them aim at sounding the inverted hierarchy region of neutrino masses.

Promising results from a number of R&D's show that significant improvements of the experimental sensitivities are possible making tellurium an excellent candidate for next generation experiments.

References

[1] Cremonesi O and Pavan M 2013 *Adv. High Energy Phys.* **2014** 951432

[2] Robertson R G H 2013 *Mod. Phys. Lett.* **28** 1350021

[3] Fiorini E and Niinikoski T 1984 *Nucl. Instrum. Meth.* **224** 83

[4] Alduino C et al (CUORE Collaboration) 2018 *arXiv:1801.05403*

[5] Fiorini E 1998 *Phys. Rep.* **307** 309

[6] Ardito R et al 2005 CUORE: A cryogenic underground observatory for rare events *arxiv:hep-ex/0501010*

[7] Alessandria F et al 2017 The CUORE cryostat: A 10 mk infrastructure for large bolometric arrays *In preparation*

[8] Bersani A et al 2017 The detector suspension system of the CUORE experiment *In preparation*

[9] Bucci C et al 2017 The faraday room of the CUORE experiment *arxiv/1710.05614*

[10] Alduino C et al (CUORE Collaboration) 2018 *Phys. Rev. Lett.* **120** 132501 *arxiv:1710.07988*

[11] Cushman J S et al 2017 *Nucl. Instrum. Meth.* A **844** 32

[12] D'Addabbo A et al2017 *Eur. Phys. J.* **77** 543; *arXiv:1704.08970*

[13] Wang G et al (CUPID Group of Interest) 2015 *arxiv:1504.03599*

[14] Wang G et al (CUPID Group of Interest) 2015 *arxiv:1504.03612*

[15] Pavan M et al (CUPID-0 Collaboration) 2017 *These proceedings*

[16] Armengaud E et al 2017 *Subm. to Eur. Phys. J. C.*

[17] D'Addabbo A et al (CALDER Collaboration) 2017 *arXiv:1705.04483*

[18] Gironi L et al 2016 *Phys. Rev.* **94** 054608; *arXiv:1603.08049*

[19] Artusa D R et al 2017 *Physics Letters* **767** 321; *arXiv:1610.03513*

[20] Berge L et al 2017 *arXiv:1710.03459*

[21] Bellini G et al 2000 *Eur.Phys.J.* **19** 43; *arXiv:0007012*

[22] Andringa S et al 2016 *Adv. High En. Phys.* **2016** 6194250

[23] Caden E et al 2017 *Proc. of the TAUP17 conf., July 2017, Sudbury, Ontario, Canada*

[24] Knoll G 2000 *Radiation Detectibn and Measurement* John Wiley & Sons Inc., Third Edition

[25] Zuber K 2001 *Phys. Lett.* **519** 1

IOP Conf. Series: Journal of Physics: Conf. Series **1056** (2018) 012014 doi:10.1088/1742-6596/1056/1/012014

The SPES radioactive ion beam facility at the Legnaro National Laboratories and the EDM search

Giacomo de Angelis for the SPES collaboration

INFN Laboratori Nazionali di Legnaro, I 35020 Legnaro (Pd), Italy

giacomo.deangelis@lnl.infn.it

Abstract. Radioactive ions can be used to amplify the sensitivity to symmetry violations. An example is the time reversal (T-) violation connected, assuming the charge conjugation, the parity and time reversal (CPT) symmetry, to the CP-violation. This is linked to the question of the role that CP violation plays in explaining the cosmological matter-antimatter asymmetry and the search for new physics beyond the Standard Model (SM). One of the most direct probes of such symmetry violation is the existence of a permanent electric dipole moment (EDM) of a particle or system. Octupole-deformed nuclei with high Z and A are expected to be among the best candidates for testing the limit of existence of EDM. The production, extraction and characterization of such unstable systems is part of the scientific program of the new SPES radioactive ion beam facility which is presently under construction at the Legnaro National Laboratories of INFN. In the following I will discuss the present status of the SPES facility and the scientific program foreseen for those studies.

1. Introduction: SPES, an acronym for Selective Production of Exotic Species, is a new infrastructure of the Italian Institute of Nuclear Physics (INFN) dedicated to forefront nuclear research and nuclear physics based applications. It is presently under construction at the Legnaro National Laboratories (LNL) and is expected to provide the first low energy radioactive ion beams in 2019. The facility will provide high-intensity and high-quality beams of unstable neutron-rich nuclei with the aim to perform forefront research in nuclear structure, reaction dynamics, nuclear astrophysics, fundamental interaction studies and interdisciplinary fields like medical, biological and material sciences. SPES is a second generation ISOL radioactive ion beam facility and represents an intermediate step toward the future European ISOL facility EURISOL. It is part of the European ISOL distributed facility (EURISOL-DF) presently under construction in Europe. The production of radioactive beams is based on a proton beam impinging on an uranium carbide (UCx) Direct Target able to sustain a maximum power of 8 kW. Neutron-rich radioactive ions are produced by Uranium fission at an expected maximum fission rate in the target of the order of 10^{13} fissions per second. The exotic isotopes after extraction and separation are re-accelerated by the ALPI superconducting linac of LNL at energies of 10 MeV/A or higher, for masses up to A=200 amu, with an expected rate on the secondary target of 10^7-10^9 pps for the most prolific species.

2. The SPES radioactive ion beam facility: The realization of SPES strongly involves two national laboratories, the LNL and the Laboratori Nazionali del Sud (LNS), and other INFN sites located all over Italy. It is presently under construction at LNL and makes use of the Isol (Isotope separation on line) technique for the production of radioactive beams. A primary proton beam provided by a driver accelerator, a cyclotron, induces nuclear reactions (mainly fissions) inside a thick, mainly uranium carbide, target. The reaction products are extracted from the target by thermal process due to the high temperature of the target-ion source system (exceeding 2000^0 C). Once reached the 1^+ source, the products are ionized and extracted. After an isotopic selection obtained using an high resolution spectrometer and a further ionization to an higher n^+ charge state, the exotic beam is injected into a re-accelerator, the ALPI superconductive linac. A pictorial sketch of the facility is shown in Fig. 1.

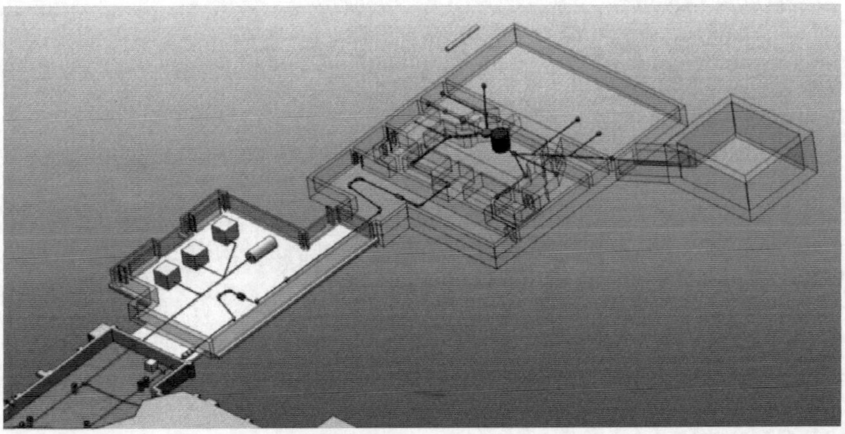

Figure 1: The Pictorial sketch of the SPES facility. The cyclotron is represented in blue. The red area is the one hosting the superconducting linac.

The driver accelerator is a 70 MeV proton cyclotron with two exit ports providing a total current of 750 μA. Two production Isol targets are planned to be installed and operated alternatively even if only one is presently under construction for budgetary reason. As the cyclotron can supply two beams at the same time, a second independent facility can be operated in parallel. In particular the high intensity proton beam will be used to produce radioisotopes for nuclear medicine as well as used as a neutron source. Concerning the latter possibility, the expected neutron production at SPES will exceed 10^{14} n/s with energy up to 70 MeV and a peak between 0.1 and 10 MeV according to target and moderator. The proton beam of the SPES cyclotron is suited for the production of new radioisotopes for medical applications. Taking advantage of the high energy and current, it opens up the possibility to produce ^{82}Sr, ^{67}Cu, and ^{47}Sc isotopes to be used in therapy and diagnostics. The SPES cyclotron is shown in Fig.2.

2.1 The Isol front-end: it is the part coupling the proton beam with the UCx target, the ion source and the first part of the transport line of the exotic beams. The whole system has been completed and is now in operation with stable beams at LNL for off-line tests. The most critical element of the SPES Isol facility is the Direct Target. It has been constructed following the basic design of the CERN ISOLDE target but represents an innovation in term of capability to sustain an higher primary beam power. The design is carefully oriented to optimize the radiative cooling, taking advantage of the target system's high operating temperature, which is of the order of 2000^0 C. The SPES target is designed in order to maximize the release efficiency and the heat dissipation. It consists of multiple thin disks housed in a cylindrical graphite box. The presently installed one is composed of 7 UCx disks (diameter and thickness of 40 and 1.3 mm, respectively), appropriately spaced in the axial direction in order to dissipate, by thermal radiation, the average power of 8 kW originated by the energy release of the proton beam which, passing through them, induces nuclear reactions. The graphite box is housed inside a tubular hollow tungsten ohmic heater. In fact, due to the intense heat exchange by radiation, the energy released by the proton beam is not sufficient to make the system reaching the operating temperature. This allows for a better thermal control of the target operation. At the basis of the target-ion-source design there is an extensive simulation study of the target behaviour for thermal and release properties. Experimental work to benchmark the simulations was carried out in collaboration with HRIBF [1], the Oak Ridge National Laboratory Isol facility (USA) and the i-Themba Nuclear Physics Laboratory in Stellenbosch (South - Africa).

Figure 2: The SPES cyclotron

2.2 The ion-source system: neutron-rich fission products are produced through the interaction of the proton beam with the UCx target. Such neutron rich nuclei are subsequently extracted by thermal motion and ionized at 1^+ charge state by the Isol source, directly connected to the production target. The hot-cavity ion source chosen for the SPES facility was designed at CERN (ISOLDE) [2] and adapted to the LNL environment. The source has the basic structure of the standard high temperature radioactive ion sources employed for on-line operation. The ionizer cavity is a tungsten tube (34 mm length, 3 mm inner diameter and 1 mm wall thickness) resistively heated to near 2000^0 C. Due to the high temperature the isotopes produced in the target diffuse in the target material and then effuse through the transfer tube into the ionizer cavity where they undergo surface or laser ionization. The surface ionization process occurs when an atom comes into contact with a hot metal surface. In the positive surface ionization, the transfer of a valence electron from the atom to the metal surface is energetically favourable for elements with an ionization potential lower than the work function of the metal. For alkalis and some rare earth elements, high ionization efficiencies can be achieved using the surface ionization technique. This kind of 1^+ source has good efficiency and selectivity for the elements like Rb, Cs, Ba. For most of the other elements it is preferable to apply laser resonant photo-ionization. It uses the same hot cavity cell and it is a powerful method to achieve sufficiently selected ionization. This technique, now under development in the laboratory, has been implemented with the aim to produce beams as pure as possible (chemical selectivity) also for metal isotopes. Very recently laser ionization of Sn ions has been achieved at LNL. Laser ionization schemes for other elements are presently under development in the newly created laser laboratory. To ionize elements with high ionization potential, as rare gasses, the plasma source is needed. This source ionizes all the elements without any selectivity. At SPES both surface, plasma and laser ion sources have been developed and are now in operation at the off line test stand.

2.3 Beam selection and transport: the ions after the extraction from the source undergo a first mass selection performed by a Wien Filter having a mass resolution of 1/100. The Wien filter is installed just after the first electrostatic quadrupole triplet inside the production bunker with the aim to confine the larger part of radioactivity inside this high shielded area. The transfer line toward the ALPI superconductive linac is equipped with several beam handling systems to purify the secondary ion beams. Purification is achieved by the combination of a Beam Cooler and an High Resolution Mass

Separator (HRMS) designed to reach 1/20000 mass resolution and installed inside the SPES building. Special efforts have been dedicated to reach such high mass resolution taking advantage of the emittance reduction provided by the Beam Cooler and of a 260 keV high voltage platform on which the mass separator is mounted. The mass separator is a scaled-up version of the separator designed for the CARIBU facility [3] in ANL (USA). In the SPES configuration the calculated resolution is of the order of 1/20000 constrained by an emittance of 3 mm mrad and an energy spread of 1.3 eV. Such high selectivity results in a strong advantage also for the safety issue, reducing the contaminations along the beam transport area, the re-accelerator and the final target location. The injection of the secondary beams in the ALPI superconducting linac requires an increase of the ion charge state from 1^+ to n^+. This is performed by means of a Charge Breeder. The SPES Charge Breeder is based on ECR method [4] and aims to produce ions with A/q up to 7 for masses of the order of A~130. The design and construction of the Charge Breeder has been performed in the framework of the SPES-SPIRAL2 collaboration at LPCS (Grenoble, France). The new design represents an upgraded version of the Phoenix booster in operation at LPSC. After the Charge Breeder a second mass separator with 1/1000 mass resolution is installed with the aim to clean the secondary beam from the contaminants introduced by the Charge Breeder itself (mainly contamination from stable gasses). The same basic configuration used for the HRMS is adopted. Finally an RFQ pre-accelerator is used to increase the secondary beam energy to match the ALPI acceptance. The layout of the radioactive beam line is shown in Fig. 1.

2.4 Exotic beam re-acceleration: the secondary beams after mass selection and n^+ ionization are injected into the ALPI superconducting linac by a new RFQ that is operating in a CW mode (100% duty factor) at a resonant frequency of 80 MHz. This frequency is the same as that of the superconducting cavities used for the low-energy section of the ALPI linear accelerator. The injection energy of the ions is set to 5.7 keV/u. This choice is a compromise between the desire to reduce the ion energy to simplify the low energy injection and the RFQ bunching section design and the need to increase the beam rigidity in the 1^+ transport line to reduce space charge effects. The extraction energy of the RFQ is set to 727 keV/u to optimize the beam dynamics of the superconducting linac ALPI. In the last year the ALPI linac has gone through a continuous upgrade in the number and performance of its accelerating cavities, and consequently in the maximum achievable transport efficiency and beam energy. In the context of such upgrade all the electromagnetic elements have been realigned and transport efficiency has been maximized.

2.5 Status of the SPES project: the SPES facility is presently in an advanced stage of construction at the LNL. The production building hosting the cyclotron and up to two Isol sources has been completed. The cyclotron, provided by the BEST company in Canada, after the low energy commissioning in the factory, has been delivered and brought to his final location - Fig.1. The commissioning and acceptance test of the accelerator was completed in 2017. The authorizations to operate the cyclotron have been obtained allowing the acceleration of the proton beam on non-fissile target at high power and the test of the UCx target up to a current of 5 μA. Very recently the proton beam has been accelerated inside the cyclotron up to the maximum energy and intensity. A further authorization request to run the UCx target at the designed power has been presented to the authorities. The physics design of the radioactive ion beam transfer line (1^+ and n^+) is completed and the reshaping of the ALPI beam line in the experimental hall has been performed to allow the installation of the Charge Breeder and of the low and high resolution mass separators. The Isol front-end is completed and under operation in the off-line laboratory to test ion sources, assembly and controls. The project is fully financed and the first reaccelerated exotic beam is expected for the end of 2019 [5].

3. Challenges of the SPES project: Our present knowledge of the structure of the nuclei is mostly limited to systems close to the valley of stability or nuclei with a deficiency of neutrons. Only recently unexplored regions of the nuclear chart, especially on the neutron-rich side, have become accessible due the availability of beams of unstable ions. Starting from a nucleus located on the stability line and

IOP Conf. Series: Journal of Physics: Conf. Series **1056** (2018) 012014 doi:10.1088/1742-6596/1056/1/012014

adding successively neutrons, the binding energy of the last neutron decreases steadily until it vanishes and the system decays by neutron emission. The position in the nuclear chart where the neutron emission starts defines the neutron drip line. Due to the absence of electrical repulsion between neutrons, the neutron drip line lies much farther away from the valley of stability than the corresponding drip line associated with protons. The location of the neutron drip line is largely unknown as experimental data are available only for light nuclei. Nuclei with large neutron excess are of interest not only for the location of the neutron drip line but also for the investigation of the density dependence of the effective interaction between the nucleons. Variations of the nuclear density and size in nuclei with increasing N/Z ratios are expected to lead to different nuclear symmetries and new excitation modes.

3.1 Structure of exotic nuclei and nuclei of astrophysical interest: The modification of nuclear properties when moving towards the neutron drip line depends on the shell structure changes for increasing neutron excess. As an example, studies of neutron-rich nuclei beyond the doubly magic ^{132}Sn are of key importance to investigate the single-particle structure above the N=82 shell closure, the stability of the shell closure and to find out how the effective interaction between valence nucleons behaves far from stability. In connection with the formation of a neutron skin new modes of collective motion are also expected, namely oscillation of the skin against the core, similar to the soft dipole mode already identified in the case of very light halo nuclei. Neither the thickness nor the detailed properties of the neutron skin of exotic nuclei are presently known. This information is essential to allow a quantitative description of compact systems like neutron stars, where exotic nuclei forming a Coulomb lattice are immersed in a sea of free neutrons, a system displaying the properties of both finite and infinite (nuclear matter) objects. At the energies of the secondary beams provided by the SPES facility, it will be possible to address important questions related to the study of neutron-rich matter such as nuclear forces, level density, viscosity, barrier, neutron pairing and collective modes. Interesting areas where new data will be collected are those in the very neutron rich regions, where shell evolution is a major issue. Investigations of the modification of the pairing interaction in the nuclear medium will receive significant inputs by measurements of multi-nucleon transfer reactions to specific nuclear states. Another important issue is the study of the basic nuclear properties, like lifetimes, masses and neutron capture probabilities, for systems involved in the element formation in the universe. Those quantities can allow to probe the predictive power of the nuclear models used to predict the element abundancies in different astrophysical scenarios (for a detailed discussion see [6].

3.2 Fundamental Interactions: Exotic nuclei can be used to amplify the sensitivity to symmetry violations. One example is CP/T violation. Parity (P), Time Reversal (T) and Charge Conjugations (C) are discrete symmetry transformations. Strong and electromagnetic interactions are symmetric under C, P and T separately and under CP and T whereas weak interaction is maximally antisymmetric under P and C involving only left-handed neutrinos and right-handed antineutrinos. CP is also violated in weak decays of kaons and b-mesons. CPT symmetry is also a requirement for any Lorenz-invariant quantum-field theory. CPT symmetry requires the Electric Dipole Moment (EDM) of a particle d and

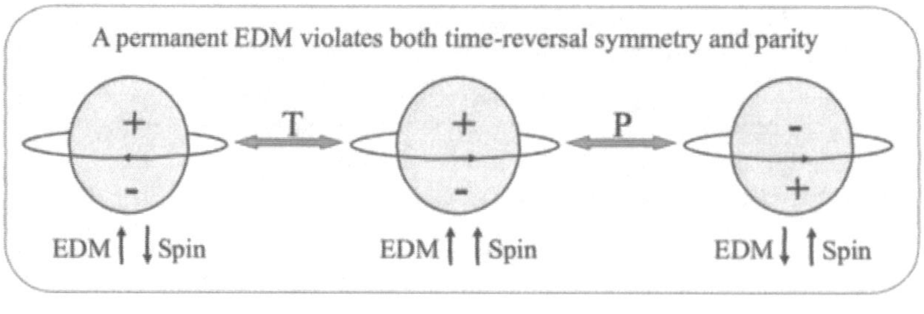

Figure 3: EDM and parity and time-reversal invariances.

the EDM of its antiparticle be equal in magnitude and opposite in sign. The determination of electric dipole moments (EDM), which violate parity (P) and time-reversal invariance (T), is one of the crucial steps to pinpoint physics beyond the Standard Model – Fig.3. Experiments can probe P-odd/T-odd observables in different systems. For the neutron and proton, the fundamental CP-violating interactions arise from a short range contribution and a long range contribution, the latter arising from the P-odd/T-odd pion-nucleon interactions. Paramagnetic atoms and molecules are most sensitive to electron EDM and the nuclear-spin independent electron-nucleus coupling. In diamagnetic atoms the dominant contributions are the nuclear-spin dependent electron-nucleus interaction and the Schiff moment, which also arises from long-range pion exchange and short range four-nucleon interactions. At present the most sensitive EDM search is performed on ^{199}Hg and the upper limits already constrain various extension of the Standard Model [7].
Nuclear structure can strongly amplify the sensitivity of nucler EDM measurements, in particular the occurrence of octupole deformation and enhanced octupole vibrations in nuclei lead to considerably larger Schiff moments inducing a contribution to an atomic EDM. Octupole collectivity is expected to arise when neutrons and protons near the Fermi surface populate states of opposite parity separated by total angular momentum 3ħ, which corresponds to proton and neutron numbers in the range Z or N ≈ 34, 56, 88 and N ≈ 134. Those systems have reflection asymmetric shapes, which can lead to permanent octupole deformation or to octupole vibration. Especially in the case of stable octupole deformation a large intrinsic dipole moment is polarized along the nuclear-spin by P-odd/T-odd interactions. This leads to the nuclear Schiff moment which is further enhanced by the electric polarizability of the nucleus. Since the Schiff moment can be expressed as:

$$S \propto \eta e\,(\beta_2\beta_3^2\, Z\, A^{2/3}\, r_0^3)/(E_+ - E_-)$$

where η represents the strength of the P-odd/T-odd interaction, β_2 and β_3 the quadrupole and octupole deformation parameters and E_{+-} the energies of the opposite parity states.
One of the most promising cases is the ^{229}Pa nucleus where the splitting was originally reported to be as small as 0.22 keV and evidence of strong octupole correlations support a ground state parity doublet with spin I=5/2. Due to the octupole collectivity for an odd nucleus an enhancement factor up to the order of 10^4 has been calculated respect to Hg nuclei [7]. Experimentally there is a strong evidence of octupole collectivity for nuclei in the A=200-226 mass region. This concerns interleaved even/odd parity states in even-A nuclei, parity doublets in odd-A nuclei and enhanced electric-dipole transition moments. One of the most stringent evidence of octupole collectivity concerns the large E3 strength found through Coulomb excitation measurements of radioactive beams of ^{220}Rn and ^{224}Ra at Isolde.
A present experimental issue is the characterization of the octupole collectivity for nuclei like ^{229}Pa and ^{225}Ac. At SPES an experimental program is presently under discussion for Ac, Th amd Pa nuclei aiming to the direct determination of the octupole strength (very challenging for odd systems) through Coulomb excitation and inelastic scattering experiments.

References

[1] D. Scarpa, L. Biasetto, S. corradetti, M. Manzolaro, A. Andrighetto, S. Carturan, G. Prete, P. Zanonato and D.W. Stracener, *Eur. Phys. J.* A (2011) 47
[2] M. Lindroos, *CERN-AB-2004-086* (2004)
[3] T. Lamy, J.L. Bouly, J.C. Curdy, P. Sole, P. Sortais, T. Thuillier, J.L. Vieux-Rochaz and D. Voulot; *Rev. Sci. Instrum* 75, 1624 (2004).
[4] G. Savard, R. Pardo, S. Baker, C. Davids, A. Levand, D. Peterson, D. Phyllips, T. Sun, R. Vondrasek, B. Zabransky, G. Zinkann *Hyperfine Interactions*, Vol. 199 Issue 1-3, (2011) 301.
[5] *http://web.infn.it/spes/*
[6] R. Surman et al., arXiv:13090059v1 (2013)
[7] T.E. Chupp, P. Fierlinger, M.J. Ramsey-Musolf, J.T. Singh, arXiv:1710.02504 [physics.atom -ph] ACFI-T17-18, ACFI-T17-18

Heavy–ion particle identification for the transfer reaction channels for the system ^{18}O + ^{116}Sn under the NUMEN Project

N Deshmukh[1*], F Cappuzzello[1,2], C Agodi[1], D Carbone[1], M Cavallaro[1], L Acosta[3], D Bonanno[4], D Bongiovanni[1], I Boztosun[5], S Calabrese[1,2], D Calvo[6], E R Chávez Lomelí[3], F Delaunay[6,7], P Finocchiaro[1], M Fisichella[6], A Foti[4], G Gallo[1,2], A Hacisalihoglu[1,8], F Iazzi[6,9], R Introzzi[6,9], G Lanzalone[1,10], R Linares[11], F Longhitano[4], D Lo Presti[2,4], N Medina[12], A Muoio[1], J R B Oliveira[12], A Pakou[13], L Pandola[1], F Pinna[6,9], S Reito[4], G Russo[2,4], G Santagati[1], O Sgouros[1,13], S O Solakcı[5], G Souliotis[13], V Soukeras[1,13], A Spatafora[1,2], D Torresi[1], S Tudisco[1], A Yildirin[5] and V A B Zagatto[11]

1. Istituto Nazionale di Fisica Nucleare, Laboratori Nazionali del Sud, Catania, Italy
2. Dipartimento di Fisica e Astronomia, Università di Catania, Catania, Italy
3. Universidad Nacional Autònoma de México, Ciudad de México, Mexico
4. Istituto Nazionale di Fisica Nucleare, Sezione di Catania, Catania, Italy
5. Akdeniz University, Antalya, Turkey
6. Istituto Nazionale di Fisica Nucleare, Sezione di Torino, Turin, Italy
7. LPC Caen, Normandie Université, ENSICAEN, UNICAEN, CNRS/IN2P3, Caen, France
8. Institute of Natural Science, Karadeniz Teknik Universitesi, Trabzon, Turkey
9. Politecnico di Torino, Turin, Italy
10. Università di Enna "Kore", Enna, Italy
11. Universidade Federal Fluminense, Niteroi, Brazil
12. Universidade de Sao Paulo, Sao Paulo, Brazil
13. University of Ioannina, Ioannina, Greece

[*]e-mail: drnikitdeshmukh@gmail.com

Abstract. The current work is a part of the NUMEN project, which aims to deduce the nuclear matrix elements (NME) of neutrinoless double beta decay by measuring the cross sections in heavy-ion induced Double Charge Exchange (DCE) reactions. The particle identification for the competing transfer reaction channels has been studied for the ^{18}O + ^{116}Sn system at 270 MeV.

1. Introduction

Neutrinoless double beta decay, 0νββ, is at the present time strongly pursued both experimentally and theoretically [1]. Its observation will determine whether the neutrino is a Dirac or Majorana particle and will provide a measurement of the average neutrino mass, which is one of the most fundamental problems in physics. The NUMEN project [2 – 4], at INFN Laboratory Nazionali del Sud in Catania (Italy), proposes to access the Nuclear Matrix Elements (NMEs) entering the expression of the life time of the 0νββ decay by measuring cross sections of DCE reactions in a wide range of incident energies. Crucial aspects of the project are the high-resolution in the energy spectra necessary to identify the transitions of interest and the accurate measurement of the absolute cross sections at very forward angles, including zero degree [5,6]. Moreover, a peculiarity of

NUMEN is the concurrent measurement of the other relevant reaction channels i.e., one- and two-neutron transfer, one- and two-proton transfer and single charge exchange. These studies are important because they allow to isolate the direct DCE mechanism from the competing multi-nucleon transfer processes. These are at least of 4th order in the nucleus-nucleus interaction and can be effectively minimized by the choice of the proper projectile-target system and incident energy [7]. In this manuscript we discuss the experimental measurement of the system $^{18}O + ^{116}Sn$, focusing in particular on the multi-nucleon transfer channels.

2. Experimental Details and Data reduction

The following competing channels were studied: 1n-transfer $^{116}Sn(^{18}O,^{17}O)^{117}Sn$, 2n-transfer $^{116}Sn(^{18}O,^{16}O)^{118}Sn$ and 1p-transfer $^{116}Sn(^{18}O,^{19}F)^{115}In$. These correspond to the almost complete net of transfer reactions for the system $^{18}O + ^{116}Sn$. Figure 1 shows the studied reaction channels.

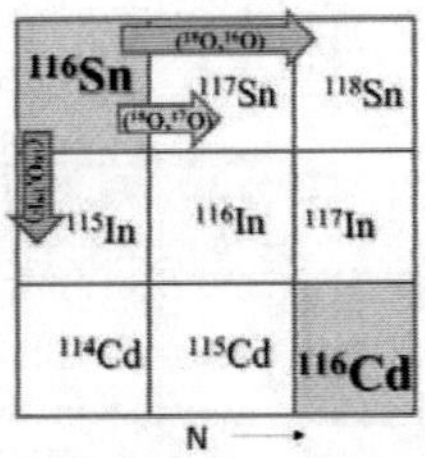

Figure 1: The net of multi-nucleon transfer reactions for the system $^{18}O + ^{116}Sn$

A beam of $^{18}O^{4+}$ ions, extracted by the K800 Superconducting Cyclotron accelerator, bombarded on 257 μg/cm² Sn target, at 270 MeV incident energy. The ejectiles produced were momentum-analyzed by MAGNEX [8,9,10], which is a quadrupole–dipole magnetic spectrometer characterized by a large angular (~ 50 mr) and momentum (–14.3%, +10.3%) acceptance and detected by its Focal Plane Detector (FPD) [11]. An angular range of $4^{o} < \theta_{lab} < 12^{o}$ was explored in the laboratory frame.

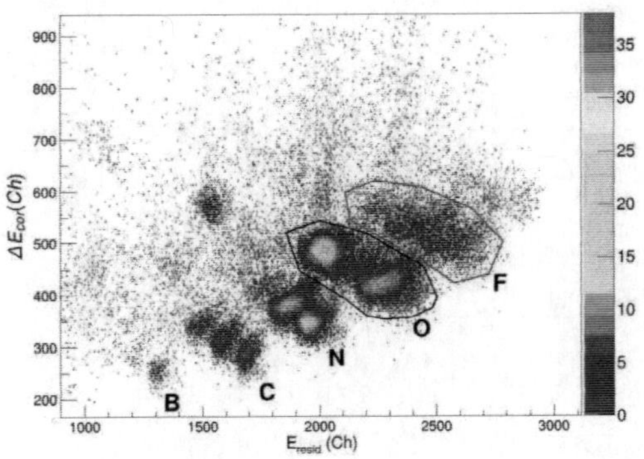

Figure 2: ΔE_{PC} - E_{resid} plot for the unselected ejectiles detected in the reaction $^{18}O + ^{116}Sn$ at 270 MeV incident energy for a single silicon detector. The other ejectiles like Boron, carbon, Nitrogen are also quite observable, apart from Oxygen and Fluorine. The graphical contours on Oxygen (black line) and Fluorine (red line) regions are also indicated.

The identification for the ejectiles were performed, firstly by identifying the atomic number (Z) using the standard ΔE – E technique based on Bethe-Bloch formula. A typical ΔE – E plot is shown in Figure 2 for a single silicon detector together with two coarse graphical contours: one includes the oxygen ejectiles (black line) and the other includes the Fluorine ejectiles (red line). The plotted parameters are the residual energy measured

by the Silicon detectors (E_{resid}) in abscissa, and in ordinate the energy loss in the gas measured by the PC counter (ΔE_{PC}) [11].
An innovative particle identification technique for the mass was introduced in ref. [12] that exploits the property of the Lorentz force, which determines the trajectory of a charged particle in a magnetic field to its momentum

$$B\rho = \frac{p}{q}$$

where p and q are the momentum and electric charge of the ion respectively, while ρ is the radius of curvature of the ion trajectory inside a dispersive element (bending magnet) with the field induction B.

In a non-relativistic approximation, the momentum p is related to the kinetic energy E and, approximately, to the residual energy measured by the silicons E_{resid}, by a quadratic relation $p = \sqrt{2mE}$, where m is the ion mass. Since the curvature ρ is related to the position at the focal plane X_{foc}, the relationship between the two measured quantities (X_{foc} and E_{resid}) is approximately quadratic with a factor depending on the ratio $\sqrt{m}/q$

$$X_{foc} = \frac{\sqrt{m}}{q}\sqrt{E_{resid}}$$

Therefore, in a X_{foc} v/s E_{resid} plot, the ions are distributed on different curves according to the ratio $\sqrt{m}/q$. The seperation between the different oxygen isotopes with the same charge state (8^+) is visible in Figure 3, where the X_{foc}-E_{resid} matrix shown is for the data selected in Figure 2, with the graphical condition on the ΔE_{PC}-E_{resid} (black line). For the present case, we chose the $^{16}O^{8+}$ ejectile (graphical cut in black), for the 2n-transfer reaction $^{116}Sn(^{18}O,^{16}O)^{118}Sn$, and $^{17}O^{8+}$ ejectile (graphical cut in red), for the 2n-transfer reaction $^{117}Sn(^{18}O,^{17}O)^{117}Sn$, as done with the graphical contour shown in Figure 3. Using the graphical condition on fluorine isotope in Figure 2 (red line), we can select the $^{19}F^{9+}$ ejectiles, for the 1p-transfer reaction $^{116}Sn(^{18}O,^{19}F)^{115}In$, as shown in Figure 4.

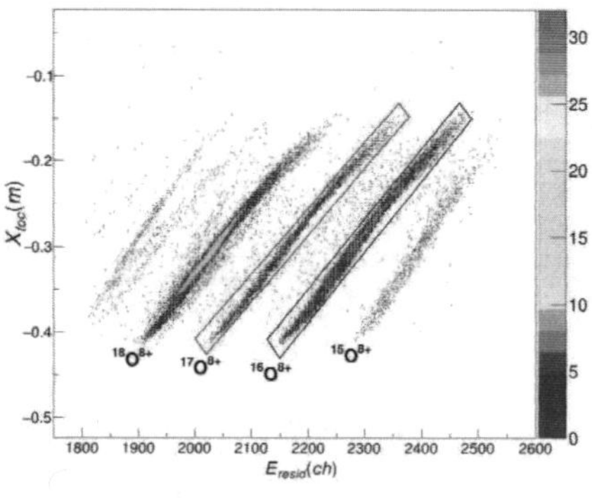

Figure 3: Typical X_{foc}-E_{resid} matrix plotted with the graphical condition of Figure 2 (black line) for the same silicon detector. The different isotopes for oxygen are seen and the graphical contours selecting the $^{16}O^{8+}$ ejectiles (black line) and the $^{17}O^{8+}$ ejectiles (red line) are also indicated.

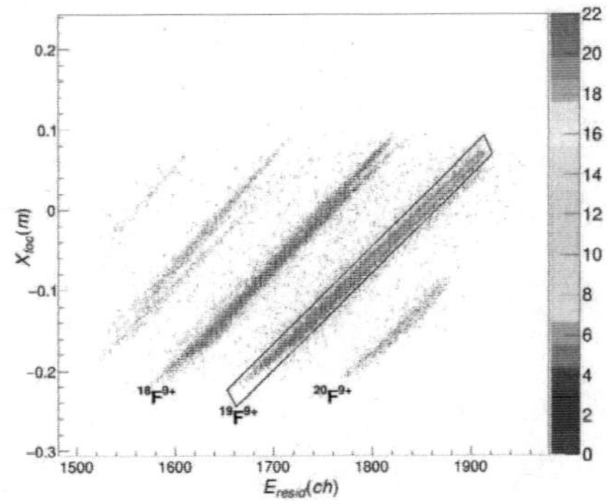

Figure 4: Typical X_{foc}-E_{resid} matrix plotted with the graphical condition of Figure 2 (red line) for the same silicon detector. The different isotopes for fluorine are seen and a graphical contour selecting the ^{19}F ejectile is also shown.

3. Summary and Conclusions

The identification procedures to separate the different ions has been carried out. Now after that the final phase space parameters has to be analyzed in detail which will provide an important information for the achieved horizontal and vertical focusing along with the aberrations [13,14]. Finally, the ray-reconstruction procedure will be applied to each set of analyzed data, according to the procedure described in refs. [15][16], in order to obtain the excitation energy spectra of the residual nuclei and the cross section angular distributions for the main transitions.

References

[1] Dell'Oro S, Marcocci S, Viel M, Vissani F 2016 *Adv. High Energy Phys.* **2162659**
[2] Cappuzzello F *et al.* 2015 *J. Phys.: Conf. Ser.* **630** 012018
[3] Agodi C *et al.* 2015 *Nucl. Part. Phys. Proceedings* **265–266** 28
[4] Cappuzzello F *et al.* 2018 *Eur. Phys. J. A* **54** 72
[5] Cappuzzello F *et al.* 2015 *Eur. Phys. J. A* **51** 145
[6] Matsubara H *et al.* 2013 *Few-Body Systems* **54** 1433
[7] Brink D M 1972 *Phys. Lett. B* **40** 37
[8] Cappuzzello F *et al.* 2016 *Eur. Phys. J. A* **52** 167
[9] Cappuzzello F *et al.* 2014 *Nucl. Instr. and Meth A* **763** 314
[10] Cunsolo A *et al.* 2007 *Eur. Phys. J. ST* **150** 343
[11] Cavallaro M *et al.* 2012 *Eur. Phys. J. A* **48** 59
[12] Cappuzzello F *et al.* 2010 *Nucl. Instr. and Meth A* **621** 419
[13] Lazzaro A *et al.* 2008 *Nucl. Instr. and Meth.* **591** 394.
[14] Lazzaro A *et al.* 2008 *Nucl. Instr. and Meth.* **585** 136
[15] Carbone D 2015 *Eur. Phys. J. Plus* **130** 143
[16] Calabrese S 2018 *Acta Phys. Pol. B* **49** 275

IOP Conf. Series: Journal of Physics: Conf. Series **1056** (2018) 012016 doi:10.1088/1742-6596/1056/1/012016

A Multi-Pixel Photon Counter detector prototype for direct detection of scintillation light in liquid xenon.

Adriano Di Giovanni

New York University Abu Dhabi, Abu Dhabi, United Arab Emirates

E-mail: adriano.digiovanni@nyu.edu

Abstract.
This work concerns the preliminary tests and characterization of a cryogenic preamplifier board for an array made of 16 S13370-3050CN (VUV4 family) Multi-Pixel Photon Counters manufactured by Hamamatsu and operated at liquid xenon temperature. The electronics for the proposed prototype is based on the use of the Analog Devices AD8011 current feedback operational amplifier, but other models can be also considered. The detector allows for single photon detection, making this device a promising choice for future generation of neutrino and dark matter detectors based on liquid xenon targets.

1. Introduction

Technology advancements in material science, electronics and devices gave a tremendous contribution not only to the industry, but also to the fundamental physics opening for the last generation astroparticle experiments.

Liquefied noble gases based detectors are at the forefront of Dark Matter search [1, 2, 3, 4]. In this scenario, the intrinsic detector background and the few photons expected in a WIMP (Weakly Interacting Massive Particle) - Nucleus interaction are carefully taken into account.

The screening of materials is a common approach to select the cleanest components (in terms of radiopurity) for the actual realisation of the cryostat containing the high purity liquefied noble gas. On the other hand, the selection of the photosensor needs to take into account many parameters (quantum efficiency in the region of interest, reliability, geometrical coverage, bulkiness, costs, etc.)

Photomultiplier tubes (PMTs) are still the most common option, however, in order to achieve optimal geometrical coverage and to scale down the total radioactivity budget of the detector, more compact cryogenic photosensors with single photon detection capability are being considered for the instrumentation of large volumes.

The requirements for a photosensor to be a good candidate are mainly driven by the limited light yields in the liquid phase (order of a few tens photons/keV): to improve the sensitivity, the ideal dark matter detector must have a high geometrical coverage, single photon counting capability, adequate photon detection efficiency (PDE, larger than 20% at the scintillation emission peak) and large gain (in the order of 10^6).

Photosensors used in liquid xenon based experiments are usually required to be sensitive in the wavelength region of vacuum ultraviolet (VUV) where the scintillation happens ($\lambda_{\text{scintillation}} \approx$ 178 nm).

IOP Conf. Series: Journal of Physics: Conf. Series **1056** (2018) 012016 doi:10.1088/1742-6596/1056/1/012016

Silicon Photomultiplier (SiPM) or Multi-pixel Photon Counter (MPPC) are the most promising candidate to replace PMTs in the next generation of large scale dark matter detectors [5].

The prototype presented in this work consists of 16 3×3 mm^2 S13370-3050CN, the fourth generation of vacuum ultraviolet (VUV) sensitive multi-pixel photon counters (VUV4-MPPC) manufactured by Hamamatsu. The VUV4-MPPC family:

(i) can be operated at low voltage (< 60 V) in LXe,
(ii) shows single photon counting capability,
(iii) has PDE close to 25% at 178 nm,
(iv) can be operated at gains larger than 2×10^6.

To get the equivalent coverage of a PMT, by keeping the same number of readout channels, several tens of MPPC needs to be readout as they were a single channel (in form of arrays or matrices of sensors with one single channel of biasing and signal output). The readout of many devices as a single channel is still challenging. A few typical readout examples are described in [6].

The aim of this work is therefore to provide a single output cryogenic amplifier readout for 16 MPPCs for liquid xenon based experiments.

2. The detector and the experimental setup

The detector shown in figure 1 is an array of 16 MPPCs soldered to an interface board that is in turn connected to the preamplifier board.

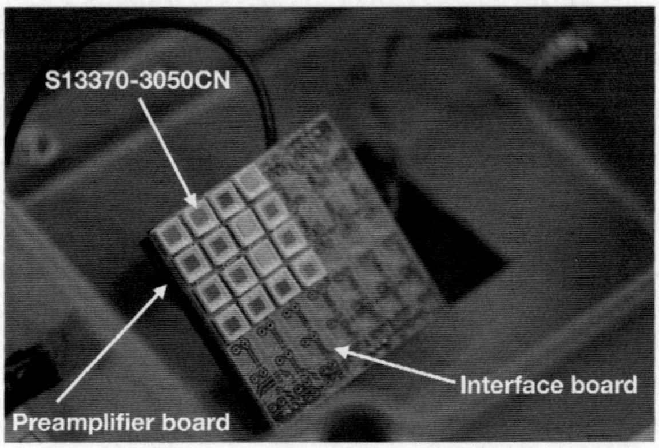

Figure 1. The 16 MPPCs array used in the experiment.

The preamplifier circuit is based on the use of the Analog Devices AD8011 operational amplifier. This latter had been already successfully test in a preamplifier circuit for the Hamamatsu PMT R11410 [7].

The array has been characterized at LXe temperature (175 K), the setup has been cooled down by using a cold finger immersed in a liquid nitrogen reservoir. A change in the liquid height reflects a variation in the operational temperature. The MPPC array and its electronics have been mounted in a light tight metal enclosure (see figure 2) instrumented with all the required IN/OUTPUT connectors along with an optical diffuser to isotropically shine light from a pulsed UV LED. A PT100 has been used for temperature measurement and consequently to compensate the biasing of the MPPCs in the array. Gas Nitrogen has been constantly flushed through the box in order to prevent water vapor condensation.

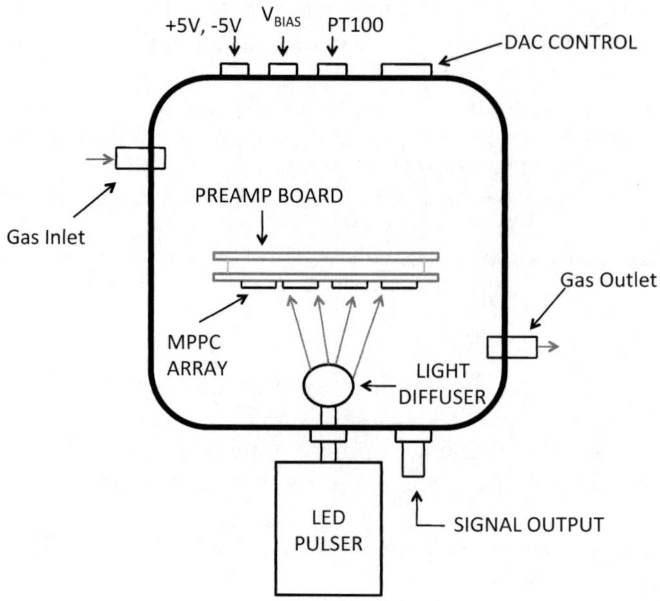

Figure 2. A sketch of the test unit used in the experiment.

A Digital to Analog Converter (DAC) has been implemented in the biasing circuit to equalize the gain of the 16 MPPCs due to their different breakdown voltages (in the range 55.56 V ÷ 55.78 V) by fine tuning the over-voltages[1]

An Agilent E3645A has been used for the biasing of the MPPCs, while for the amplifier unit a linear DC Elind 32DP8 power supply has been used. The temperature measurements has been acquired through a Keithley 2100 digital multimeter. A LeCroy HDO6104 oscilloscope has been used to collect waveforms in 1 μs time window and sampled with 2500 points at 12 bits at full bandwidth.

Typical waveforms corresponding to single and double photon event families are shown in figure 3.

3. The Electronics

An MPPC can be represented by a matrix of independent channels connected in parallel [8], each one consisting of a series of one avalanche photodiode (APD) and a quenching resistor. The equivalent circuit of a MPPC is reported in figure 4.

An avalanche photodiode can be modeled through a junction capacitance Cj, a voltage source (breakdown voltage Vbd), a junction resistance Rj and a switch S that closes when a photon hits the sensor. The resistor Rq quenches the signal and restore S to the open position. Current limiting resistors (R_a), bypass capacitances (C_b) and the decoupling resistors (R_S) are all wired outside the MPPC.

The resistor R_S is used to decouple the MPPC equivalent parasitic capacitance C_S of any non-fired photosensor from the operational amplifier.

figure 5 is an effective approximation of the array being exposed to a single photon.

The signal is represented by the current generator I_S connected in parallel with C_S towards ground. Since all the quenching resistors are connected in parallel, the equivalent resistance is negligible with respect to R_S.

[1] The over-voltage (V_{OV}) is the voltage above the breakdown

IOP Conf. Series: Journal of Physics: Conf. Series **1056** (2018) 012016 doi:10.1088/1742-6596/1056/1/012016

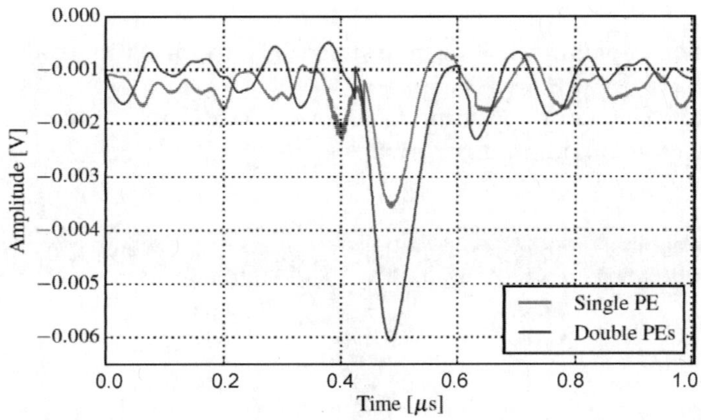

Figure 3. Example of typical waveforms corresponding to a single photon and to 2 photons event taken at 175 K, V_{OV}=2 V (50 Ω termination).

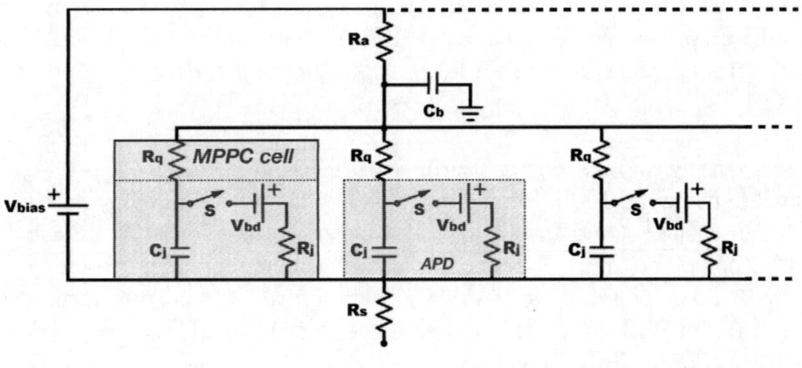

Figure 4. Equivalent circuit of a MPPC.

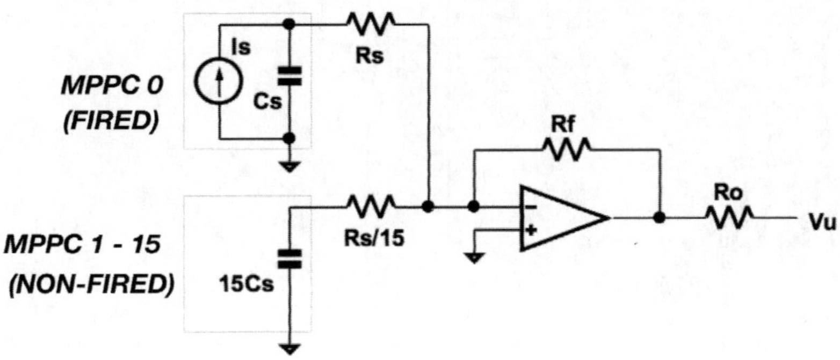

Figure 5. Scheme of the equivalent circuit of a single fired MPPC.

The operational amplifier output voltage is the product between the value of the feedback resistance (R_f= 1 kΩ) and the current I_S (the effect of C_S and R_S is negligible).

From the measurements, the average amplitude of a typical single photon signal (see figure 3) is ~2.5 mV corresponding to I_S= 5 μA (R_f= 1 kΩ and R_0= 50 Ω).

IOP Conf. Series: Journal of Physics: Conf. Series **1056** (2018) 012016 doi:10.1088/1742-6596/1056/1/012016

4. Results

Single photon detection capability has been assessed by exposing the array at pulsed UV light and operated at T = 175 K, $V_{OV} = 3$ V.

The measurements have been performed at lowest achievable intensity to maximize the probability of having only one photon per event (see figure 6).

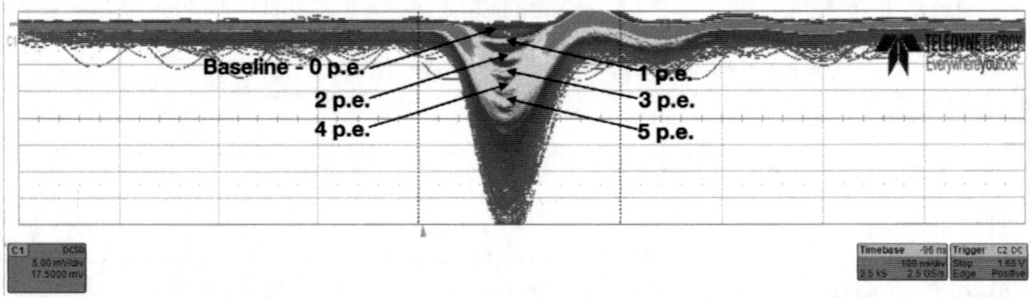

Figure 6. Waveforms taken in persistence mode at 175 K, $V_{OV} = 3$ V, by illuminating the detector with a LED pulser (used as trigger too). The spacing between signal families is the consequence of the preserved single photon counting capability after summing up the 16 individual MPPCs.

The separation among signals corresponding to different number of photons detected in a single event is visible in figure 6. The first family, below the baseline band, corresponds to single photon events (1 p.e.), the second family to a double photons events (2 p.e.), etc.

The typical photoelectron charge spectrum, obtained by integrating the acquired waveforms, has been fitted with a set of multiple gaussians (see figure 7): the measured pedestal (0 p.e.) charge is (1.47 ± 0.16) pC, while for the 1 p.e. is (3.21 ± 0.26) pC, corresponding to an overall gain of $\sim 2.0 \times 10^7$ (175 K, V_{OV}= 3 V).

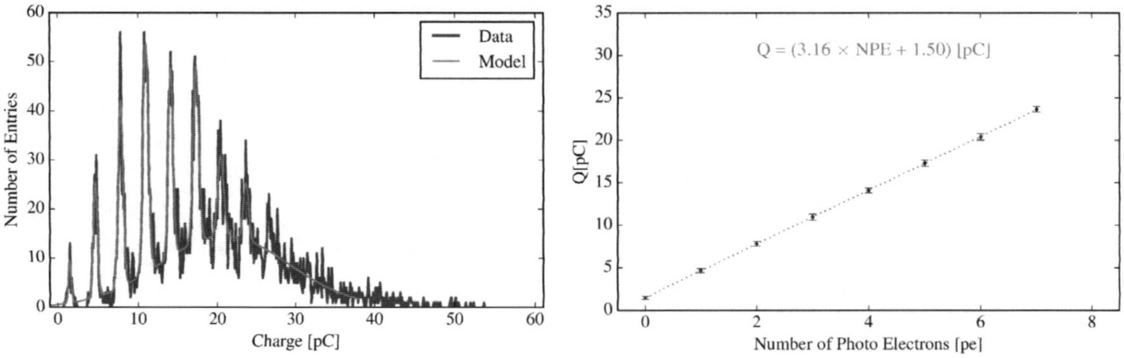

Figure 7. Photoelectron charge distribution with the array operated at 175 K, $V_{OV} = 3$ V. The data have been fitted by using a set of multiple gaussians. The measured charge of the 0 p.e. peak (baseline) is of the order of 1.5 pC, while the measured charge of the single photoelectron peak is about 3.2 pC.

The sigma of the charge pedestal showed in figure 7 is smaller than the average charge separation between two consecutive peaks (0.16 pC versus 3.16 pC); therefore a distinctive photoelectron peak charge distribution can be observed. Assuming no statistical effects, the main contributions to the sigma of each photoelectron peak are in fact due to noise (σ_{ELE}), dark counts (σ_{DC}), afterpulses (σ_{AP}), crosstalk (σ_{CT}) and gain fluctuation (σ_{GF}). The equation:

$$\sigma^2_{p.e.} = \sigma^2_{ELE} + \sigma^2_{DC} + \sigma^2_{AP} + \sigma^2_{CT} + \sigma^2_{GF} \quad (1)$$

Since $\sigma_{\mathrm{p.e.}}$ of the first photoelectron peak is 0.26 pC and σ_{ELE} is 0.16 pC, the contributions due to the detector (AP+DC+CT+GF) is dominant (0.20 pC).

5. Conclusions

This work concerns the design and tests of a photo detector prototype based on the use of 16 S13370-3050CN (VUV4 family) Multi-Pixel Photon Counters manufactured by Hamamatsu sensitive to VUV light.

The array is readout as a single detector through an electronics based on the use of an operational amplifier suitable for cryogenic environments (AD8011).

The measurements show an excellent photo detection capability, that is one of the most stringent requirements for the new generation of dark matter experiments. Overall, the performance of this device encourages for further studies for its use in liquid xenon based detectors.

References

[1] Akerib DS et al. 2017 Phys. Rev. D **95** 012008
[2] Abe K et al. 2013 NIM A **716** 78-85
[3] Agnes P et al. 2015 Phys. Lett. B **743** 456-466
[4] Aprile E et al. 2016 JCAP04 **027**
[5] Yamamoto K et al. 2006 IEEE Nucl. Sci. Symp. Conf. Rec. **N30-102** 1094-1097
[6] Sawada R 2015 POS (TIPP2014) **033**
[7] Arneodo F et al. 2016 NIM A **824** 275276
[8] Piatek S 2014 Internal Note www.Hamamatsu.com

IOP Conf. Series: Journal of Physics: Conf. Series **1056** (2018) 012017 doi:10.1088/1742-6596/1056/1/012017

A NaI-based cryogenic scintillating calorimeter: status and results of the COSINUS project

N Di Marco[a], G Angloher[b], P Carniti[c,d], L Cassina[c,d], L Gironi[c,d], C Gotti[c,d], M Mancuso[b], L Pagnanini[g,a], G Pessina[c], F Petricca[b], S Pirro[a], F Pröbst[b], R Puig[e,f], F Reindl[e,f], K Schäffner[g,a], J Schieck[e,f] and M Stahlberg[e,f]

[a] INFN - Laboratori Nazionali del Gran Sasso, I-67010 Assergi - Italy
[b] Max-Planck-Institut für Physik, D-80805 München - Germany
[c] INFN - Sezione di Milano Bicocca, I-20126 Milano - Italy
[d] Dipartimento di Fisica, Università di Milano-Bicocca, I-20126 Milano - Italy
[e] Institut für Hochenergiephysik der ÖAW, A-1050 Wien - Austria
[f] Atominstitut, Technical University Vienna, A-1020 Wien - Austria
[g] Gran Sasso Science Institute, I-67100 L'Aquila - Italy

E-mail: natalia.dimarco@lngs.infn.it

Abstract. The COSINUS (Cryogenic Observatory for SIgnals seen in Next-generation Underground Searches) project aims to provide a model independent cross-check of the long-standing DAMA/LIBRA claim on the observation of dark matter, by using the same target material (NaI) with a different experimental approach. The use of sodium iodide (NaI) crystals, operated at cryogenic temperature as scintillating calorimeters, provides both a low energy threshold for nuclear recoil events as expected from dark matter particle interactions, and the possibility to perform particle discrimination. Indeed, the dual read-out of phonon and light allows to perform signal-to-background discrimination on an event-by-event basis, a unique feature in comparison to other NaI-based dark matter searches. In this paper we will discuss in detail the COSINUS detector concept and we will present the performances of our first prototypes together with the results of the first measurements.

1. Introduction

Although the evidence for the existence of Dark Matter (DM) is nowadays well-established on the basis of several astrophysical measurements [1], its nature still remains unknown. Among several DM theories, the Weakly Interacting Massive Particles (WIMPs) are one of the most promising DM candidates. WIMPs, with a mass in the (GeV-TeV)/c^2 range, are expected to interact with ordinary matter with cross sections similar or smaller than the weak processes.

Direct DM search techniques are currently based on the detection of WIMPs scattering off the target nuclei and, since the expected recoil energies are well below 100 keV, an ultra-low background and highly sensitive detection apparatus is needed. At present several experimental efforts are ongoing all over the word using a variety of experimental approaches exploiting single or dual channel read-out based on the detection of scintillation light, phonon and/or charge signals induced by particles interacting in the detector target.

The analysis of present DM experiments is based on the detection of an excess of events over the expected background or on the detection of an annual modulation of the event rate

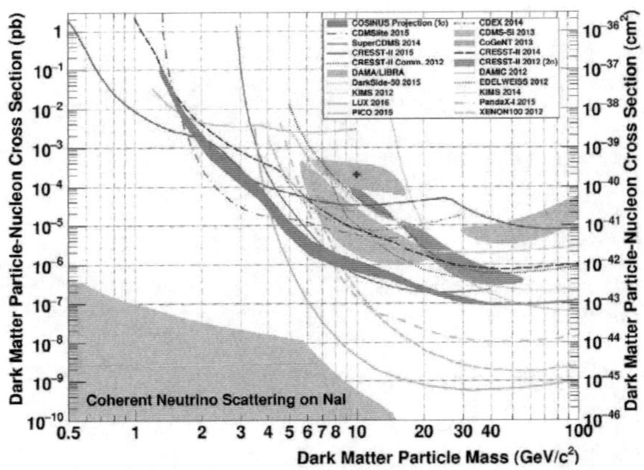

Figure 1. Projected sensitivity for a NaI-based experiment using COSINUS detector technology (solid blue band, 1 σ confidence level for spin-independent elastic dark matter nucleus scattering. Recent results from experiments using silicon and germanium targets are drawn in green, results from CRESST-II ($CaWO_4$) are depicted in red and exclusion limits using liquid noble gases in magenta. Limits drawn in cyan correspond to bubble chamber technology and experiments with CsI target. The light blue shaded regions correspond to the interpretation of the DAMA/LIBRA (NaI(Tl)) modulation signal by Savage et al. [2]. The benchmark point (blue cross) indicates the mass and cross-section chosen for the simulated WIMP contribution presented in section III of [3]. Gray-shaded regions in parameter space will be affected by coherent neutrino nucleus scattering on NaI mainly originating from solar neutrinos [4]. Figure from [3],references therein.

caused by the seasonal variation of the Earth's velocity with respect to the dark matter halo [5]. While in the former approach the number of expected recoil events depends on the DM model assumptions, as the mass of dark matter particle, the interacting cross-section and the detector properties, a pure modulation signal can be considered model-independent.

Nonetheless, in the experimental panorama of the DM direct search the situation is strongly controversial: the DAMA/LIBRA collaboration reports a statistically robust modulation signal ($>9\sigma$) over 14 annual cycles [6] using a detector based on 250 kg of highly radiopure sodium iodide (NaI(Tl)) crystals operated at room temperature in the underground site of the Laboratori Nazionali del Gran Sasso (LNGS) in Italy. In the standard elastic scattering scenario, the DAMA/LIBRA result is not consistent with the null results of other direct search experiments (see Figure 1). Further studies are therefore needed to understand the origin of the modulation signal observed by the DAMA/LIBRA experiment and, in particular, the target dependence of the result; several R&D projects and experiments with alkali halide as target are currently running or in construction [7, 8, 9]. Among them, the COSINUS project [3] aims to develop a NaI-based detector operated at cryogenic temperature, thus exploiting both the scintillation light signal of NaI crystals and the phonon signal in a cryogenic detector. The dual channel read-out allows to perform particle identification, thus resulting in an effective suppression of the dominant β/γ background.

2. NaI-based cryogenic scintillating calorimeter

In the DM sector, the use of cryogenic scintillating calorimeters is successfully implemented in the CRESST experiment [10] while it is used by LUCIFER/CUPID−0 [11] for the search for neutrinoless double-beta decay.

A particle interacting in a cryogenic scintillating calorimeter produces two coincident signals:

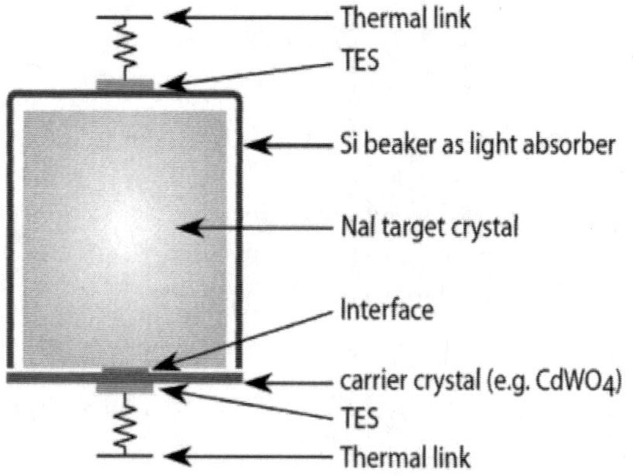

Figure 2. Scheme of a COSINUS detector module.

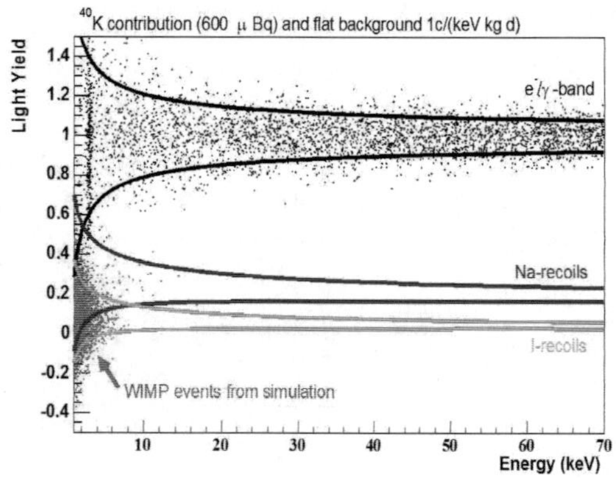

Figure 3. Result of simulated data for an exposure of 100 kg days with the assumption detailed in [3] in the light yield - energy plane.

the major part of the deposited energy is converted into phonons (heat signal) while a small fraction (O(few percent)) goes into the production of scintillation light. The heat signal is therefore exploited to precisely measure the energy deposited in the target crystal and is almost independent from the particle type. The amount of scintillation light, instead, strongly depends on the particle type and can be used to perform particle identification, thus allowing the suppression of the dominant β/γ background with respect to nuclear recoil signal events. The light to phonon ratio, the so-called *light yield* (LY), is characteristic for each type of event: in particular β and γ particles produce the most light and get assigned a LY$=1$ by definition, while α-particles and neutrons show a lower LY quantified by the so-called *quenching factor* (QF).

Figure 2 shows the concept of a COSINUS detector module: a NaI target crystal is coupled via an interface (typically silicon oil) to a crystal (e.g. $CdWO_4$) carrying the Transition Edge temperature Sensor (TES) which reads the phonon signal. The target is inserted in a beaker-shaped light detector made from high-purity silicon. The light detector is equipped with a second TES reading the heat rise of the beaker caused by the absorption of scintillation photons. The whole detector module is operated at few milli-Kelvin. The need for the use of the $CdWO_4$ carrier crystal, is due to the hygroscopic nature and the low melting point of the NaI crystal making it difficult to evaporate the TES directly onto the NaI surface. By covering a large fraction of the NaI crystal surface, the beaker-shaped light detector exhibits a very high light yield while simultaneously serving as an active veto for external backgrounds and radioactive decays close to the surfaces of the NaI crystal and its surrounding material.

Figure 3 shows the result of simulated data for an exposure of 100 kg days (with the assumption detailed in [3]) in the light yield - energy plane: black, blue and green bands correspond to electron, Na and I recoils respectively. Black dots indicate events originating from a flat background contribution of 1 c/(keV kg d) and a ^{40}K contamination with an activity of 600 μBq while red dots indicate Na and I scattered off by a hypothetical WIMP with a mass of 10 GeV/c^2 and a cross section of 2×10^{-4} pb (following the DAMA interpretation of Savage et al. [2]). Magenta boxes indicate the regions which may contribute to the DAMA/LIBRA energy range of positive modulation detection corresponding to 2-6 keV$_{ee}$ (where ee = electron equivalent).

IOP Conf. Series: Journal of Physics: Conf. Series **1056** (2018) 012017 doi:10.1088/1742-6596/1056/1/012017

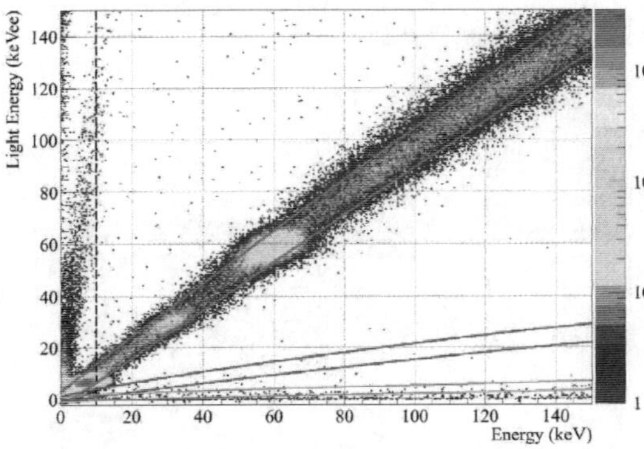

Figure 4. Data corresponding to a net exposure of 0.46 kg days of the first detector prototype in the light vs.(phonon) energy plane [12].

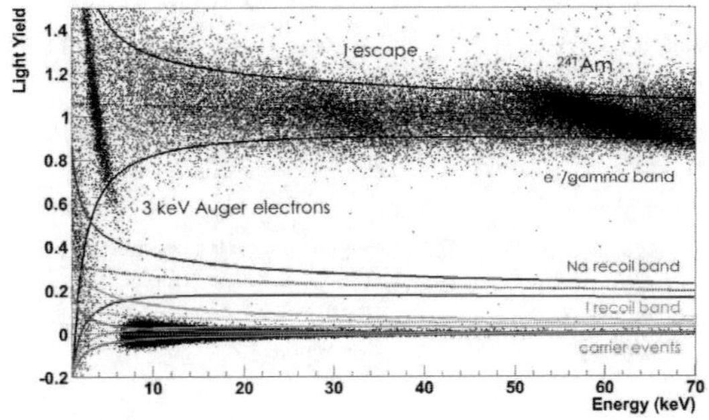

Figure 5. Data from the second COSINUS prototype. Lines indicate the central 80% bands for electron (black), Na (blue) and I (green) recoils respectively while the events without associated light signal are described by the magenta band [13].

3. Status and results of the COSINUS project

The first COSINUS detector prototype was realized using an undoped/pure NaI crystal with dimensions of 30x30x20 mm^3 and a weight of 66 g. The NaI is attached through a very thin layer of silicon oil to a carrier crystal made of $CdWO_4$ (diameter = 39.2mm, thickness = 1.6mm) equipped with a TES made by a thin film (200 nm) of superconducting tungsten (W-TES) produced at the Max-Planck-Institute for Physics in Munich. The cryogenic light detector consists of a sapphire wafer (diameter = 40mm, thickness $\simeq$460 μm) equipped with a TES optimized for light measurement. To increase the absorption efficiency of the blue scintillation light a 1-μm-thick layer of silicon is epitaxially grown onto the sapphire disc (SOS = Silicon On Sapphire). The whole setup is surrounded by a reflective foil [12].

The measurement was carried out in the test facility of the Max-Planck-Institute for Physics located in the underground site of LNGS, consisting of a dilution refrigerator allowing detector operation at temperatures as low as 7 mK. The TESs were read out with commercial dc-SQUID electronics (Applied Physics Systems company). The hardware-triggered signals were recorded in a 328 ms window at a sampling rate of 25 kHz. Both detectors were always read out simultaneously, independent of which one triggered. The detectors were irradiated with γ-rays from an ^{41}Am source with an energy of 59,541 keV. After applying quality cuts to reject unstable detector operation, pile-up and bad pulse-shape events, a final exposure of 0.46 kg days was collected [12].

Figure 4 shows the events surviving quality cuts in the light-energy plane. A linear relation between light output and the energy deposited in the NaI crystal is clearly visible. Magenta solid lines confine the central 80% of the β/γ-band. Accounting for the QFs given in [14], red and green bands correspond to recoils off Na and I respectively, while the almost horizontal event population at very small light energies originates from particle events taking place in the $CdWO_4$ carrier crystal. The black dashed line indicates the hardware trigger threshold of the phonon detector at 10 keV, essentially limited by the high pile-up rate observed. The energy resolution at the $\sim$60 keV calibration line is found to be 5.0 keV showing a significant improvement towards lower energies. By comparing the amplitude of a scintillation light pulse induced from a 59.541 keV x-ray in the NaI, originating from the ^{241}Am calibration source, to a direct hit of the light

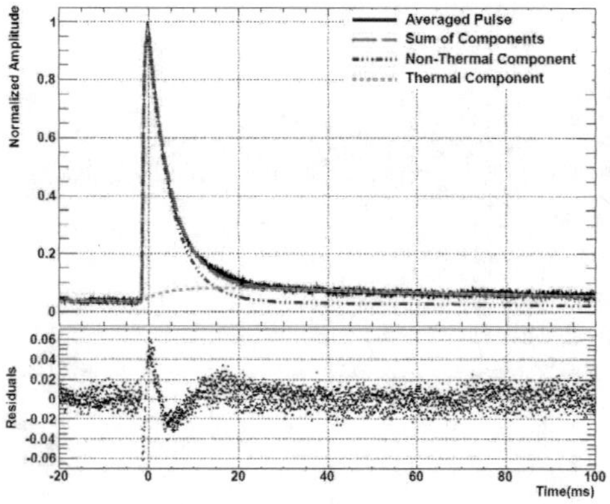

Figure 6. Top: Normalized average pulse (black) and Fit of the two-component model. Bottom: Fit residuals [13].

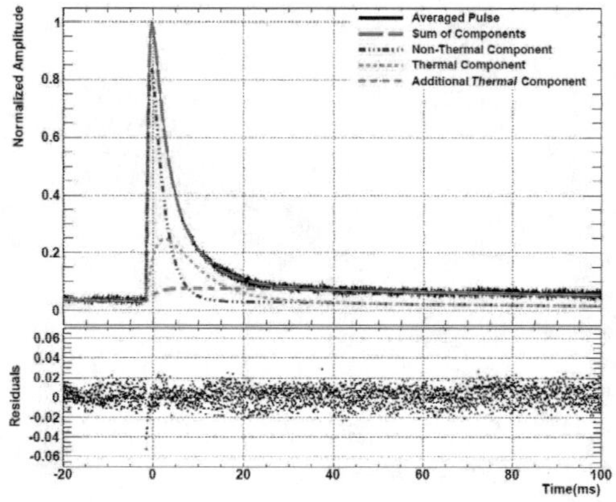

Figure 7. Top: Normalized average pulse (black) and Fit of the three-component model. Bottom: Fit residuals [13].

detector from a ^{55}Fe X-ray calibration source, we find that 3.7% of the energy deposited in the NaI crystal is measured as light [12].

In the second detector prototype realized, the beaker-shaped light detector, foreseen for the final detector design (Figure 2), was implement. The prototype was tested in the same facility used for the first protoytpe and, after applying the quality cuts, a total exposure of 1.32 kg days was collected. The result is shown in 5: black solid lines indicate a fit of the β/γ-band to the data with 80% bands. The mean line of the band is well modeled by a simple linear function, the width of the band includes contributions from the baseline resolutions of phonon and light detector, a Poisson-statistics term accounting for the quantization of the scintillation light photons and a empiric term scaling with light energy (e.g. potential position dependencies). Moreover, three characteristic structures are clearly evident: the line at ~60 keV originating from gammas from the ^{241}Am source, the so-called escape peaks from I at ~30 keV and a line at around 3 keV caused by ^{40}K Auger electrons. Na (blue) and I (green) recoil bands are calculated using the QFs of [14]. The magenta band is populated by events originating in the carrier crystal and showing a different pulse-shape than events in the NaI absorber [13].

The phonon detector was operated with a hardware trigger threshold of (8.26±0.02(stat.))keV, still significantly above the COSINUS design goal of 1 keV. A baseline resolution for the phonon detector of 1.01 keV was achieved increasing to 4.5 keV at the ^{241}Am-peak at about 60 keV. With the second prototype an absolute light yield of 13.1 %, corresponding to an average of ~40 photons/keV (given a mean scintillation photon energy of 3.3 eV), was obtained. This value is roughly three times higher than that obtained with the first prototype and has been achieved thanks to the geometry of the beaker-shaped light detector [13].

Both in the measurement of the first and second COSINUS prototypes, we found that NaI exhibits a different pulse shape compared to other materials (Al_2O_3, $CaWO_4$, $ZnWO_4$, $CdWO_4$, CaF, and TeO_2) and, thus, cannot be convincingly described by the model for cryogenic particle detectors of [15]. We performed a pulse shape analysis by averaging O(100) events of the same energy deposition. We fit the averaged pulse both with the two-component model described in [15] and with a three-component fit. In the latter case, to the standard thermal and non-thermal components described in [15], we added a second thermal component. The results are shown in Figure 6 and 7 respectively: the residuals of both fits clearly show that the additional third component signifiantly improves the fit result and almost perfectly describes the measured pulse, while the two-component fit does not [12, 13]. As a result of this analysis we can conclude that the peculiarity of NaI is a long decay time of the pulse. As the decay time exceed the maximally available record window of the present DAQ we will switch to a dead-time free DAQ in the future, thus allowing to record the full tail which is a immensely useful tool to understand the origin of this third pulse component.

4. Conclusions and perspectives

The COSINUS R&D project aims at developing a NaI-based detector to solve the long-standing tension of the DAMA/LIBRA result in the panorama of direct DM searches. By using the same target material of the DAMA experiment, the target dependencies can be ruled out. Moreover, the use of a cryogenic apparatus and of a dual channel read-out, will allow to perform a signal to background discrimination on an event-by-event basis unique to NaI DM searches.

The first prototype developed by the COSINUS collaboration was the first successful measurement of a NaI crystal operated as cryogenic scintillating calorimeter measuring simultaneously the phonon and light signal from a particle interaction. The implementation of the second prototype has been the proof-of-principle measurement of the final detector design. We measured 13.1% (~40 photons/keV) of the energy deposited in the NaI as scintillation light thus exceeding the original COSINUS design goal (4 %) by more than a factor of three and proving that undoped NaI exhibits an outstandingly high light output at low temperatures. Moreover we observed a peculiar phonon pulse-shape previously not seen in numerous other materials.

At present, we are carrying out the optimization of the performances of a new detector prototype in a dedicated low-background cryostat in the underground Hall C at LNGS while the realization of further prototypes, by using high-purity both doped and undoped crystals, is foreseen. These tests together with the planned neutron test beam at the Tandem accelerator at TUM/LMU in Munich, to directly measure for the first time the QFs of a NaI crystal operated at cryogenic temperatures, will open the way for the realization of the COSINUS DM detector.

Acknowledgements

This work was carried out in the frame of the COSINUS R&D project funded by the INFN (CSN5). We want to thank the LNGS mechanical workshop team E. Tatananni, A. Rotilio, A. Corsi, and B. Romualdi for continuous help in the overall set-up construction and M. Guetti for his cryogenic expertise and his constant support.

References

[1] Olive K A *et al.* (Particle Data Group) 2014 *Chin. Phys.* **C38** 090001
[2] Savage C, Gelmini G, Gondolo P and Freese K 2009 *JCAP* **0904** 010 (*Preprint* 0808.3607)
[3] Angloher G *et al.* 2016 *Eur. Phys. J.* **C76** 441 (*Preprint* 1603.02214)
[4] Gütlein A *et al.* 2015 *Astropart. Phys.* **69** 44–49 (*Preprint* 1408.2357)
[5] Drukier A K, Freese K and Spergel D N 1986 *Phys. Rev.* **D33** 3495–3508
[6] Bernabei R *et al.* 2013 *Eur. Phys. J.* **C73** 2648 (*Preprint* 1308.5109)
[7] Thompson W G (COSINE-100) 2017 Current status and projected sensitivity of COSINE-100 *TAUP 2017* (*Preprint* 1711.01488)
[8] Amare J *et al.* 2017 The ANAIS-112 experiment at the Canfranc Underground Laboratory *TAUP 2017* (*Preprint* 1710.03837)
[9] Tomei C (SABRE) 2017 *Nucl. Instrum. Meth.* **A845** 418–420
[10] Angloher G *et al.* (CRESST) 2016 *Eur. Phys. J.* **C76** 25 (*Preprint* 1509.01515)
[11] Artusa D R *et al.* 2017 *Phys. Lett.* **B767** 321–329 (*Preprint* 1610.03513)
[12] Angloher G *et al.* (COSINUS) 2017 *JINST* **12** P11007 (*Preprint* 1705.11028)
[13] Reindl F *et al.* (COSINUS) 2017 (*Preprint* 1711.01482)
[14] Tretyak V I 2010 *Astropart. Phys.* **33** 40–53 (*Preprint* 0911.3041)
[15] Pröbst F 1995 *Journal of Low Temperature Physics* **100** 69–104 ISSN 1573-7357

IOP Conf. Series: Journal of Physics: Conf. Series **1056** (2018) 012018 doi:10.1088/1742-6596/1056/1/012018

NEWSdm: an emulsion-based directional Dark Matter experiment

N. Di Marco for the NEWSdm Collaboration
INFN - Laboratori Nazionali del Gran Sasso, I-67010 Assergi - Italy
E-mail: natalia.dimarco@lngs.infn.it

Abstract. In the field of direct Dark Matter search, a variety of experiments have been developed over the past decades aiming at detecting Weakly Interactive Massive Particles (WIMPs) via their scattering in a detector medium and, in the last years, several experimental efforts are concentrated on the directionality approach: the observation of the incoming apparent direction of WIMPs would in fact provide a new and unambiguous signature and the proof of the galactic origin of Dark Matter. Furthermore the directionality appears as the only way to overcome the neutrino background that is expected to finally prevent standard techniques to further lower cross-section limits. The NEWSdm experiment has been proposed to measure the direction of WIMP-induced nuclear recoils by using a novel emulsion technology with improved spatial resolution. In this paper we will present the status of the experiment, the performances of the newly developed read-out systems reaching sub-micrometric resolution and we will discuss the expected sensitivity and discovery potential.

1. Introduction

The evidence of the existence of Dark Matter (DM) is nowadays well supported by several astrophysical measurement [1]. In the sector of direct DM search, the experimental signature is the detection of the nuclear recoil scattered-off by a WIMP. The analysis is based on the measurement of an excess of events over the expected background or on the detection of an annual modulation of the event rate due to the motion of the Earth-based detector in the DM Halo. A different experimental approach would consist in the detection of the direction of the recoiled nuclei: the motion of the Sun inside the galaxy causes in fact a peak in the nuclear recoils induced by WIMPs in the direction opposite to the motion itself (i.e. toward the Cygnus costellation). First proposed by Spergel [2] in 1988, the detection of the anisotropy of the recoiled nucleus direction, with respect to the isotropically distributed background, would provide a strong signature and an unambiguous proof of the galactic origin of DM particles.

Current directional DM projects are mainly based on the use of a large TPC [3]: in low pressure gaseous detectors the nuclear recoils are long enough (O(mm)) to be reconstructed as tri-dimensional tracks, but the target mass is rather small being limited by the low gas density. The sensitivity of such experiments is therefore quite poor. The use of a solid target would allow to explore low cross section sectors in the phase space covered by recent direct search experiments, the challenge being the shorter track length (O(100 nm)) resulting in the WIMP-nucleus scattering. The ground-breaking approach of the NEWSdm project, using very high resolution nuclear emulsions as solid target and novel technologies for their read-out, would overcome this problem.

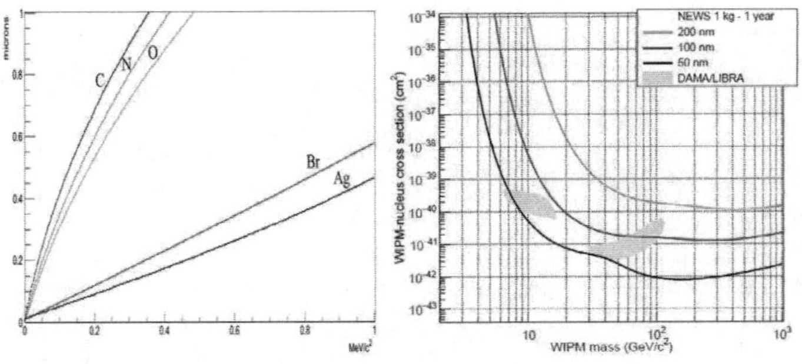

Figure 1. Left: correlation between the track length of the recoiled nuclei and their kinetic energy, for the different target nuclei contained in NIT emulsions. Right: the 90% C.L. upper limits for a NIT detector with an exposure of 1 kg × year, a threshold ranging from 200 nm down to 50 nm, in the zero background hypothesis. The directionality information is not included. For a detailed study on the NEWSdm sensitivity the reader is referred to [4].

2. Directional Dark Matter Searches: the NEWSdm approach

In the NEWSdm (Nuclear Emulsion for WIMP Search, directional measurement) project, the detector is conceived as a bulk of nuclear emulsions acting both as a target and as a tracking device surrounded by a shield to reduce the external background. The target is placed on an equatorial telescope in order to absorb the earth rotation, thus keeping fixed the detector orientation with respect to the incoming apparent WIMP flux [5]. The angular distribution of the WIMP-scattered nuclei is therefore expected to be strongly anisotropic with a peak centered in the forward direction.

In order to reach the extremely high spatial resolution needed to detect the submicrometric track left by a WIMP-scattered nucleus in a solid medium, the target is planned to be instrumented with the so-called Nano Imaging Tracker (NIT) [6] consisting of silver halide crystals with diameters down to a few tens of nm, an order of magnitude smaller than conventional emulsions (like the ones used in the OPERA [7] experiment). The silver halide crystals are immersed in an organic gelatin: NIT therefore contains both heavy (Ag, Br) and light nuclei such as C, N and O with weight fractions of ~ 80% for AgBr and ~ 20% for CNO. The correlation between the track length of the recoiled nucleus and its kinetic energy is shown in Figure 1 for the different target nuclei.

The read-out of the target emulsion films is based on a two-step approach and is performed by using an improved version [8, 9, 10] of the optical microscope used for the scanning of the OPERA films [11] with an intrinsic spatial resolution of the order of 200 nm. In the first step a fast pre-selection of candidate signal tracks is performed on the basis of the so-called shape-analysis [12]: a signal-like cluster, made by several grains, tends indeed to have an elliptical shape with the major axis coincident with the direction of the trajectory, while a background cluster, produced by a single grain, tends to have a spherical shape. An elliptical fit of the cluster shape allows to perform a first separation between background grains and signal tracks.

In the second read-out step, a pin-point check of preselected candidates is needed in order to resolve the grains belonging to a track. A microscope system using polarized light and with the capability of extending the nanometric resolution to all 3 axes, has been designed and tested [13]. The basic idea is to exploit the resonance effect occurring when nanometric metal grains are dispersed in a dielectric medium [14]. The polarization dependence of the resonance frequencies strongly reflects the shape anisotropy and can be used to infer the presence of non-spherical nanometric grains. Images of the same cluster taken with different polarization angles will then

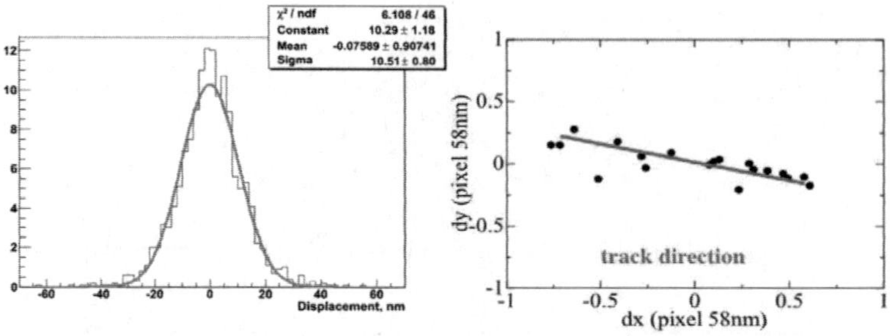

Figure 2. Right: study of the displacement of the barycenter position of a real track image on NIT obtained exploiting the resonant light scattering and polarized light. Left: the spatial resolution achieved is ~10 nm in both x and y coordinates.

show a displacement of the position of its barycenter. The analysis of the displacements allows to distinguish clusters made of a single grain from those made of two (or more) grains thus resulting in a clear signal to background track discrimination.

In order to measure efficiency and resolution of the new scanning system, a test beam with low velocity ions was performed. Kr ion beam with energies of 200 and 400 keV [15] and a C ion beam with energies of 60, 80 and 100 keV were used. The results are reported in [12] and [5]. As shown in Figure 2 the unprecedented accuracy of about 10 nm has been achieved. Moreover, in order to evaluate the intrinsic angular resolution of the scanning system we exposed a NIT sample to a 2.8 MeV neutron beam at the Fusion Neutronics Source (FNS-JAEA) in Japan. An intrinsic angular resolution of 230 mrad was achieved, by far the best resolution achieved with direction sensitive detectors in this energy range [5].

Background sources for dark matter searches are α and β particles, γ-rays and neutron induced recoils, while NIT are essentially not sensitive to minimum ionizing particles (MIP). α-particles, originating from U and Th radioactive chains, can be identified by measuring their range in emulsion ($O(10)\mu$m), by far longer than WIMP-induced nuclear recoils ($< 1\mu$m). The γ radiation due to environmental radioactivity constitutes a non-negligible contribution to the total background budget as well as β-rays produced in ^{14}C decay. This kind of background is anyway less critical for NIT emulsion with respect to other sources being possible to rejected it by properly regulating the emulsion sensitivity through a chemical treatment, exploiting the response of β-rays to the polarized light scattering or performing a cryogenic exposure and by exploiting the phonon effect.

Neutron induced recoils are the main background source. While the external neutron flux can be reduced to a reasonable level with an appropriate shielding, the intrinsic emulsion radioactivity would be responsible of an irreducible neutron yield through (α, n) and ^{238} U spontaneous fission reaction. In order to estimate this contribution, a detailed study including evaluation of the activities of U and Th in the emulsion components via spectroscopic methods and a SOURCES [16]/GEANT4 MC simulation, was performed. We found that the detectable neutron-induced background would be $0.02 \div 0.03$ per year per kilogram. The reader is referred to [17] for further details.

3. Status and perspectives

At present, a technical test with a 10 g NIT sample was performed in a shielded environment in the underground Hall B of the Laboratori Nazionali del Gran Sasso (LNGS). In parallel, the construction of the underground infrastructure for emulsion production and development is ongoing together with the studies of the final detector design.

We plan to perform a pilot exposure with a target mass of 1 kg and the corresponding analysis of the data on a time scale of five years: in Fig. 1 the 90% C.L. upper limit, in case of null observation for an exposure of 1 kg year of NIT emulsions, with a minimum detectable track length ranging from 200 nm down to 50 nm and in the hypothesis of zero background, is shown. Even not including the directionality discrimination of the signal and assuming to reach a negligible background level, such an experiment would cover a large part of the parameter space indicated by the DAMA/LIBRA results with a small detector mass, using a powerful and complementary approach.

A detailed simulation was performed to take into account the scattering effect of nuclear recoils together with a study of the NEWSdm potentialities in terms of sensitivity and discovery potential asssuming a given background budget and including the directionality information. The reader is referred to [4] for details.

4. Conclusions

The aims of the NEWSdm project is to realize the first directional DM experiment with high sensitivity in the spin-independent case. A Letter of Intent [17] was presented in 2015 to INFN and to the Scientific Committee of the National Gran Sasso Laboratories. The project was founded for the first three years of operation.

The idea to exploit fine-grained nuclear emulsions and an innovative read-out system has to be proven successful thanks to a campaign of measurements carried out by implanting low energy ions simulating nuclear recoils with track lengths of the order of few hundred of nm and with neutron test beams. We demonstrated to be able to reconstruct track as long as the one expected for WIMP-scattered nuclei ($< 1\mu$m even for large (O(TeV)) WIMP masses) with the impressive resolution of 10 nm in position and 13° in angle. These results constitute a breakthrough in the sector of tracking detectors.

A technical test with a low-mass (10g) sample has been already realized and a pilot experiment with 1 kg mass is foreseen on a time scale of 5 years. The actual intrinsic radioactive level allows to scale the target mass and exposure time up to one order of magnitude. A careful selection of the emulsion components and a better control of their production could further increase the radiopurity and reduce the thermal background (typically referred as fog), thus allowing larger detector mass. Recent developments in optical microscopy envisage the possibility to lower the threshold down to 50 nm and achieve scanning speeds suitable for the analysis of ton scale emulsion detectors in the near future.

References

[1] Bertone G, Hooper D and Silk J 2005 *Phys. Rept.* **405** 279–390 (*Preprint* hep-ph/0404175)
[2] Spergel D N 1988 *Phys. Rev.* **D37** 1353
[3] Battat J B R *et al.* 2016 *Phys. Rept.* **662** 1–46 (*Preprint* 1610.02396)
[4] Agafonova N *et al.* (NEWSdm) 2017 (*Preprint* 1705.00613)
[5] Aleksandrov A *et al.* (NEWS) 2016 (*Preprint* 1604.04199)
[6] Natsume M *et al.* 2007 *Nucl. Instrum. Meth.* **A575** 439–443
[7] Acquafredda R *et al.* 2009 *JINST* **4** P04018
[8] Alexandrov A *et al.* 2015 *JINST* **10** P11006
[9] Alexandrov A *et al.* 2016 *JINST* **11** P06002
[10] Alexandrov A *et al.* 2017 *Nature Scientific Reports* **7**
[11] Armenise N *et al.* 2005 *Nucl. Instrum. Meth.* **A551** 261–270
[12] Kimura M and Naka T 2012 *Nucl. Instrum. Meth.* **A680** 12–17
[13] Alexandrov A *et al. patent N. 102016000132813*
[14] Tamaru H, Kuwata H, Miyazaki H T and Miyano K 2002 *Applied Physics Letters* **80** 1826–1828
[15] Naka T *et al.* 2012 *EAS Publ. Ser.* **53** 51–58 (*Preprint* 1109.4485)
[16] Wilson W B *et al. Submitted to: American Nuclear Society/Radiation Protection and Shielding Division, 12th Biennial Topical Meeting, Santa Fe, NM, April 14-18, 2002*
[17] Alexandrov A *et al.* 2016 *Astropart. Phys.* **80** 16–21 (*Preprint* 1507.03532)

IOP Conf. Series: Journal of Physics: Conf. Series **1056** (2018) 012019 doi:10.1088/1742-6596/1056/1/012019

Experimental studies of neutrino nuclear responses and nuclear structures for neutrino nuclear physics

Hiroyasu Ejiri
Research Center for Nuclear Physics, Osaka University, Ibaraki, Osaka, Japan
E-mail: ejiri@rcnp.osaka-u.ac.jp

Abstract. Neutrino nuclear responses associated with astro neutrinos and double beta decays are crucial to extract neutrino properties of astro particle physics interests. The present report reviews briefly recent experimental studies of the neutrino nuclear responses. Single beta decays and electron captures, charge exchange nuclear reactions, muon and photon nuclear reactions, neutrino nuclear reactions and nucleon transfer reactions are used to study neutrino nuclear responses. Discussions are made on nucleonic and nonnucleonic spin isospin correlations and renormalization (quenching) effects on axial weak responses.

1. Introduction

Neutrino is a key particle for astro nuclear physics, particle physics and cosmology. It is the elementally particle that has only the weak charge, and has no electric and color charges. Neutrino interactions with nuclei are so weak that experimental studies of the neutrino and the weak interaction are very hard.

Several basic questions on the neutrino remain yet unsolved. Some of them are i the nature of the neutrino if it is the Majorara or the Dirac particle, ii the absolute mass scale and the mass hierarchy (spectrum), iii the lepton sector CP phases, iv solar neutrino sources and the fluxes, and iv supernova neutrino fluxes, spectra and nucleosyntheses.

These questions are studied well also by investigating neutrino weak processes in nuclei such as neutrino nuclear interactions, single beta decays (SBD) and electro captures (EC), inverse beta decays (IBD) induced by neutrinos and neutrino-less double beta decays (DBDs). The neutrino nuclear responses are crucial for the SBD/EC, IBD and DBD neutrino studies in nuclei. Historical reviews and previous works on the neutrino nuclear responses are given in reviews [1, 2] and references therein.

The present report reviews briefly recent experimental studies of the neutrino nuclear responses. The details of the review on neutrino nuclear responses for astro neutrino interactions and double beta decays will be given elsewhere [3]. The astro neutrino, DBD, SBD/EC and IBD processes are schematically shown in Figure 1. Astro neutrino interactions are studied by measuring astro neutrino interactions in nuclei. The interaction (IBD) rate $R(\alpha)$ with α being the interaction mode is expressed as

$$R_(\nu)(\alpha) = g^2 G_\nu B(\nu) I(\nu) \quad B(\nu) = (2Ji+1)^{-1} |M_\nu|^2, \tag{1}$$

where g is the weak coupling constant, G_ν is the phase space (kinematic) factor, and $I(\nu)$ is the astro neutrino flux. The nuclear response $B(\nu)$ is given by the square of the nuclear matrix element (NME M^{ν}) with $2J_i$+1 being the initial state spin factor.

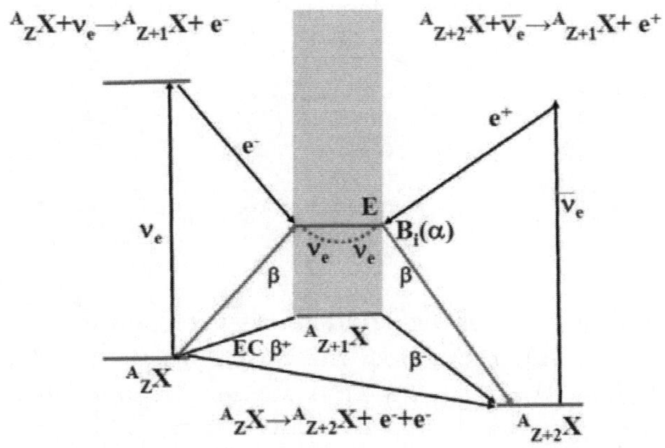

Figure 1. Schematic neutrino CC interactions and transition processes in nuclear femto laboratories for IBD, SBD and DBD. p: proton, n: neutron, e: electron, ν and $\bar{\nu}$: neutrino and anti neutrino, E: excitation energy, $B_i(\alpha)$: nuclear response.

The neutrino-less DBD rate for the neutrino mass mechanism is expressed as

$$R(0\nu) = g^4 G^{0\nu} B(0\nu)(m^{eff})^2 \quad B(0\nu) = (2J_i + 1)^{-1}|M^{0\nu}|^2, \tag{2}$$

where $G^{0\nu}$ is the phase space (kinematic) factor and m^{eff} is the effective mass. The nuclear response $B(0\nu)$ is given by the square of the DBD NME $M^{0\nu}$ with $2J_i + 1$ being the initial state spin factor. The energy sum of the DBD electrons is given by the DBD Q value.

2. Experimental probes for neutrino nuclear responses

Neutrino nuclear responses are neutral current (NC) and charged current (CC) weak responses for nucleons (neutron and proton) in the nucleus. The Z and $W^{\pm}$ weak bosons are involved in the NC and CC interactions, respectively. The weak responses with weak couplings of g_A and g_V for astro neutrinos and DBD neutrinos are very sensitive to nucleonic and non-nucleonic correlations. Therefore they have been studied experimentally by using various kinds of weak, electro magnetic and nuclear probes to help theoretical evaluations for the responses and the NMEs. They are schematically shown in Figure 2.

The weak processes with astro ν and $\bar{\nu}$ are ${}^A_Z$X + $\nu \rightarrow {}^A_Z$X' + ν, ${}^A_Z$X + $\bar{\nu} \rightarrow {}^A_Z$X' + $\bar{\nu}$ for NC process, and ${}^A_Z$X + $\nu \rightarrow {}^A_{Z+1}$X' + β^- and ${}^A_{Z+1}$X + $\bar{\nu} \rightarrow {}^A_Z$X' + β^+ for CC process. NC process is a nucleon excitation process of N$\rightarrow$N' with N and N' being nucleons in the nucleus, while CC is a charge exchange process of p$\leftrightarrow$n with p and n being proton and neutron in the nucleus.

The neutrino-less DBD process is ${}^A_Z X \rightarrow {}^A_{Z\pm2}$X + $2\beta^{\mp}$ with the two nucleon double charge exchange of (n_1, n_2)$\leftrightarrow$(p_1, p_2) in the nucleus. The nuclear response is given by the product of the initial spin factor $1/(2J_i + 1)$ and the square of the NME for (n_1,n_2) $\leftrightarrow$ (p_1,p_2). Here the DBD NME is associated indirectly with the $p_1 \leftrightarrow n_1$ and the $p_2 \leftrightarrow n_2$ via the neutrino potential in case of the neutrino-less DBD ($0\nu\beta\beta$) and directly with them in case of the two neutrino DBD ($2\nu\beta\beta$).

The weak NMEs involved in the neutrino nuclear responses are essentially the NMEs for ${}^A_Z X(N) \leftrightarrow {}^A_Z$X'(N) and ${}^A_Z X(n) \leftrightarrow {}^A_{Z+1}$X(p) transitions via the NC and CC weak interaction operators. They are studied by using weak interaction probes and also by using electro magnetic (EM) probes with the interaction operators similar to the weak interaction ones as discussed in [4, 5, 6].

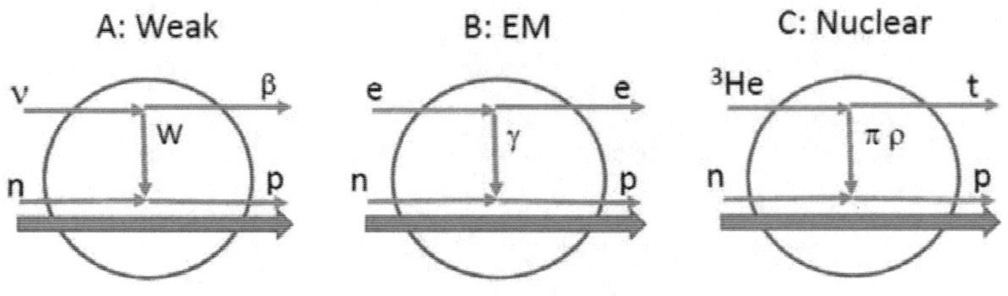

Figure 2. Experimental probes for neutrino nuclear responses. A: leptons with weak interaction. B: photons with EM interaction. C: nuclei with strong/nuclear interaction. p: proton, n: neutron, e: electron, W: weak boson, γ: gamma ray, and $\pi\rho$: mesons.

Weak interaction probes of ν and $\bar{\nu}$ beams are used to study directly the weak (neutrino) responses. The neutrino cross section, however, is extremely small because of the weak interaction. It is of the order of $10^{-(40-44)}$ cm^2. Then one needs high flux neutrinos of the order of 10^{13-15}/sec and multi-ton scale detectors for the ν beam experiment in order to get adequate event rates. Medium energy neutrinos obtained from stopped π decays are used to study the neutrino reactions. The intense GeV protons from SNS and J-PARC may be used to produce the intense pions [4, 7, 8].

SBDs ($\beta^{\pm}$) and ECs are used to get CC neutrino n$\leftrightarrow$p responses. They are limited mostly to allowed and first forbidden transitions from the ground and isomeric states. SBD responses are given in review articles [1, 2]. Recently, K-forbidden SBD NMEs of astro nucleosynthesis interests are studied [9].

Negative muons trapped in atomic orbits are captured mostly into nucleus by the weak interaction in case of medium heavy nuclei with the atomic number $Z \geq 20$. Then the ordinary muon capture reaction of ${}^A_ZX + \mu \rightarrow {}^A_{Z-1}\mathrm{X} + \nu_\mu$ is used to study the anti neutrino p$\rightarrow$n response in the medium energy (10-70 MeV) and momentum (10-70 MeV/c) regions. Recently the anti neutrino response on ^{100}Mo was studied by measuring RIs from muon captures [10]. The response shows for the first time a μ-capture giant resonance (GR) around 10-15 MeV [10].

Photons with EM interactions are used to study neutrino nuclear responses because EM interactions have similar spin isospin and multipole transition operators as the weak ones. The electric and magnetic transitions correspond to the vector and axial-vector weak transitions, respectively. In particular, photo nuclear reactions via IASs (Isobaric Analogue States) provide EM NMEs corresponding to the analogous $\beta^+(\bar{\nu})$ NMEs [11].

Nuclear reactions with nuclear/strong interactions are very useful for studying neutrino nuclear responses because of the large reaction/interaction cross section. The nuclear (strong) interaction itself is quite different from the weak interaction, but the strong interaction includes the spin, isospin and multipole operators as the weak interaction. The spin-flip and spin non-flip inelastic scatterings of p,n,d, and light ions are used to study vector and axial vector NC responses, respectively. Charge exchange reactions (CERs) are used for the CC response study, (p,n), (^{3}He,t) and other CERs for (n$\rightarrow$p) responses and (n,p), (d,^{2}He), (t,^{3}He), (^{7}Li,^{7}Be) and other CERs for p$\rightarrow$ n ones. They are discussed in review articles [2, 3].

High energy-resolution (^{3}He,t) CERs with the medium energy 0.42-0.45 GeV are used to study the n$\rightarrow$p axial vector responses because the spin isospin interaction gets predominant at this energy. Since both the projectile of ^{3}He and the out-going particle of t are charged

IOP Conf. Series: Journal of Physics: Conf. Series **1056** (2018) 012019 doi:10.1088/1742-6596/1056/1/012019

particles, high precision energy-analyses of them are possible to perform high energy-resolution measurements required to separate individual states. The high energy-resolution system with the energy resolution of $\Delta E/E \approx 10^{-5}$ at RCNP Osaka University has extensively been used to study neutrino nuclear responses [4, 5].

The (^{3}He,t) CER on ^{71}Ga was measured to study the solar neutrino responses. The measured responses for the low-lying states in ^{71}Ge are found to be consistent with the calculated ones. This shows that the deviations of the measured neutrino capture rates from the estimations based on the calculated responses are not due to the responses, but something new(sterile neutrino)/unknown effects [12].

Spin dipole (SD 2^-) NMEs play major roles for neutrino-less DBD NMEs. They are for the first time studied by using the (^{3}He,t) CERs on DBD nuclei [13]. The observed SD cross sections are shown to be consistent with the Fermi surface quasi particle model (FSQP) NMEs [14]. Thus it is shown that the SD NMEs are well studied by means of the (^{3}He,t) CERs as GT NMEs.

Double charge exchange reactions (DCERs) are used to get strength distributions for DCER responses relevant to DBDs. The (^{11}B,^{11}Li) DCER at E/A=80 MeV shows that most double spin-isospin strengths are pushed up to the high excitation region, leaving little strengths at the low-lying states [15]. Heavy ion (HI) double CERs for low-lying states were studied [16, 17]. The NUMEN project there aims at extensive studies of DCERs for DBD neutrino nuclear responses.

Nucleon transfer reactions provide experimentally single quasi-particle properties of nucleons in the nucleus, which are used to help evaluate the neutrino nuclear responses [18]. The single particle and hole probabilities are smaller by a factor ≈ 0.55 than the shell model expectation [19], suggesting uniform reduction of the single nucleon nature in the nuclear medium.

3. Axial vector responses and quenching of the axial vector coupling

Experimental GT and SD NMEs $M_{EX}(\alpha)$ derived from SBD/EC and CER data are uniformly reduced with respect to the single quasi-particle NMEs $M_{QP}(\alpha)$ and pnQRPA NMEs $M_{QR}(\alpha)$, where the transition modes are given by α=GT and SD [1, 2, 20, 21]. They are expressed as

$$M_{EX}(\alpha) = k(\alpha)M_{QP}(\alpha) = k_{NM}(\alpha)M_{QR}(\alpha), \quad M_{QR}(\alpha) = k_{\tau\sigma}(\alpha)M_{QP}(\alpha). \tag{3}$$

Here the reduction coefficient $k(\alpha) \approx 0.25$ is expressed as $k(\alpha) = k_{\tau\sigma}(\alpha) \times k_{NM}(\alpha)$ with $k_{\tau\sigma}(\alpha) \approx 0.4$ and $k_{NM}(\alpha) \approx 0.6$ standing for the reduction coefficients due to the nucleonic $\tau\sigma$ correlation and non-nucleonic (nuclear medium) effect, respectively [20, 21]. The nucleonic correlation is taken into accounts by the pnQRPA, while the Δ isobar and other non-nucleonic correlation may be incorporated with the effective axial vector coupling $g_A^{eff}/g_A \approx 0.6$.

Neutrino nuclear responses associated with medium energy supernova neutrinos and neutrino-less DBDs involve medium momentum and angular momentum transfers of q=20-200 MeV/c and $l^{-}h$=1-5. Nuclear and muon CERs provide opportunities to study neutrino nuclear responses in the wide momentum-transfer region. The (^{3}He,t) CERs on DBD nuclei were measured for F(IAS 0^+), GT(1^+) and SD(2^-) states in the wide angular range of θ=0-4 deg., corresponding to the momentum-transfer range of q=5-100 MeV/c in order to study the momentum dependence of the neutrino nuclear responses. The q dependent cross section is expressed as

$$\frac{d\sigma_i}{d\Omega} = K_i(\alpha)F_i(\alpha,q)J_i(\alpha)^2 k^{eff}(q)^2 B_i(\alpha), \tag{4}$$

where $K_i(\alpha)$ and $J_i(\alpha)$ with α =F,GT, SD are the kinematic factor and the volume integral of the interaction, respectively, and $B_i(\alpha)$ is the nuclear response at q=0. The kinematic q-dependence is given by $F_i(\alpha,q) \approx |J_L(qR)|^2$ and the q dependence of the nuclear response is by $k^{eff}(q)^2$. The kinematic q dependence $F_i(\alpha,q)$ is given by the DWBA (Distorted Wave Born Approximation) calculation, and the q dependent coupling $k^{eff}(q)$ manifests as deviation of the observed q (angular) distribution from the DWBA calculation. Actually, the observed q-dependence (angular distribution) is well reproduced by the constant $k^{eff}(q)^2$ as shown in Figure 3.

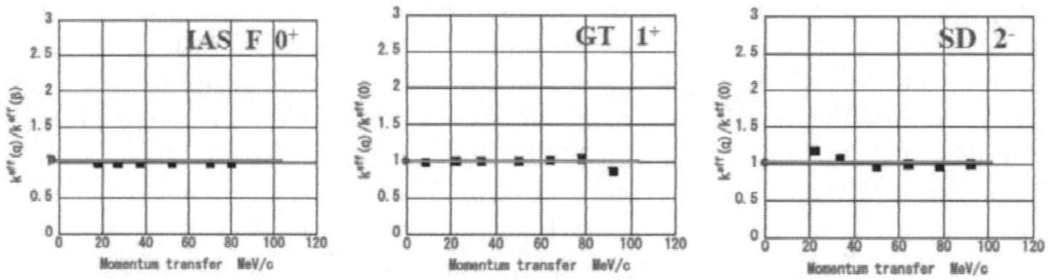

Figure 3. Momentum dependence of the ^{76}Ge(^{3}He,t) CER cross sections [22]. The ratio of $k^{eff}(q)/k^{eff}(q=0)$ for α =F (IAS), GT (1st GT), and SD (ground) states in ^{76}Ge. The red point is the normalization point at q=0.

The (^{3}H,t) CER spectra for medium heavy nuclei show that i. the Fermi strength is concentrated in the sharp IAS (Fermi GR) and no other Fermi strengths are in the low excitation region, ii. the GT and SD strengths are mainly concentrated, respectively, in the broad GT and SD GRs, and thus some strengths are in the low excitation region. In other words, the IAS is a good eigen state because of the good isospin symmetry, while GT and SD GRs are kinds of quasi-eigen states because of the quasi (incomplete) spin-isospin (super-multiplet) symmetry.

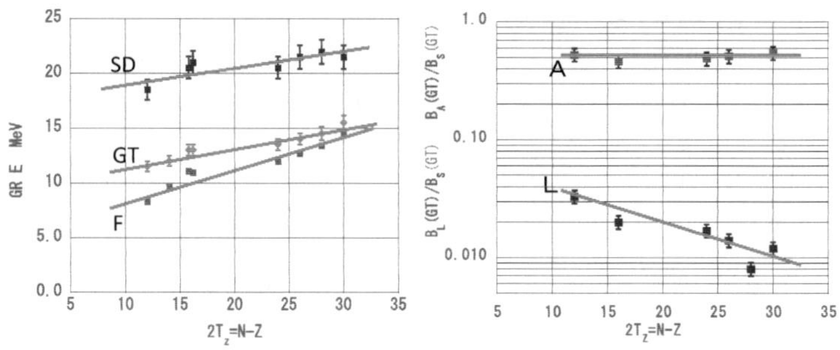

Figure 4. Left hand side. Fermi (IAS), GT and SD GR energies in units of MeV for DBD nuclei as a function of $2T_z = N - Z$. Right hand side: Summed GT strengths, $\sum_L GT$ for the low-lying states (L) and the $\sum GT$ for all GT states (A) in units of the sum rule limit B_S as a function of $2T_z = N - Z$. The GT GR tail at E=3-4 MeV in the ^{76}Ge is corrected for. Ref. [3].

The Fermi, GT and SD GRs are expressed as coherent (in-phase) $\tau^-, \tau^-\sigma$ and $\tau^-\sigma r Y_1$ excitations of the all neutron-hole proton-particle states, respectively. Their excitation energies are pushed up to the high excitation region due to the repulsive $\tau, \tau\sigma$ and $\tau\sigma r Y_1$ interactions. The GR energies are shown as a function of $N - Z = 2T_z$ with T_z being the isospin z component. The IAS, GT GR, and SD GR energies are expressed in units of MeV as

$$E(IAS) = 5 + 0.6T_z, \ E(GT) = 9.0 + 0.4T_z, \ E(SD) = 16.5 + 0.4T_z. \tag{5}$$

Note that the GT GR and SD GR energies increase as T_z increases, and the SD GR is higher in energy than the GT GR by $\hbar\omega \approx 8$ MeV, reflecting the transition operator r involved in the

SD. The energies of IAS with the upper T increase more with T_z than the GT and SD GRs with the lower T.

The sum $\sum B_i(GT)$ for all GT states including the GT GR is derived from the measured (^{3}He,t) CERs by assuming the Lorentzian shape of the GR and the quasi-free scattering shape at the higher excitation region beyond $E=$ 30 MeV. The sum is around 0.50-0.55 of the sum rule limit, indicating the large reduction of the GT strengths, as seen in other CERs.

The reduction of the GT strengths suggests some nuclear medium and non-nucleonic NΔ isobar effects [23]. The isobar effect is discussed for the first forbidden β transitions [24]. The reduction (quenching) of the B(GT) sum is very interesting in views of the reduced effective g_A suggested for low-lying GT and SD states [1, 20, 21, 23, 24].

The author thank Prof. D. Frekers and Prof. J. Suhonen for valuable discussions.

References

[1] Ejiri H and Fujita J.I 1978 *Phys. Rep.* C **38** 85
[2] Ejiri H 2000 *Phys. Rep.* C **338** 265
[3] Ejiri H, Suhonen J and Zuber K 2018 *Phys. Rep.* C submitted
[4] Ejiri H 2005 *J. Phys. Soc. Jpn.* **74** 2101
[5] Vergados J, Ejiri H and Simkovic F 2012 *Rep. Prog. Phys.* **75** 106301
[6] Vergados J, Ejiri H and Simkovic F 2016 *Int. J. Mod. Phys.* E **25** 1630007
[7] Avignone III F, ORLaND Collaboration 1999 *ORLaND proposal*
[8] Ejiri H 2003 *Nucl. Instr. Methods Phys. Research* A **503** 276
[9] Ejiri H and Shima T 2017 *J. Phys. G: Nucl. Part. Phys.* **44** 065101
[10] Hashim I and Ejiri H et al. 2018 *Phys. Rev.* C **97** 014617
[11] Ejiri H, Titov A, Boswell M and Young A 2013 *Phys. Rev.* C **88** 054610
[12] Frekers D, Ejiri H, et al. 2011 *Physics Letters* B **706** 134
[13] Ejiri H and Frekers D 2016 *J. Phys. G: Nucl. Part. Phys.* **43** 11LT01
[14] Ejiri H 2017 *J. Phys. G: Nucl. Part. Phys.* **44** 115201
[15] Takahisa K, Ejiri H, et al. 2017 *arXiv* 1703.08264 nucl-ex
[16] Cappuzzello F et al. 2015 *Eur. Phys. J.* A **51** 145
[17] Cappuzzello F et al. 2018 *Eur. Phys. J.* A **54** 72
[18] Freeman S J and Schiffer J. P 2012 *J. Phys. G: Nucl. Part. Phys.* **39** 124004
[19] Kay B.P, Schiffer J.P and Freeman S.J 2013 *Phys. Rev. Lett.* **111** 042502
[20] Ejiri H, Soukouti N and Suhonen J 2015 *J. Phys. G: Nucl. Part. Phys.* **42** 055201
[21] Ejiri H and Suhonen J 2014 *Phys. Letters* B **729** 27
[22] Thies J H, et al. 2012 *Phys. Rev.* C **86** 014304
[23] Bohr A and Mottelson B.R 1975 *Nuclear Structure* **II** (Benjamin, New York)
[24] Ejiri H 1982 *Phys. Rev.* C **26** 2618

IOP Conf. Series: Journal of Physics: Conf. Series **1056** (2018) 012020 doi:10.1088/1742-6596/1056/1/012020

Determination of the neutrino mass in ^{163}Ho.

Amand Faessler

Institute of Theoretical Physics, University of Tuebingen, Germany,

E-mail: faessler@uni-tuebingen.de

Abstract. KATRIN plans to determine the electron anti-neutrino mass in the Tritium decay. Electron capture in ^{163}Ho can measure the electron neutrino mass supported by the small decay energy Q = 2.8 keV $^{163}_{67}Holmium + electron \rightarrow ^{163}_{66}Dysprosium + \nu$. The decay energy of 2.8 keV is used to emit a neutrino and to excite the Dy atom. The neutrino mass is then given by the difference between the Q value and the upper end of the Dy deexcitation spectrum. The Dirac-Hartree-Fock approach is used for the bound and the continuum states in Holmium and in Dysprosium. The present background must be reduced by at least two orders to extract the neutrino mass from the spectrum. The present work discuses the one- and the two-hole excitations in Dy. The two-hole states are excited by shake-up of an electron into a free bound state and the shake-off into the continuum. The shake-off contribution can practically be neglected. It seems not to influence the neutrino mass determination.

1. Introduction

Although the neutrino is an essential particle for the weak interaction many of its properties are unknown. Unknown are the mass of the three neutrinos and three anti-neutrinos, the hierarchy and the nature of the neutrinos, Dirac (neutrino different from the antineutrino) or Majorana (neutrino identical with the corresponding antineutrino).

The topic of this contribution is the theoretical side of the determination of the electron neutrino mass [1, 2, 3, 4, 5, 6, 7, 9] by electron capture in $^{163}_{67}Ho$ by the Tuebingen group and coworkers [5, 15, 16, 18]. For the capture

$$^{163}Holmium + bound\ Electron \rightarrow excited\ ^{163}Dysprosium^{*} + Neutrino \tag{1}$$

the decay energy Q was obtained by a Penning trap measurement by Eliseev et al. [19] in the MPI in Heidelberg ([20, 21, 22]).

$$Decay\ energy\ Q = 2.833(30_{stat})(15_{sys})\ keV \tag{2}$$

Electron capture in ^{163}Ho is only possible from electron orbits bound less than Q = 2.8 keV, thus from $3s_{1/2}, M1$ with a binding energy 2.128 keV and from $p_{1/2}$, $M2$ with binding energy of 1.8 keV and less bound electron states. In addition the electron orbits in Ho, from which an electron can be captured must have an overlap with the nucleus to react with a proton. This is only the case for $(n > 2)s_{1/2}$ and the lower relativistic amplitude of the $(n > 2)p_{1/2}$ states. The decay energy of 2.8 keV is shared by the excitation energy of the Dy atom (one-, two-, three-hole and higher excitations) and by the emitted neutrino. The excited Dy decays by X-rays and

Auger electron emissions. The sum of the deexcitation energy can be measured by a bolometer. The maximum bolometer energy yields then the minimum neutrino energy, which is the electron neutrino rest mass. Thus the neutrino mass can be determined by the difference of the decay energy and the upper end of the Dy deexcitation spectrum. The excitation of the Dysprosium*, calculated in the sudden approximation, is then given by the overlap with Holmium* with an electron hole in the capture orbit with the different excited configurations $|D>$ in the daughter Dysprosium*.

A publication [9], which studies also shake-up and shake-off in Holmium capture, uses non-relativistic pure Coulomb wave functions for the electrons in Ho, which do not allow capture from $np_{1/2}$. The effect of the partial screening of the nuclear Coulomb potential by the electron cloud is considered there by effective charges. Since in reference [9] the perturbation approach is used to calculate the Dy* wave functions, this requires orthonormal basis functions. Thus ref. [9] uses for each capture process, e. g. from $3s_{1/2}$, the same screening factor for all single electron orbits needed in the perturbation approach for this special capture process independent of the binding energy of the single electron orbits. Starting from these "bad" wave functions for Ho ref. [9] calculates the bound and continuum electron wave functions for Dy* with the perturbation:

$$H'(r) = \; +1/r \; - \; \int d^3r_1 \; |\varphi_{3s/4s}(\vec{r}_1)|^2/|\vec{r} - \vec{r}_1|. \tag{3}$$

The first term takes into account, that in Dy one has one proton less compared to Ho. The second term originates from the electron hole in the $3s_{1/2}$ or the $4s_{1/2}$ state, which are the only hole states considered in ref. [9].

We use here selfconsistent Dirac-Hartree-Fock wave functions calculated specifically for Ho* and Dy* [23, 24, 25]. The continuum electron wave functions in Dy* are determined [26] in the selfconsistent Z-1 = 65 electron potential orthogonal to all 65 bound electrons in Dy.

2. The capture process

The decay spectrum of the excited $^{163}Dy^*$ after electron capture in ^{163}Ho refs. [4, 9] and [5] assuming Lorentzian shapes for the decay resonances are:

$$\begin{aligned}
\frac{d\Gamma}{dE_c} \propto & \sum_{i=1,\ldots N_\nu} (Q - E_c) \cdot U_{e,i}^2 \cdot \sqrt{(Q - E_c)^2 - m_{\nu,i}^2} \; * \\
& \Big(\sum_{f=f'} \lambda_0 B_f \frac{\Gamma_{f'}}{2\pi} \frac{1}{(E_c - E_{f'})^2 + \Gamma_{f'}^2/4} + \qquad\qquad (4) \\
& \sum_{f=f';p'<F;q_b'>F} \lambda_0 B_{f,p'<F;q_b'>F} \frac{\Gamma_{f',p'}}{2\pi} \frac{1}{(E_c - E_{f',p'})^2 + \Gamma_{f',p'}^2/4} + \\
& \int dk_{q'} \lambda_0 B_{f,p'<F;q_c'>0} \frac{\Gamma_{f',p',q'}}{2\pi} \frac{1}{(E_c - E_{f',p',q'})^2 + \Gamma_{f',p',q'}^2/4}\Big)
\end{aligned}$$

The factor in front of the bracket is the same as for the single beta decay. It reflects the phase space of the emitted neutrino (1). The three expressions in the bracket of eq. (4) describe the decay of the 1-hole f' excitations in Dy*, of the 2-hole excitations of the bound occupied electron orbits f', p' with shake-up of p' to q' into an empty bound orbit q' above the Fermi surface F and of 2-holes in originally occupied electron orbits f', p' with one electron p' moved to (an unoccupied) q' into the continuum (shake-off). The last term has as the others a dimensionless strength $dk_{q'} \cdot B_{f',p'<F,q'>o}$. The transformation to an integral over the energy yields the factor in front of the third term, which can be deduced from ref. (10). Different neutrino mass eigenstates $|i>$ are mixed into the electron neutrino flavor states by $U^2_{e,i}$. The decay energy Q is given by eq. (2) from the ECHo collaboration [20, 21, 27, 28, 29]. The excitation energy of the final Dy* is E_c. The energy $Q - E_c$ is carried away by the neutrino. B_f, $B_{f,p'<F;q'_b>F}$ and $B_{f,p'<F;q'_c>0}$ are the overlap and exchange corrections for the 1-hole, the bound 2-hole and the shake-off 2-hole states. λ_0 contains the nuclear matrix element squared [32]. Eq. (4) yields the spectrum in arbitrary units and must be normalized at one energy to the data. We use the N1, $4s_{1/2}$ peak for the normalisation. $E_{f'}$, $E_{f',p'}$ and $E_{f',p';q'>0}$ are the 1-hole, the 2-hole shake-up and the 2-hole shake-off excitation energies in Dysprosium (see refs. [7] and [10]). $\Gamma_{f'}$, $\Gamma_{f',p'}$ and $\Gamma_{f',p';q'>0}$ are the widths of the one- and two-hole states and the two-hole states with shake-off in Dysprosium [15, 16, 18].

The overlap and exchange corrections B_f, $B_{f,p'<F;q'_b>F}$ and $B_{f,p'<F;q'_c>0}$ are calculated in second quantization with Dirac-Hartree-Fock (DHF) electron wave functions. This yields automatically the antisymmetrization and the exchange terms. The antisymmetrized two-hole state in Dy with shake-off is for example:

$$|A'_{p',f':q'>0}>= a_1^{'\dagger} a_2^{'\dagger}...a_{f'-1}^{'\dagger} a_{f'+1}^{'\dagger}...a_{p'-1}^{'\dagger} a_{p'+1}^{'\dagger}...a_Z^{'\dagger} a_{q'>0}^{'\dagger}|0> \tag{5}$$

The probability to form a two-hole shake-off state is proportional to:

$$P_{f',p';q'>0} = | < A'_{f',p';q'}|a_i|G> |^2 \tag{6}$$

f' has in Dysprosium the same quantum numbers as the electron state in Holmium f, from which the electron in captured. f' and p' are here originally occupied electron orbitals in Dy below the Fermi surface, while $q' > 0$ is in Dy an empty state in the continuuum with an energy larger 0. As everywhere in this article the primes indicate electron orbits in the daughter atom Dy and unprimed states indicate electrons in Holmium.

3. The Dy Continuum Wave Functions.

The selfconsistent Coulomb field for the shake-off electron is in atomic units:

$$V_{shake-off}(r) = \frac{-66}{r} + \sum_{k\ occupied\ e} g'_{k'} \int d^3\, r \cdot \frac{|\varphi_{k'}(\vec{r'})|^2}{|\vec{r} - \vec{r'}|} \tag{7}$$

$g'_k s$ are the number of bound electrons in the selfconsistent occupied orbits $|k> = |n, \ell, j>$ of Dy. To determine the potential for the shake-off electrons in Dy one needs the occupied selfconsistent relativistic upper and lower Dirac-Hartree-Fock orbitals in Dysprosium with the single electron quantum numbers k': $P_{k'}(r)$ and $Q_{k'}(r)$.

$$\varphi_{k'}(\vec{r}) * r = (\ P_{k'}(r); Q_{k'}(r)) \tag{8}$$

The normalization of the continuum Dirac wave function is treated in M. E. Rose "Relativistic electron theory " [30] or Walter Greiner "Relativistic Quantum Mechanics" [31]. We follow here this recommendation of Perger and Karrighattam on their page 394 [34]. The electron energy in the continuum can be due to energy conservation not larger than the Q-value of 2.8 keV minus the excitation energy of the two hole state.

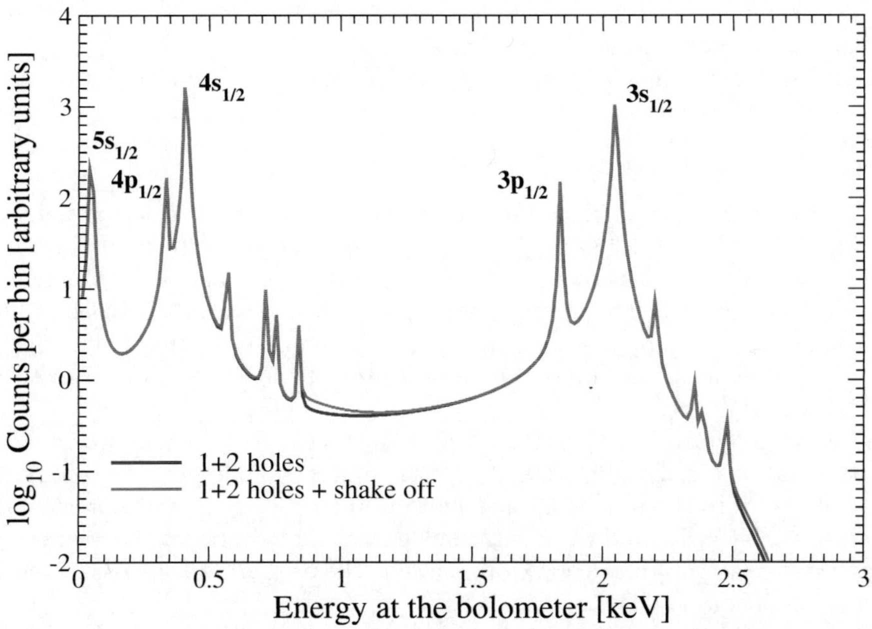

Figure 1. Sum of the one- and two-hole deexcitations compared to the sum of the one-, two-hole and the shake-off deexcitation as measured by the bolometer spectrum (4) in arbitrary units. The units are adjusted to the experimental N1, $4s_{1/2}$ 1-hole peak. For the one hole states the quantum numbers are indicated. The two-hole peaks are by about two orders of magnitudes smaller than the one hole peaks. Shake-off can almost not been seen.

$$\int_{r=0\ to\ \infty} dr \cdot 2 \cdot \sin(kr - \ell \cdot \frac{\pi}{2} - \eta \ln 2kr + \Delta + \delta) \cdot \tag{9}$$
$$2 \cdot \sin(k'r - \ell \cdot \frac{\pi}{2} - \eta \ln 2kr + \Delta + \delta) \approx 2\pi \cdot \delta(k - k').$$

The wave number is connected with the relativistic and non-relativistic energies by the equations:

$$E_{rel}^2 = c^{2-}\hbar k^2 + m^2c^4 \;\rightarrow\; c^2k^2 + c^4 \;\; (in\ atomic\ units); \;\; E_{n-rel} = \frac{1}{2}k^2;$$
$$k = \alpha\sqrt{E_{n-rel}(E_{n-rel} + 2 \cdot c^2)}; \quad with\ c \;=\; 1/\alpha \;=\; 137.035999\ in\ [atomic\ units]. \tag{10}$$

4. Electron Shake-off into the Dy Continuum.

The overlap between the bound Ho states $ns_{1/2}$ and $np_{1/2}$ with $n \geq 3$ and the continuum wave functions in Dy squared gives the probability for shake-off. We restrict this work to s-wave shake-off, which is expected to be the largest contribution. The s-wave shake-off is given by the overlap $< n \geq 3, s_{1/2}, Ho|E,\ s, Dy >$ squared. In the spectrum with 1-hole, 2-hole and shake-off in figure 1 the shake-off contribution is hardly visible. Here and also in ref. [9] only the decay width of the 2-hole states are included. Three hole states can be neglected. The electron in the continuum has an escape width, which is not included.

Figure 1 shows the logarithmic spectrum of the 1-hole, the 2-hole and the s-wave shake-off contributions. The shake-off spectrum can hardly be seen on this scale in the total spectrum.

5. Summary

This publication treats the theoretical side of our determination of the electron neutrino mass by electron capture [5, 7, 15, 16, 18] in ${}^{163}_{67}Ho$ to ${}^{163}_{66}Dy$. The energy of the $Q - value = 2.80 \pm 0.08\ keV$ is distributed to the excitation of Dy and the emission of a neutrino. The recoil of ${}^{163}Ho$ can be neglected. The deexcitation as X-rays and Auger electrons of Dy can be measured in a bolometer. The upper end of the spectrum requires the minimum energy of the neutrino. Thus the difference of Q minus this upper limit of the deexcitation spectrum is the rest mass of the electron neutrino.

We describe the bound states in Ho and Dy by the Dirac-Hartree-Fock approach [25, 23, 24] even including occupations in Dy due to different hole states. The s-wave continuum wave functions in Dy are determined with the Dirac equation in the selfconsistent potential [26]. The energy of the capture orbit in Ho and the excitations of Dy* including also the continuum states are limited by energy conservation to the Q-value minus the excited one-hole states in Ho* and Q minus the ecitation energies in Dy*.

Remaining discrepancies between theory and experiment could be due to configuration mixing not included here. Finally we want to stress, that the accuracy needed to extract the neutrino mass can not be obtained by theoretical calculation alone. One must fit simultaneously the neutrino mass, the Q-value, the excitation energy of the highest resonance, their width and their strength at the upper end of the spectrum to extremely accurate data. For this the present background of the experimental spectrum has to be drastically reduced. The measurement must probably be done in an underground lab to protect against cosmic rays. This reduces the background as a first measurement in the Modane underground lab in the Fréjus tunnel has shown [35]. If the background can be reduced by one to two orders, it should be enough to adjust only the two parameters, the Q-value and the neutrinos mass, to the upper end of the spectrum for the determination of the electron neutrino mass.

6. Acknowledgment

I want to thank the ECHo collaboration and specifically my coauthors in earlier works [5, 15, 16, 18] on this topic.

References

[1] Drexlin G, Hannen V, Mertens S and Weinheimer C 2013 *Adv. High Energy Phys.* **2013** 293986 (*Preprint* physics.ins-de/13077.0101v1)
[2] Faessler A, Huster E, Kraft O and Krahn F 1970 *Z. Physik* **238** 352
[3] Otten E W and Weinheimer C 2008 *Rep. Prog. Phys.* **71** 086201
[4] De Rujula A 2013 Two old ways to measure the electron-neutrino mass *Preprint* hep-ph/1305.4857v1; De Rujula A and Lusignoli M 1982 *Phys. Lett. B* **118** 429
[5] Faessler A, Gastaldo L and Šimkovic F 2015 *J. Phys. G* **42** 015108 (*Preprint* nucl-th/1407.6504)
[6] Robertson R G H 2015 *Phys. Rev. C* **91** 035504 (*Preprint* nucl-th/411.2906v1)
[7] Faessler A and Šimkovic F 2015 *Phys. Rev. C* **91** 045505 (*Preprint* hep-hp:/1501.04338)
[8] Intemann R L and Polock F 1967 *Phys. Rev.* **157** 41
[9] De Rujula A and Lusignoli M 2016 *JHEP* 1605 015 (*Preprint* hep-ph/1601.04990)
[10] Faessler A, Hodak R, Kovalenko S, and Šimkovic F 2013 arXiv:1304.5032v3, and *European Physical Journal Web of Conferences*
[11] Faessler A 2009 *Nucl. Phys. Proc. Suppl.* **188** 20
[12] Šimkovic F, Faessler A, Rodin V, Vogel P and Engel J 2008 *Phys. Rev. C* **77** 045503

IOP Conf. Series: Journal of Physics: Conf. Series **1056** (2018) 012020 doi:10.1088/1742-6596/1056/1/012020

[13] Vogel P and Engel J 2008 *Phys. Rev. C* **77** 045503
[14] Avignone F, Elliott S R and Engel J 2008 *Rev. Mod. Phys.* **80** (2008) 481
[15] Faessler A, Gastaldo L and Šimkovic F 2017 *Phys. Rev. C* **95** 045502 (*Preprint* nucl-th/1611.00325)
[16] Faessler A and Simkovic F 2016 *Phys. Scripta* **91** 043007
[17] Faessler A, Enns Ch, Gastaldo L and Šimkovic F 2015 *Phys. Rev. C* **91**, 064302
[18] Faessler A and Šimkovic F 2015 *Phys. Rev. C* **91** 045505
[19] Eliseev S, Blaum K, Block M, Chenmarek S, Dorrer H, Duellmann Ch E, Enss C, Filiamin P E, Gastaldo L, Goncharov M, Koester U, Lautenschlaeger F, Novikov Yu N, Rischka A, Schuessler R X, Schweighard L and Tuerler A 2015 *Phys. Rev. Lett.* **115** 062501
[20] Ranitzsch P C -O *et al.* 2012 *Jour. of Low Temp. Phys.* **167** 1004, and private communication
[21] Blaum K, Doerr A, Duellmann C E, Eberhardt K, Eliseev S, Enss C, Faessler A, Fleischmann A, Gastaldo L, Kempf S, Krivoruchenko M, Lahiri S, Matai M, Novikov Yu N, Ranitzsch P C, Šimkovic F, Scusc Z and Wegner M 2013 The Electron Capture ^{163}Ho Experiment ECHo *Preprint* physics.ins-det/1306.2655v1
[22] Ranitzsch P C, Hassel C, Wegner M, Kempf S, Fleischmann A, Enss C, Gastaldo L, Herlert A and Johnston K 2014 First Calorimetric Measurement of OI-line in the Electron Capture Spectrum of ^{163}Ho *Preprint* physics.ins-det/1409.007v1
[23] Grant I P 1970 *Adv. Phys.* **19** 747
[24] Desclaux J P 1975 *Comp. Phys. Com* **9** 31
[25] Ankudinov A L, Zabinsky S I and J. J. Rehr 1996 *Comp. Phys. Com.* **98** 359
[26] Salvat F, Fernandez-Varea J M and Williamson W 1995 *Comp. Phys. Com.* **90** 151
[27] Anderson J U *et al.* 1982 *Phys. Lett. B* **398** 72
[28] Gatti F *et al.* 1997 *Phys. Lett. B* **398** 415
[29] Audi G, Wapstra A H and Thibault C 2003 *Nucl. Phys. A* **279** 337
[30] M.E. Rose, Relativistic Electron Theory. (John Wiley and Sons, New York, 1961).
[31] W. Greiner, Relativistische Quantenmechanik. (Band 6; Harri Deusch, Frankfurt, 1987).
[32] Bambynek W, Behrens H, Chen M H, Crasemann B, Fitzpatrick M L, Ledingham K W D, Genz H, Mutterer M and Intemann R L 1977 *Rev. Mod. Phys.* **49** 109
[33] Williams G P 2001 *X-Ray Data Booklet* (Lawrence Berkeley National Laboratory) Table 1-1
[34] Perger W F, Vasan Karighattam 1991 *Comp. Physics Com.* **66** 392
[35] Gastaldo L private information

IOP Conf. Series: Journal of Physics: Conf. Series **1056** (2018) 012021 doi:10.1088/1742-6596/1056/1/012021

First results of the KM3NeT/ARCA detector

G Ferrara[1,2] and S Biagi[1] on behalf of the KM3NeT Collaboration
[1] INFN, Laboratori Nazionali del Sud, Catania, Italy
[2] Università degli Studi di Catania, Italy

E-mail: gferrara@lns.infn.it

E-mail: simone.biagi@lns.infn.it

Abstract. The KM3NeT collaboration started to build a multi-km^3 neutrino telescope in the Mediterranean Sea. The telescope is composed of two parts: the ARCA detector, optimised for searches for high energy neutrino sources in the Universe, and ORCA, for the determination of the mass hierarchy of neutrinos. ARCA is under construction at the Capo Passero site, Italy, 80 km offshore at a depth of 3500 m while ORCA is located in the Toulon area, France, 40 km offshore at a depth of 2500 m. The basic detection element of the KM3NeT detector is the Digital Optical Module. The module is a pressure resistant glass sphere, containing 31 photo-multiplier tubes. Eighteen modules are arranged in the detection unit, a vertical structure anchored on the sea floor. The detection units are deployed on the sea bed to form a three-dimensional array of optical modules to detect Cherenkov light produced by neutrino-induced muons. In these proceedings preliminary results obtained with the first two detection units of ARCA are presented. The capability to select and reconstruct atmospheric muons is discussed. The dependence of the muon flux with the sea depth is derived, showing that the detector is well calibrated and the systematics are kept under control.

1. Introduction

KM3NeT (km^3-scale Neutrino Telescope) [1] is a research infrastructure in the Mediterranean Sea which will host two neutrino detectors: ORCA (Oscillation Research with Cosmics in the Abyss), located in the Toulon area, France, 40 km offshore at a depth of 2500 m, with the main objective of neutrino mass hierarchy determination [2] and ARCA (Astroparticle Research with Cosmics in the Abyss) optimised for searches for high energy neutrino sources in the Universe. The ARCA neutrino telescope is under construction at the Capo Passero site, Italy, about 90 km offshore at a depth of 3500 m and it will reach a volume of cubic kilometre scale. The first two ARCA-DUs were deployed in December 2015 and in May 2016. The detector was taking data continuously from December 2015 up to April 2017. In the first period of data taking the finalization of data acquisition system has been carried on and on-line trigger algorithms have been optimized. The calibration procedures allow to obtain sub-nanosecond accuracy for the timing of PMTs and uniform tuning for their high voltages.

The basic detection element of the detector is the Digital Optical Module (DOM) [3], a pressure resistant glass sphere of 43 cm diameter, containing 31 3" photo-multipliers tubes. 18 DOMs are arranged with an interspace of 36 m in the Detection Unit (DU), a vertical string anchored on the sea floor. The DUs are 700 m tall and are deployed on the sea bed to form a three-dimensional array of DOMs. The DUs are connected to submarine Junction Boxes and through them to shore for power feed and data transmission.

The events of interest for a neutrino telescope are the upward-going cosmic neutrinos in the energy range $10^2 \div 10^8$ GeV. The neutrinos interact weakly with matter, producing detectable secondary particles. Muon neutrinos are considered the best candidates for a neutrino telescope. Neutrino-induced muons, in fact, can travel in water over distances of the order of kilometers and can be tracked with high accuracy. At energies $E \geq 1$ TeV the muon direction is almost collinear with the generating neutrino, it is therefore possible to infer the neutrino direction allowing for neutrino astronomy. The detection of high energy muons is done exploiting the Cherenkov effect, which consists of a coherent radiation emitted by charged particles when cross a medium with a velocity greater than the speed of light in that medium. A large volume of a natural medium optically transparent (water or ice) is instrumented with optical sensors to detect the Cherenkov radiation.
A neutrino telescope has three main components of background: the optical background, due to the decay of ^{40}K in the sea water, the atmospheric muons and the atmospheric neutrinos, produced by the interactions of cosmic rays in the atmosphere. The optical background is responsible for uncorrelate signals on PMTs. It can be reduced looking for time-space correlation between PMTs in the same DOMs. The atmospheric muon and neutrino fluxes produce detectable events like cosmic neutrinos. The background of atmospheric muon can be reduced applying cuts on the reconstructed zenith angle and on quality parameters. For point-like source search, the contribution of cosmic neutrinos over the atmospheric neutrino background is investigated searching for a statistical excess of events around the source position in the sky. In the next sections the preliminary results obtained with the first two detection units of ARCA are presented. The capability to select and reconstruct atmospheric muons is discussed and the dependence of the muon flux with the sea depth is derived.

2. Study of atmospheric muons and data-MonteCarlo comparison

The multi-PMT DOM structure of the ARCA telescope allows for selecting atmospheric muons over the optical background looking for local coincidences between PMTs. In the KM3NeT framework, a local coincidences (L1) is defined as a signal of two PMTs in the same DOM in a time window less than 25 ns. The PMT multiplicity (m) is the number of PMTs in the same DOM with a L1 local coincidence.
The study of the detector response to the optical and atmospheric muon backgrounds is done through MonteCarlo (MC) simulations. MC simulations of the uncorrelated optical background have been performed in order to simulate the decay of K^{40} in sea water, the light production and signals on PMT. The Atmospheric muons are simulated through the muon generation and propagation in the sea water, the emission of Cherenkov light and the generation of PMT signals. Data-MC comparison of the rates of local coincidences in a time window of 25 ns, for the lowest and the highest DOMs of the two ARCA-DUs, are shown in Fig. 1. It can be seen that low PMT multiplicities are related with optical background. Atmospheric muon events produce local coincidences on a great number of PMTs in the same DOM, with a consequent high PMT multiplicity. The data-MC comparison shows that the MC well reproduces the data and indicates that for a multiplicity $m \geq 8$ the coincidences rate due to optical background from ^{40}K decay is negligible [4]. Cuts on PMT multiplicity allows to selects atmospheric muons over the optical background.

3. Atmospheric muon flux dependence with the sea depth

The stability of the PMTs and the homogeneity of the ARCA detector performances can be verified through the dependence of atmospheric muon flux rate with the sea depth. For the study of the atmospheric muon flux dependence a data sample corresponding to a detector livetime of $\sim$10 day has been considered. A full MC simulation, with the PMT efficiencies equal to 1, has been produced. Then the same trigger algorithms have been applied both for data

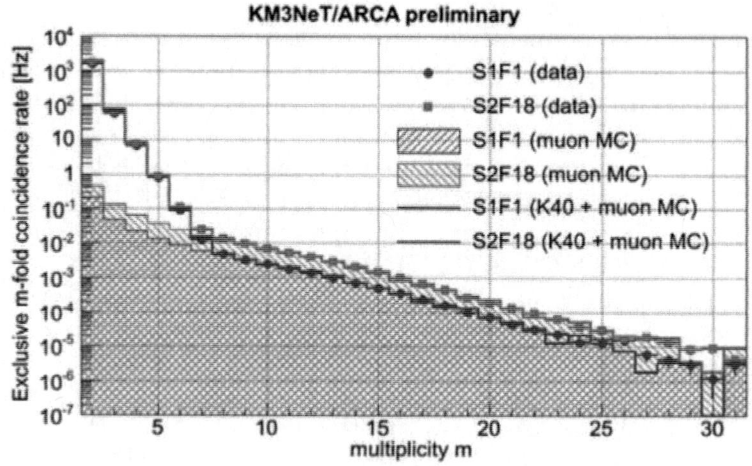

Figure 1. Rate of multiple coincidences for the DOM1 of ARCA-DU1 and DOM18 of ARCA-DU2 as a function of PMT multiplicity. Data are reported as blue circles and red squares. The MC simulations of the atmospheric muon background are represented by full areas. The sum of the atmospheric muons and optical background are indicated by full lines. For a multiplicity $m \geq 8$ the coincidences rate due to optical background is negligible.

and MC. To obtain a pure atmospheric muon sample, events with more than $m > 7$-fold local coincidences have been selected. In Fig. 2 the multiplicity rates in single DOM induced by muon flux as a function of the sea depth is shown. The coincidence rate shows an exponential decrease of the PMT rates according to the decrease of atmospheric muon flux with the depth both for data and MC. From the analysis of the ^{40}K coincidences, a relative efficiency of each PMT has been obtained and include in a new MC production (blue square in Fig. 2). The MC simulation with the relative efficiencies of PMTs well reproduces the characteristics of the selected data [4], indicating that the detector is well calibrated and the systematics are kept under control.

4. Data reconstruction

The KM3NeT/ARCA telescope allows for the reconstruction of muons trajectories looking for time-space correlation among the PMT signals of different DOMs. The reconstruction procedure consists of different steps. At first, a track reconstruction algorithm [5] performs a scan over the full solid angle, selecting PMT signals correlated with candidate reconstructed tracks. After that, a likelihood-maximisation procedure, based on multi-dimensional probability distribution of the arrival times of the Cherenkov light from events, is applied. Finally, the track with the best likelihood is chosen as the reconstructed track. To investigate the quality of the reconstruction procedure a data sample corresponding to a detector livetime of ~20 days has been selected. For this analysis only the MC simulation of atmospheric muons has been considered. In Fig. 3 the rate of events as a function of the reconstructed zenith angle is presented both for data and MC. The data and MC comparison shows a good agreement, proving the reliability of MC simulation codes and that the detector is well calibrated in time. Even if a small fraction of downward-going atmospheric muons is wrongly reconstructed as upward-going events, they overcome the atmospheric neutrino flux. In Fig. 3 the data-MC ratio is also reported. An excess of data over the MC event has been found in the rage of reconstructed zenith angle of very upward-going tracks. Further studies of this excess of upward-going tracks found in the data are ongoing in order to search for neutrino-induced events.

IOP Conf. Series: Journal of Physics: Conf. Series **1056** (2018) 012021 doi:10.1088/1742-6596/1056/1/012021

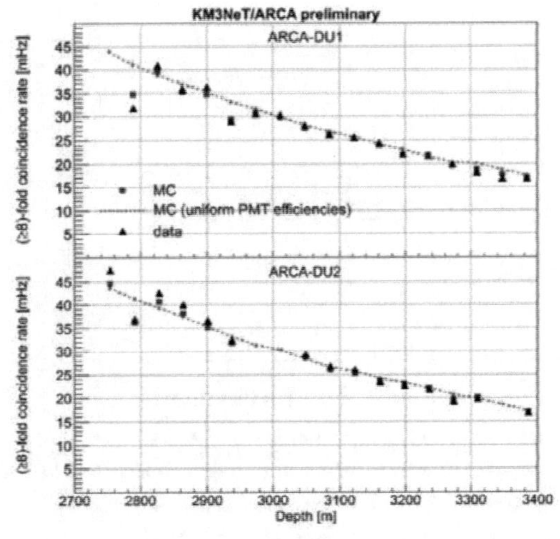

Figure 2. Comparison between data (triangle), MC with uniform PMTs efficiencies (dotted line) and MC with relative PMTs efficiencies (blue square) of the rates of $m \geq 8$ PMT coincidences in the DOMs as a function of depth for ARCA-DU1 and ARCA-DU2. DOM1 is the lowermostDOM of ARCA-DU1 (3390 m), DOM18 is the uppermost DOM of ARCA-DU2 (2750 m), see Fig. 1

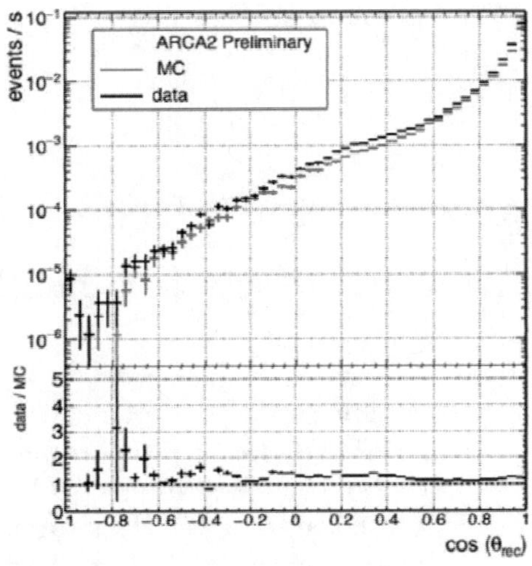

Figure 3. Comparison between the rate of reconstructed events for data and MC as a function of the cosine of the reconstructed zenith angle. In the bottom, the data-MC ratio is shown. No quality cuts are applied to the data sample. Only atmospheric muons are simulated.

5. Conclusion

In these proceedings preliminary results obtained with the first two ARCA-DUs have been presented. The capability of the ARCA telescope for selecting atmospheric muons over the optical background has been verified. The detector performances have been investigated by means of the study of the atmospheric muon flux dependence with the sea depth. Finally, the data-MC comparison of the rate of events as a function of the reconstructed zenith angle demonstrates that the simulation fits well with the data and shows the reconstruction algorithm performances.

References

[1] Adrian-Martínez S et al. 2016 *J. Phys. G: Nucl. Part. Phys.* **43** 084001.
[2] Van Elewyck V (for the KM3NeT Coll.) these proceedings.
[3] Leonora E (for the KM3NeT Coll.) these proceedings.
[4] Jongen M (for the KM3NeT Coll.) **PoS(ICRC2017)**1018.
[5] Melis K et al. (for the KM3NeT Coll.) **PoS(ICRC2017)**950.

IOP Conf. Series: Journal of Physics: Conf. Series **1056** (2018) 012022 doi:10.1088/1742-6596/1056/1/012022

Characterization of the microwave multiplexing readout and TESs for HOLMES

E Ferri[1,2], B K Alpert[3] , D T Becker[3] , D A Bennett[3] , M Biasotti[4,5] , V Ceriale[4,5], M De Gerone[5], M Faverzani[1,2], J W Fowler[3] , G Gallucci[5] , J D Gard[3] , A Giachero[1,2], J P Hays-Wehle[3] , G C Hilton[3] , J A B Mates[3] , A Nucciotti[1,2], A Orlando[2] , G Pessina[2] , A Puiu[1,2] , C D Reintsema[3] , D R Schmidt[3] , D S Swetz[3] , J N Ullom[3] and L R Vale[3]

[1] Dipartimento di Fisica, Università di Milano-Bicocca, Milano, Italy
[2] Istituto Nazionale di Fisica Nucleare, Sezione di Milano-Bicocca, Milano, Italy
[3] National Institute of Standards and Technology, Boulder, CO, USA
[4] Dipartimento di Fisica, Università di Genova, Genova, Italy
[5] Istituto Nazionale di Fisica Nucleare, Sezione di Genova, Genova, Italy

E-mail: `elena.ferri@mib.infn.it`

Abstract.
A powerful tool to determine the effective electron-neutrino mass is the calorimetric measurement of the energy released in a nuclear beta decay. Performing a precision measurement of the end point of the Electron Capture decay spectrum of ^{163}Ho, HOLMES aims at pushing down the sensitivity on the neutrino mass below 1 eV. In its final configuration HOLMES will deploy an array of 1000 microcalorimeters based on Transition Edge Sensors with gold absorbers in which the ^{163}Ho will be ion implanted. The best technique to easily read out such a number of detector with a common readout line is the microwave frequency domain multiplexing. Therefore, the TESs are coupled to multiplexed rf-SQUIDS operated in flux ramp modulation for linearization purposes. The rf-SQUIDS are then coupled to superconducting quarter wavelength resonators in the GHz range, from which the modulating signal is finally recovering using software defined radio techniques. In the last two years an extensive R&D activity has been carried out in order to maximize the multiplexing factor while preserving the performances of each detector which fulfil the HOLMES requirements (i.e. an energy resolution of few eV and a time-resolution of a few microseconds). We report here the progress made towards the characterization of the multiplexing system together with the results of the characterization of the HOLMES detectors.

1. Introduction

One of the most compelling issue in particle physics is the neutrino mass measurement. To date, a model-independent method to determine the neutrino mass is the calorimetric measurement of the energy released in a nuclear decay involving neutrino. Performing a calorimetric measurement of the ^{163}Ho Electron Capture decay spectrum, as proposed in 1982 by De Rújula and Lusignoli [1], the HOLMES [2] experiment aims at reaching a sensitivity on neutrino mass as low as 1 eV. For reaching such target, HOLMES will deploy an array of 1000 microcalorimeters characterized by an energy and time resolution of $\sim 1\,\mathrm{eV}$ at the end point and of $\sim 1\,\mu\mathrm{s}$,

IOP Conf. Series: Journal of Physics: Conf. Series **1056** (2018) 012022 doi:10.1088/1742-6596/1056/1/012022

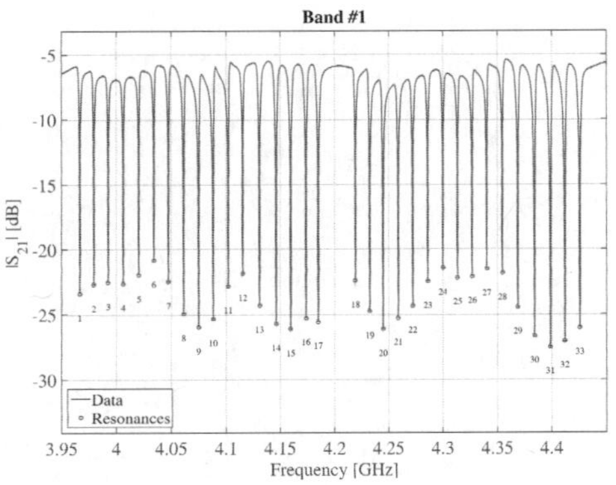

Figure 1. The total measured transfer function $|S_{21}|$ of the resonators on the multiplexing chip model 17A.

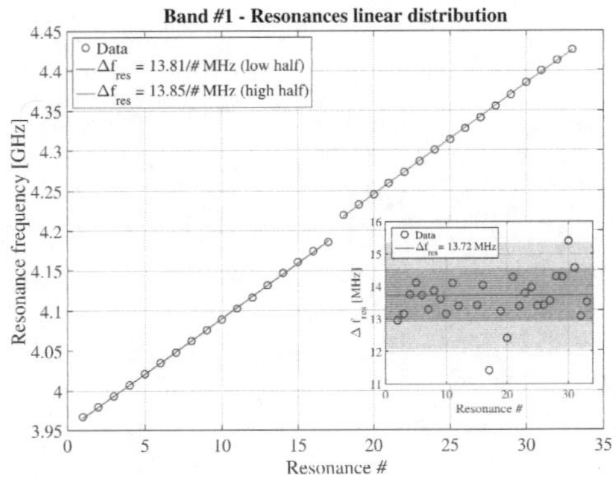

Figure 2. The resonances linear distribution and the resonances spacing on the multiplexing chip model 17A.

respectively. A good time resolution is an important element in this kind of experiment since the sensitivity is strongly dependent on the unresolved pile-up fraction in the region of interest. The 1 μs time resolution is obtained applying pile-up resolving algorithms to pulses with 20 μs rise time sampled at 500 kHz [3, 4].

A complete description of the current status of the HOLMES experiment is reported in [5].

2. The microwave multiplexing readout

A promising approach to read out a large number of detectors using a common transmission line is the microwave multiplexng read out. This can be leveraged to multiplex the signals from many detectors with large signal bandwidth and high energy resolution. Therefore, the HOLMES detectors are read out with the microwave multiplexing system (μmux) [7], which is based on the use of the rf-SQUIDs as input devices. Their response is linearised using a flux ramp modulation [8]: a sawtooth signal is applied to a common line inductively coupled to each rf-SQUIDs. The modulated rf-SQUID signals are read out coupling the rf-SQUID to a superconducting $\lambda/4$ wave resonators in the GHz range. By tuning the resonators at different frequencies it is possible to read many SQUIDs with a set of RF carries. The modulated signal is finally recovered using Software Defined Radio (SDR) technique: in this scheme frequency tones in the MHz range are generated by a digital to analog converter, up-converted in the GHz range by means of an IQ mixer and combed together to probe the resonators. Then, the output signal is down converted by another IQ mixer and digitized by an analog to digital converter. Finally, the individual channel recovery is performed via software. The SDR technique in HOLMES exploits the Reconfigurable Open Architecture Computing Hardware (ROACH2) [9] board with a Xilinx Virtex6 FPGA. The SDR is implemented in the firmware of the ROACH2 board by NIST [10]. We are currently using a version of the firmware which is able to multiplex 16 channels. The ADC sampling frequency f_{ADC} sets the total bandwidth available for each ROACH2 board. The bandwidth of the ADC boards currently available for ROACH2 is 550 MHz. Considering a 500 MHz ADC bandwidth BW_{ADC}, the multiplexing factor is given by:

$$N_{TES} = \frac{BW_{ADC}}{N_S \times f_{ramp} \times N_{\Phi_0}} \tag{1}$$

where f_{ramp} is the sampling frequency of the TES signal and N_{Φ_0} is the number of rf-SQUID oscillation per ramp, which is set to 2. To properly reconstruct the TES signal while providing

IOP Conf. Series: Journal of Physics: Conf. Series **1056** (2018) 012022 doi:10.1088/1742-6596/1056/1/012022

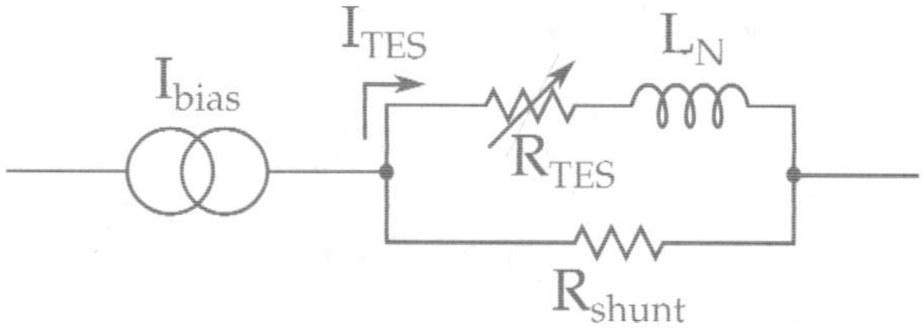

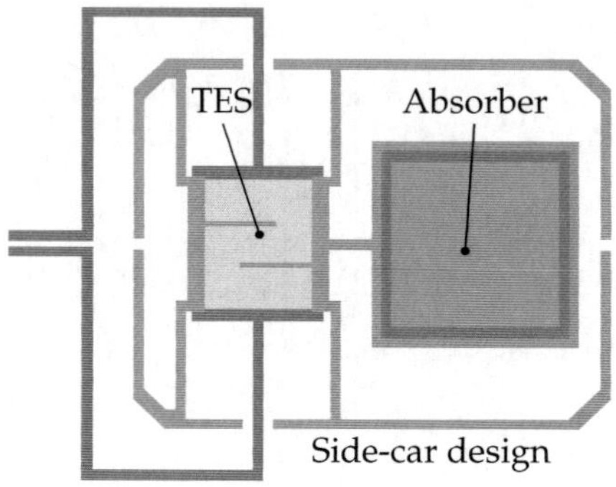

Figure 3. Left side: the TES bias circuit . Right side: a sketch of a pixel with the gold absorber beside the TES sensor (side-car design).

a reasonable multiplexing factor, the frequency separation and the bandwidth of the resonators in the multiplexing chip need to be precisely controlled. A spacing between resonances (N_S) of at least 14 MHz assures a negligible cross-talk, while a bandwidth of 2 MHz for each resonators allows to sample the TES signal at 500 kHz. In this configuration, the number of TES detectors which can be multiplexed is 36.

We have characterized a preliminary version of the HOLMES multiplexing chip (μmux17A) which was developed and produced at NIST. The chip has 33 resonances in 500 MHz frequency interval with an average separation of 14 MHz. The resonance width is around 2 MHz. The transfer function $|S_{21}|$ of the resonators on the multiplexing chip is shown in figure 1, while the resonances linear distribution and spacing are displayed in figure 2. The big frequency gap between resonance 17 and 18 corresponds to the centre of the multiplexing chip and its presence is required by the read out system.

3. TES performances

The requirement of low pile-up probability for reaching a sensitivity on neutrino mass as low as 1 eV sets strict constraints on the detector design. A fast rise time requires a large electrical bandwidth which is limited by the ratio between the stray inductance L and the total resistance R in the bias circuit (left side of figure 3). Therefore, L must be kept as small as possible. The decay time is set by the ratio between the thermal capacity C of the absorber and the thermal conductance G toward the heat sink. Since the value of C is set by the request of full containment of the energy in the absorber, the only way to have fast decay time is to increase G. The selected design is $125\times125\,\mu m^2$ Molybdenum-Copper bilayer TESs on Si_3N_3 membrane with 2 μm thick gold absorber in which the ^{183}Ho will be ion implanted [11, 12]. The $200\times200\,\mu m^2$ gold absorber lays aside of the sensor (side-car design) in order to avoid proximity effects between the sensor and absorber itself. The thermal conductance G is increased by the addition of a Cu thermal radiating perimeter. C is around $0.8\,$pJ/K while G is about $600\,$pW/K. A sketch of a single pixel is reported in right side of figure 3.

The rise time of $20\,\mu$s is obtained by tuning the stray inductance in the TES bias circuit. A study on the variation of the rising time at different working point is reported in [13]. The decay time is of the order of $100\,\mu$s.

We have studied the energy resolution at different energies illuminating the detectors with a X-ray fluorescence source. The source is based on ^{55}Fe as primary source which irradiates targets of Al, Cl, Ca. As a consequence, the detectors are exposed to the Rayleigh scattered K_α and K_β X-rays of Mn and to the fluorescence K_α X-rays at 1.5, 2.6, 3.7 keV excited in Al, Cl, Ca. The results are reported in figure 4.

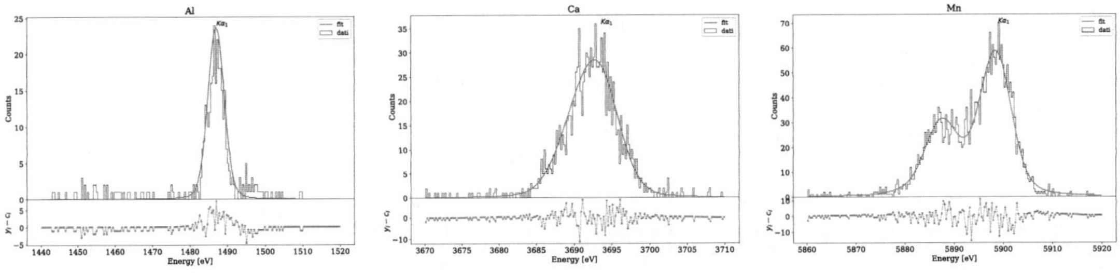

Figure 4. Energy spectra acquired in the presence of the fluorescence calibration source. The FWHM of the fitted peaks is 4.7 eV at 1.5 keV; 5.7 eV at 3.7 keV and 6.3 eV at 5.8 keV.

The results proved that the HOLMES read out based on ROACH-2 combined with the multiplexing system allow us to properly sample fast pulses without significant energy resolution degradation.

4. Conclusion

HOLMES is a calorimetric experiment aiming at reaching a sensitivity on neutrino mass as low as 1 eV by means the study of the EC decay of ^{163}Ho. In order to reach such target there are strict requirements not only on the multiplexing system for reading out 1000 detectors but also on the single detector. In fact, as many detectors as possible must to be read in parallel without deteriorating the energy and time resolutions. In this contribution we have presented the characterization of the microwave multiplexing read out and we have shown that a multiplexing factor of 33 preserves the performances of each detectors in term of energy and time resolutions.

Acknowledgments

The HOLMES experiment is funded by the European Research Council under the European Union Seventh Framework Programme (FP7/2007-2013)/ERC Grant Agreement no. 340321. We also acknowledge support from INFN and from the NIST Innovations in Measurement Science program.

References

[1] De Rújula A and Lusignoli M 1983 *Nucl. Phys.* B **219** 277
[2] Alpert B et al 2015 *Eur. Phys. J.* C **75** 112
[3] Ferri E et al 2016 *J. Low Temp. Phys.* **184** 405
[4] Alpert B et al 2016 *J. Low Temp. Phys.* **184** 263
[5] Ferri E et al 2018 *J. Physics: Conf. Series* this issue
[6] Faverzani M et al 2016 *J. Low Temp. Phys.* **184** 922
[7] Mates J A B et al 2008 *Appl. Phys. Lett.* **92** 023514
[8] Mates J A B et al 2012 *J. Low Temp. Phys.* **167** 707
[9] Hickish J et al 2016 *J. Astron. Instrum.* **05** 1641001
[10] Mates J A B et al 2017 *Appl. Phys. Lett.* **111** 062601
[11] Hays-Wehle J P et al 2016 *J. Low Temp. Phys.* **184** 492
[12] Orlando A et al 2016 *J. Low Temp. Phys.* **184** 892
[13] Puiu A et al submitted to *J. Low Temp. Phys.*

IOP Conf. Series: Journal of Physics: Conf. Series **1056** (2018) 012023 doi:10.1088/1742-6596/1056/1/012023

Focal plane detector optical readout

G Gallo[1,2], D L Bonanno[3], D G Bongiovanni[2],F Cappuzzello[1,2] M Cortesi[4], F Longhitano[2], D Lo Presti[1,3], L Pandola[2], S Reito[3], for the NUMEN collaboration

[1]Dipartimento di Fisica e Astronomia, Università di Catania, Italy
[2]INFN Laboratori Nazionali del Sud, Catania, Italy
[3]INFN Sezione di Catania, Catania, Italy
[4]National Superconducting Cyclotron Laboratory, MSU, East Lising, USA

E-mail: giuseppe.gallo@ct.infn.it

Abstract. A preliminary study of an alternative solution for the optical detection and tracking of the ions in the MAGNEX focal plane detector for the NUMEN project is presented. The tracks of the ions are sampled by means of the light that they produce through a scintillating gas, which is collected by arrays of Silicon Photo Multiplier suitably arranged. A complete Geant4 simulation is under development in order to correlate detector geometry, characteristics of the gas and light collection efficiency. Fast timing performance could be achieved through a completely digital (on-off) read-out system of the light sensors and a position reconstruction algorithm.

1. Introduction

The nature of the neutrino, if it is a Dirac or Majorana particle, is an open question in nuclear and particle physics. The NUMEN project aims to answer this question looking for lepton number violation in double beta decay [1, 2, 3].

The already obtained results indicate that suitable information can be extracted from heavy-ion induced Double Charge Exchange (DCE) reactions [4]. The MAGNEX large acceptance spectrometer [5, 6] of the INFN-Laboratori Nazionali del Sud (LNS) is of fundamental importance for high resolution measurement of reaction channels with very low cross section (in order of few nb). But, in order to explore cases of interest for $\beta\beta$ decay, an upgrade of the LNS Superconducting Cyclotron (CS) is planned to increase beam current about two orders of magnitude [7, 8].

In order to keep track of the foreseen high rate, a major upgrade of the present detection system is also mandatory. The main limitations come from the MAGNEX Focal Plane Detector (FPD) [9], with a sustainable rate of about 5 kHz. The NUMEN project R&D phase sets for the substitution of the present FPD gas tracker with a tracker system based on micro-patterns gas detectors (MPGD).

This paper describes a preliminary exploration of a solution based on purely optical tracking of the ions at focal plane and digital readout of the signals produced in the light sensors, alternative to the MPGD system.

IOP Conf. Series: Journal of Physics: Conf. Series **1056** (2018) 012023 doi:10.1088/1742-6596/1056/1/012023

2. The optical readout

2.1. Design of the detectors

The detector is schematically designed as a series of light sensitive rings that surround the gas chamber of the FPD, see figure 1. The chamber is filled with a suitable scintillating gas or gas mixture which produces scintillation light during the passage of charged particles. Each ring is composed of a sequence of four linear arrays of square Silicon Photo Multipliers (SiPMs). In front of each SiPM is placed a collimator in order to reduce its field of view as shown in figure 2. The internal walls of the collimator could be made reflective to improve the photon collection efficiency.

Figure 1. Sketch of the detector. The inner box represents the gas chamber whereas the elements that surround it stand for the light sensors arrays. The dimensions of the gas chamber are 150(l) $\times$ 20(h) $\times$ 15(w) cm^3.

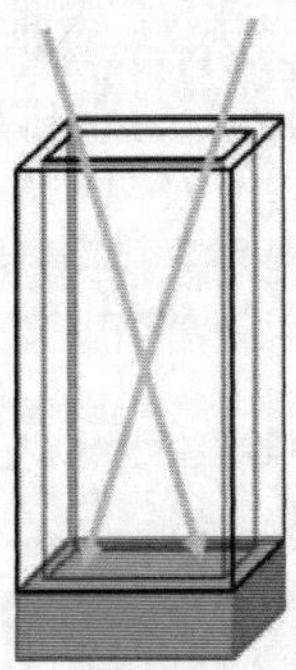

Figure 2. The box on the bottom represents a SiPM coupled with a collimator of which only the edges are drawn to show the light rays that reach the sensitive surface of the sensor.

2.2. The digital readout

The digital readout procedure consists in recording only light sensors which are hit by the scintillation light, without counting how many photons reach each sensors. In practice, it could be considered equivalent to a measure of the width of the light cone over a sensitive array. Starting from this measurement on two facing arrays, a geometric reconstruction algorithm can find the coordinate of the ion on the plane passing through the two arrays.

The optical readout solution is under investigation because it could give some advantages, compared with standard solutions for ions tracking like drift chambers, such as:

- low operating voltages limited to SiPMs bias;
- high uniformity due to no electric field distortion;
- high spatial resolution of a few hundreds microns;
- high counting rate up to a few hundreds of MHz;
- reduced data stream.

These last two features will be achieved only if the information needed to accurately reconstruct the position of the ions can be obtained by a fully digital readout of the SiPMs.

IOP Conf. Series: Journal of Physics: Conf. Series **1056** (2018) 012023 doi:10.1088/1742-6596/1056/1/012023

3. Geant4 simulation

To verify the feasibility of an optical tracker with these characteristics and to study its performance, a detailed Monte Carlo simulation is under development by means of the Geant4 toolkit [10, 11]. Figure 3 shows some screenshots acquired during the debugging phase.

The parameters of the simulation are: (*i*) type and pressure of scintillating gas; (*ii*) number, size and position of the SiPMs; (*iii*) collimator size and its optical properties. These parameters determine the luminescence spectrum and the photon yield of the gas, the energy loss and angular straggling of the ions and the geometric photon collection efficiency at the light sensitive surfaces.

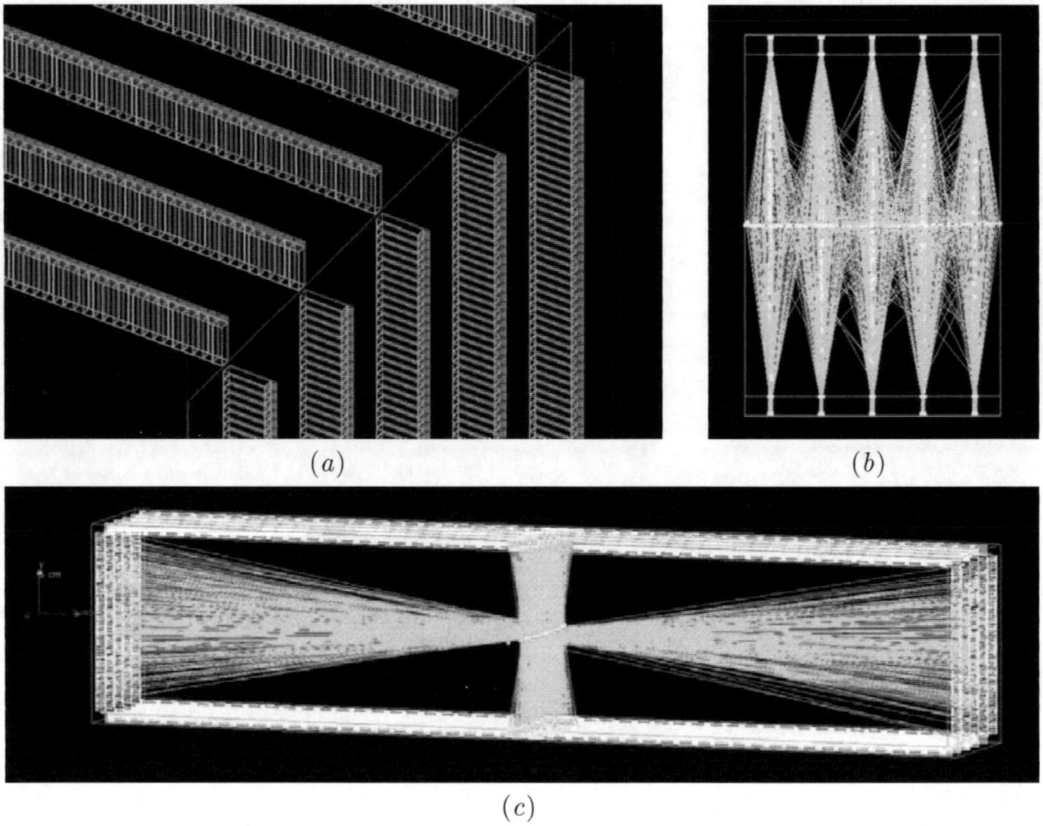

Figure 3. Figure *a* shows the arrangement of light sensors (orange) and collimators (white frame) on a corner of the gas box that represents the FPD volume. Figures *b* and *c* show the scintillation photons produced by an α particle of 5 MeV traversing the FPD filled with CH_4 at 25 torr. Photons are radiated isotropically, but it is possible to display only that ones which hit any of the SiPM.

3.1. Preliminary results

Once the number of rings has been chosen, the first step of the simulation is to calculate the angular straggling and the energy lost by the ions. Figure 4 shows an example for 1000 α particles of 5 MeV passing through 150 mm of CH_4 at a pressure of 25 torr.

4. Conclusions and outlooks

The idea of a new detector with optical readout is described as alternative for the FPD of the NUMEN experiment and preliminary results of the simulations are reported. The next

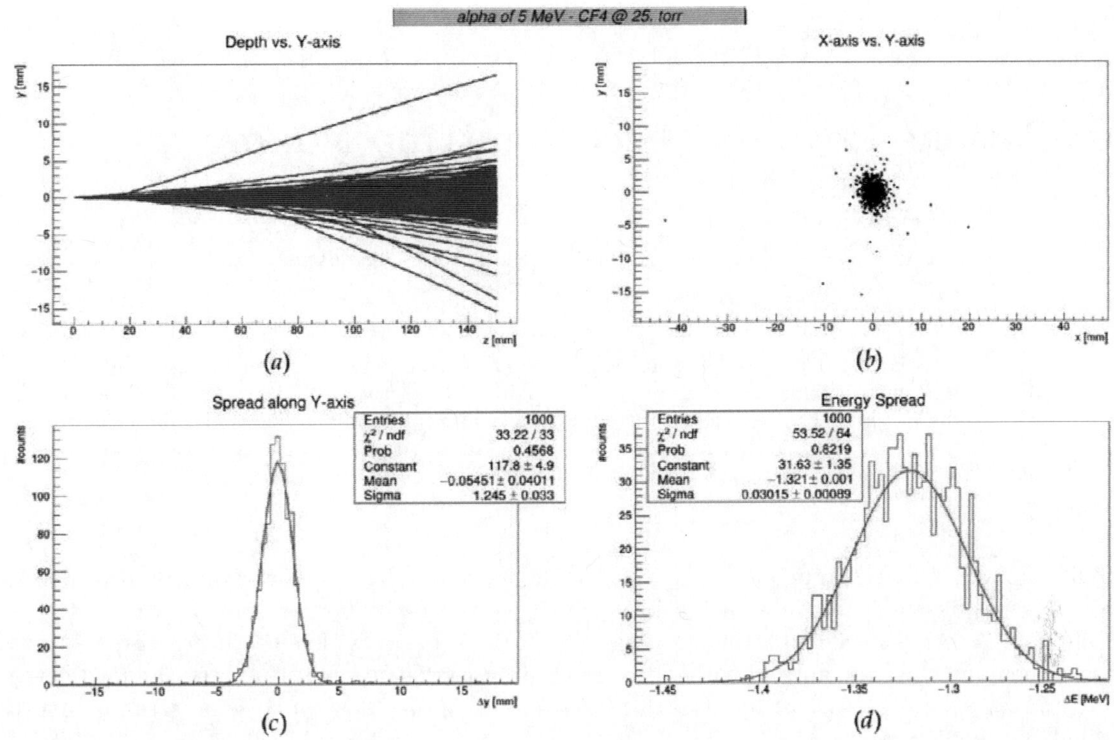

Figure 4. The upper part of the figure shows as the beam spread traversing the gas along z direction: (a) is a side projection showing the spread along vertical direction y; (b) is a plot of the beam spot on a plane at z = 150 mm perpendicular to the propagation direction of the ions. The bottom part of the figure shows an histogram of the beam spread along y direction at the exit of the beam (c) and an histogram of the energy loss after traversing the gas chamber (d).

steps of this work include the study of detector performance with ions involved in the reaction channels of interest for the NUMEN project, such as double β decay, finding the best parameters to maximize the number of photons collected, and the implementation of the algorithm to reconstruct the position of the ions through the detector. This simulation platform will help to design other detectors with similar requirements which could be of interest in many field of particle identification. If the first promising results of the simulations are confirmed, a prototype will be built to confirm performance.

References

[1] Cappuzzello F *et al.* 2015 *Journal of Physics: Conference Series* **630** 012018
[2] Agodi C *et al.* 2015 *Nuclear and Particle Physics Proceedings* **265-266** 28 – 30 ISSN 2405-6014
[3] Cappuzzello F *et al.* 2018 *Eur. Phys. J. A* **54** 72
[4] Cappuzzello F *et al.* 2015 *Eur. Phys. J. A* **51** 145
[5] Cappuzzello F *et al.* 2016 *Eur. Phys. J. A* **52** 167
[6] Cappuzzello F *et al.* 2014 *Nucl. Instrum. Methods Phys. Res. A* **763** 314 – 319 ISSN 0168-9002
[7] Cappuzzello F *et al.* 2017 *AIP Conference Proceedings* **1894** 020004
[8] Calabretta L *et al.* 2017 *Modern Physics Letters A* **32** 1740009
[9] Cavallaro M *et al.* 2012 *The European Physical Journal A* **48** 59 ISSN 1434-601X
[10] Agostinelli S *et al.* 2003 *Nucl. Instrum. Methods Phys. Res. A* **506** 250 – 303 ISSN 0168-9002
[11] Allison J *et al.* 2016 *Nucl. Instrum. Methods Phys. Res. A* **835** 186 – 225 ISSN 0168-9002

IOP Conf. Series: Journal of Physics: Conf. Series **1056** (2018) 012024 doi:10.1088/1742-6596/1056/1/012024

Oscillations Beyond Three-Neutrino Mixing

Carlo Giunti

INFN, Sezione di Torino, Via P. Giuria 1, I–10125 Torino, Italy

E-mail: giunti@to.infn.it

Abstract. The LSND, Gallium and reactor neutrino anomalies can be explained by short-baseline neutrino oscillations due to the mixing of the active neutrinos with sterile neutrinos at the eV scale. I review the results of a 3+1 global fit of short-baseline neutrino oscillation data that includes the recent measurements of the MINOS, IceCube, and NEOS experiments, and I discuss the implications for neutrinoless double-beta decay.

The LSND [1, 2], Gallium [3–7] and reactor [8–10] anomalies are intriguing indications in favor of short-baseline neutrino oscillations due to sterile neutrinos at the eV scale. Here, I consider 3+1 active-sterile neutrino mixing (see Ref. [11]), in which there are three sub-eV massive neutrinos ν_1, ν_2, and ν_3 which are the main constituents of the three standard active neutrinos ν_e, ν_μ, ν_τ, and there is a fourth massive neutrino ν_4 at the eV scale which is mainly sterile ($\nu_4 \simeq \nu_s$).

In the framework of 3+1 active-sterile mixing, the effective oscillation probabilities of the flavor neutrinos in short-baseline experiments are given by [12]

$$P_{\alpha\beta}^{(\mathrm{SBL})} \simeq \left| \delta_{\alpha\beta} - \sin^2 2\vartheta_{\alpha\beta} \sin^2\left(\frac{\Delta m^2_{41} L}{4E}\right) \right| , \tag{1}$$

where $\alpha, \beta = e, \mu, \tau, s$, L is the source-detector distance and E is the neutrino energy. The short-baseline oscillation amplitudes depend only on the absolute values of the elements in the fourth column of the mixing matrix:

$$\sin^2 2\vartheta_{\alpha\beta} = 4|U_{\alpha 4}|^2 \left| \delta_{\alpha\beta} - |U_{\beta 4}|^2 \right| . \tag{2}$$

Here I review the results of the 3+1 global fit of short-baseline neutrino oscillation data presented in Ref. [13], which includes the recent measurements of the MINOS [14], IceCube [15], and NEOS [16] experiments, and I discuss the implications for neutrinoless double-beta decay.

It is well known (see Ref. [11]) that the global fits of short-baseline data are affected by the so-called "appearance-disappearance" tension, that is present [17] for any number N_s of sterile neutrinos in 3+N_s mixing schemes which are perturbations of the standard three-neutrino mixing required for the explanation of the observation of solar, atmospheric and long-baseline neutrino oscillations. In Ref. [18], we proposed a "pragmatic approach" in which the appearance-disappearance tension is alleviated by excluding from the global fit the low-energy bins of the MiniBooNE experiment [19, 20], which have an anomalous excess of $\overset{(-)}{\nu}_e$-like events that is widely considered to be suspicious and is under investigation in the MicroBooNE experiment at Fermilab. In this paper I assume the pragmatic approach from the beginning.

	PrGlo16A	PrGlo16B	PrGlo17
$\chi^2_{\min}$	262.0	530.3	595.1
NDF	244	519	579
GoF	20%	36%	31%
Δm^2_{41}	1.6	1.6	1.7
$\lvert U_{e4}\rvert^2$	0.026	0.030	0.020
$\lvert U_{\mu4}\rvert^2$	0.013	0.011	0.015
$\sin^2 2\vartheta_{e\mu}$	0.0014	0.0013	0.0012
$\sin^2 2\vartheta_{ee}$	0.10	0.12	0.079
$\sin^2 2\vartheta_{\mu\mu}$	0.053	0.042	0.058
$\Delta\chi^2_{\rm NO}$	48.3	47.3	47.4
$\mathrm{NDF}_{\rm NO}$	3	4	4
$n\sigma_{\rm NO}$	6.4	6.1	6.1
$\Delta\chi^2_{\rm PG}$	3.8	4.7	7.2
$\mathrm{NDF}_{\rm PG}$	2	2	2
$\mathrm{GoF}_{\rm PG}$	15%	9.7%	2.7%

Table 1. Results of the pragmatic 3+1 global PrGlo16A, PrGlo16B, and PrGlo17 fits of SBL data. The first group of rows gives: the minimum χ^2 ($\chi^2_{\min}$), the number of degrees of freedom (NDF), the goodness of fit (GoF), the best fit values of the mixing parameters Δm^2_{41}, $|U_{e4}|^2$, $|U_{\mu4}|^2$, and of the oscillation amplitudes $\sin^2 2\vartheta_{e\mu}$, $\sin^2 2\vartheta_{ee}$, $\sin^2 2\vartheta_{\mu\mu}$. The second group of rows gives the χ^2 difference $\Delta\chi^2_{\rm NO}$ between the χ^2 of no oscillations and $\chi^2_{\min}$ and the resulting number of σ's ($n\sigma_{\rm NO}$) for $\mathrm{NDF}_{\rm NO}$ degrees of freedom corresponding to the number of fitted parameters. The third group of rows gives the results for the appearance-disappearance parameter goodness of fit [21]: the χ^2 difference $\Delta\chi^2_{\rm PG}$ and the resulting goodness of fit $\mathrm{GoF}_{\rm PG}$ for $\mathrm{NDF}_{\rm PG}$ degrees of freedom.

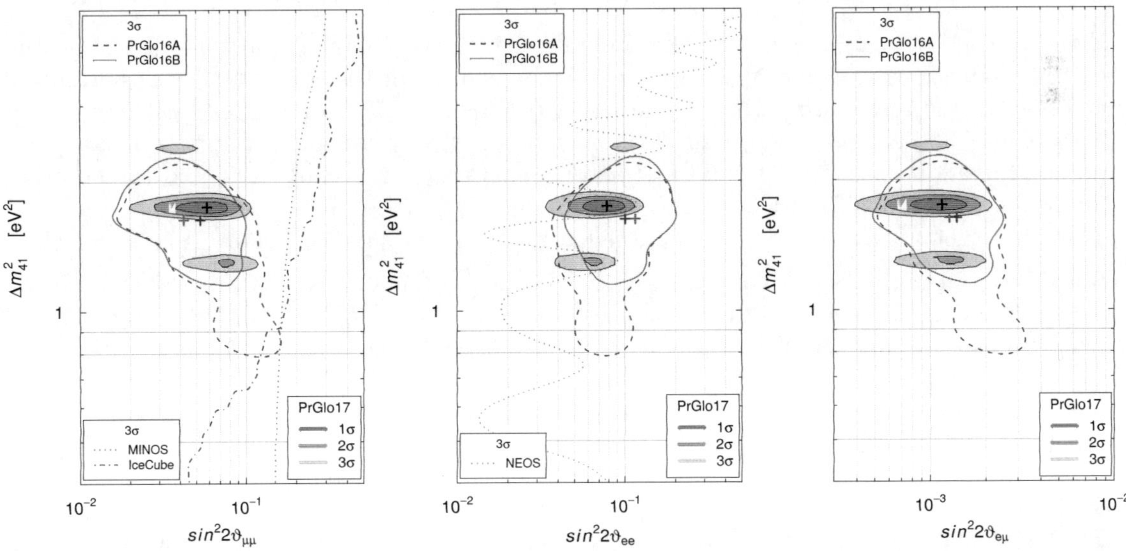

Figure 1. Allowed regions in the $\sin^2 2\vartheta_{\mu\mu}$–$\Delta m^2_{41}$, $\sin^2 2\vartheta_{ee}$–Δm^2_{41}, and $\sin^2 2\vartheta_{e\mu}$–Δm^2_{41} planes obtained in the pragmatic 3+1 global PrGlo16A, PrGlo16B, and PrGlo17 fits of SBL data.

The pragmatic global fit of short-baseline neutrino oscillation data [13] considers the following three groups of experiments:

(A) The $\overset{(-)}{\nu}_\mu \to \overset{(-)}{\nu}_e$ appearance data of the LSND [2], MiniBooNE [19, 20] (without the anomalous low-energy bins), BNL-E776 [22], KARMEN [23], NOMAD [24], ICARUS [25] and OPERA [26] experiments.

(B) The following $\overset{(-)}{\nu}_e$ disappearance data: 1) The ratios of measured and predicted [8–10] $\bar{\nu}_e$ rates of the short-baseline reactor experiments listed in Table 1 of Ref. [13]); 2) The $\bar{\nu}_e$ spectra measured in the Bugey-3 [27] and NEOS [16] short-baseline reactor experiments; 3) the data of the GALLEX [28–30] and SAGE [3, 31–33] Gallium radioactive source experiments with the statistical method discussed in Ref. [6]; 4) the solar neutrino constraint on $\sin^2 2\vartheta_{ee}$ [13]; 5) the KARMEN [34, 35] and LSND [36] $\nu_e + {}^{12}\text{C} \to {}^{12}\text{N}_{\text{g.s.}} + e^-$ scattering data [37], with the method discussed in Ref. [38].

(C) The constraints on $\overset{(-)}{\nu}_\mu$ disappearance obtained from the data of the CDHSW experiment [39], from the analysis [40] of the data of atmospheric neutrino oscillation experiments, from the analysis of the SciBooNE-MiniBooNE neutrino [41] and antineutrino [42] data, and the recent constraints of the MINOS [14] and IceCube [15] experiments.

Table 1 and Figure 1 summarize the results of the following three global fits:

PrGlo16A. In this analysis we considered all the appearance and disappearance SBL data available in 2016 (see Ref. [13]), except MINOS [14] and IceCube [15]. The PrGlo16A fit is an update of the PrGLO fit presented in Ref. [11], with a similar set of data.

PrGlo16B. In this analysis we added the MINOS [14] and IceCube [15] data to the data considered in the PrGlo16A fit, in order to clarify their effects on the results of the analysis.

PrGlo17. In this analysis we added the NEOS [16] data, which have been available to us in the beginning of 2017.

From Table 1 one can see that all the three fits have an acceptable goodness of fit and the case of no oscillations is excluded at the level of about 6σ. On the other hand, the parameter goodness of fit, decreases from 15% in the PrGlo16A fit to 9.7% in the PrGlo16B fit to 2.7% in the PrGlo17 fit. This is a symptom of the increase of the appearance-disappearance tension caused by the addition of the MINOS and IceCube data in the PrGlo16B fit and the addition of NEOS data in the PrGlo17 fit.

The left panel in Fig. 1 shows the allowed regions in the $\sin^2 2\vartheta_{\mu\mu}$–$\Delta m^2_{41}$ plane, which is relevant for $\overset{(-)}{\nu}_\mu$ disappearance. One can see that the constraints on $\overset{(-)}{\nu}_\mu$ disappearance given by the MINOS and IceCube data disfavor the low-Δm^2_{41}–high-$\sin^2 2\vartheta_{\mu\mu}$ part of the region allowed by the PrGlo16A fit. This effect was expected [43] and is consistent with the results of the 3+1 global fit presented in Ref. [44], which updated Ref. [45] with the addition of the IceCube data. There is an increase of the appearance-disappearance tension caused by the shrinking if the allowed range of $|U_{\mu4}|^2$ from $0.0050 - 0.033$ at 3σ in the PrGlo16A fit to $0.0048 - 0.023$ in the PrGlo16B fit, which is quantified by the decrease of the parameter goodness of fit from 15% in the PrGlo16A fit to 9.7% in the PrGlo16B.

The center panel in Fig. 1 shows the allowed regions in the $\sin^2 2\vartheta_{ee}$–Δm^2_{41} plane, which is relevant for $\overset{(-)}{\nu}_e$ disappearance. One can see that the inclusion of the NEOS constraints has the dramatic effect of fragmenting the allowed region in three islands with narrow Δm^2_{41} widths. The best-fit island is at $\Delta m^2_{41} \approx 1.7\,\text{eV}^2$. There is an island allowed at 2σ at $\Delta m^2_{41} \approx 1.3\,\text{eV}^2$, and an island allowed at 3σ at $\Delta m^2_{41} \approx 2.4\,\text{eV}^2$. Moreover, the NEOS constraints shifts the allowed range of $|U_{e4}|^2$ from $0.013 - 0.050$ at 3σ in the PrGlo16B fit to $0.0098 - 0.031$ in the

IOP Conf. Series: Journal of Physics: Conf. Series **1056** (2018) 012024 doi:10.1088/1742-6596/1056/1/012024

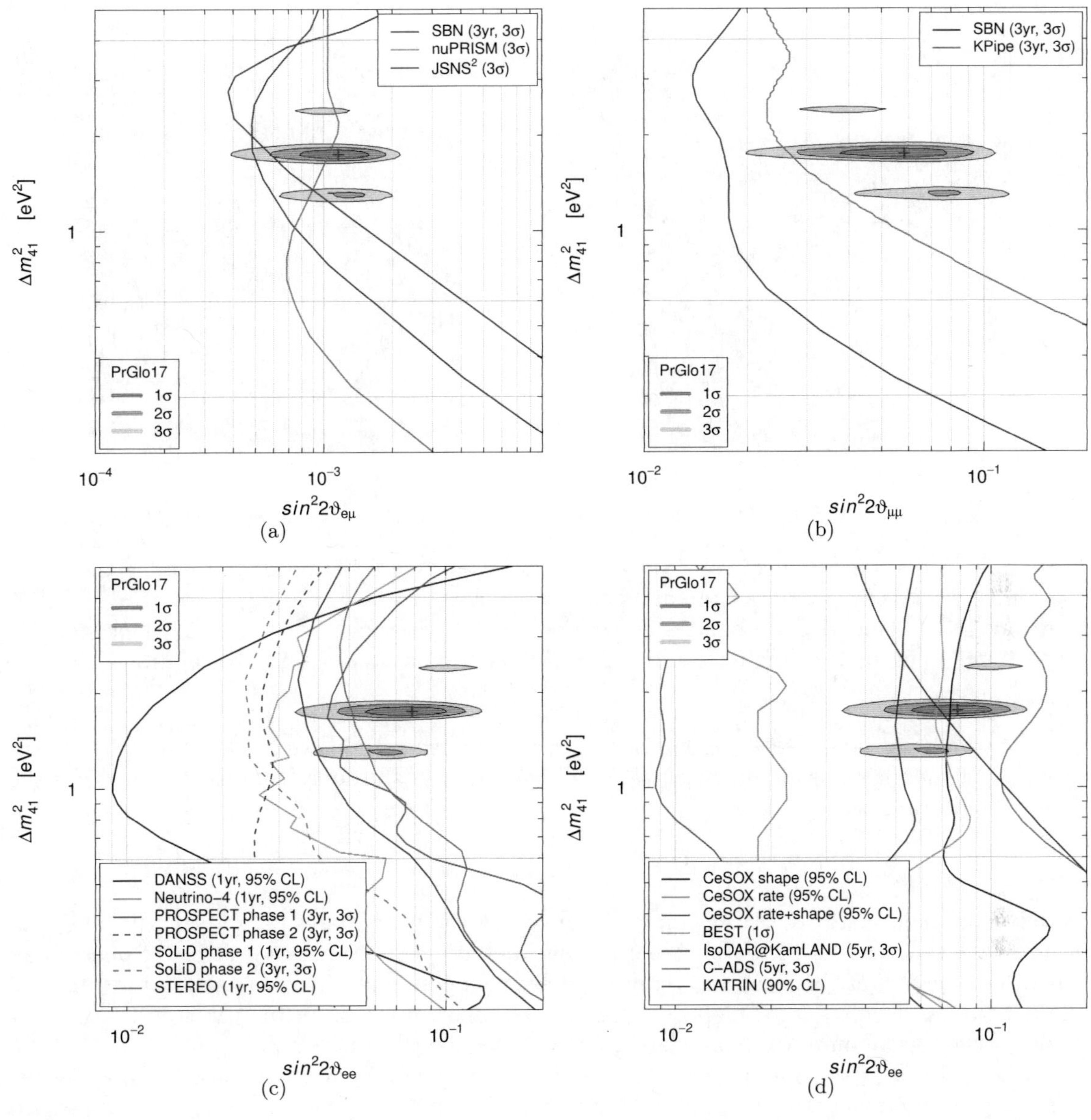

Figure 2. Sensitivities of future experiments compared with the PrGlo17 allowed regions in Fig. 1.

PrGlo17 fit. The corresponding increase of the appearance-disappearance tension is quantified by the decrease of the parameter goodness of fit from 9.7% in the PrGlo16B fit to 2.7% in the PrGlo17 fit.

Figure 2 shows a comparison of the sensitivities of future experiments with the PrGlo17 allowed regions for: (a) $\overset{(-)}{\nu}_\mu \to \overset{(-)}{\nu}_e$ transitions; (b) $\overset{(-)}{\nu}_\mu$ disappearance; (c),(d) $\overset{(-)}{\nu}_e$ disappearance. It is clear that these experiments will give definitive information on the existence of active-sterile short-baseline oscillations connected with the LSND, Gallium and reactor anomalies.

The determination of active-sterile neutrino mixing is of interest also for the phenomenology of neutrinoless double-β decay experiments [7, 46–57].

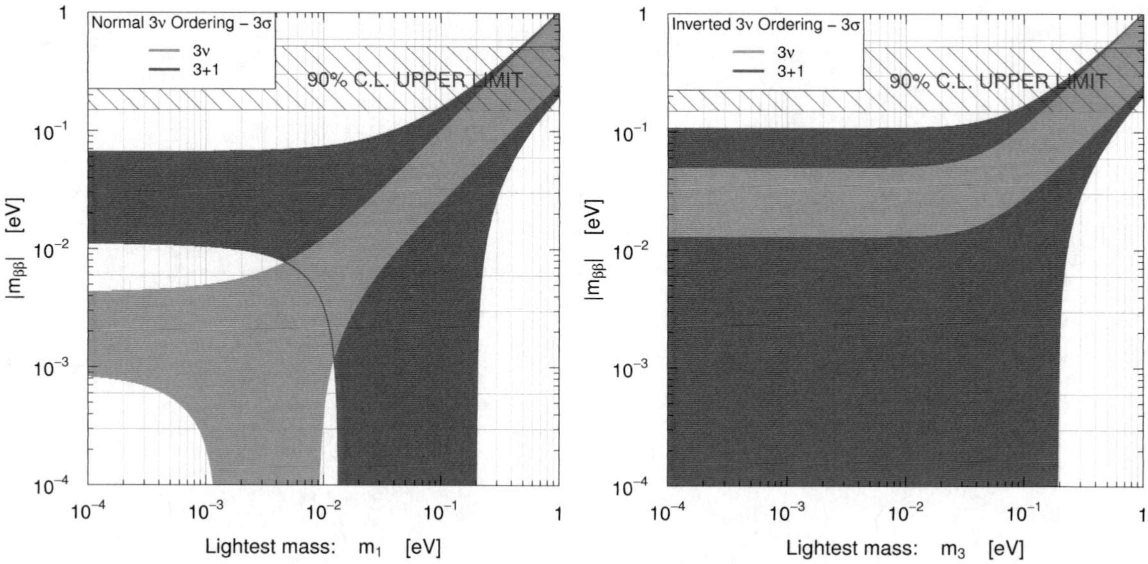

Figure 3. Value of the effective Majorana mass $|m_{\beta\beta}|$ as a function of the lightest neutrino mass in the cases of 3ν and 3+1 mixing with normal and inverted ordering of the three lightest neutrinos [55]. The horizontal band is an estimate of the experimental 90% C.L. upper limit for $|m_{\beta\beta}|$ taking into account the uncertainties of the nuclear matrix element calculations [58].

If massive neutrinos are Majorana particles (see the recent reviews in Refs. [56, 58]), in the case of 3+1 mixing the rate of neutrinoless double-β decay is proportional to the square of the effective Majorana mass

$$|m_{\beta\beta}| = \left||U_{e1}|^2 m_1 + |U_{e2}|^2 e^{i\alpha_2} m_2 + |U_{e3}|^2 e^{i\alpha_3} m_3 + |U_{e4}|^2 e^{i\alpha_4} m_4\right| . \tag{3}$$

In this expression there are three completely unknown complex phases α_2, α_3, α_4 which depend on the Majorana phases in the neutrino mixing matrix. These unknown complex phases can generate cancellations between the different mass contributions. Figure 3 shows the range of allowed values of $|m_{\beta\beta}|$ as a function of the lightest neutrino mass in the cases of 3ν and 3+1 mixing with normal and inverted ordering of the three lightest neutrinos [55].

From Fig. 3 one can see that the presence of an additional massive neutrinos at the eV scale can change dramatically the predictions for the possible range of values of $|m_{\beta\beta}|$. In the case of a normal 3ν mass hierarchy ($m_1 \ll m_2 \ll m_3$) the value of $|m_{\beta\beta}|$ is dominated by the contribution of ν_4, which implies that $1 \times 10^{-2} \lesssim |m_{\beta\beta}| \lesssim 7 \times 10^{-2}$ eV. This range of values of $|m_{\beta\beta}|$ is larger than that predicted by the standard 3ν mixing in the case of a normal hierarchy and similar to that predicted in the case of an inverted hierarchy in the standard 3ν mixing scheme. On the other hand, in the case of an inverted 3ν mass ordering there can be a complete cancellation between the contribution of ν_4 and those of the three standard light neutrinos, leading to the disappearance of the lower limit for $|m_{\beta\beta}|$ predicted by the standard 3ν mixing scheme.

The next generation of neutrinoless double-beta decay experiments (see Refs. [59–64]) is planned to explore the range of $|m_{\beta\beta}|$ between about 1×10^{-2} and 5×10^{-2} eV predicted by the standard 3ν mixing in the case of an inverted hierarchy. They are not expected to reach the range of $|m_{\beta\beta}|$ between about 8×10^{-4} and 5×10^{-3} eV predicted by the standard 3ν mixing in the case of a normal hierarchy. From Fig. 3 it is clear that the predictions are dramatically changed in the 3+1 neutrino mixing scheme and a positive result in these experiments is guaranteed in the case of a normal mass hierarchy, whereas in the case of an inverted mass hierarchy the allowed range of $|m_{\beta\beta}|$ goes from zero to about 0.1 eV.

Let me finally emphasize that the confirmation of the existence of sterile neutrinos would be a major discovery which would have a profound impact not only on neutrino physics, but on our whole view of fundamental physics, because it would prove that there is new physics beyond the Standard Model at the low-energies accessible in laboratory experiments. The measurement of the properties of the sterile neutrinos can give important information on this new physics.

References

[1] Athanassopoulos C *et al.* (LSND) 1995 *Phys. Rev. Lett.* **75** 2650–2653 (*Preprint* nucl-ex/9504002)
[2] Aguilar A *et al.* (LSND) 2001 *Phys. Rev.* **D64** 112007 (*Preprint* hep-ex/0104049)
[3] Abdurashitov J N *et al.* (SAGE) 2006 *Phys. Rev.* **C73** 045805 (*Preprint* nucl-ex/0512041)
[4] Laveder M 2007 *Nucl. Phys. Proc. Suppl.* **168** 344–346
[5] Giunti C and Laveder M 2007 *Mod. Phys. Lett.* **A22** 2499–2509 (*Preprint* hep-ph/0610352)
[6] Giunti C and Laveder M 2011 *Phys. Rev.* **C83** 065504 (*Preprint* arXiv:1006.3244)
[7] Giunti C, Laveder M, Li Y, Liu Q and Long H 2012 *Phys. Rev.* **D86** 113014 (*Preprint* arXiv:1210.5715)
[8] Mueller T A *et al.* 2011 *Phys. Rev.* **C83** 054615 (*Preprint* arXiv:1101.2663)
[9] Mention G *et al.* 2011 *Phys. Rev.* **D83** 073006 (*Preprint* arXiv:1101.2755)
[10] Huber P 2011 *Phys. Rev.* **C84** 024617 (*Preprint* arXiv:1106.0687)
[11] Gariazzo S, Giunti C, Laveder M, Li Y and Zavanin E 2016 *J. Phys.* **G43** 033001 (*Preprint* arXiv:1507.08204)
[12] Bilenky S M, Giunti C and Grimus W 1998 *Eur. Phys. J.* **C1** 247–253 (*Preprint* hep-ph/9607372)
[13] Gariazzo S, Giunti C, Laveder M and Li Y 2017 *JHEP* **1706** 135 (*Preprint* arXiv:1703.00860)
[14] Adamson P *et al.* (MINOS) 2016 *Phys. Rev. Lett.* **117** 151803 (*Preprint* arXiv:1607.01176)
[15] Aartsen M G *et al.* (IceCube) 2016 *Phys. Rev. Lett.* **117** 071801 (*Preprint* arXiv:1605.01990)
[16] Ko Y *et al.* (NEOS) 2017 *Phys.Rev.Lett.* **118** 121802 (*Preprint* arXiv:1610.05134)
[17] Giunti C and Zavanin E M 2016 *Mod. Phys. Lett.* **A31** 1650003 (*Preprint* arXiv:1508.03172)
[18] Giunti C, Laveder M, Li Y and Long H 2013 *Phys. Rev.* **D88** 073008 (*Preprint* arXiv:1308.5288)
[19] Aguilar-Arevalo A A *et al.* (MiniBooNE) 2009 *Phys. Rev. Lett.* **102** 101802 (*Preprint* arXiv:0812.2243)
[20] Aguilar-Arevalo A *et al.* (MiniBooNE) 2013 *Phys. Rev. Lett.* **110** 161801 (*Preprint* arXiv:1303.2588)
[21] Maltoni M and Schwetz T 2003 *Phys. Rev.* **D68** 033020 (*Preprint* hep-ph/0304176)
[22] Borodovsky L *et al.* (BNL-E776) 1992 *Phys. Rev. Lett.* **68** 274–277
[23] Armbruster B *et al.* (KARMEN) 2002 *Phys. Rev.* **D65** 112001 (*Preprint* hep-ex/0203021)
[24] Astier P *et al.* (NOMAD) 2003 *Phys. Lett.* **B570** 19–31 (*Preprint* hep-ex/0306037)
[25] Antonello M *et al.* (ICARUS) 2013 *Eur.Phys.J.* **C73** 2599 (*Preprint* arXiv:1307.4699)
[26] Agafonova N *et al.* (OPERA) 2013 *JHEP* **1307** 004 (*Preprint* arXiv:1303.3953)
[27] Achkar B *et al.* (Bugey) 1995 *Nucl. Phys.* **B434** 503–534
[28] Anselmann P *et al.* (GALLEX) 1995 *Phys. Lett.* **B342** 440–450
[29] Hampel W *et al.* (GALLEX) 1998 *Phys. Lett.* **B420** 114–126
[30] Kaether F, Hampel W, Heusser G, Kiko J and Kirsten T 2010 *Phys. Lett.* **B685** 47–54 (*Preprint* arXiv:1001.2731)
[31] Abdurashitov J N *et al.* (SAGE) 1996 *Phys. Rev. Lett.* **77** 4708–4711
[32] Abdurashitov J N *et al.* (SAGE) 1999 *Phys. Rev.* **C59** 2246–2263 (*Preprint* hep-ph/9803418)
[33] Abdurashitov J N *et al.* (SAGE) 2009 *Phys. Rev.* **C80** 015807 (*Preprint* arXiv:0901.2200)
[34] Bodmann B E *et al.* (KARMEN.) 1994 *Phys. Lett.* **B332** 251–257
[35] Armbruster B *et al.* (KARMEN) 1998 *Phys. Rev.* **C57** 3414–3424 (*Preprint* hep-ex/9801007)
[36] Auerbach L B *et al.* (LSND) 2001 *Phys. Rev.* **C64** 065501 (*Preprint* hep-ex/0105068)
[37] Conrad J and Shaevitz M 2012 *Phys. Rev.* **D85** 013017 (*Preprint* arXiv:1106.5552)
[38] Giunti C and Laveder M 2011 *Phys. Lett.* **B706** 200–207 (*Preprint* arXiv:1111.1069)
[39] Dydak F *et al.* (CDHSW) 1984 *Phys. Lett.* **B134** 281
[40] Maltoni M and Schwetz T 2007 *Phys. Rev.* **D76** 093005 (*Preprint* arXiv:0705.0107)
[41] Mahn K B M *et al.* (SciBooNE-MiniBooNE) 2012 *Phys. Rev.* **D85** 032007 (*Preprint* arXiv:1106.5685)
[42] Cheng G *et al.* (SciBooNE-MiniBooNE) 2012 *Phys. Rev.* **D86** 052009 (*Preprint* arXiv:1208.0322)
[43] Giunti C 2017 *J. Phys. Conf. Ser.* **888** 012019 (*Preprint* arXiv:1609.04688)
[44] Collin G, Arguelles C, Conrad J and Shaevitz M 2016 *Phys. Rev. Lett.* **117** 221801 (*Preprint* arXiv:1607.00011)
[45] Collin G H, Arguelles C A, Conrad J M and Shaevitz M H 2016 *Nucl. Phys.* **B908** 354–365 (*Preprint* arXiv:1602.00671)
[46] Goswami S and Rodejohann W 2006 *Phys. Rev.* **D73** 113003 (*Preprint* hep-ph/0512234)

[47] Goswami S and Rodejohann W 2007 *JHEP* **10** 073 (*Preprint* `arXiv:0706.1462`)
[48] Barry J, Rodejohann W and Zhang H 2011 *JHEP* **07** 091 (*Preprint* `arXiv:1105.3911`)
[49] Li Y and Liu S 2012 *Phys. Lett.* **B706** 406–411 (*Preprint* `arXiv:1110.5795`)
[50] Rodejohann W 2012 *J. Phys.* **G39** 124008 (*Preprint* `arXiv:1206.2560`)
[51] Girardi I, Meroni A and Petcov S T 2013 *JHEP* **1311** 146 (*Preprint* `arXiv:1308.5802`)
[52] Pascoli S, Mitra M and Wong S 2014 *Phys. Rev.* **D90** 093005 (*Preprint* `arXiv:1310.6218`)
[53] Meroni A and Peinado E 2014 *Phys. Rev.* **D90** 053002 (*Preprint* `arXiv:1406.3990`)
[54] Abada A, Romeri V D and Teixeira A 2014 *JHEP* **1409** 074 (*Preprint* `arXiv:1406.6978`)
[55] Giunti C and Zavanin E M 2015 *JHEP* **07** 171 (*Preprint* `arXiv:1505.00978`)
[56] Pas H and Rodejohann W 2015 *New J. Phys.* **17** 115010 (*Preprint* `arXiv:1507.00170`)
[57] Barea J, Kotila J and Iachello F 2015 *Phys. Rev.* **D92** 093001 (*Preprint* `arXiv:1509.01925`)
[58] Bilenky S and Giunti C 2015 *Int.J.Mod.Phys.* **A30** 0001 (*Preprint* `arXiv:1411.4791`)
[59] Gomez-Cadenas J, Martin-Albo J, Mezzetto M, Monrabal F and Sorel M 2012 *Riv.Nuovo Cim.* **35** 29–98 (*Preprint* `arXiv:1109.5515`)
[60] Giuliani A and Poves A 2012 *Adv.High Energy Phys.* **2012** 857016
[61] Schwingenheuer B 2013 *Annalen Phys.* **525** 269–280 (*Preprint* `arXiv:1210.7432`)
[62] Cremonesi O and Pavan M 2014 *Adv. High Energy Phys.* **2014** 951432 (*Preprint* `arXiv:1310.4692`)
[63] Artusa D R *et al.* (CUORE) 2014 *Eur.Phys.J.* **C74** 3096 (*Preprint* `arXiv:1404.4469`)
[64] Gomez-Cadenas J and Martin-Albo J 2015 *PoS* **GSSI14** 004 (*Preprint* `arXiv:1502.00581`)

IOP Conf. Series: Journal of Physics: Conf. Series **1056** (2018) 012025 doi:10.1088/1742-6596/1056/1/012025

Search for solar dark matter annihilation with JUNO

Wan-Lei Guo

Institute of High Energy Physics, Chinese Academy of Sciences, P.O. Box 918, Beijing 100049, China

E-mail: guowl@ihep.ac.cn

Abstract. We explore the electron neutrino signals from light dark matter (DM) annihilation in the Sun for the large liquid scintillator detector JUNO. In terms of the spectrum features of three typical DM annihilation channels $\chi\chi \to \nu\bar{\nu}, \tau^+\tau^-, b\bar{b}$, we take two sets of selection conditions to calculate the expected signals and atmospheric neutrino backgrounds based on the Monte Carlo simulation data. Then the JUNO sensitivities to the spin independent DM-nucleon and spin dependent DM-proton cross sections are presented. It is found that the JUNO projected sensitivities are much better than the current spin dependent direct detection experimental limits for the $\nu\bar{\nu}$ and $\tau^+\tau^-$ channels. In the spin independent case, the JUNO will give the better sensitivity to the DM-nucleon cross section than the LUX and CDMSlite limits for the $\nu\bar{\nu}$ channel with the DM mass lighter than 6.5 GeV.

1. Introduction

The existence of dark matter (DM) is by now well confirmed [1]. The current cosmological observations have helped to establish the concordance cosmological model where the present Universe consists of about 69.3% dark energy, 25.8% dark matter and 4.9% atoms [2]. Understanding the nature of dark matter is a prime open problem in particle physics and cosmology. Here we focus on the neutrino signals from the DM annihilation in the Sun. As well as in the DM direct detection experiments, the halo DM particles can also elastically scatter with nuclei in the Sun. Then they may lose most of their energy and are captured in the Sun [1]. These trapped DM particles will be accumulated in the core of the Sun due to repeated scatters and the gravity potential. Therefore the Sun is a very interesting place for us to search the DM annihilation signals. Due to the interactions of the DM annihilation products in the Sun, only the neutrino can escape from the Sun and reach the Earth. Therefore the terrestrial neutrino detectors can detect these neutrinos.

The Jiangmen Underground Neutrino Observatory (JUNO) [3], a 20 kton multi-purpose underground liquid scintillator (LS) detector, are constructing in China to primarily determine the neutrino mass hierarchy by detecting reactor antineutrinos. The JUNO central detector as a LS calorimeter has an excellent energy resolution and a very low energy threshold. It is found that the LS detector has capability to reconstruct the track direction of the energetic charged particle by use of the timing pattern of the first-hit on the photomultiplier tubes (PMTs) since the energetic particle travels faster than light in the LS [4]. Therefore the JUNO LS detector can detect the neutrinos from the DM annihilation in the Sun. In Ref. [3], the JUNO has calculated the $\nu_\mu/\bar{\nu}_\mu$ signals from the DM annihilation in the Sun. Some authors have analyzed the $\nu_e/\bar{\nu}_e$ and $\nu_\mu/\bar{\nu}_\mu$ signals for the other LS detectors [5, 6]. Based on the Ref.[7], here we shall briefly discuss the $\nu_e/\bar{\nu}_e$ signals from the solar DM annihilation in JUNO.

IOP Conf. Series: Journal of Physics: Conf. Series **1056** (2018) 012025 doi:10.1088/1742-6596/1056/1/012025

In this paper, we shall explore the $\nu_e/\bar{\nu}_e$ signals in JUNO from the light DM annihilation $4\,\mathrm{GeV} \le m_D \le 20\,\mathrm{GeV}$ and consider three typical DM annihilation channels $\chi\chi \to \nu\bar{\nu}, \tau^+\tau^-, b\bar{b}$. The JUNO sensitivities to the spin-independent (SI) DM-nucleon and spin-dependent (SD) DM-proton elastic scattering cross sections will be given for the three DM annihilation channels. This paper is organized as follows: In Sec. 2, we outline the main features of the DM captured by the Sun and give the produced $\nu_e/\bar{\nu}_e$ fluxes. In Sec. 3, we present the selection conditions and numerically calculate the corresponding $\nu_e/\bar{\nu}_e$ event numbers in JUNO. In Sec. 4, we analyze the expected atmospheric neutrino background and calculate the JUNO sensitivities to the DM direct detect cross sections. Finally, a conclusion will be given in Sec. 5.

2. Neutrinos from the DM annihilation in the Sun

A halo DM particle via elastic scattering with the solar nuclei may lose most of its energy and is trapped by the Sun [1]. On the other hand, each DM annihilation in the Sun will deplete two DM particles. The evolution of the DM number N in the Sun can be written as:

$$\dot{N} = C_\odot - C_A N^2 \,, \tag{1}$$

where the dot denotes differentiation with respect to time. The DM solar capture rate $C_\odot$ in Eq. (1) is proportional to the DM-nucleon (DM-proton) elastic scattering cross section σ_n^{SI} (σ_p^{SD}) in the SI (SD) interaction case. In the next paragraph, we shall give the corresponding formulas to calculate $C_\odot$. The last term $C_A N^2$ in Eq. (1) controls the DM annihilation rate in the Sun. The coefficient C_A depends on the the thermally averaged annihilation cross section times the relative velocity $\langle\sigma v\rangle$ and the DM distribution in the Sun. To a good approximation, one can obtain $C_A = \langle\sigma v\rangle/V_{eff}$, where $V_{eff} = 5.8 \times 10^{30}$ $\mathrm{cm}^3(1\mathrm{GeV}/m_D)^{3/2}$ is the effective volume of the core of the Sun. In Eq. (1), we have neglected the evaporation effect since this effect is very small when the DM mass $m_D \ge 4$ GeV. One can easily solve the evolution equation and derive the DM solar annihilation rate [1]

$$\Gamma_A = \frac{1}{2} C_A N^2 = \frac{1}{2} C_\odot \tanh^2(t_\odot \sqrt{C_\odot C_A}) \,, \tag{2}$$

where $t_\odot \simeq 4.5$ Gyr is the solar age. If $t_\odot\sqrt{C_\odot C_A} \gg 1$, the DM annihilation rate reaches equilibrium with the DM capture rate. In this case, one may derive the maximal DM annihilation rate $\Gamma_A = C_\odot/2$ which means that the DM annihilation signals will only depend on σ_n^{SI} or σ_p^{SD}.

In the following parts, we shall use $C_\odot^{\mathrm{SI}}$ and $C_\odot^{\mathrm{SD}}$ to represent the solar capture rate $C_\odot$ in the SI and SD interaction cases, respectively. $C_\odot^{\mathrm{SI}}$ and $C_\odot^{\mathrm{SD}}$ may be approximately written as [1]

$$C_\odot^{\mathrm{SI}} \approx 4.8 \times 10^{24}\mathrm{s}^{-1} \frac{\rho_0}{0.3\,\mathrm{GeV/cm^3}} \frac{270\,\mathrm{km/s}}{\bar{v}} \frac{1\mathrm{GeV}}{m_D} \sum_i F_i \frac{\sigma_{N_i}^{\mathrm{SI}}}{10^{-40}\mathrm{cm}^2} f_i \phi_i S\left(\frac{m_D}{m_{N_i}}\right) \frac{1\mathrm{GeV}}{m_{N_i}}, \tag{3}$$

$$C_\odot^{\mathrm{SD}} \approx 1.3 \times 10^{25}\mathrm{s}^{-1} \frac{\rho_0}{0.3\,\mathrm{GeV/cm^3}} \frac{270\,\mathrm{km/s}}{\bar{v}} \frac{1\mathrm{GeV}}{m_D} \frac{\sigma_p^{\mathrm{SD}}}{10^{-40}\mathrm{cm}^2} S\left(\frac{m_D}{m_p}\right), \tag{4}$$

where the local DM density $\rho_0 = 0.3\,\mathrm{GeV/cm^3}$ and the local DM root-mean-square velocity $\bar{v} = 270\,\mathrm{km/s}$. f_i and ϕ_i describe the mass fraction and the distribution of the element N_i in the Sun, respectively. f_i, ϕ_i, the form-factor suppression $F_i(m_D)$ and the kinematic suppression $S(x)$ can be found in Ref. [1]. We shall sum over the following elements in the Sun: ^{1}H, ^{4}He, ^{12}C, ^{14}N, ^{16}O, ^{20}Ne, ^{24}Mg, ^{28}Si, ^{32}S and ^{56}Fe. The SI DM-nucleus elastic scattering cross section $\sigma_{N_i}^{\mathrm{SI}}$ is related to the SI DM-nucleon scattering cross section σ_n^{SI} by the formula $\sigma_{N_i}^{\mathrm{SI}} = \sigma_n^{\mathrm{SI}} A_{N_i}^2 M^2(N_i)/M^2(n)$ [8], where A_{N_i} is the mass number of the nucleus N_i and $M(x) = m_D m_x/(m_D + m_x)$. Here we have assumed that the DM couplings to protons and neutrons are isospin-invariant. Assuming $\sigma_n^{\rm SI} = \sigma_p^{\rm SD} = 10^{-40}$ cm^2, we calculate $C_\odot^{\rm SI}$ and $C_\odot^{\rm SD}$ as shown in Fig. 1. It is found that $C_\odot^{\rm SI}$ from other elements in the Sun is much larger than that from the hydrogen element although it has the maximal mass fraction. In the SD case, the hydrogen element plays the dominant role.

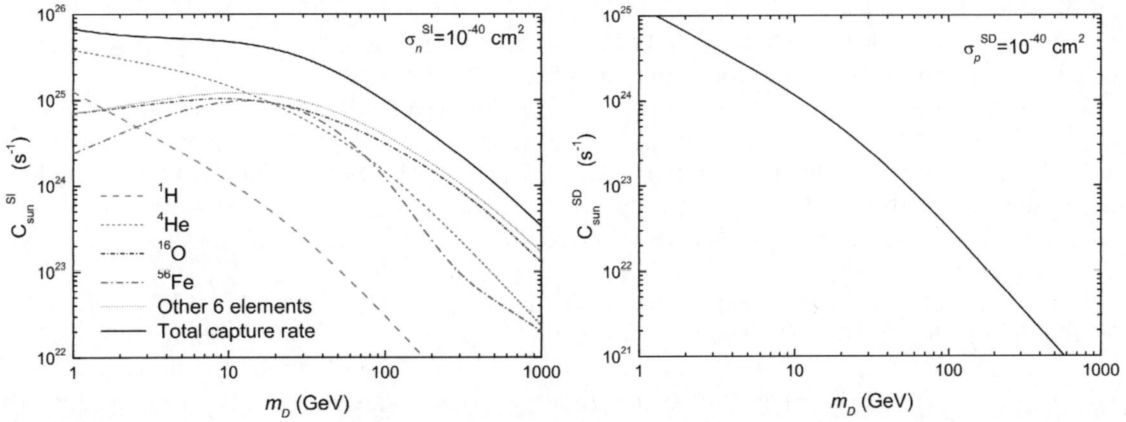

Figure 1. The DM solar capture rates $C_\odot^{\rm SI}$ (left) and $C_\odot^{\rm SD}$ (right) as a function of m_D with $\sigma_n^{\rm SI} = \sigma_p^{\rm SD} = 10^{-40}$ cm^2. The related contributions to $C_\odot^{\rm SI}$ from ^{1}H, ^{4}He, ^{16}O, ^{56}Fe and other 6 elements have also been shown.

Considering the usual $\langle\sigma v\rangle \approx 3\times10^{-26}$ cm^3 s^{-1} induced from the observed DM relic density, we can obtain $C_\odot \geq 8.6\times10^{22}/(m_D/1{\rm GeV})^{3/2}{\rm s}^{-1}$ [9] from $t_\odot\sqrt{C_\odot C_A} \geq 3.0$ which means $\tanh^2[t_\odot\sqrt{C_\odot C_A}] \geq 0.99$ in Eq. (2). In terms of the results in Fig. 1, we may safely assume that the DM annihilation rate reaches equilibrium with the DM capture rate. Then the electron neutrino flux at the surface of the Earth from the solar DM annihilation can be written as:

$$\frac{d\Phi_{\nu_e}}{dE_\nu} = \frac{\Gamma_A}{4\pi R_{\rm ES}^2}\frac{dN_{\nu_e}}{dE_\nu} \simeq \frac{C_\odot}{8\pi R_{\rm ES}^2}\frac{dN_{\nu_e}}{dE_\nu}\,, \tag{5}$$

where $R_{\rm ES} = 1.496\times10^{13}$ cm is the Earth-Sun distance. dN_{ν_e}/dE_ν is the differential electron neutrino energy spectrum from per DM pair annihilation in the Sun. In order to calculate dN_{ν_e}/dE_ν, one must consider the hadronization, interactions and decay processes of the DM annihilation final states in the core of the Sun. In addition, we should consider the neutrino interactions in the Sun and neutrino oscillations. Here we use the program package WimpSim [10] and the neutrino oscillation parameters in Eq.(2.8) of Ref. [7] to calculate dN_{ν_e}/dE_ν and $dN_{\bar\nu_e}/dE_\nu$ for three typical DM annihilation channels $\chi\chi \to \nu\bar\nu, \tau^+\tau^-, b\bar b$ and 4 GeV $\leq m_D \leq$ 20 GeV. Note that $\nu_e\bar\nu_e$, $\nu_\mu\bar\nu_\mu$ and $\nu_\tau\bar\nu_\tau$ have the same contributions in the $\nu\bar\nu$ channel.

3. Electron neutrinos and antineutrinos in JUNO

Here we shall discuss the $\nu_e/\bar\nu_e$ signals in JUNO from the DM $\nu\bar\nu$, $\tau^+\tau^-$ and $b\bar b$ annihilation channels. The JUNO central detector holds 20 kton LS which will be in a spherical container of radius of 17.7 m [3]. According to the detector properties, we have made a Monte Carlo (MC) simulation based on the GENIE generator [11] and the Geant4 detector simulation [12]. For the $\nu_e/\bar\nu_e$ charged current (CC) interactions in the JUNO detector, the MC simulation can provide many useful information which includes the event visible energy E_{vis}, the $e^\pm$ visible energy E_{vis}^e, the initial neutrino direction, the final state $e^\pm$ direction, etc. The JUNO can reconstruct E_{vis} with a very excellent energy resolution $\sigma_{E_{vis}}$. Note that the $\sigma_{E_{vis}}$ will be neglected in the following analysis since it has the very slight effects. On the other hand, the JUNO can also reconstruct the single muon direction with the angular resolution better than 1° if its track length $L_\mu > 5$ m and intrinsic PMT timing resolution better than 4 ns [3]. For the single electron/positron track, the 50 kton LS detector LENA find that the angular resolution is a few degrees [4]. Note that the hadronic final states in the ν_e and $\bar{\nu}_e$ CC interactions will affect the $e^\pm$ angular resolution and the identification of electron shower. The related studies are under way. Here we assume the $\nu_e/\bar{\nu}_e$ CC events with $E^e_{vis} > 1$ GeV and $Y_{vis} \equiv E^e_{vis}/E_{vis} > 0.5$ can be identified and reconstructed very well. For these selected events, the $e^\pm$ angular resolution will be assumed to be 10° in the following analysis.

For the DM $\nu\bar{\nu}$ annihilation channel, the WimpSim gives an approximate monoenergetic $\nu_e/\bar{\nu}_e$ spectrum. This is because that the initial monoenergetic $\nu_e/\bar{\nu}_e$ spectrum ($E_\nu = m_D$) will be slightly modified by the neutrino interactions in the Sun. The DM $\tau^+\tau^-$ and $b\bar{b}$ annihilation channels have the continuous spectra. In order to suppress the atmospheric neutrino backgrounds, we shall take different selection conditions for a given $\nu_e/\bar{\nu}_e$ energy E_ν in the monoenergetic and continuous $\nu_e/\bar{\nu}_e$ spectrum cases. Then two selection efficiencies $\epsilon(E_\nu)$ will be obtained from the MC simulation data. Except for $E^e_{vis} > 1$ GeV and $Y_{vis} \equiv E^e_{vis}/E_{vis} > 0.5$, we only select the events with $\theta_{\rm sun} < 20^\circ\sqrt{10{\rm GeV}/E_\nu}$ [1] and $1 \geq E_{vis}/E_\nu > 0.9$ for the monoenergetic spectrum case. Here $\theta_{\rm sun}$ denotes the angle between the initial neutrino direction (the Sun direction) and the reconstructed $e^\pm$ direction. Based on the $e^\pm$ initial direction and 10° angular resolution, one can easily calculate the angle $\theta_{\rm sun}$ for every MC simulation event. For the continuous spectrum case, $E^e_{vis} > 1$ GeV, $Y_{vis} > 0.5$, $\theta_{\rm sun} < 30^\circ$ and $1\,{\rm GeV} < E_{vis} < E_\nu$ will be chosen. Then we calculate the corresponding selection efficiencies $\epsilon(E_\nu)$ for a given energy $\nu_e/\bar{\nu}_e$ in the monoenergetic and continuous spectrum cases. It is found that $\epsilon(E_\nu)$ will not obviously change for $E_\nu > 3$ GeV as shown in Fig. 2 of Ref. [7]. The selection efficiencies are $\sim 40\%$ and $\sim 60\%$ for the monoenergetic and continuous spectrum cases, respectively.

For a given DM mass m_D, the expected $\nu_e/\bar{\nu}_e$ CC event numbers from the DM annihilation in the Sun can be expressed as

$$N_S = N_n\, t \int_{E_{th}}^{m_D} \left[\frac{d\Phi_{\nu_e}}{dE_\nu}\sigma_{\nu_e} + \frac{d\Phi_{\bar{\nu}_e}}{dE_\nu}\sigma_{\bar{\nu}_e}\right] \epsilon(E_\nu)\, dE_\nu \,, \tag{6}$$

where the total nucleon number $N_n \simeq 20\,{\rm kton}/m_n$ and m_n is the nucleon mass. In the following parts, we shall take the JUNO exposure time $t = 10$ years. Here we extract ν_e ($\bar{\nu}_e$) per nucleon CC cross section σ_{ν_e} ($\sigma_{\bar{\nu}_e}$) from the GENIE [11] and consider that the LS target includes 12% ^{1}H and 88% ^{12}C. σ_{ν_e} and $\sigma_{\bar{\nu}_e}$ can also be found in the left panel of Fig. 7-5 in Ref. [3]. In Eq. (6), the integral lower limit $E_{th} = 0.9 m_D$ ($E_{th} = 1$ GeV) for the DM $\nu\bar{\nu}$ channel ($\tau^+\tau^-$ and $b\bar{b}$ channels) will be chosen. With the help of Eq. (5), Eq. (6) and the efficiencies $\epsilon(E_\nu)$ in Fig. 2 of Ref. [7], one can calculate the expected event numbers N_S for different DM masses and annihilation channels. For illustration, we plot N_S as a function of m_D in the left panel of Fig. 2 with $\sigma^{\rm SI}_n = 10^{-40}$ cm^2 and $\sigma^{\rm SD}_p = 10^{-40}$ cm^2.

4. Sensitivities to the DM direct detection cross sections

For the $\nu_e/\bar{\nu}_e$ signals from the DM annihilation in the Sun, the related backgrounds are due to the atmospheric neutrino CC and neutral current (NC) interactions in the JUNO. The atmospheric $\nu_e/\bar{\nu}_e$ CC events are the irreducible background. The atmospheric neutrino NC background mainly origins from the π^0 misidentification as $e^\pm$. As shown in Fig. 7-5 of Ref. [3], the NC event rate is about half of the $\nu_e/\bar{\nu}_e$ CC event rate for $E_{vis} > 1$ GeV. Note that the $e^\pm$ takes on average about 60% of the initial $\nu_e/\bar{\nu}_e$ energy in the CC interaction. For the NC interaction, several hadronic particles will usually share E_{vis}. Considering the requirement $E^e_{vis} > 1$ GeV and the π^0 misidentification rate, we shall neglect the atmospheric neutrino NC background and only calculate the atmospheric $\nu_e/\bar{\nu}_e$ CC background.

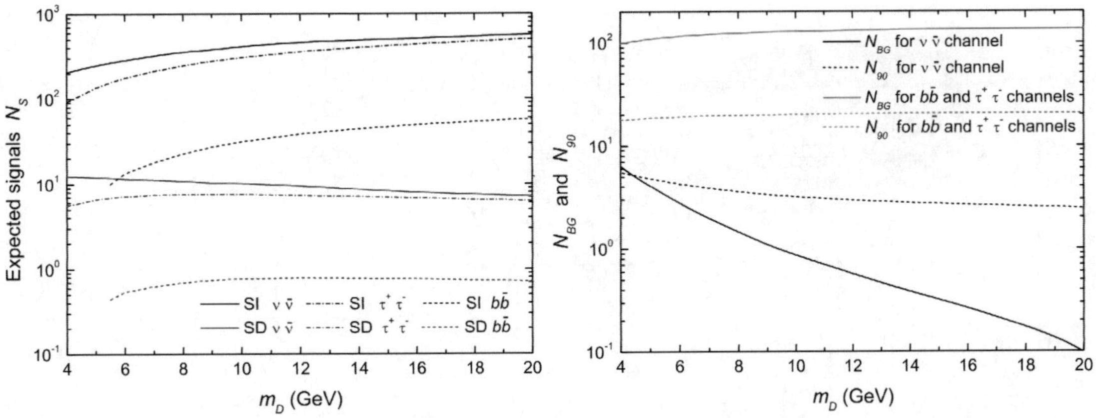

Figure 2. Left panel: the expected signals N_S from the DM $\nu\bar{\nu}$, $\tau^+\tau^-$ and $b\bar{b}$ channels for the $\sigma_n^{\rm SI} = 10^{-40}$ cm^2 and $\sigma_p^{\rm SD} = 10^{-40}$ cm^2 cases. Right panel: the atmospheric $\nu_e/\bar{\nu}_e$ CC backgrounds N_{BG} and the deduced 90% CL upper limit N_{90} to N_S for the monoenergetic and continuous spectrum cases.

With the help of the GENIE generator and Geant4 detector simulation, 1.5 million $\nu_e/\bar{\nu}_e$ CC events in JUNO detector have been simulated [3]. On the other hand, we calculate the expected atmospheric $\nu_e/\bar{\nu}_e$ CC event numbers in every energy and zenith angle bin by use of the atmospheric neutrino fluxes [13] and the oscillation parameters in Eq. (2.8) of Ref. [7]. Comparing the event numbers per bin of the MC simulation and the corresponding theoretical values, we determine the weight value for every MC event. Then the expected atmospheric $\nu_e/\bar{\nu}_e$ CC sample can be obtained. For the DM $\nu\bar{\nu}$ channel, we apply the selection conditions $E^e_{vis} > 1$ GeV, $Y_{vis} > 0.5$ and $1 \geq E_{vis}/m_D > 0.9$ to calculate the corresponding atmospheric $\nu_e/\bar{\nu}_e$ CC event numbers from all directions. Then we average the all direction result and obtain the atmospheric $\nu_e/\bar{\nu}_e$ CC background N_{BG} within the cone half-angle $\theta_{\rm sun} < 20^\circ\sqrt{10{\rm GeV}/m_D}$. For the $\tau^+\tau^-$ and $b\bar{b}$ channels, $E^e_{vis} > 1$ GeV, $Y_{vis} > 0.5$, $1\,{\rm GeV} < E_{vis} < m_D$ and $\theta_{\rm sun} < 30^\circ$ will be used. In the right panel of Fig. 2, we plot the atmospheric $\nu_e/\bar{\nu}_e$ CC background N_{BG} as a function of m_D. It is clear that the $\nu\bar{\nu}$ channel has the smaller N_{BG} than the $\tau^+\tau^-$ and $b\bar{b}$ channels. This is because that we take $1\,{\rm GeV} < E_{vis} < m_D$ and a larger cone half-angle $\theta_{\rm sun}$ for the DM $\tau^+\tau^-$ and $b\bar{b}$ channels.

To estimate the JUNO sensitivities to the direct detection cross sections $\sigma_n^{\rm SI}$ and $\sigma_p^{\rm SD}$, we assume the observed event number $N_{obs} = N_{BG}$. Then the 90% confidence level (CL) upper limit N_{90} to the expected $\nu_e/\bar{\nu}_e$ signals N_S can be derived through the Poisson-based likelihood functions in Eqs. (4.1-4.2) of Ref. [7]. In the right panel of Fig. 2, we plot N_{90} as a function of m_D for the $\nu\bar{\nu}$, $\tau^+\tau^-$ and $b\bar{b}$ channels. It is found that N_{90} decreases from 5.6 to 2.5 as the DM mass m_D increases in the $\nu\bar{\nu}$ channel case. For the $\tau^+\tau^-$ and $b\bar{b}$ channels, N_{90} only slightly changes. The JUNO sensitivities to $\sigma_n^{\rm SI}$ and $\sigma_p^{\rm SD}$ can be derived from $N_S = N_{90}$ and Eq. (6). In Fig. 3, we plot the JUNO sensitivities as a function of m_D for the DM $\nu\bar{\nu}$, $\tau^+\tau^-$ and $b\bar{b}$ channels with 10 year running. Here the current experimental results from the direct detection experiments have also been shown. It is clear that the JUNO projected sensitivity to $\sigma_n^{\rm SI}$ is better than the direct detection experiments for the $\nu\bar{\nu}$ channel with $m_D < 6.5$ GeV. If the $\nu\bar{\nu}$ or $\tau^+\tau^-$ channel is dominant, the future JUNO results are very helpful for us to understand the tension between the DAMA annual modulation signal and other direct detection exclusions. In the SD case, the JUNO sensitivities to $\sigma_p^{\rm SD}$ from the $\nu\bar{\nu}$ and $\tau^+\tau^-$ channels are much better than the current PICO-2L, PICASSO and SIMPLE-II experimental limits.

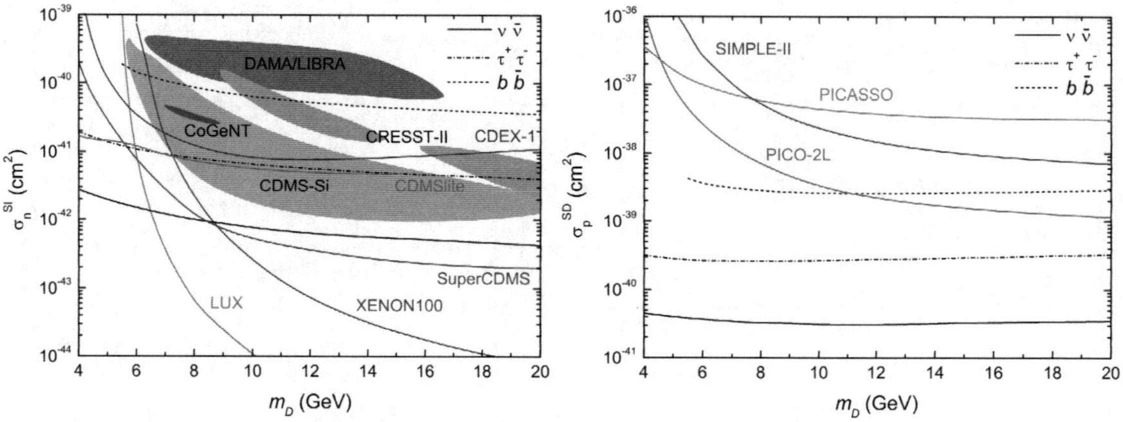

Figure 3. The JUNO sensitivities to $\sigma_n^{\rm SI}$ (left) and $\sigma_p^{\rm SD}$ (right) as a function of m_D for the DM $\nu\bar{\nu}$, $\tau^+\tau^-$ and $b\bar{b}$ annihilation channels with 10 year data.

5. Conclusions

In conclusion, we have investigated the ν_e and $\bar{\nu}_e$ CC signals in JUNO from the DM annihilation in the Sun. Three typical DM annihilation channels $\chi\chi \to \nu\bar{\nu}, \tau^+\tau^-, b\bar{b}$ have been analyzed for the light DM mass 4 GeV $\leq m_D \leq$ 20 GeV. In terms of the spectrum features, we take two sets of selection conditions for the monoenergetic ($\nu\bar{\nu}$ channel) and continuous ($\tau^+\tau^-$ and $b\bar{b}$ channels) spectra. The corresponding selection efficiencies $\epsilon(E_\nu)$ can be derived for a given $\nu_e/\bar{\nu}_e$ energy from the MC simulation data and 10° angular resolution. Then we numerically calculate the expected ν_e and $\bar{\nu}_e$ CC event numbers for the SI and SD interactions. On the other hand, we calculate the irreducible atmospheric $\nu_e/\bar{\nu}_e$ CC background which is far larger than the atmospheric neutrino NC background. Finally, we present the JUNO sensitivities to $\sigma_n^{\rm SI}$ and $\sigma_p^{\rm SD}$ for the DM $\nu\bar{\nu}$, $\tau^+\tau^-$ and $b\bar{b}$ annihilation channels with 10 year running. It is found that the JUNO projected sensitivities to $\sigma_p^{\rm SD}$ are much better than the current experimental upper limits for the $\nu\bar{\nu}$ and $\tau^+\tau^-$ channels. For the SI case, the JUNO will give the better sensitivity to $\sigma_n^{\rm SI}$ than the current direct detection experiments for $m_D < 6.5$ GeV in the $\nu\bar{\nu}$ channel case.

Acknowledgments

This work is supported in part by the National Nature Science Foundation of China (NSFC) under Grants No. 11575201 and the Strategic Priority Research Program of the Chinese Academy of Sciences under Grant No. XDA10010100.

References

[1] Jungman G, Kamionkowski M and Griest K 1996 *Phys. Rept.* **267** 195
[2] Ade P A R *et al.* 2016 *Astron. Astrophys.* **594** A13
[3] An F *et al.* 2016 *J.Phys. G* **43** 030401
[4] Wurm M *et al.* 2012 *Astropart. Phys.* **35** 685
[5] Kumar J, Learned J G, Sakai M and Smith S 2011 *Phys. Rev. D* **84** 036007
[6] Kumar J and Sandick P 2015 *J. Cosmol. Astropart. Phys.* JCAP06(2015)035
[7] Guo W L 2016 *J. Cosmol. Astropart. Phys.* JCAP01(2016)039
[8] Guo W L and Wu Y L 2010 *J. High Energy Phys.* JHEP10(2010)083
[9] Guo W L, Liang Z L and Wu Y L 2014 *Nucl. Phys. B* **878** 295
[10] Blennow M, Edsjo J and Ohlsson T 2008 *J. Cosmol. Astropart. Phys.* JCAP01(2008)021
[11] Andreopoulos C *et al.* 2010 *Nucl. Instrum. Meth. A* **614** 87
[12] Agostinelli S *et al.* 2003 *Nucl. Instrum. Meth. A* **506** 250
[13] Honda M, Athar M S, Kajita T, Kasahara K and Midorikawa S 2015 *Phys. Rev. D* **92** 023004

Study Of EoS Dependence of SNe via Relic Supernova Neutrino Spectrum

J Hidaka[1,2], T Kajino[2,3,4] and G J Mathews[2,5]

[1] Mechanical Engineering Department, Meisei University, 2-1-1 Hodokubo, Hino, Tokyo 191-8506, Japan
[2] National Astronomical Observatory of Japan 2-21-1 Osawa, Mitaka, Tokyo, 181-8588, Japan
[3] International Research Center for Big-Bang Cosmology and Element Genesis, and School of Physics and Nuclear Energy Engineering, Beihang University, Beijing 100191, China
[4] Department of Astronomy, School of Science, the University of Tokyo, 7-3-1 Hongo, Bunkyo-ku, Tokyo, 113-0033, Japan
[5] Center for Astrophysics, Department of Physics, University of Notre Dame, Notre Dame, IN 46556, USA

E-mail: [1]jun.hidaka@meisei-u.ac.jp

E-mail: [2]kajino@nao.ac.jp

E-mail: [5]gmathews@nd.edu

Abstract.
The energy spectrum of relic supernova neutrinos (RSNs) tells us valuable information about neutrinos emitted from the core-collapsed SNe and their cosmological evolution. Recent astronomical observations and theoretical studies of SNe give new insights on stellar evolution, which also influence supernova explosion mechanism. The RSN energy spectrum is an important tool to investigate these findings. We show the results of the RSN spectrum based on a variety of astronomical scenarios, which include different supernova occurance, the cosmic star formation history, and metallicity dependent initial mass function. They reveal the signature of nuclear EoS dependence, which appears robustly in spite of the different scenarios.

1. Introduction

Recent progress of the numerical simulation provides insight into the SN explosion mechanism. This includes the complicated picture about the criteria for successful explosions in terms of their progenitor mass [1, 2]. We have previously assumed that the successful explosions occur within certain mass range. Many numerical investigations, however, have shown the density profile of progenitor is essential to determine their final outcome, and so-called compactness parameter is proposed to be a key to understand physics behind this situation[3]. We have been studying the relic supernova neutrino spectrum for different types of SNe, and showed that the clear EoS dependence in the spectrum[4, 5]. According to these new finding, it is important to re-investigate the EoS dependence of the RSN spectrum.

Moreover, the RSN spectrum highly depends upon the cosmic star formation rate (SFR). The observational estimation of the SFR relies mostly on the UV light from galaxies. It is, however, pointed out by the recent study that the SFR based on UV underestimates the contribution from the starburst galaxies, especially at high-z, and a new cosmic SFR is proposed. Therefore it is interesting to see the consequence in RSN spectrum with this SFR.

Star formation rate also depends upon the choice of the initial mass function (IMF), which is usually assumed universal. There is another possibility that IMF varies according to the metallicity. In general the metal works as a coolant, and metal-poor molecular cloud tends to form more massive stellar objects selectively leading to the top-heavy IMF. This could be the case at high-z, and it is worth studying the effect of z-dependent IMF in the RSN spectrum.

In this article we investigate the EoS dependence in the RSN spectrum by applying a variety of scenarios described above and show that the EoS dependence is robust enough to be a useful tool to get insight of the nuclear EoS for the supernova explosion.

2. Detection Rate of RSN

The detection rate of the RSN is given as follows. For each type of SNe, the neutrino number flux per unit energy is written as

$$\frac{dN_\nu}{dE_\nu} = \frac{c}{H_0}\int_0^{z_{max}} R_{SN}(z)\frac{dN_\nu(E'_\nu)}{dE'_\nu} \times \frac{dz}{\sqrt{\Omega_m(1+z)^3+\Omega_\Lambda}}, \tag{1}$$

where we adopt $z_{max} = 5$ as the redshift at which star formation begins, and R_{SN} is the cosmic supernova rate. $dN_\nu(E'_\nu)/dE'_\nu$ is the neutrino spectrum emitted at the source. $E'_\nu = (1+z)E_\nu$ is the energy at emission, and E_ν is the redshifted energy observed in the detector. We adopt $H_0 = 70$ km s^{-1} Mpc^{-1}, $\Omega_m = 0.3$, $\Omega_\Lambda = 0.7$ [3, 6]. Fermi-Dirac distribution for all neutrino species is assumed and parametrized by the effective neutrino temperature and the total energy of emitted neutrinos as summarized in Table 1. We choose two models of fSNe with the different EoS, a soft EoS (LS-EoS [7]) and a stiff EoS (Shen-EoS [8, 9]). The supernova rate R_{SN} is given

Table 1. Parameters for the various neutrino sources.

Detail	ONeMg	SNe(optically luminous)	fSNe (Shen EoS)	fSNe (LS EoS)
mass ($M_\odot$)	8-10	10-40(74%)	10-40(26%), 40-125	10-40(26%), 40-125
$T_{\nu_e}(MeV)$	3.0 ± 0.3	3.2 ± 0.32	7.25 ± 0.73	5.58 ± 0.56
$T_{\bar{\nu}_e}(MeV)$	3.6 ± 0.36	4.0 ± 0.50	8.25 ± 0.83	6.67 ± 0.67
$T_{\nu_x}(MeV)$	3.6 ± 0.36	6.0 ± 0.60	9.27 ± 0.93	8.22 ± 0.82
$E^{total}_{\nu_e}(erg)$	3.3×10^{52}	5.0×10^{52}	13.5×10^{52}	4.4×10^{52}
$E^{total}_{\bar{\nu}_e}(erg)$	2.7×10^{52}	5.0×10^{52}	12.7×10^{52}	3.7×10^{52}
$E^{total}_{\nu_x}(erg)$	1.1×10^{52}	5.0×10^{52}	5.3×10^{52}	1.9×10^{52}

by

$$R_{SN}(z) = \Psi_*(z) \times \frac{\int_{M_{SN}} dM\, \phi_0(M)}{\int_{M_{min}}^{M_{max}} dM\, M\phi_0(M)}\ , \tag{2}$$

where $\Psi_*(z)$ is the cosmic SFR, and $\phi_0(M)$ is the initial mass function (IMF). M_{SN} is the range of progenitor masses for a type of SN. We assume $M_{min} = 0.1M_\odot$ and $M_{max} = 125M_\odot$. According to these formulae, we can recognize the neutrino flux depends not only upon the cosmic SFR and the IMF, but also upon the mass range of the progenitors for each type of SNe. The event rate at the detector is given by

$$\frac{dN_{\rm event}}{dE_{e^+}} = N_{\rm target}\frac{1}{c}\frac{dN_\nu}{dE_\nu}\sigma(E_\nu)\frac{dE_\nu}{dE_{e^+}}, \tag{3}$$

where N_{target} is the number of target particles in the detector. $\sigma(E_\nu)$ is the cross section for neutrino interactions within the detector [10]. We assume a water Čerenkov detector, and $\bar{\nu}_e + p \rightarrow e^+ + n$ is considered as the dominant reaction. $E_\nu = E_{e^+} + 1.3$ MeV is used.

As noticed above, the RSN detection rate depends upon SFR, IMF, and M_{SN}. In this work, we adopt a cosmic SFR parametrized by Madau & Dickinson[11], and it is given by

$$\Psi_*(z) = 0.015 \frac{(1+z)^{2.7}}{1+[(1+z)/2.9]^{5.6}} \ \mathrm{M_\odot year^{-1} Mpc^{-3}}. \tag{4}$$

We use this SFR for the feducial case, in which Salpeter-A IMF is also adoped and written as

$$\phi_0(M) = M^{-\zeta} = M^{-(1+\Gamma)} \tag{5}$$

with $\zeta = 2.35$ for stars with $M \geq 0.5 M_\odot$ and $\zeta = 1.5$ ($\Gamma = 0.5$) for stars with $0.1 < M < 0.5 M_\odot$[12]. Since each type of SNe emits neutrinos differently, it is crucial to determine the mass range of progenitor for a type of SNe. Most studies of the RSN have been assuming some definitive mass ranges within which a certain type of SNe occurs. Recent numerical simulations, however, show more complicated pictures about the relationship between the progenitors and SN explosions. Here for the fiducial investigation, we adopt a case such that the fraction of fSN, $f_{\mathrm{fSN}} = 0.26$ for the progenitors in the mass range of $10-40\mathrm{M_\odot}$ and $f_{\mathrm{fSN}} = 1.00$ for $\mathrm{M} > 40\mathrm{M_\odot}$ based on the study by Pejcha & Thompson[1] (see Table 1).

3. Results with a variety of the astronomical scenarios

Figure 1 shows the detector event rate of the fiducial case by assuming 10 years run of the Hyper-Kamiokande detector. No neutrino oscillation is considered. The uncertainty bands are based upon the observed cosmic SFR. Rectangular regions represent the peak of the spectrum and the locations where the neutrino signals start overwhelming the atmospheric neutino noise. As noticed from the figure, the EoS dependence is clearly seen in the high energy tail of the spectrum as the well-separated rectangles. The locations of these rectangular simbols are used in order to characterize the EoS dependence for other cases described below.

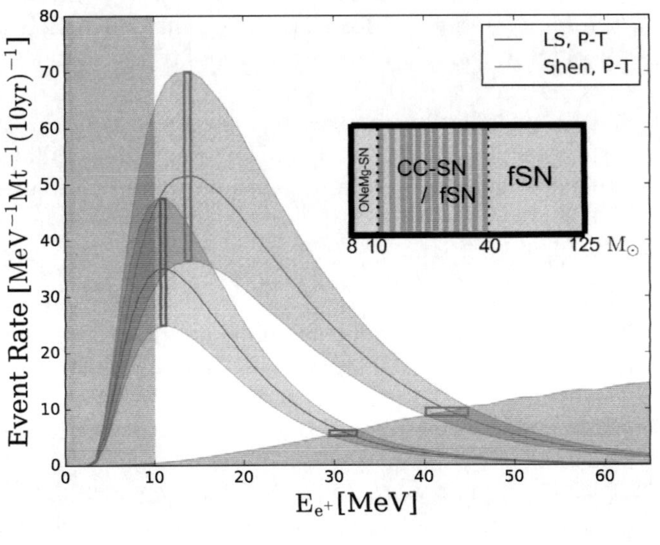

Figure 1. Predicted e^+ energy spectra and uncertainties in the total SRN detections for the fiducial case. The shaded energy range below 10 MeV indicates the region where the background noise due to reactor $\bar{\nu}_e$ may dominate. The shaded energy range that intersects the spectrum at ~ 30 to 46 MeV indicates the region where the background may be dominated by noise from atmospheric neutrinos. The figure includes the schematic representation for the the progenitor-SNe relationship.

Besides the fiducial case explained above, we investigate many astronomical possibilities, which include the different SFR, starburst and quiescent star formation phase, and metallicity-dependent IMF. Generally SFR is obtained based on the assumption of an IMF. The correction on SFR is, therefore, necessary when we apply the different IMF. In the followings we consider four different cases. All results are included in figure 2 with only rectangular simbols at both the peaks and tails. The fiducial case is labeled by "PT".

First we investigate the senario which adopts the different progenitor-SNe relationship which is suggested by so-called the red supergiant problem. In this case, we consider $10 - 18M_\odot$ progenitor stars to become CCSNe, and progenitors in $18 - 125M_\odot$ are taken to be as fSNe. They are labeled by "RSG" in the figure.

As the second case, we consider two phase star formation history. Recently N-body cosmological simulation identifies starburst and quiescent star formation phases, and it reveals that the starburst is dominant phase at high-z while the quiescent one becomes dominant at low-z[13]. Moreover the study shows that top-heavy IMF ($\zeta = 2$ (or $\Gamma = 1$)) is required in order to recover the SFR by Madau & Dickinson. This result is physically reasonable. The environment of star formation at high-z could be metal poor and lack of coolant for the molecular clouds to be filamented into small scale. This might lead to the top-heavy IMF at high-z. The result of this cosmologically motivated scenario is labeled by "PT, Lacey-SB-Q" in figure 2.

Next we enploy another cosmic SFR, which has higher star formation rate comparing with the SFR by Madau & Dickinson especially at high-z. As mentioned earlier, the RSN spectrum is affected by the SFR very much, and we expect to see the difference in the spectrum. This SFR is derived by newly interpreted data based on the idea that the UV luminosity from the starburst galaxies could underestimate the SFR[14]. The analysis of this work also identifies the starburst and quiescent phases. Thus we apply the same IMF used in the second case for each star formation phase. The results are denoted by "PT, IMF-Lacey, RR-SB-Q" in figure 2.

As the last case, we adopt a varying IMF, which is metallicity dependent. Recent observational study of nearby galaxies shows the clear metallicity dependence in the IMF[15], and it is formulated by

$$\Gamma = 2.2(\pm 0.1) + 3.1(\pm 0.5) \times [\mathrm{M/H}]. \tag{6}$$

We apply this varying IMF, instead of Salpeter-A IMF, to the fiducial case. We also adopt the galaxy mass function[16] and the mass-metallicity relation[17] so that we have a z-dependent IMF. Figure 2 includes the results labeld by "PT, varying IMF".

According to all cases in figure 2, we can see the robustness of EoS dependence in the RSN spectrum. The locations of rectangles in the high energy tail for cases of LS and Shen EoS are well-separated each other. Thus, the RSN spectrum could become a good tool to investigate the nuclear EoS in the supernova explosion.

In addition to the cases without the neutrino oscillation (see figure 2), we also consider the RSN spectrum with two possible cases of the neutrino ocillation[4, 5]. The neutrino oscillation of a normal or inverted mass hierarchy associated with complete non-adiabatic mixing is given by

$$\phi_{\bar{\nu}_e} = 0.7 \times \phi^0_{\bar{\nu}_0} + 0.3 \times \phi^0_{\bar{\nu}_x}, \tag{7}$$

and the RSN spectra of this case are shown in figure 3. Figure 4 shows the results with another cases of the neutrino oscillation, in which an inverted mass hierarchy associated with complete adiabatic mixing through the MSW high-density resonance is considered. It is given by

$$\phi_{\bar{\nu}_e} = \phi^0_{\bar{\nu}_x}, \tag{8}$$

where the effect from shock wave propagation is ignored. Figure 2-4 clearly show the EoS dependence in the RSN spectrum in the high energy tail, though the robustness for the case of inverted mass hierarchy becomes slightly weaker since the separation of the locations of rectangles for LS and Shen EoS is smaller than those of two other cases.

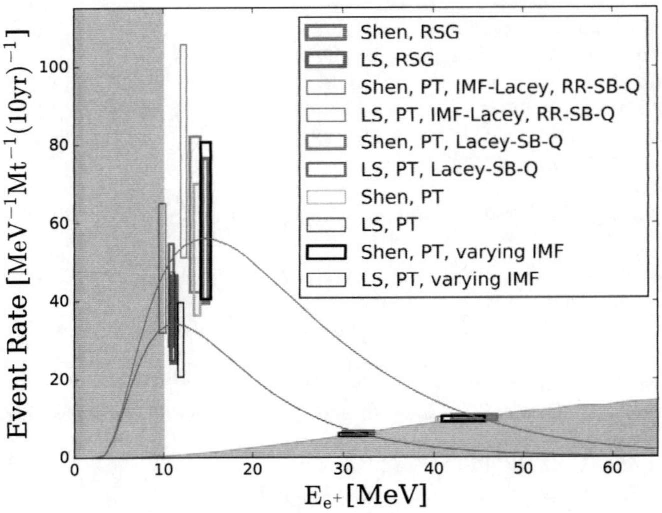

Figure 2. Predicted e^+ energy spectra for all cases without neutrino oscillation. The figure shows only rectangules for the characteristic locations of both peaks and tails for each cases. As the eye-guide, the lines are included only for the fiducial cases with the best fitted SFR.

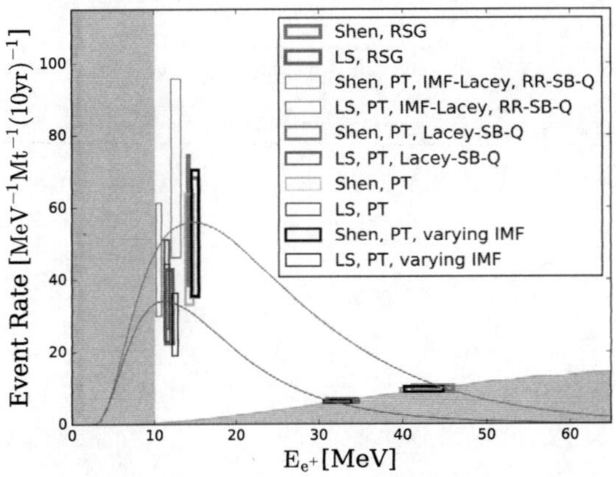

Figure 3. Same as figure 2 but with the neutrino oscillation by equation 7.

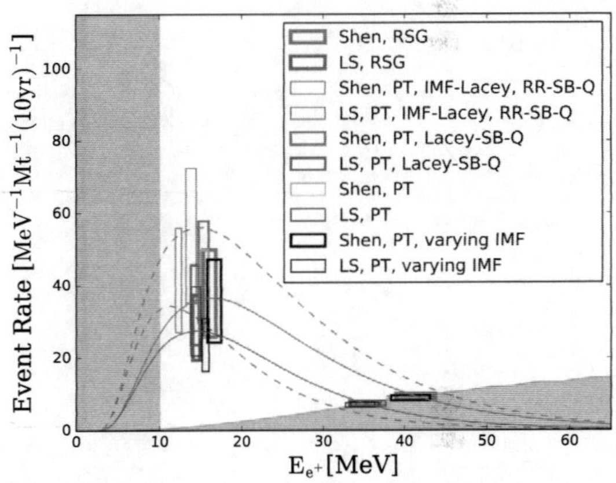

Figure 4. Same as figure 2 but with the neutrino oscillation by equation 8. The dashed lines, the fiducial cases in figure 3 are also included for comparison.

4. Conclusion

We investigate the RSN spectrum with a variety of astronomical scenarios. We enploy the different progenitor-SNe relationship, the cosmic SFR, the star formation phase of starburst and quiescent, and the metallicity-dependent IMF. We found that the appearance of EoS dependence in the spectrum is clearly seen in the high energy tail. We also consider the effects of the neutrino oscillations. Although we have slightly weaker EoS dependence for the case of the inverted mass hierarcy, the EoS dependence still robust to be a good tool in order to get valuable knowledge about the nuclear EoS in the supernova explosion.

References

[1] Pejcha O and Thompson T A 2015 *Astrophys. J.* **801** 90

IOP Conf. Series: Journal of Physics: Conf. Series **1056** (2018) 012026 doi:10.1088/1742-6596/1056/1/012026

[2] Ugliano M, Janka H T, Marek A and Arcones A 2012 *Astrophys. J.* **757** 69
[3] Horiuchi S, Beacom J, Kochanek C *et al.* 2011 *Astrophys. J.* **738** 154
[4] Mathews G, Hidaka J, Kajino T and Suzuki J 2014 *Astrophys. J.* **790** 115
[5] Hidaka J, Kajino T and Mathews G J 2016 *Astrophys. J.* **827** 85
[6] Hinshaw Gand Larson D, Komatsu E *et al.* 2013 *Astrophys. J. Suppl.* **208** 19
[7] Lattimer J M and Swesty F D 1991 *Nucl. Phys. A* **535** 331
[8] Shen H, Toki H, Oyamatsu K and Sumiyoshi K 1998 *Nucl. Phys. A* **637** 435
[9] Sumiyoshi K, Yamada S and Suzuki H 2008 *Astrophys. J.* **688** 1176
[10] Strumia A and Vissani F 2003 *Phys. Lett. B* **564** 42
[11] Madau P and Dickinson M 2014 *Annual Review of Astronomy and Astrophysics* **52** 415
[12] Baldry I K and Glazebrook K 2003 *Astrophys. J.* **593** 258
[13] Lacey C G, Baugh C M, Frenk C S, Benson A J, Bower R G, Cole S, Gonzalez-Perez V, Helly J C, Lagos C D P and Mitchell P D 2016 *MNRAS* **462** 3854
[14] Rowan-Robinson M, Oliver S, Wang L, Farrah D, Clements D L, Gruppioni C, Marchetti L, Rigopoulou D and Vaccari M 2016 *MNRAS* **461** 1100
[15] Martín-Navarro I, Vazdekis A, La Barbera F, Falcón-Barroso J, Lyubenova M, van de Ven G, Ferreras I, Sánchez S F, Trager S C, García-Benito R, Mast D, Mendoza M A, Sánchez-Blázquez P, González Delgado R, Walcher C J and Team T C 2015 *Astrophys. J.* **806**
[16] Lopes A R, Iribarrem A, Ribeiro M B and Stoeger W R 2014 *A&A* **572**
[17] Ma X, Hopkins P F, Faucher-Giguère C A, Zolman N, Muratov A L, Kereš D and Quataert E 2016 *MNRAS* **456** 2140

IOP Conf. Series: Journal of Physics: Conf. Series **1056** (2018) 012027 doi:10.1088/1742-6596/1056/1/012027

Open problems in neutrino physics

Francesco Iachello

Center for Theoretical Physics, Sloane Laboratory
Yale University, New Haven, Connecticut 06520-8120, USA

E-mail: francesco.iachello@yale.edu

Abstract. Three open problems in neutrino physics are mentioned. A brief review of neutrino-less double beta decay (DBD) is given. The question of quenching of g_A is discussed. Other possible scenarios, Majoron emission, sterile neutrinos and non-standard mechanisms, are presented and their implication to DBD discussed.

1. Introduction

Although neutrinos were introduced over 80 years ago, their properties remain to a large extent unknown. Three unanswered questions in neutrino physics are:

- What is the absolute mass scale of neutrinos? [1]
- Are neutrinos Dirac or Majorana particles? [2]
- How many neutrino species are there? [3]

An answer to the first question can be obtained by a measurement of:

(1) The end point of the electron spectrum in single beta decay (KATRIN) [4] ${}^{3}_{1}H_2 \rightarrow {}^{3}_{2}He_1^{+} + e^{-} + \overline{\nu}_e$.

(2) The end-point of the spectrum of single electron capture (ECHO) [5] ${}^{163}_{67}Ho_{96} + e^{-} \rightarrow {}^{163}_{66}Dy_{97}^{*i} + \nu_e \rightarrow {}^{163}_{66}Dy_{97} + E_i + \nu_e$.

An answer to the third question can be obtained by a study of neutrino oscillations in the range <1Km (Short-Baseline Neutrino, SBN, Program at FERMILAB).

An answer to all three questions can be obtained from neutrino-less double-beta decay (DBD) and related processes ${}^{A}_{Z}X_N \rightarrow {}^{A}_{Z+2}Y_{N-2} + 2e^{-}$.

2. Double beta decay

Double beta decay is a process in which a nucleus X decays into a nucleus Y with the emission of two electrons or positrons and eventually other particles

$$ {}^{A}_{Z}X_N \rightarrow {}^{A}_{Z\pm2}Y_{N\mp2} + 2e^{\mp} + anything \ . \qquad (1) $$

For processes not allowed by the standard model, the half-life is given by

$$ \left[\tau_{1/2}^{0\nu\beta\beta}\right]^{-1} = G_{0\nu}\left|M_{0\nu}\right|^2 \left|f(m_i,U_{ei})\right|^2 \qquad (2) $$

where $G_{0\nu}$ is a phase-space factor, PSF, which depends on atomic physics, $M_{0\nu}$ are the nuclear matrix elements, NME, which depend on nuclear physics, and the function f contains physics beyond the standard model. For processes allowed by the standard model, the half-life can be, to a good approximation, factorized in the form

$$\left[\tau_{1/2}^{2\nu\beta\beta}\right]^{-1} = G_{2\nu}\left|M_{2\nu}\right|^2 . \tag{3}$$

For all processes and to extract physics beyond the standard model, one needs to calculate the phase space factors (PSF) and the nuclear matrix elements (NME).

2.1. *Phase space factors*

PSF were calculated in the 1980's by Doi *et al.* [6]. Also a calculation of phase-space factors is reported in the book of Böhm and Vogel [7]. These calculations use an approximate expression for the electron wave functions at the nucleus. PSF have been recently recalculated [8] with exact Dirac electron wave functions and including screening by the electron cloud. These new PSF are available from jenni.m.kotila@jyu.fi and are on the webpage nucleartheory.yale.edu.

2.2. *Nuclear matrix elements*

NME for neutrino-less DBD can be written as:

$$M_{0\nu} = g_A^2 M^{(0\nu)}$$
$$M^{(0\nu)} \equiv M_{GT}^{(0\nu)} - \left(\frac{g_V}{g_A}\right)^2 M_F^{(0\nu)} + M_T^{(0\nu)} . \tag{4}$$

Several methods have been used for evaluation of $M_{0\nu}$: Quasiparticle Random Phase Approximation (QRPA), Shell Model (ISM), Interacting Boson Model (IBM-2), Density Functional Theory (DFT) among others. The same methods have also been used for evaluation of $M_{2\nu} = g_A^2 M^{(2\nu)}$.

For 0ν processes, two scenarios have been extensively considered: (1) Emission and re-absorption of a light ($m_{light} \ll 1 keV$) neutrino for which the f function and the neutrino potential are $f = \frac{\langle m_\nu \rangle}{m_e}$, $\langle m_\nu \rangle = \sum_{k=light} \left(U_{ek}\right)^2 m_k$ and $v(p) = \frac{2}{\pi}\frac{1}{p(p+\tilde{A})}$, where $\tilde{A}$ is the closure energy, $\tilde{A} = 1.12 A^{1/2} MeV$, also called long-range, and (2) Emission and absorption of a heavy neutrino ($m_{heavy} \gg 1 GeV$) with $f_h = m_p \left\langle \frac{1}{m_{\nu_h}} \right\rangle$, $\left\langle \frac{1}{m_{\nu_h}} \right\rangle = \sum_{k=heavy} \left(U_{ek_h}\right)^2 \frac{1}{m_{k_h}}$, and $v(p) = \frac{2}{\pi}\frac{1}{m_p m_e}$, also called long-range. Recent results for NME for some of the models are summarized in figures 1 and 2. By combining the PSF with the NME, one can set limits on neutrino masses. Figure 3 shows current limits on neutrino masses (light neutrino exchange) for KI-PSF [8], IBM-2 NME [9] and g_A=1.269.

2.3. *Quenching of* g_A

Results in figures 1 and 2 are obtained with g_A=1.269. It is well known from single beta decay and electron capture that g_A is renormalized in models of nuclei. Two reasons for this renormalization are: (i) the limited model space in which the calculation is done, giving rise to a quenching factor $q_{N_{ex}}$ and (ii) the omission of non-nucleonic degrees of freedom, giving rise to a quenching factor q_Δ .

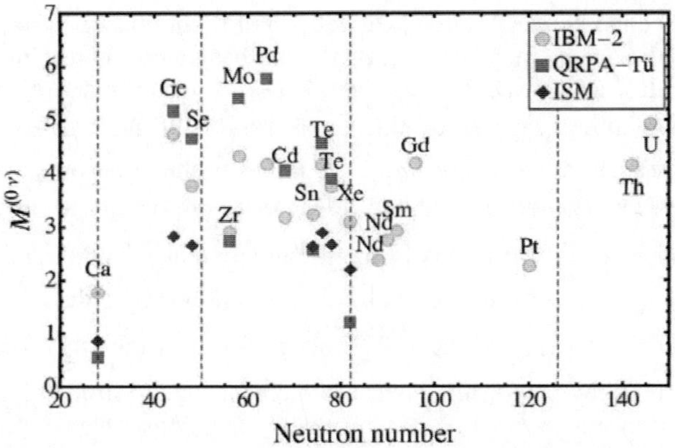

Figure 1. NME for light neutrino exchange in IBM-2 [9], QRPA-Tü [10], ISM [11]. IBM-2 and QRPA-Tü results are with isospin restoration and Argonne short-range correlations.

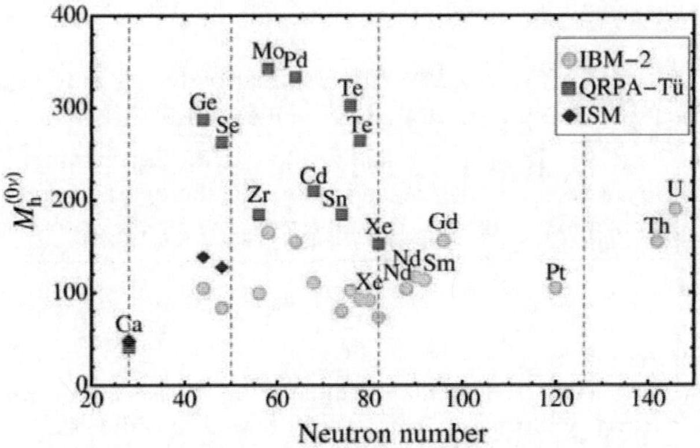

Figure 2. Same as fig. 1 for heavy neutrino exchange.

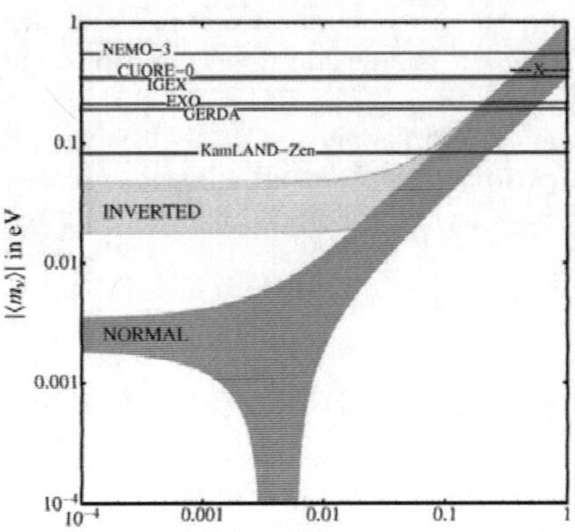

Figure 3. Current limits on neutrino masses from NEMO-3 [12], CUORE-0 [13], IGEX [14], EXO [15], GERDA [16] and KamLAND-Zen [17].

Since g_A appears to the second power in the NME, equation (5), and hence to the fourth power in the half-life, its quenching will have a dramatic effect on double-beta decay. Quenching of g_A in 2ν double beta decay, consistent with single-beta decay, has been observed (see figure 5 of [9]). The question of whether or not g_A in $0\nu\beta\beta$ is renormalized as much as in $2\nu\beta\beta$ is of much debate. The two processes differ by the momentum transferred to leptons. In $2\nu\beta\beta$ this is or the order of few MeV, while in $0\nu\beta\beta$ it is of the order of 100 MeV. The current view (2017) is that both factors, q_Δ and $q_{N_{ex}}$, contribute to $2\nu\beta\beta$, while only q_Δ contributes to $0\nu\beta\beta$. This is due to the fact that, while the missing non-nucleonic degrees of freedom have a scale of excitation set by $m_\Delta - m_p \cong 300MeV$, the missing nuclear degrees of freedom have a scale set by $\langle m_{N_{ex}} \rangle \cong 10MeV$. This problem is currently being addressed from various sides. Experimentally, by measuring the matrix elements to and from the intermediate odd-odd nucleus in $2\nu\beta\beta$ decay by means of single charge exchange reactions (^{3}He, t) [18], and theoretically, by using effective field theories (EFT) to estimate the effect of non-nucleonic degrees of freedom (two-body currents) [19]. Very recently, an experimental program (NUMEN) [20] has been set up at LNS in Catania, Italy to measure both single and double charge exchange reaction intensities with heavy ions. This program will also provide useful information on the Fermi and Gamow-Teller matrix elements of interest in $0\nu\beta\beta$ and $2\nu\beta\beta$ decay.

Results in figure 3 are with g_A=1.269. Three scenarios for g_A are [21]: (i) g_A= 1.269 (free value obtained from neutron decay), g_A=1 (quark value), and (iii) $g_A = 1.269A^{-0.18}$ (maximal quenching obtained from $2\nu\beta\beta$ [9]). If g_A is renormalized to maximal quenching, all estimates for half-lives should be increased by a factor of ~4-34 and limits on the average neutrino mass should be increased by a factor ~1.6-6, making it very difficult to reach in the foreseeable future even the inverted region.

3. Other possible scenarios

Possibilities to escape this negative conclusion are: (1) neutrino masses are degenerate and large. This possibility will be in tension with the cosmological bound on the sum of neutrino masses, $\sum_i m_i \leq 0.230eV$ [22]. (2) Other scenarios (Majoron emission, sterile neutrinos, ...) must be considered. Here Majoron means a massless neutral boson [23] and sterile means a neutrino with no standard model interaction [3]. (3) Other non-standard mechanisms contribute [24]. Here non-standard means mechanisms other than V-A.

The scenario (2) of sterile neutrinos is currently being extensively investigated both experimentally, with planned experiments at FERMILAB and CERN-LHC, and theoretically. NME for sterile neutrinos of arbitrary mass can be calculated by using a transition operator as in Section 2, but with $f = \frac{m_{\nu I}}{m_e}$ and

$$v(p) = \frac{2}{\pi} \frac{1}{\sqrt{p^2 + m_{\nu I}^2}\left(\sqrt{p^2 + m_{\nu I}^2} + \tilde{A}\right)}, \tag{5}$$

where $m_{\nu I}$ is the effective mass of sterile neutrinos. IBM-2 matrix elements for this scenario have been calculated [25]. PSF are the same as in Section 2. Contribution of all hypothetical neutrinos can be written in the general form [25]

$$\left[\tau_{1/2}^{0\nu}\right]^{-1} = G_{0\nu}\left|\begin{array}{l}\left[\frac{1}{m_e}\sum_{k=1}^{3}U_{ek}^2 m_k + \frac{1}{m_e}\sum_{i}U_{ei}^2 m_i + \frac{1}{m_e}\sum_{j}U_{ej}^2 m_j\right]M^{(0\nu)} \\ +\left[m_p\sum_{N}U_{eN}^2\frac{m_N}{\langle p^2\rangle + m_N^2} + m_p\sum_{k_h=1}^{3}U_{ek_h}^2\right]M^{(0\nu_h)}\end{array}\right|^2, \quad (6)$$

where the first three terms represent the contribution of light neutrinos (known with m_ν<0.1eV, unknown with m_i~1eV, unknown with m_j~1keV) and the last two terms represent contributions of heavy neutrinos (unknown with mass m_N~1MeV , unknown with mass $m_{\nu h}$>1GeV).
Several types of sterile neutrinos have been suggested. (i) In scenario a, heavy sterile neutrinos with masses in the keV-GeV range have been suggested in [26]. (ii) In scenario b, Giunti and Laveder have suggested light sterile neutrinos with masses in the eV range to account for the reactor anomaly in oscillation experiments [27]. The presence of sterile neutrinos changes completely the picture, as shown in figure 4. With sterile neutrino (with properties of scenario b [27]) and g_A=1.269, the inverted hierarchy is reachable by GERDA-PHASE II and CUORE.

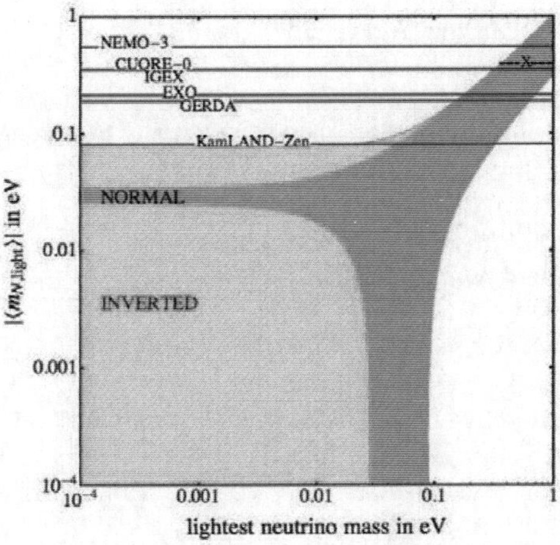

Figure 4. Limits on neutrino masses when sterile neutrinos are included. Figure courtesy of J. Kotila adapted from [25].

Scenario 3 of non-standard mechanisms is also being investigated both long-range, with Lagrangian [28]

$$L_{long} = \frac{G_F}{\sqrt{2}}\left[J_{V-A,\mu}^{\dagger} j_{V-A}^{\mu} + \sum_{\alpha,\beta}\varepsilon_{\alpha\beta}J_{\alpha}^{\dagger}j_{\beta}\right], \quad (7)$$
$$\alpha,\beta = S \pm V, V \pm A, T \pm T_5$$

and short-range, with Lagrangian [29]

$$L_{short} = \frac{G_F^2}{\sqrt{2}}\left[\varepsilon_1 JJj + \varepsilon_2 J^{\mu\nu}J_{\mu\nu}j + \varepsilon_3 J^{\mu}J_{\mu}j + \varepsilon_4 J^{\mu}J_{\mu\nu}j^{\nu} + \varepsilon_5 J^{\mu}Jj_{\mu}\right]. \quad (8)$$

Predictions and limits for these mechanisms will be available in early 2018.

4. Conclusions

No matter what the mechanism of neutrino-less DBD is, its observation will answer the fundamental questions posed in the introduction. Indeed, if observed, neutrino-less DBD may provide evidence for

IOP Conf. Series: Journal of Physics: Conf. Series **1056** (2018) 012027 doi:10.1088/1742-6596/1056/1/012027

physics beyond the standard model other than the mass mechanism. Conversely, its non-observation will set stringent limits on other scenarios (sterile, ...), and on non-standard mechanisms. In this sense, neutrino-less DBD is a *search for lepton number violation* rather than a measurement of the neutrino mass.

Acknowledgements
This work was supported in part by U.S.D.O.E. Grant DE-FG02-91ER40608.

References
[1] Fermi E 1934 *Z. Physik* **88** 161
[2] Majorana E 1937 *Nuovo Cimento* **14** 171
[3] Pontecorvo B 1968 *Sov. Phys. JEPT* **26** 984
[4] Drexlin J *et al.* 2013 *Adv. High Energy Phys.* **2013** 293986
[5] Faessler A, Gastaldo L and Šimkovic F 2015 *J. Phys. G: Nucl. Part. Phys.* **42** 015108
[6] Doi M, Kotani T, Nishiura N, Okuda K and Takasugi E 1981 *Prog. Theor. Phys.* **66** 1739
[7] Böhm F and Vogel P 1987 *Physics of massive neutrinos* (Cambridge: Cambridge University Press)
[8] Kotila J and Iachello F 2012 *Phys. Rev.* C **85** 034316
[9] Barea J, Kotila J and Iachello F 2015 *Phys. Rev.* C **91** 034304
[10] Šimkovic F, Rodin V, Faessler A and Vogel P 2013 *Phys. Rev.* C **87** 045501
[11] Menendez J, Poves A, Caurier E and Nowacki F 2009 *Nucl. Phys.* A **818** 139
[12] Arnold R *et al.* (NEMO-3) 2006 *Nucl. Phys.* A **765** 483
[13] Alfonso K *et al.* (CUORE-0) 2013 *Phys. Rev. Lett.* **111** 122503
[14] Aalseth *et al.* (IGEX) 2002 *Phys. Rev.* D **65** 092007
[15] Auger M *et al.* (EXO) 2014 *Nature* **510** 229
[16] Agostini M *et al.* (GERDA) 2017 *Nature* **544** 47
[17] Gando A *et al.* (KamLAND-Zen) 2016 *Phys. Rev. Lett.* **117** 082503
[18] Puppe P *et al.* 2012 *Phys. Rev.* C **86** 044603
[19] Menendez J, Gazit D and Schwenk A 2011 *Phys. Rev. Lett.* **107** 062501
[20] Cappuzzello F *et al.* 2015 *Journal of Physics Conference Series* **630** 012018
Cappuzzello F, Agodi C *et al.* 2017 *The NUMEN Project at LNS:Status and Perspectives* Proceedings MEDEX17, in press
[21] Dell'Oro S, Marcocci S and Vissani F 2014 *Phys. Rev.* D **90** 033005
[22] Ade PAR *et al.* (PLANCK) arXiv:1303.5076
Mattarese S (PLANCK) 2015 *Proc. XVI Int. Workshop NEUTEL 2015*
[23] Kotila J, Barea J and Iachello F 2015 *Phys. Rev.* C **91** 064310
[24] Deppisch F, Hirsh M and Päs H 2012 *Neutrinoless Double Beta Decay and Physics Beyond the Standard Model* arXiv:1208.0727v1[hep-ph]
[25] Barea J, Kotila J and Iachello F 2015 *Phys. Rev.* D **92** 093001
[26] Asaka T and Shaposhnikov M 2005 *Phys. Lett.* B **620** 17
Asaka T, Blanchet S and Shaposhnikov M 2005 *Phys. Lett.* B **631** 151
[27] Giunti C and Laveder M 2010 *Phys. Rev.* D **82** 053005
Giunti C, Laveder M, Li YF and Long HW 2013 *Phys. Rev.* D **88** 073008
Giunti C 2015 *Proc. XVI Int. Workshop NEUTEL 2015*
[28] Ali A, Borisov AV and Zhuridov DV 2008 *Probing New Physics in the Neutrinoless Double Beta Decay Using Electron Angular Correlation* arXiv:0706.4165v3[hep-ph]
[29] Graf L and Iachello F 2017 in preparation

IOP Conf. Series: Journal of Physics: Conf. Series **1056** (2018) 012028 doi:10.1088/1742-6596/1056/1/012028

New results on the ^{8}Be anomaly

A J Krasznahorkay*, M Csatlós, L Csige, Z Gácsi, J Gulyás, Á Nagy, N Sas, J Timár, T G Tornyi, I Vajda
Institute for Nuclear Research, Hungarian Academy of Sciences, MTA Atomki

A Krasznahorkay
CERN, Geneva, Switzerland

E-mail: * kraszna@atomki.hu

Abstract.
Recently, we observed anomalous internal pair creation for the M1 transition depopulating the 18.15 MeV isoscalar 1^+ state in ^{8}Be. We observed a significant (σ = 7.37) peak-like deviation from the predicted angular correlation of the e^+e pairs at $\theta = 140^o$. To the best of our knowledge no nuclear physics related description of such deviation can be made. However, the deviation can be described by assuming the creation and subsequent decay of a boson with mass of ≈17 MeV. In order to clarify the interpretation, we re-investigated the ^{8}Be anomaly with an improved, and independent setup. We have confirmed the signal of the assumed X(17) particle and constrained its mass (m_0c^2 = 17.01(16) MeV) and branching ratio compared to the γ-decay ($B_x = 6(1) \times 10^{-6}$).

1. Introduction

Our current knowledge of Nature at the fundamental level is successfully explained by the Standard Model (SM) of elementary particles, but this theory describes only five percent of the entire content of the Universe. We named the overwhelming unknown constituents as dark matter and dark energy. The nature of dark matter is currently one of the greatest unsolved mysteries in physics.

The search for new physics beyond the SM can be divided into two categories. The first one is the search for new heavy particles and interactions at high energies, the so-called energy frontier research. A complementary and vital role is played by low-energy, precision and/or high-intensity experiments, which require a joint effort of the particle, atomic, and nuclear physics communities [1, 2]. One of the main targets of the intensity frontier in particle physics is a new force carrier which is much lighter than the weak scale and which interacts with the Standard Model particles very weakly. Observations of rare nuclear transitions can be used to search for new hidden force-carrier particles at the MeV scale. In this approach, a fixed target is bombarded with a hadron beam to produce excited states of a nucleus. The excited state then decays also by emitting new particles.

Recently, we have observed an anomaly in the nuclear decay of ^{8}Be [3]. The ^{7}Li(p,γ)^{8}Be reaction was used to populate the excited states in ^{8}Be selectively and the differential internal pair conversion coefficients were studied for the 17.6 MeV, and 18.15 MeV ($J^\pi = 1^+ \rightarrow 0^+$) M1 transitions in ^{8}Be. Significant, peak-like enhancement of the internal pair creation was observed at large angles in the angular correlation of the 18.15 MeV transition, but not in the 17.6 MeV one [3]. This observation was interpreted as the creation and the subsequent decay of an X boson with mass m_0c^2=16.70±0.35(stat)±0.5(sys) MeV. The branching ratio of the e^+e^- decay of such a boson to the γ decay of the 18.15 MeV level of ^{8}Be is found to be 5.8×10^{-6} [3].

Zhang and Miller [4] have discussed in detail a possible nuclear physics origin of the anomaly. Therefore, they have improved the existing model of e$^+$e production in the current experimental context, by including the interferences between E1, E2, and M1 multipoles and by considering two different angular dependencies, and introducing also important constraints from the photon production measurements. These improvements have to be taken into account in extracting new particle properties from such experiments, but can not explain the observed anomaly [4].

J. Feng and his group [5, 6] studied our data along with other previous experiments and showed that our observation strongly disfavors dark photons. Instead, they proposed a new theory and explained the observation by introducing a fifth fundamental force [5]. If confirmed by further experiments, this discovery of a possible fifth force would completely change our understanding of the universe, by taking a step towards the unification of forces and dark matter.

The possible relation of the X boson to the dark matter problem and the fact that it might explain the $(g\text{-}2)_\mu$ puzzle, triggered an enhanced theoretical and experimental interest in the particle and hadron physics community [7]. Inspired by this enhanced interest, we re-measured the effect in different and independent experimental conditions.

2. Experiments

To populate the 18.15 MeV 1^+ state in ^{8}Be selectively, we used the ^{7}Li(p,γ)^{8}Be reaction at the E_p=1030 keV resonance [8]. The average energy loss of protons in the target was 70 keV, so the actual proton energy was 1100 keV. The experiment was performed at the new 2-MV Tandetron accelerator in Debrecen. A proton beam with a typical current of 1.0 μA impinged on 300 μg/cm^2 thick Li target evaporated on 20 $\mu g/cm^2$ thick carbon foil. In contrast to our previous experiment [3, 9], we used a much thinner ^{12}C backing and we increased the number of telescopes (from 5 to 6), which resulted in different electron detection efficiency as a function of the correlation angle. As a considerable improvement, we replaced the gas-filled MWPC detectors with a double-sided silicon strip detector (DSSSD) array.

The e^+e^- pairs were detected by six plastic scintillator + DSSSD detector telescopes placed perpendicularly to the beam direction at azimuthal angles of 0°, 60°, 120°, 180° 240° and 300°. The size of the scintillators is 82×86× 80 mm^3. The positions of the hits were registered by the double-sided silicon strip detectors having strip widths of 3 mm. The telescope detectors were placed around the vacuum chamber made of a carbon fibre tube with a wall thickness of 1 mm.

γ rays were also detected for monitoring purposes. A ϵ_{rel}=100% HPGe detector was used at 25 cm from the target to detect the 18.15 MeV γ rays produced in the ^{7}Li(p,γ)^{8}Be reaction.

3. Efficiency calibration of the e^+e^- spectrometer

The well-known, strong 6.05-MeV IPC transition ($0^+ \rightarrow 0^+$, E0) following the ^{19}F$(\mathrm{p},\alpha)^{16}$O reaction was applied to perform the energy calibration of the spectrometer.

The efficiency calibration of the telescopes was made by using the same dataset but with uncorrelated e^+e^- pairs of consecutive events. Accordingly, an energy independent approximation of the efficiency curve could be extracted.

Such an approximation is accurate when only the central sections of the telescopes are in use. This condition was automatically fulfilled in our previous experiments, when MWPC detectors were used with an effective area of 30x30 mm^2.

However, the size of the DSSSD detectors used in the present experiment is 50x50 mm^2 resulting a high probability of event loss when one of the particle escapes from the scintillator.

IOP Conf. Series: Journal of Physics: Conf. Series **1056** (2018) 012028 doi:10.1088/1742-6596/1056/1/012028

It results in an efficiency reduction near the surface of the scintillator causing minor deviations in the efficiency curve. Thus, the energy dependence of the efficiency calibration was simulated by the GEANT3 code (for the same e^+e^- sum-energy gate as we used in the experimental data reduction) and taken into account as a correction for the experimentally determined efficiency curve.

The efficiency curve differs considerably for the present and previous setups, as shown in figure 1. Therefore, the present results could be considered as an independent measurement in the sense that any geometry-related systematic effect is eliminated from the measured data.

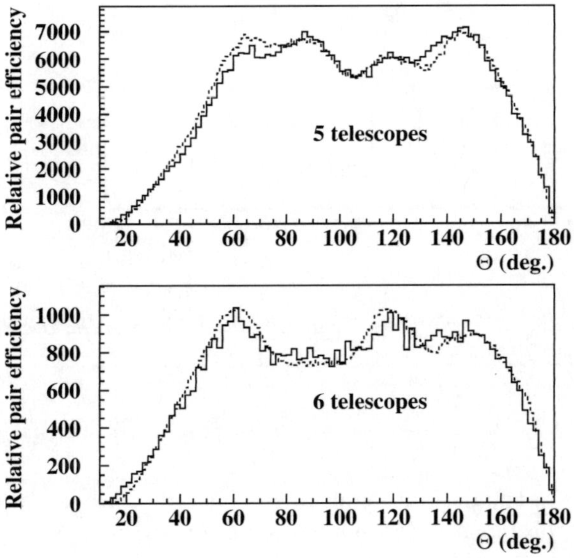

Figure 1. *Measured relative e^+e^- pair detection efficiency for the spectrometer with 5 telescopes (upper histogram) and 6 telescopes (lower histogram) compared to the results of the corresponding Monte Carlo simulations (dotted lines).*

4. Subtraction of the background caused by cosmic rays

Since the Q-value of the ^{7}Li(p,γ)^{8}Be reaction is exceptionally large (Q= 17.25 MeV [8]), we were not expecting any background transitions up to 18 MeV. This expectation was also supported by our experimental data obtained for the high energy γ spectra of this energy region, which was completely empty.

However, in an investigation of such rare processes, the cosmic ray background has to be been taken into account. The background was measured for two weeks, before and after the experiment, and was subtracted with an experimentally determined factor from the results by using the same gates and conditions as for the in-beam data. The subtraction factor was derived by setting a high-energy gate (E(sum)=25-50 MeV) on the cosmic rays for both cases (in-beam and off-beam). The cosmic ray background subtraction was then performed until eliminating all events within the high-energy gate.

The shape of the cosmic-ray background angular correlations determined for the 18 MeV gate is shown in figure 2 for the 5 detector and 6 detector configurations.

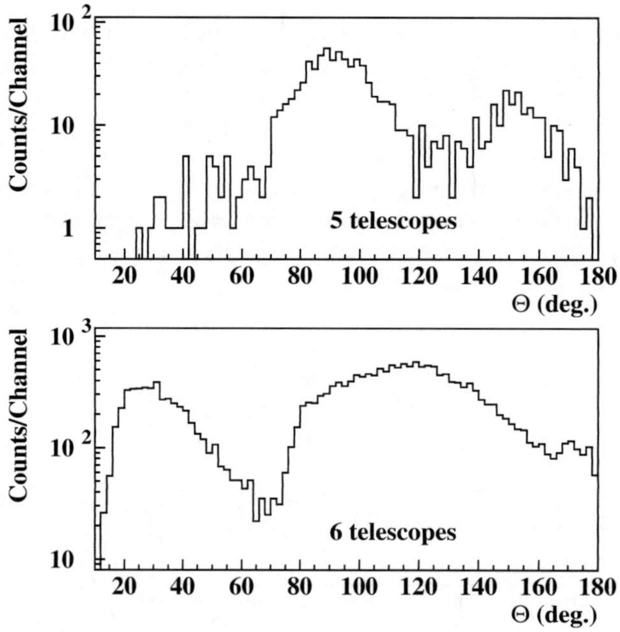

Figure 2. *Relative e^+e^- angular correlations measured for the cosmic rays, analysed in the same way as the 18.15 MeV transition for the spectrometer with 5 telescopes (upper histogram) and 6 telescopes (lower histogram).*

The yield of the cosmic-rays in the present experiment was two times larger compared to the previously published one. The previous experiment was performed in a laboratory situated in the basement of a building with three levels above, while the present one in a lab at ground level, having practically no shielding above.

In order to get a reduction of the cosmic-ray background an active shield were installed above the e^+e^- spectrometer, which was constructed from 12 peaces of 1.0 cm thick, 4.5 cm wide and 100 cm long plastic scintillators. Half of the yield of the cosmic rays could be suppressed this way.

5. Results

Figure 3 shows our experimental results (red dots with error bars) for the recent angular correlation of e^+e^- pairs together with our previous results (blue dots with error bars) [3] measured at the proton absorption resonance at E_p= 1030 keV. There is a very good agreement between the two independent sets of experimental data.

6. Fitting the measured angular correlations

The e^+e^- angular correlation distribution is described as a sum of an exponentially falling background distribution modeled by the IPC simulation, and the signal distribution modeled by the simulation of a boson decaying to e^+e^- pairs.

The fit was performed with RooFit [10] by describing the e^+e^- angular correlation

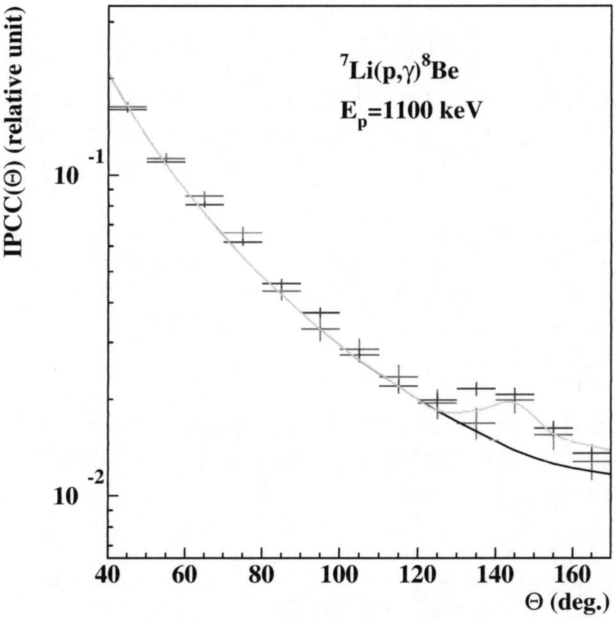

Figure 3. *Measured angular correlations published previously [3] (blue) and the present results (red) of the e^+e^- pairs originated from the decay of the 18.15 MeV ground state transition in ^8Be. The black line represents the background, while the green one is the sum of the signal and background.*

distribution with the following probability density function (PDF):

$$PDF(e^+e^-) = N_{Bkgd} * PDF(IPC) + N_{Sig} * PDF(signal) \ , \tag{1}$$

where N_{Bkgd} and N_{Sig} are the fitted number of background and signal events, respectively.

The signal PDF was constructed as a 2-dimensional model as a function of the e^+e^- opening angle and the mass of the simulated particle. To construct the mass dependence, the PDF linearly interpolates the e^+e^- opening angle distributions simulated for discrete particle masses.

The composite PDF, described in Equation 1, was used to perform a simultaneous fit of N_{Sig} , N_{Bkgd} and the simulated boson's mass. Which were used to calculate the branching ratios for the best fit as well. The results of the two fits are summarized in table 1.

Table 1. Results of the new fit for Exp1 [3] and for the present experiment Exp2.

	Previous res. [3]	Exp1	Exp2	Average
$m_0c^2(MeV)$	16.7(51)	16.86(6)	17.17(7)	17.01(16)
B_x	5.8	$6.8(10) \times 10^{-6}$	$4.7(21) \times 10^{-6}$	$6(1) \times 10^{-6}$
Significance	6.8σ	7.37σ	4.90σ	

The discrepancy in the particle masses of the two data sets could be a result of the unstable beam position in our previous experiment. According to MC simulations, such a mm order of beam position variation can cause a systematic uncertainty that cannot be neglected.

7. Conclusion

We have remeasured the e^+e^- angular correlation for the M1 transition depopulating the 18.15 MeV state in ^{8}Be. We could reproduce the peak-like deviation from the predicted IPC, confirming the signal of the new X(17) particle and constraining its mass ($m_0c^2 = 17.01(16)$ MeV) and branching ratio compared to the γ-decay ($B_x = 6(1) \times 10^{-6}$).

8. Acknowledgements

This work has been supported by the Hungarian NKFI Foundation No. K124810, by the GINOP-2.3.3-15-2016-00034 and by the János Bolyai Research Fellowship of the Hungarian Academy of Sciences (L. Csige).

References

[1] Pospelov M, Ritz A, and Voloshin M B 2008 *Phys. Lett.* B **662** 53
[2] Kitahara T and Yamamoto Y 2017 *Phys. Rev.* D **95** 015008
[3] Krasznahorkay A J et al., 2016 *Phys. Rev. Lett.* **116** 042501
[4] Zhang Xilin, Miller Gerald A 2017 *Phys. Lett.* B **773** 159
[5] Feng J et al., 2016 *Phys. Rev. Lett.* **117** 071803
[6] Feng J et al. 2017 *Phys. Rev.* D **95** 035017
[7] https://inspirehep.net/search?ln=en&p=refersto%3Arecid%3A1358248&jrec=26&sf=earliestdate
[8] Tilley D R et al. 2004 *Nucl. Phys.* A **745** 155
[9] Gulyás J et al. 2016 *Nucl. Instr. and Meth. in Phys. Res.* A **808**, 21
[10] Verkerke W and Kirkby D P 2003 *eConf* C **0303241** MOLT007 [physics/0306116]

Double charge-exchange reactions and the effect of transfer

J A Lay[1,†], S Burrello[2], J I Bellone[2,3], M Colonna[2] and H Lenske[4] within the NUMEN project

[1] Dpto. de Física Atómica, Molecular y Nuclear, Universidad de Sevilla, Sevilla, Spain
[2] Istituto Nazionale di Fisica Nucleare, Laboratori Nazionali del Sud, Catania, Italy
[3] Dipartimento di Fisica e Astronomia, Universit di Catania, Italy
[4] Institut für Theoretische Physik, University of Giessen, Giessen, Germany

E-mail: †lay@us.es

Abstract. There is a recently renewed interest on the study of single and double charge exchange reactions with heavy ions. We report here a preliminary theoretical study of double charge exchange reactions in terms of two successive charge exchanges in 2nd order Distorted Wave Born approximation. This allows us to include the effect of the different transfer channels in an approximate way. Evidences show that the charge exchange process is dominant. The following one in importance is the combination of one single charge exchange, one neutron pick-up and one proton stripping. This particular process calls for further investigation.

1. Introduction

Nuclear reactions are one of the most important tools to unravel the nature of the interaction between neutrons and protons inside the atomic nucleus. In principle, we can extract a lot of information by making collide a nucleus of interest with a light ion or with a heavy ion. In the latter case, the theoretical description of the reaction is largely more involved, since we have to deal with the nucleon-nucleon interaction of two different many-body systems. In spite of this extra difficulty, we find in the literature different cases in which we have improve our understanding of certain problems thanks to particular characteristics not available with light nuclei. As an example, we can cite the study of breakup of halo nuclei under extreme electric fields supplied by a heavy target [1] or the search of the Giant Pairing Resonance, where a large Q-value is needed and only reachable with a heavy projectile [2].

For the case of charge exchange reactions, we can study the single charge exchange (SCE) in the nucleus of interest with both light and heavy ions. However, performing double charge exchange (DCE) with light ions presents certain difficulties like dealing with the measurement and reconstruction of the trajectory of three different final projectile particles as we would have with a light ion like ^{3}He or t. We can use heavy ions to avoid such problems since both the target-like and projectile-like products of the reactions will be bound in most of the cases. In addition, the final cross section for double charge exchange will depend on the Fermi and Gamow Teller response of both the nucleus of interest (the target) and the heavy ion used as experimental probe in the reaction. By choosing carefully a heavy ion with a large Fermi or Gamow Teller response, we can increase the final cross section, helping to the extraction of the particular response of the nucleus of interest.

Nowadays, there is a large experimental interest on double charge-exchange reactions with heavy ions which has given origin to different campaigns mainly at the radioactive beam facilities of RCNP at Osaka and RIKEN in Japan [3, 4] and at the LNS-INFN of Catania in Italy [5, 6]. The former ones are focused on studying the Gamow-Teller response of nuclei and in the production of exotic nuclei through double charge exchange reactions. The latter is devoted to the study of double charge-exchange reactions in connection with the neutrinoless double-beta decay.

In the present contribution we will focus in one of the reactions performed at the LNS-INFN: 15 MeV/A ^{18}O projectiles impinging on a ^{40}Ca target [5]. After the double charge-exchange we will measure ^{18}Ne and ^{40}Ar in different states. Here, we do not consider any direct process leading from the initial to the final counterparts. This means that the nucleus-nucleus interaction cannot change two charges acting only once. Such a possibility is under investigation [7]. Therefore, the double charge-exchange is produced by a part of the effective nucleus-nucleus interaction able to change one charge at a time but acting twice, thus being called a second order process.

The final situation can be reached also through different combinations of neutron pick-up, proton stripping, and single charge-exchange processes. Any of these so-called competing processes is of a higher order than the double charge-exchange, so that this latter process should dominate. Nonetheless, at low bombarding energies, the differences can be too small to neglect the contribution of the competing processes. In any case, if we manage to control them we can have access to a complementary structure information of the nucleus considered which can be of great interest when studying double-beta decay candidates.

The study of the competition between charge exchange with heavy ion and the corresponding transfer at different bombarding energies has only been properly done for the case of single charge-exchange [8]. In single charge-exchange, the transfer consists of one neutron (proton) pick-up followed by a one proton (neutron) stripping (or vice versa) what means that can be solved in terms of 2nd order Distorted Wave Born Approximation (DWBA). For double charge-exchange we can have transfer channels involving four particles, which means the need of a 4th order DWBA. Non-orthogonality between prior and post representation in trasnfer generates a non-orthogonality contribution. This term, together with simultaneous and succesive contributions, contributes coherently to the final total cross section. Full 2nd order DWBA including simultaneous, succesive and non-orthogonality terms has been recently fully implemented, helping us to calculate exact cross sections for these cases, whereas for 4th order DWBA there is no implementation yet for the corresponding non-orthogonalities.

Therefore, our purpose is to preliminary asses the importance of the different contributions in order to study the possibility of extracting the desired response from the experimental data. We will estimate the transfer contribution for the ^{18}O + ^{40}Ca reaction at 15 MeV/A and we will compare with the double charge-exchange experimental data from [5].

2. Double charge-exchange and multinucleon transfer cross sections

On one hand, we only consider double charge exchange as a second order charge exchange, so that the form factors needed for the reaction calculation are those for the reactions: ^{18}O+^{40}Ca $\rightarrow$ ^{18}F+^{40}K and ^{18}F+^{40}K $\rightarrow$ ^{18}Ne+^{40}Ar. For the calculation of such form factors we can rely on a double folding procedure of a residual nucleon-nucleon interaction and the corresponding transition densities for target and projectile.

On the other hand, corresponding transfer contributions must be considered. The formalism of multinucleon transfer in DWBA has been known for several decades but only recently we have started to calculate second order DWBA transfer due to its computational demands. For the case of double charge exchange we deal with the transfer of two protons and two neutrons. That would mean the need of a 4th order DWBA which has no implementation yet. Since our purpose is to estimate the order of magnitude and its relative importance with respect to the charge-exchange, we can approximate the full cross section by the one obtained considering a second order DWBA two-neutron transfer to the ground states of ^{16}O+^{42}Ca followed by a second order DWBA two-proton transfer to the final products ^{18}Ne+^{40}Ar.

We have to add here the possibility of having one neutron and one proton transfer followed by a single charge exchange and vice versa. This process is of third order. Therefore, even though it can be against our intuition to mix two different responses of nuclei, it could be larger than the full fourth order transfer.

The structure input needed for the transfer are the different overlaps between the different target-like products in the different initial, intermediate and final situations considered. The same reads for the projectile like ones. For the sake of completeness, these overlaps should be provided by the same structure calculation providing transition densities for the charge exchange calculation. However, they are usually approximated by the overlap obtained with single particle wavefunctions multiplied by the corresponding spectroscopic factor. An alternative option is to look for different spectroscopic factors extracted from experimental transfer reactions on the considered nuclei.

The last ingredient for calculating the reaction cross section are the optical potential between the different counterparts of the initial, intermediate, and final situations. All of them can be obtained again by a double folding procedure with the corresponding central densities of projectile-like and target-like nuclei.

Finally, the total cross section is the coherent sum of all these contributions. Therefore, we cannot neglect the processes whose cross sections are small when computed separately. All of them interfere and, through this interference, their final impact on the total cross section can be larger than expected.

3. Preliminary Results

In order to tackle the ^{18}O + ^{40}Ca reaction at 15 MeV/A we combine two programs HIDEX [9] and FRESCO [10]. With HIDEX we calculate the single charge exchange form factor by folding QRPA transition densities of the different nuclei with the Love and Franey nucleon-nucleon interaction [11]. We approximate the ^{18}F+^{40}K $\rightarrow$ ^{18}Ne+^{40}Ar part by using the same form factors as in the first single charge exchange, i.e. ^{18}O+^{40}Ca $\rightarrow$ ^{18}F+^{40}K. However, including all the charge exchange strength for all possible multipolarities within FRESCO is very demanding computationally, so that in this preliminary approach we have constrain ourselves to the multipolarity leading to the ground states of the intermediate counterparts: ^{18}F(1^+) and ^{40}K(4^-). Then we increase the total strength of this specific part of the total response of the nuclei in order to be of the order of the experimental cross section at zero degrees. In this way the result should be approximate but still representative of the strength of the double charge exchange process and its comparison with the competing channel remains meaningful.

We use the same interaction to calculate the corresponding optical potentials. We include the form factors and optical potentials into FRESCO where we can add the different transfer contributions. For this last part, we rely on experimentally extracted spectroscopic factors [12, 13, 14, 15] for the different cases. Only ground states are considered for even-even or odd-odd nuclei. For the odd-even or even-odd ones, we consider all the single particle levels in the vicinity of the Fermi energy. Therefore, all these levels provide a large basis and we only lack of completeness in the even-even and odd-odd nuclei.

A final remark has to be done regarding the choice of prior or post interactions for each steps. We only consider the successive part of the transfer choosing post interactions for all the steps. In order to have the full transfer cross section, we will need to consider the different non-orthogonalities.The transfer leading to the same counterparts as DCE is a 4th order DWBA calculation. The corresponding non-orthoganlity terms have not been implemented so far. A calculation including only succesive terms with post interactions, as the calculation presented here, should be a good estimation and must reach the exact result if our basis is increased towards completeness [10].

In order to assess the relative importance of pure four particle transfer, the combination of transfer and a single charge exchange and the successive double charge exchange, we have performed three different calculation. In the first one we include all contributions. In the second one, only charge exchange and the combination of charge exchange and transfer. Finally, in a third calculation, only charge exchange is included.

We show the preliminary results in table 1. When only double charge exchange is considered, we get a differential cross section of 18.3 μb/sr at zero degrees. This value is an 8% larger than the calculation that includes all possible competing channels. This last calculation is entitled to be the most correct one and therefore this 8% can be understood as the level of accuracy found when neglecting the effect of transfer. However, if we consider the double charge exchange in combination with one-neutron and one-proton transfer channels, the difference with respect to the full calculation is reduced to less than a 5%.

Table 1. Preliminary calculations of the differential cross section at zero degrees and the total cross section for the different processes considered.

Processes considered	$\frac{d\sigma}{d\Omega}$ (0) (μb/sr)	Integrated σ (μb)
DCE only	18.3	0.060
DCE and (np transfer + SCE)	16.2	0.051
DCE and all transfers considered	17.0	0.054
Experimental [5]	11.1	0.072

This reduced difference is only present in the case of the differential cross section at zero degrees. The angular distribution also changes depending on the processes considered. In fact, the results for the total integrated cross sections show a larger difference as can also be found in table 1.

4. Conclusions

We have studied the double charge exchange reaction ^{18}O + ^{40}Ca→ ^{18}Ne+^{40}Ar at 15 MeV/A measured at the LNS-INFN [5] focusing in the relative importance of the different competing channels, i.e. double charge exchange and the different possible transfer combinations leading to the same final nuclei.

We have found that double charge-exchange is the leading process since there is only a difference of an 8% between the differential cross section at zero degrees calculated including all processes considered and the one calculated by considering only the double charge-exchange. This result was as expected according to the fact that this process is the lowest order one considered.

Among the processes involving transfer of particles, the combinations of one single charge exchange, one proton stripping and one neutron pick up is the following in importance. Considering this process and the double charge-exchange we obtain a differential cross section at zero degrees just a 5% different from the one obtained including all processes.

Finally, the 4th order multinucleon transfer is the lowest in importance. This is important since there is not a complete derivation and implementation of the corresponding non-orthogonality terms needed to properly calculate this contribution. This is also consistent with what found for the ^{20}Ne + ^{116}Cd reaction [16] where different calculations for the transfer of two neutron and two protons produce strongly suppressed cross sections.

The combinations of one single charge exchange, one proton stripping and one neutron pick up is a third order process. In order to do a full DWBA calculation one has to include non-orthogonality terms for the part that includes transfer. However, these non-orthogonalities are parallel to those of the second order process and its derivation should be much simpler than those for the 4th order transfer.

In any case, these conclusions are based on preliminary, not complete, calculations which should be just representative of the order of magnitude. Further investigations are demanded in order to corroborate these preliminary results. Moreover, a full derivation of the third order process arising from the combination of transfer and charge exchange could be of great interest in order to improve accuracy when comparing our structure models for neutrinoless double-beta decay candidates with the double charge-exchange reaction experimental data. Another issue to investigate is whether the choice of the heavy ion could reduce even more this transfer contribution. As shown in [17], the cross section for transfer of a particle depends of the alignment of its spin with its angular momentum in the initial and final occupied orbits. This feature is not seen in transfer with light ions since the particles transferred are initially in $s-$orbits. However, the relevance of the transfer can be very different if we use a heavy ion with valence neutrons and protons occupying a spin-aligned or a spin-antialigned orbit.

Acknowledgments

The research leading to these results received funding from Spanish Ministerio de Economía y Competitividad and FEDER funds under Project No. FIS2014-53448-C2-1-P, by the Ayudas a Consolidación de Grupos from Junta de Andalucía, and by the European Union's Horizon 2020 research and innovation program under Grant Agreement No. 654002.

References

[1] Pesudo V *et al.* 2017 *Phys. Rev. Lett.* **118** 152502 URL `https://link.aps.org/doi/10.1103/PhysRevLett.118.152502`
[2] Cappuzzello F *et al.* 2015 *Nature Communications* **6** 6743
[3] Kisamori K *et al.* 2016 *Phys. Rev. Lett.* **116** 052501 URL `https://link.aps.org/doi/10.1103/PhysRevLett.116.052501`
[4] Takaki M *et al.* 2014 *Heavy-Ion Double-Charge Exchange Study via a 12C(18O,18Ne)12Be Reaction* vol 6 p 02038 (*Preprint* `http://journals.jps.jp/doi/pdf/10.7566/JPSCP.6.020038`) URL `http://journals.jps.jp/doi/abs/10.7566/JPSCP.6.020038`
[5] Cappuzzello F, Cavallaro M, Agodi C, Bond M, Carbone D, Cunsolo A and Foti A 2015 *Eur. Phys. J. A* **51** 145 URL `https://doi.org/10.1140/epja/i2015-15145-5`
[6] Cappuzzello F *et al.* 2018 *Eur. Phys. J. A* **54** 72 URL `https://doi.org/10.1140/epja/i2018-12509-3`
[7] Lenske H 2018 *see this volume* **ibid**
[8] Lenske H, Wolter H H and Bohlen H G 1989 *Phys. Rev. Lett.* **62**(13) 1457–1460 URL `https://link.aps.org/doi/10.1103/PhysRevLett.62.1457`
[9] Lenske H, Bellone J I, Colonna M and Lay J A 2018 *Phys. Rev. C* (Submitted)
[10] Thompson I 1988 *Comp. Phys. Rep.* **7** 167
[11] Franey M A and Love W G 1985 *Phys. Rev. C* **31**(2) 488–498 URL `https://link.aps.org/doi/10.1103/PhysRevC.31.488`
[12] Vernotte J, Berrier-Ronsin G, Kalifa J, Tamisier R and Wildenthal B 1994 *Nuclear Physics A* **571** 1 – 42 ISSN 0375-9474 URL `http://www.sciencedirect.com/science/article/pii/0375947494903395`
[13] Evers D, Ley C, Spindler E, Assmann W, Rudolph K, Konrad P and Sperr P 1977 *Nuclear Physics A* **275** 363 – 380 ISSN 0375-9474 URL `http://www.sciencedirect.com/science/article/pii/0375947477904584`
[14] Becha M B, Blyth C O, Pinder C N, Clarke N M, Ward R P, Hayes P R, Pearce K I, Watson D L, Ghazarian A, Cohler M D, Thompson I J and Nagarajan M A 1997 *Phys. Rev. C* **56** 1960–1971 URL `https://link.aps.org/doi/10.1103/PhysRevC.56.1960`
[15] Terakawa A, Orihara H, Oura M, Hosaka M, Suzuki H, Kumagai K, Kikuchi Y, Tohei T, Nakagawa T, Takamatsu J, Narita A, Hosomi K, Ishii K, Jon G C, Miura K and Ohnuma H 2002 *Phys. Rev. C* **66** 064313 URL `https://link.aps.org/doi/10.1103/PhysRevC.66.064313`
[16] Lubian J *et al.* 2018 *see this volume* **ibid**
[17] Lay J A, Fortunato L and Vitturi A 2014 *Phys. Rev. C* **89** 034618 URL `https://link.aps.org/doi/10.1103/PhysRevC.89.034618`

IOP Conf. Series: Journal of Physics: Conf. Series **1056** (2018) 012030 doi:10.1088/1742-6596/1056/1/012030

Probing Double Beta-Decay by Heavy Ion Charge Exchange Reactions

Horst Lenske[1,2]

[1]Institut für Theoretische Physik, Justus-Liebig-Universität Gießen, D-35392 Gießen
[2]NUMEN Collaboration, LNS Catania

E-mail: `horst.lenske@physik.uni-giessen.de`

Abstract. A special class of nuclear double charge exchange (DCE) reactions proceeding as a one-step reaction through a two-body process are shown to involve nuclear matrix elements of the same diagrammatic structure as in $0\nu2\beta$ decay. These hadronic Majorana-type DCE reactions (MDCE) have to be distinguished from second order DCE reactions, given by double single charge exchange (DSCE) processes resembling $2\nu2\beta$ decay. The theoretical concepts of MDCE are discussed. First results show that ion-ion DCE reactions are the ideal testing grounds for investigations of second order nuclear matrix elements.

1. Introduction

There is a broad consensus that neutrino-less nuclear double beta-decay will be a highly promising gateway to physics beyond the standard model. However, $0\nu2\beta$ double-beta decay is still a hypothetic process awaiting experimental confirmation. If detected, those decays would give direct evidence on the Majorana-nature of neutrinos with far reaching implications for neutrino masses, neutrino-matter interactions and flavour mixing up to the question of the matter-antimatter asymmetry in the universe [1, 2, 3, 4, 5]. Such a signal has to be distinguished from the two-neutrino $2\nu2\beta$ beta-decay [6, 7] which is fully compatible with the standard model. Although both decays correspond to second order nuclear processes, they are dynamically distinct. Double beta-decay with neutrino emission is a sequential decay process where the leptons are emitted subsequently in an uncorrelated manner. A few nuclei are known to decay by this already rather rare process, as discussed e.g. in Ref. [8, 9]. While the matrix elements are accessible by the observed $2\nu2\beta$ transitions such a check against data does not yet exist for $0\nu2\beta$ processes. Thus, estimates of life times and transition probabilities are relying on theoretical investigations, notoriously showing an uncomfortably large spread of values as discussed in this volume e.g. by Iachello [10]. Independent tests of the nuclear structure input under controllable dynamical conditions are highly necessary, allowing to evaluate and gauge the theoretical results by an independent process. Hence, the field will profit tremendously if a surrogate process could be identified which is technically and physically easily accessible. While SCE reactions with light and heavy ions have been studied intensively, including our own work [11, 12, 13, 14], close to nothing is known about double charge exchange reactions. Only very recently, the NUMEN project has been initiated [15], using heavy ion reactions to explore that unknown territory, also aiming at establishing the relation to double beta decay. In section 2 the physical concept of

IOP Conf. Series: Journal of Physics: Conf. Series **1056** (2018) 012030 doi:10.1088/1742-6596/1056/1/012030

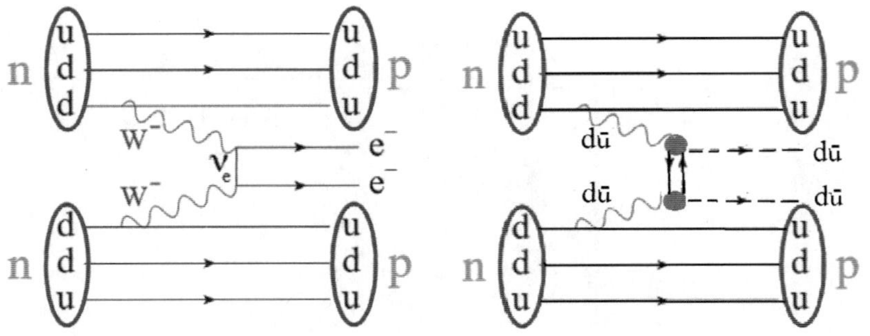

Figure 1. The elementary weak interaction process mediating nuclear $0\nu2\beta$ decay (left) and a corresponding strong interaction process (right) are depicted schematically. The QCD counterpart is given by the simultaneous emission of two $[d\bar{u}]$ pairs in an isovector-vector, e.g. 1^-, configuration (wavy lines), decaying into a pionic $[d\bar{u}]$ configuration and a charge-neutral $q\bar{q}$ pair.

Majorana DCE reactions is briefly introduced. First results are discussed in section 3 and in section 4 an outlook will be given.

2. Majorana Double Charge Exchange Reactions

Second order quantal processes like heavy ion double charge exchange reactions are of genuine reaction theoretical interest. First of all, until now heavy ion DCE reactions have not been studied, neither experimentally nor theoretically. Some attempts were made on (π^+, π^-) reactions but their notoriously bad energy definition of the incoming beams is unfavorable for spectroscopic work. Thus, double charge exchange reactions with heavy ions are much better suited for explorations of weakly populated transitions. Here, we consider collisional charge exchange processes given by elementary interactions between target and projectile nucleons. In accordance with explicit calculations, the mean-field driven transfer contributions are neglected because they are at least of 4th order for DCE reactions considered here. Thus, only processes with changes of the charge partitions but leaving the projectile-target mass partition unaltered will be discussed.

A central question is whether we can identify on the elementary level a correspondences between strong and weak interaction processes. The answer is yes, as illustrated in Fig. 1. Under nuclear structure aspects, the $0\nu2\beta$ decay of a nucleus is nothing but special class of two-body correlation, sustained by the exchange of a (pair of) Majorana neutrino(s) between two nucleons where the interaction vertices are given by the emission of virtual $W^\pm$ gauge bosons. The strong interaction counterpart is a two-nucleon correlation built up by the exchange of a virtual charge-neutral quark-antiquark ($q\bar{q}$) pair accompanied by the emission of a charged $q\bar{q}$ component, thus changing at the same time the nucleonic charges. Similar to the weak process, the strong vertices are originating from gauge bosons, here given by the initial emission of gluons which materialize into two $q\bar{q}$ pairs. At the end, the highly off-shell $q\bar{q}$ compounds will decay into mesons, preferentially into pions but also multi-pion configurations like the scalar and vector mesons.

Such interaction process may occur frequently in nuclei, both on the weak as well as on the strong interaction scale. They remain unobserved if the emitted electrons or charged mesons are reabsorbed by the same nucleon. This will lead to vertex and propagator correction and, as such, contributes to the nucleon in-medium mass operator of a, however, negligibly small strength.

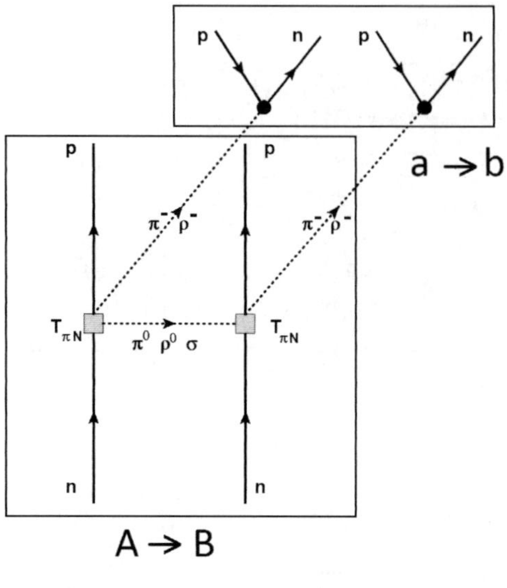

Figure 2. Generic diagram illustrating the hadronic surrogate process for $0\nu2\beta$ decay. A virtual $nn \to pp\pi^-\pi^-$ scattering process, causing the $\Delta Z = +2$ target transition $A \to B$, is accompanied by $nnp^{-1}p^{-1}$ double-CC excitation in the projectile. As indicated, other isovector mesons as e.g. the rho-meson isotriplet will contribute, too. Contributions of crossed diagrams are not displayed.

Nevertheless, it is worthwhile to keep in mind that we are dealing here with phenomena belonging to the large class of nuclear ground state correlations beyond the commonly studied mean-field sector [16, 17, 18, 19]. Short-range correlations are known to modify nuclear momentum contributions on the level of up to 20%.

Both processes become of interest if they reveal their existence and nature in observable signals. In this respect, we encounter a fundamental difference between $0\nu2\beta$ decay and the hadronic process: Only the former may occur in an isolated nucleus while the latter one is forbidden by energy conservation. Thus, in order to observe the double-meson emission by a nucleon pair a partner nucleus is required which takes care of the virtuality of the process by absorbing the two charged virtual mesons. For that purpose, heavy ion double charge exchange reactions are the ideal tool. Diagrammatically, the microscopic dynamics of such a reaction are indicated in Fig. 2. The target undergoes a correlated double-meson pair decay $nn \to pp+\pi^-\pi^-$ (in fact a $2p^{-1}2n$ two particle-two hole transition) and the projectile absorbs the pions by the simultaneous excitation of two np^{-1} -type configurations. Other meson configurations will contribute as well.

The whole reaction will proceed as a one-step reaction via a special kind of two-body interaction generated by the correlation diagram. Denoting the (in-medium) pion-nucleon T-matrix by $T_{\pi N,\pi' N'}$, the target-part of the interaction is in a somewhat symbolic notation

$$V^{(MDCE)}(\mathbf{13},\mathbf{24}) \sim T_{\pi^- p,\pi^0 n}(\mathbf{1},\mathbf{3})D_{\pi^0}(\mathbf{1}-\mathbf{2})T_{\pi^0 n,\pi^- p}(\mathbf{2},\mathbf{4}) \tag{1}$$

where the $n \to p$ target transitions are denoted by **1** and **2**, respectively. The coordinates **3** and **4** indicate the outgoing charged pions, inducing the complementary transitions in the projectile. The correlation built up by the neutral pion is described by the propagator D_{π^0}. A decomposition into irreducible tensors gives rise in particular to an effective rank-2 iso-tensor projectile-target interaction of operator structure $[\boldsymbol{\tau}_{1A} \otimes \boldsymbol{\tau}_{2A}]_2 \cdot [\boldsymbol{\tau}_{3a} \otimes \boldsymbol{\tau}_{4a}]_2$. We recognize immediately the

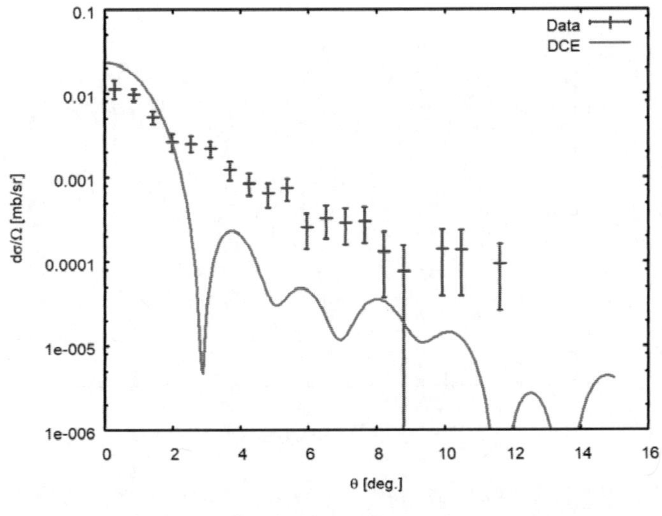

Figure 3. Angular distribution of the DCE reaction $^{18}O + {}^{40}Ca \to {}^{18}Ne + {}^{40}Ar$ at $T_{lab} =$ 270 MeV. Theoretical results are compared to the data of Ref. [29]. The theoretical cross section includes MDCE and DSCE contributions, added up coherently.

similarity to the nuclear matrix element of $0\nu2\beta$ decay, justifying the name Majorana-DCE. At present, the strengths of the nucleon-nucleon and the pion-nucleon T-matrices are taken from data. More refined description will be scrutinized in future work by an effective field theoretical description [20] and by referring to the data base available for free space $nn \to pp\pi^-\pi^-$ reactions. Previously, the charge-conjugated reaction $pp \to nn\pi^+\pi^+$ reaction and other double-pion production channels were investigated at CELSIUS and COSY [21, 22, 23, 24, 25] and more recent also at HADES [26]. Theoretical studies of the on-shell reaction combining meson exchange and resonance excitation have been performed by the Valencia group [27] and in somewhat extended form by Xu Cao et al. [28].

If we consider, on the other hand, the effective operator underlying the conventional double-SCE two-step reaction mechanism, we find

$$V^{(DSCE)}(\mathbf{13},\mathbf{24}) \sim \sum_{cC} T_{NN}(\mathbf{3},\mathbf{4})\mathcal{G}_{cC}(\mathbf{2}-\mathbf{4},\mathbf{1}-\mathbf{3})T_{NN}(\mathbf{2},\mathbf{1}) \tag{2}$$

where T_{NN} is the isovector nucleon-nucleon T-matrix and $\mathcal{G}_{bB}$ denotes the (full many-body) propagator of the intermediate nuclei reached in the first SCE reaction step. The differences are obvious, visible in particular in the different operator structures.

3. Heavy Ion DCE Reactions and Data

The full DCE reaction amplitude is given by the coherent sum of the MDCE and the DSCE amplitudes:

$$M_{\alpha\beta} \sim \langle \chi_\beta^{(-)\dagger}, bB|V^{(MDCE)} + V^{(DSCE)}|aA, \chi_\alpha^{(+)}\rangle = M_{\alpha\beta}^{(MDCE)} + M_{\alpha\beta}^{(DSCE)} \tag{3}$$

A caveat of heavy ion scattering is the strong role played by initial (ISI) and final (FSI) state ion-ion interactions. The are typically well described by optical potentials. Above, those effects are taken care off by the distorted waves $\chi_{\alpha,\beta}^{(\pm)}$. As discussed elsewhere [30, 31] a momentum

IOP Conf. Series: Journal of Physics: Conf. Series **1056** (2018) 012030 doi:10.1088/1742-6596/1056/1/012030

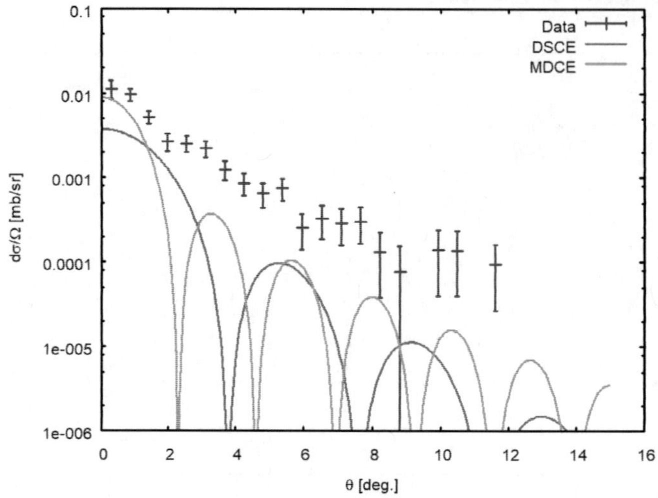

Figure 4. Angular distribution of the DCE reaction $^{18}O + {}^{40}Ca \rightarrow {}^{18}Ne + {}^{40}Ar$ at $T_{lab} =$ 270 MeV. The MDCE and DSCE partial cross sections are shown separately in comparison to the data of Ref. [29]. Both processes support large momentum transfers which is even more pronounced for the MDCE component.

space representation allows to separate the ISI/FSI contributions and the nuclear transition form factors,

$$M_{\alpha\beta} \sim \int d^3p N_{\alpha\beta}(\mathbf{p}) M_{\alpha\beta}(\mathbf{k}_\alpha - \mathbf{k}_\beta - \mathbf{p}) \quad ; \quad N_{\alpha\beta}(\mathbf{p}) = \frac{1}{(2\pi)^3} \langle \chi_\beta^{(-)\dagger} | e^{i\mathbf{p}\cdot\mathbf{r}} | \chi_\alpha^{(+)} \rangle. \qquad (4)$$

The distortion coefficient $N_{\alpha\beta}$ acts in the strong absorption limit mainly as a scaling factor, typically reducing the forward cross section by several orders of magnitudes compared to the plane wave limit. In the calculations discussed below, ion-ion interactions are treated in the strong absorption black disk limit.

First results of a DCE calculation along the line discussed above are shown on Fig. 3 and compared to recent NUMEN data [29] for the reaction $^{18}O + {}^{40}Ca \rightarrow {}^{18}Ne + {}^{40}Ar$ at $T_{lab} = 15$ AMeV. The transition strengths are taken from QRPA calculations. The transition potentials were approximated by Gaussians. Only the pionic contributions were included. The forward peak of the angular distribution is dominated, in fact, by the MDCE component. Overall, the data are described decently well in view of the exploratory character of the calculations.

The MDCE and, to a slightly lesser degree also the DSCE process, support large momentum transfers, as it is expected also for the $0\nu2\beta$ -decay. This is illustrated in Fig. 4 where the partial MDCE and DSCE cross sections are shown separately. By comparison to Fig. 3 it is seen that the coherent sum of both amplitudes is leading to a complex interference pattern, especially at large momentum transfers. From Fig. 4 is evident that for the $^{18}O + {}^{40}Ca$ system the MDCE process dominates at small scattering angels, at least for the considered case of purely pionic Majorana-correlations.

4. Outlook

A new theoretical scenario for heavy ion double charge exchange reactions was introduced. At the diagrammatic level, structures similar to $0\nu2\beta$ matrix elements have been identified. The hadronic Majorana-DCE process is accessible only by reactions of composite nuclei. The lightest possible system is the reaction $^3H + ^3He \to ^3p + ^3n$. Here, we have discussed explicitly the case of a DCE reaction with medium mass ions at relatively low incident energy. ISI and FSI ion-ion interactions were taken into account and the quantum mechanical coherence of the MDCE and the DSCE reaction mechanism was treated properly. The strongly forward peaked measured angular distributions indicate a direct mechanism which indeed is confirmed by the calculations. These first results are very promising by indicating a new way of accessing second order nuclear matrix elements of charge changing interactions. Together with the much better studied SCE reactions and their established usefulness for spectroscopic work, heavy ion DCE reactions are opening a new window to high-precision spectroscopy. Although it will not be possible to insert the extracted matrix elements directly into a $0\nu2\beta$ analysis, DCE reactions provide an unique way to validate nuclear structure models under controllable laboratory conditions by comparison to data on processes of comparable physical content. New impact on theoretical investigations in both reaction and nuclear structure theory is demanded for a quantitative understanding of these special reactions. Although the present calculations do not yet include the full spectrum of contributions, they are establishing the hadronic Majorana-DCE reaction mechanism. The refinements may lead to changes in detail but will not alter the overall picture. An exciting and encouraging result is that the MDCE process is clearly visible, even dominating the cross section at extreme forward angles.

References

[1] Haxton W C and Stephenson G J 1984 *Prog. Part. Nucl. Phys.* **12** 409
[2] Tomoda T 1991 *Rept. Prog. Phys.* **54** 53
[3] Suhonen J and Civitarese O 1998 *Phys. Rept.* **300** 123
[4] Vergados J D, Ejiri H and Šimkovic F 2012 *Rep. Prog. Phys.* **75** 106301
[5] Cremonesi O and Pavan M 2014 *Adv High Energy Phys.* 951432 (*Preprint* arXiv:1310.4692 [physics.ins-det])
[6] Goeppert M 1935 *Phys. Rev.* **48** 512
[7] Stefanik D, Simkovic F and Faessler A 2015 *Phys. Rev. C* **91** 064311 (*Preprint* arXiv:1506.00835 [nucl-th])
[8] Grewe E W and Frekers D 2006 *Prog. Part. Nucl. Phys.* **57** 260
[9] Barabash A S 2015 *Nucl. Phys. A* **935** 52 (*Preprint* arXiv:1501.05133 [nucl-ex])
[10] Iachello F 2018 *see this volume* ibid.
[11] Lenske H, Wolter H H and Bohlen H G 1989 *Phys. Rev. Lett.* **62** 1457
[12] Brendel C, von Neumann-Cosel P, Richter A, Schrieder G, Lenske H, Wolter H H, Carter J and Schll D 1988 *Nucl. Phys. A* **477** 162
[13] Bohlen H G, Lenske H *et al.* 1988 *Nucl. Phys. A* **488** 89
[14] Lenske H and Schrieder G 1998 *Eur. Phys. J. A* **2** 41
[15] Cappuzzello F *et al.* 2018 *Eur. Phys. J. A* **54** 72
[16] Froemel F, Lenske H and Mosel U 2003 *Nucl. Phys. A* **723** 544 (*Preprint* [nucl-th/0301038])
[17] Konrad P and Lenske H 2007 *Eur. Phys. J. A* **33** 291 (*Preprint* [nucl-th/0612078])
[18] Rocco N, Lovato A and Benhar O 2016 *Phys. Rev. Lett.* **116** 192501 (*Preprint* arXiv:1512.07426 [nucl-th])
[19] Benhar O 2016 *Nucl. Phys. News* **26** 15
[20] Lenske H and Santopinto E *work in progress*
[21] Johanson J *et al.* 2002 *Nucl. Phys. A* **712** 75
[22] Clement H *et al.* 2005 *Int. J. Mod. Phys. A* **20** 1747
[23] El-Bary S A *et al.* 2008 *Eur. Phys. J. A* **37** 267
[24] Tsuboyama T, Sai F, Katayama N, Kishida T and Yamamoto S S 2000 *Phys. Rev. C* **62** 034001
[25] Sarantsev V V *et al.* 2007 *Phys. At. Nucl.* **70** 1885
[26] Agakishiev G *et al.* 2015 *Phys. Lett. B* **750** 184 (*Preprint* arXiv:1503.04013 [nucl-ex])
[27] Alvarez-Ruso L, Oset E and Hernandez E 199 *Nucl. Phys. A* **633** 519
[28] Cao X, Zou B S and Xu H S 2010 *Phys. Rev. C* **81** 065201 (*Preprint* [arXiv:1004.0140 [nucl-th]])
[29] Cappuzzello F *et al.* 2015 *Eur. Phys. J. A* **51** 145 (*Preprint* arXiv:1511.03858 [nucl-ex])
[30] Bellone J *et al.* 2018 *see this volume* ibid.
[31] Lenske H, Bellone J, Colonna M and Lay J A 2018 *Phys. Rev. C* (submitted)

The Digital Optical Module of KM3NeT

E Leonora[1], on behalf of the KM3NeT Collaboration

[1] INFN-sezione di Catania, Via S. Sofia, 95123, Catania, Italy

emanuele.leonora@ct.infn.it

Abstract. The KM3NeT Collaboration is currently building neutrino telescopes with a total volume of about a cubic kilometre at the bottom of the Mediterranean Sea. The detectors consist of matrices of photodetectors, called digital optical modules, suspended in the depths of the sea by means of vertical structures, called detection units. For the optical modules, KM3NeT uses a new design that hosts 31 3-inch photomultiplier tubes and all the associated electronics inside a pressure-resistant glass vessel with a 17-inch diameter. This solution represents an innovative design with several advantages with respect to optical modules that comprise single large photomultipliers, currently used in all other operational neutrino telescopes. The digital optical module mass production phase has already started in several integration sites, together with the installation of detection units in the depths of the Mediterranean Sea.

1. Introduction

KM3NeT [1] started building a marine research infrastructure hosting new generation cubic kilometer sized neutrino telescopes located at the bottom of the Mediterranean Sea. The detector can be described as a three dimensional matrix of photosensors that are sensitive to the Cherenkov radiation emitted by products of neutrino interactions in the deep water. The arrival time of the Cherenkov light at each sensor, together with the knowledge of their spatial position, are used to reconstruct the trajectory of the particle that produces the Cherenkov light and the direction of the neutrino. The amount of the detected light can provide in addition information about the energy of the neutrino. The photosensors are called digital optical modules (DOMs), and they are suspended in the sea in vertical structures, called detection units (DUs). The modular design allows different spacing between DUs and DOMs in order to cover different energy ranges according to the physics targets. Following its first phase of construction, KM3NeT 2.0 has two scientific objectives: the discovery and observation of high-energy neutrino sources in the Universe and the determination of the mass hierarchy of neutrinos. Two goals for two detectors: ARCA (Astroparticle Research with Cosmics in the Abyss), which is optimized for the observation of astrophysical neutrinos above TeV energies, and ORCA (Oscillation Research with Cosmics in the Abyss), a denser detector that targets atmospheric neutrinos oscillations in few-GeV range [1]. Two goals, two detectors, located at two different marine sites: ARCA located at the KM3NeT-It site, at the South of the Sicilian coast in Italy, 100-km offshore from Capo Passero, at 3400 m of depth, and ORCA located at the KM3NeT-Fr site, 40-km offshore of Toulon, France, at 2475 m of depth.

2. KM3NeT detection units

In the depths of the Mediterranean Sea, the optical modules are suspended in vertical structures named detection units (DU), composed of two parallel vertical Dyneema® ropes, anchored to the seabed and kept taut with a system of submerged buoys at the top, to reduce the horizontal displacement of the top relative to the base on the bottom [2].
Each DU is equipped with 18 DOMs, and contains also all the devices required for data transmission and power supply. In deep water, all the DOMs are attached to the ropes by an external titanium collar. Attached to the ropes, there is also the vertical electro-optical cable (VEOC), which contains two copper wires for power transmission and 18 optical fibres for data transmission. In figure 1, an artistic view of the KM3NeT detection unit is shown. At the position of each DOM, two power conductors and a single fibre are branched out by means of the breakout box (BoB) positioned externally to each DOM. For marine deployments, the DU string-like structure is coiled around a large spherical frame, the launcher vehicle, in which the DOMs are positioned in cavities. A surface vessel is used to deploy the launcher vehicle at its designated position on the seabed with a 1-m accuracy. In the depths of the marine sites, a mechanical system activated by a remote operating vehicle (ROV) triggers the vertical unfurling of the string. The launcher vehicle starts to rise to the surface while slowly rotating and releasing the DOMs. At the end, the empty launcher vehicle floats to the surface and is recovered for following deployments, leaving behind the DU in vertical position. In ARCA configuration, with target energies above 1 TeV, each DU has a height of about 700 m, with DOMs 36-m vertically spaced. The DU horizontal spacing is about 95 m. For the ORCA configuration, which targets lower energies in the few GeV range, each DU has a height of about 200 m with DOMs vertically spaced 9 m. The DU horizontal spacing is about 20 m.

3. KM3NeT digital optical modules

The digital optical modules (DOMs), which represent the sensitive part of the neutrino telescope, are composed of pressure-resistant 17-inch diameter transparent glass vessels, each hosting 31 photomultiplier tubes (PMTs) with a 3-inch photocathode diameter and all associated front-end and readout electronics. In figure 2 a picture of a DOM. The multi-PMT solution represents an innovative design, considering that the optical modules of all the other neutrino telescopes, ANTARES, Baikal and IceCube, have optical modules with a single large photomultiplier, typically with a photocathode diameter of 10 inch. The multi-PMT DOM offers several advantages over traditional optical modules using single large PMT: it reaches almost three times the photocathode area of a single glass sphere equipped with a 10 inch PMT and has an almost uniform angular coverage. The influence of the Earth's magnetic field on small size PMTs is negligible, and thus a magnetic shield is not required. Since each PMT works independently, accurate photon counting can be performed, because the arrival of more than one photon at the DOM can be identified with high efficiency. The segmented photocathode layout also offers directional information with an almost isotropic field of view and provides rejection of optical background in water at the DOM detection level. The proposed DOM design matches with all the main requirements imposed by the project: a low rate of noise signals; a timing accuracy of the order of 1 ns, necessary for an accurate particle-track reconstruction and an efficient background rejection. It has also to withstand to environmental and handling conditions such as hydrostatical pressure up to 400 bar, corrosion, shocks and vibration and to ensure a lifetime of over 10 years.

3.1. DOM main components

The DOM is composed of a transparent spherical glass vessel that contains the 31 PMTs, the associated frontend and readout electronics, calibration components and accessory systems for monitoring and controls. An artistic view of the DOM inside is shown in the figure 2. The vessel is a 17-inch pressure resistant glass sphere produced by Nautilus, composed of two separate hemispheres

IOP Conf. Series: Journal of Physics: Conf. Series **1056** (2018) 012031 doi:10.1088/1742-6596/1056/1/012031

of 14-mm thick borosilicate glass (Vitrovex®). The glass provides an adequate resistance against the water and the extreme hydrostatical pressure of the deep-sea environment, while maintaining a transparency of more than 95% at a wavelength of 350 nm. The two hemispheres comprise the two different parts of the DOM, usually called top and bottom parts. The 31 PMTs are arranged in 5 rings of 6 PMTs plus a single PMT at the bottom pointing vertically downwards. The bottom hemisphere contains 19 PMTs, the top hemisphere 12 PMTs.

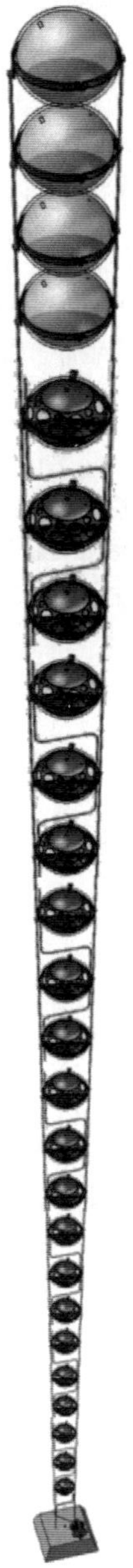

Figure 1. Artistic view of a detection unit

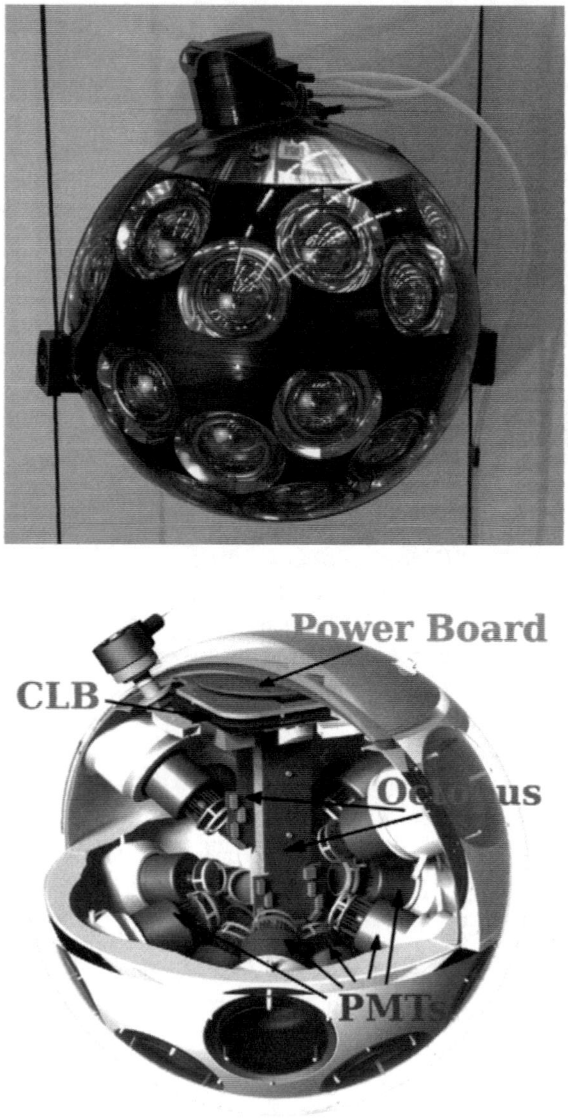

Figure 2. On top: a picture of a DOM with the breakout box and the fixation on the two parallel ropes. On bottom: a sketch of the inner components of the DOM. CLB, power board, octopus boards and PMTs are indicated

In the first phase of the project, a batch of 15000 3-inch Hamamatsu PMT R12199-02 was used for DOM production. The PMTs have a convex bialkali photocathode, with a diameter of 80 mm and a 10-stage dynode structure [3]. Inside the DOM, all the PMTs are kept in place by a nylon 3D-printed support structure, which ensures great accuracy, strength and lightness. In order to increase the photon collection efficiency without increasing the PMT diameter, a reflector ring surrounds the face of each PMT, with an angle of 45° with respect to the PMT axis. The efficiency is thus increased by 20–40%. The space between the support structure, the PMT windows and the glass sphere is filled with an optical silicone gel, which ensures optical and mechanical coupling. It is the Silgel 612 A/B produced by Wacker. The support structure and the gel are sufficiently flexible to be resistant to vibration and shocks during handling and deployment and allow for the deformation of the glass sphere under the hydrostatic pressure in deep-sea sites. An aluminum cooling structure provides heat conduction between the components inside and the environment outside of the sphere. The main warm parts are mounted inside the DOM in thermal contact with this aluminum structure. The power consumption of the DOM with all operating systems is around 7 W. A custom-designed titanium penetrator is mounted on the top hemisphere of the DOM, and contains two copper wires for power transmission and one optical fibre for data communication. On the top hemisphere is also installed a vacuum valve, to evacuate the DOM with a 0.2 Bar under pressure during the closure of the two hemispheres. A pressure gauge is installed inside to monitor the relative pressure into the DOM. Figure 3 shows an exploded view of the whole DOM. The optical module also contains three accessory sensor systems. A LED beacon [4] is mounted on the top hemisphere, to inject calibrated light pulses in water to illuminate the neighboring optical modules for time calibration. An acoustic piezo sensor [5] is glued to the inner surface of the bottom hemisphere for position measurements, in conjunction with the acoustic positioning system at the seabed. A compass system, installed as a mezzanine on the main electronic board, provides compass, tilt- and accelerometer data used to reconstruct the DOM orientation in water.

3.2. DOM electronics

All the electronic devices operative in DOMs have been developed to allow for a readout based on the "All-data-to-shore" concept, in which all the analogue signals from each PMT that pass a preset threshold are digitized and all the digital data are sent to shore via an Ethernet network of optical fibres. Each PMT works as an individual optical sensor, with an individual low-power high-voltage base with integrated amplification and tunable discrimination [6]. The photon arrival time and the time-over-threshold (ToT) of each PMT, are recorded by an individual time-to-digital converter implemented in a field programmable gate array (FPGA). The threshold is set at the level of 0.3 of the mean single photon pulse height and the high voltage base is set to provide a gain of 3×10^6 for each PMT. The FPGA is mounted on the central logic board (CLB) [7], which transfers the data to shore via an Ethernet network of optical fibres by means of a laser receiver/transceiver. The data provided by the PMT bases is collected and distributed to the CLB by means of two so-called octopus boards. The power board [8] provides the DC power for all the systems. This board is monitored and controlled by the FPGA. It is powered with 12V DC supplied via the penetrator mounted in the upper hemisphere of the DOM and connected to a DC/DC converter hosted in the breakout box external to the glass sphere (see figure 2). Each DOM in a detection unit communicates at a dedicated wavelength, which is multiplexed with the other DOM wavelengths for transfer via a single optical fibre to the shore. The broadcast of the onshore clock signal, needed for time stamping in each DOM, is embedded in the Gb-Ethernet protocol. The White Rabbit protocol [9] has been modified to implement the broadcast of the clock signal, and allows for synchronization with 1-ns resolution of all the PMTs inside the DOM and all the DOMs installed in the KM3NeT DUs.

IOP Conf. Series: Journal of Physics: Conf. Series **1056** (2018) 012031 doi:10.1088/1742-6596/1056/1/012031

3.3. DOM production

Each digital optical is composed of over 80 different elements: mechanical objects, electronics and sensors. The production is then a delicate process performed following a well-defined procedure that contains over 20 different phases.

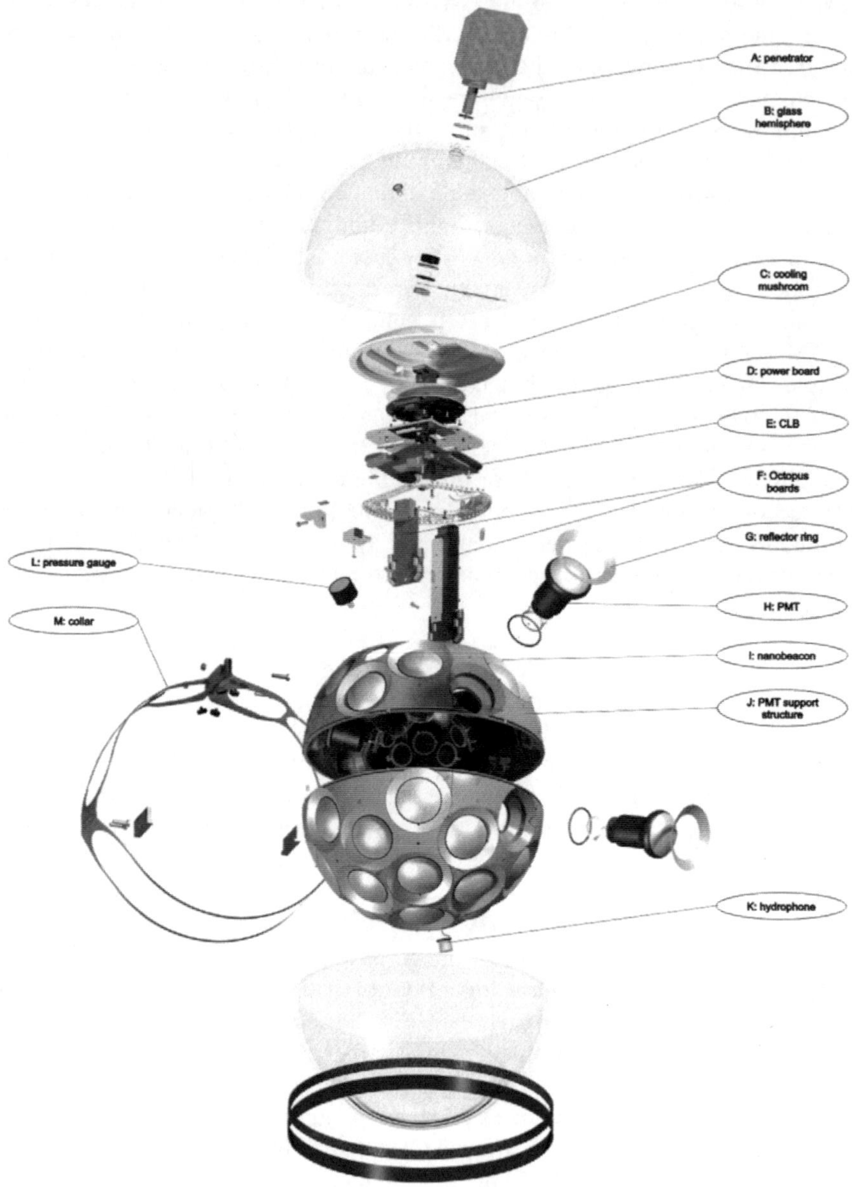

Figure 3. Exploded view of the DOM. The main components are indicated and described in the text.

All the components of each DOM are identified by means of a unique code, named UPI (Unique Product Identifier). During the production, all the components are properly registered in the KM3NeT database by means of dedicated assembly software.

The glass vessel is composed of two halves, thus the two parts of the DOM, top and bottom, are constructed separately. The support structures for the PMTs are also segmented in two halves, and each one made of two parts. The white 3-D printing quarters are black painted at the beginning of the assembly phase.
In the top hemisphere, the aluminum cooling system is glued to the inner glass sphere by means of a special tool that define its height and the distance from the glass. Power board and CLB are mounted on the inner part of the aluminum system. The penetrator is mounted in the top glass sphere, and its water tightness is tested by means of a helium leakage test. The optical fibre coming from the penetrator is connected to the laser transceiver/receiver operating an optical splice on the fibre inside. The resulting optical power and attenuation are checked, and results are stored in the database. The piezo sensor is glued to the inner surface of the bottom glass hemisphere. All the PMTs are mounted in the support structures, and their positions are stored in the database. After the positioning of the PMTs, the equipped structures are placed into the glass hemispheres, all the PMT bases are connected to the octopus boards, and a functional test of all systems is performed before applying the optical gel. Whether the test failed, is still possible replace the failing components. The pouring of the gel definitely couples optically and mechanically all the PMTs and the structures to the glass hemispheres of each half DOM. After the proper time of gel polymerization, the two halves are firstly connected electrically, and then mechanically, joining the two surfaces of the glass hemispheres. A 0.2 bar under-pressure inside the DOM is achieved by means of the external vacuum port. The application of a sealant and an anti-marine-corrosion tape on the junction of the two halves close definitely the sphere. Finally, the titanium collar is mounted externally on the DOM.
Each produced digital optical module is submitted to an accurate acceptance test. The results are stored in the KM3NeT database and used during the underwater operative conditions.

Conclusion
An innovative multi-PMT digital optical module was designed as the sensitive part of the cubic-kilometer sized KM3NeT neutrino telescope. The mass production of the KM3NeT digital optical modules is currently underway at several integration sites. Installation of detection units at the two KM3NeT sites in the deep Mediterranean Sea has already started. Two detection units have been operational for more than a year at the KM3NeT-It site. Another detection unit is operational at the KM3NeT-Fr site.

References
[1] Letter of Intent for KM3NeT 2.0, 2016 *J. Phys. G* **43** (8) 084001
[2] Kooijman P M, 2015 The mechanical structure and deployment procedure of the KM3NeT detection unit *PoS (ICRC2015)* 1173
[3] Aiello S et al., 2014 Characterization of the 80-mm diameter Hamamatsu PMTs for the KM3NeT project *AIP Conf. Proc.* 1630
[4] Calvo D et al., 2014 Nanobeacon: a low-cost time calibration instrument for the KM3NeT neutrino telescope *AIP Conf. Proc.* 1630
[5] Viola S et al., 2015 Acoustic positioning system for KM3NeT *PoS (ICRC2015)* 1169
[6] Timmer P et al., 2010 Very low power, high voltage assembly base for a Photo Multiplier Tube for the KM3NeT deep sea neutrino telescope, *JINST* **5** C12049
[7] Biagi S and Orzelli A, 2014 The Central Logic Board and its auxiliary boards for the optical module of the KM3NeT detector, JINST **9** C12033
[8] Anassontzis E. et al., Design and development of the Power Converter Board within the Digital Optical Module in KM3NeT, *PoS (TIPP2014)* 188
[9] Bouwhuis M, 2015 Time synchronization and time calibration in KM3NeT *PoS (ICRC2015)* 1170

IOP Conf. Series: Journal of Physics: Conf. Series **1056** (2018) 012032 doi:10.1088/1742-6596/1056/1/012032

Silicon Carbide detectors for nuclear physics experiments at high beam luminosity

G Litrico[1], S Tudisco[1], F La Via[2], C Altana[1], C Agodi[1], L Calcagno[4], F Cappuzzello[1,4], D Carbone[1], M Cavallaro[1], G A P Cirrone[1], G Lanzalone[1,3], A Muoio[1], S Privitera[2] and V Scuderi[1]

[1] INFN - Laboratori Nazionali del Sud, via S. Sofia, 62, 95123 Catania, Italy
[2] IMM-CNR, VIII Strada, 5, 95121 Catania, Italy
[3] Università Kore Enna, via delle Olimpiadi, 94100 Enna, Italy
[4] Physics Department, Catania University, Via S.Sofia 64 Catania, Italy

Corresponding author's e-mail: grazia.litrico@lns.infn.it

Abstract. Silicon carbide is a very promising material for next generation nuclear physics experiments at high beam luminosity. Such activities require devices able to sustain high fluxes of particles (up to 10^{14} ions/cm^2) in order to determine the cross sections of very rare nuclear phenomena. One of these activities is the NUMEN project, which aims, through the double charge exchange reactions, to impact in the determination of nuclear matrix elements entering in the expression of half-life of the neutrino-less double beta decay. Due to the very low cross sections, these features can just be explored at fluences which exceed by far those tolerated in state of the art solid state detectors, typically used in this kind of experiments. The SiC technology offers today an ideal response to such challenges, giving the opportunity to cope the excellent properties of silicon detectors with the radiation hardness, thermal stability and visible blindness of SiC material.

Introduction

Silicon carbide (SiC) is one of the compound semiconductor which has been considered as a potential alternative to Silicon (Si) for the realization of charge particles detectors and dosimeters in high energy physics. The chemical and physical material properties are promising for high temperature and high radiation operation conditions [1]. SiC diodes are predicted to be radiation harder than Si due to the high displacement threshold energy and potentially used as radiation detectors in high radiation conditions. The recent progresses in the material growth [2] and device technology, allowed to realize high performances SiC detectors [3, 4]. The potentialities of SiC as detectors are many, they have been used to detect neutrons, X-rays, protons, alpha particles and heavy-ions [5, 6]. High collection efficiency and energy resolution in the range 0.29 % - 1.7 % was reported for the detection of alpha-particle in the energy range 3.0 - 5.5 MeV [7-9].
The radiation-hardness of the material make it particularly interesting for all those activities where high flux must be detected [10, 11, 12]. One of these activities is the NUMEN project [13]. NUMEN aims to get information on the nuclear matrix elements entering the expression of the half-life of the double beta decay by relevant cross sections measurements of heavy-ion induced double charge exchange (DCE) reactions, (^{18}O,^{18}Ne), (^{20}Ne,^{20}O) and (^{12}C,^{12}Be). The DCE reactions will be investigated with incident energies ranging from 10 to 70 MeV/A and the MAGNEX large acceptance magnetic spectrometer [14-

16] will be used to detect the ejectiles produced in the nuclear collisions with its Focal Plane Detector (FPD) [17]. The operating conditions of NUMEN project require an upgrade of the actually used FPD with the replacement of silicon detectors with a new generation stopping wall telescope, which could be realized with Silicon Carbide. Simulation studies [18] have been realized in order to define the appropriate characteristics of the telescope, especially in term of thickness of the two stages (ΔE and E detectors) composing the telescope. The best solution requires the use of ΔE stage with a thickness of the order of 100 μm and a thicker E stage (~ 1000-2000 μm in order to identify ions in the range between 10 to 70 MeV/A.

The Schottky junctions represent today the state of art for SiC devices while p^+-n^- junctions are instead a novel solution for SiC detectors, very promising in analogy to similar junctions based on silicon devices. The first step for the development of new devices is the realization of some 4H-SiC prototype detectors with performances suitable to monitor heavy ions at high energy and at high dose. SiC is considered to be radiation harder than Si, however a direct comparison between the two type of detectors is not reported in literature. So in the present paper we have studied the effects of very high energy irradiation on SiC and Si detector electrical characteristics. We performed energy resolution analysis on SiC Schottky diode using alpha particles beams to evaluate the detectors performances and studythe effects of very high energy irradiation on SiC and Si detector electrical characteristics.

2. Results and discussion

The detectors were irradiated with 740 MeV Carbon beam at a dose of 94 KGy, and current-voltage (I-V) measurements of not irradiated and irradiated diodes were compared. We used 4H-SiC Schottky diode detectors realized on a 38 μm thick n-type epitaxial layer with a doping concentration of 4.8×10^{14} cm^{-3}. Schottky diodes with an active contact area of 1 mm^2 and 2.25 mm^2 were used with a nickel silicide (Ni_2Si) as metal Schottky contact. The Silicon detectors were Micron Semiconductor diodes, with an intrinsic ($1.3\times10^{11}cm^{-3}$) epitaxial layer, 300μm thick and active area of 1x1 cm^2.

In Figure 1 the energy resolution analysis obtained on SiC Schottky diode detectors under the exposition at a three peaks alpha source particles (^{239}Pu – ^{241}Am – ^{244}Cm) is shown and in Table 1 the evaluated energy resolution is reported. An Ascom preamplifier (90 mV/MeV gain), a 572 Ortec amplifier (shaping-time 2 μs) and a multichannel system were used. The obtained resolution results are not very far from the values of Silicon commercial detectors resolution (0.3%), demonstrating a good performance of prototype SiC detectors.

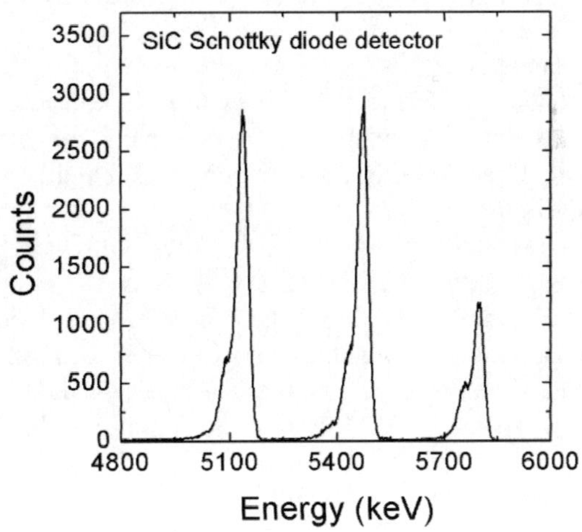

Figure 1 Response of SiC detector to a three peak α-source

Peak	Resolution (%)
^{239}Pu	0.74
^{241}Am	0.67
^{244}Cm	0.64

Table 1 SiC detector energy resolution

The reverse electrical characteristics measured in not irradiated and irradiated SiC detectors are shown in Figure 2 (a). The leakage current density in the not irradiated diode is about 2×10^{-6} A/cm^2 and it increases of about a factor two after a dose of 94 KGy. Similar measurements were performed on Si detector and the reverse I-V characteristics are reported in Figure 2 (b). The leakage current density is about 10^{-5} A/cm^2 in the not irradiated diode and it increases of more than a factor 50 after a dose of 94 KGy. The comparison between the electrical measurements on SiC detectors and Si one reveals that at the same dose the effect on the SiC is one order of magnitude lower. Energy resolution analysis are in progress to evaluate the effect of irradiation on detector resolution.

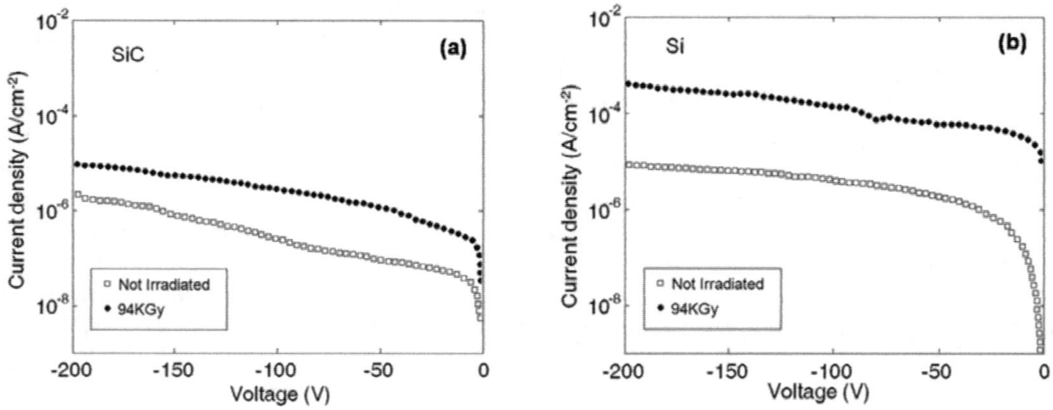

Figure 2 (a) SiC and (b) Si detector current-voltage characteristic in reverse bias

3. Conclusions

The effects of heavy ions irradiation on SiC detector was studied and compared with Si device, in particular the electrical characteristics after high dose irradiation have been compared. The high radiation hardness of SiC was demonstrated. Such results are preparatory for the development of innovative thick and large area SiC detectors useful for specific nuclear scopes.

References

[1] F. Nava, G. Bertuccio, A. Cavallini, E. Vittone, 2008 *Meas. Science Tech.* **19** 102010.
[2] F. La Via, M. Camarda, A. La Magna, 2014 *Appl. Phys. Rev.* **1** 031301.
[3] S. K. Chaudhuri, R. M. Krishna, K. J. Zavalla, K. C. Mandal, 2013 *Nucl. Instr. Meth. In Phys. Res. A* **701** 214.
[4] F. Moscatelli, A. Scorzoni, A. Poggi, M. Bruzzi, S. Lagomarsino, S. Mersi, S. Sciortino, R. Nipoti, 2005 *Nucl. Instr. Meth A* **546** 218.
[5] J. E. Lees, D. J. Bassford, E. J. Bunce, M. R. Sims, A. B. Horsfall, 2009 *Nuc. Instr. Meth. A* **604** 174.
[6] G. Bertuccio, R. Casiraghi, A. Cetronio, C. Lanzieri and F. Nava, 2004 *Nucl. Instr. Meth. A* **522** 413.
[7] S. K. Chaudhuri, K. J. Zavalla, K. C. Mandal, 2013 *Nucl. Instr. Meth. In Phys. Res. A* **728** 97.
[8] Y. B. Gurov, S. V. Rozov, E. A. Yakushev, L. Hrubcin,and B. Zat'ko, 2015 *Instr. and Exper. Tech.* **58** 22.
[9] L. Liu, J. L. Liu, L. Chen, Z. B. Zhang, P. Jin, J. L. Ruan, G. Chen, A. Liu, S. Bai, X. P. Ouyang, 2016 *Diamond and Related Materials* http://dx.doi.org/10.1016/j.diamond.2016.09.011
[10] F. Negoita, M. Roth, P. G. Thirlof, S. Tudisco, S. Mirfayzi, S. Kar, F. Hannachi, S. Moustaizis, I. Pomerantz, P. McKenna, J. Fuchs, K. Sphor, G. Acbas, A. Anzalone, P. Audebert, S. Balascuta, 17.05.2016 *Romanian Reports in Physics* **68** No. Supplement, p. S37-S144.
[11] F. Cappuzzello, C. Agodi, D. Carbone, M. Cavallaro, 2016 *Eur. Phys. T. A.* **52** 167
[12] F. Cappuzzello, C. Agodi, M. Bondì, D. Carbone, M. Cavallaro, A. Foti, 2015 *Jour. of Phys. Conf. Ser.* **630** 012018
[13] F. Cappuzzello et al 2018 *Eur. Phys. J.* A **54** 72
[14] F. Cappuzzello et al 2016 *Eur. Phys. J.* A **52** 167
[15] F. Cappuzzello et al 2014 *Nucl. Instr. and Meth.* A **763** 314

IOP Conf. Series: Journal of Physics: Conf. Series **1056** (2018) 012032 doi:10.1088/1742-6596/1056/1/012032

[16] A. Cunsolo et al. 2007 *Eur. Phys. J. A ST* **150** 343
[17] M. Cavallaro et al 2012 *Eur. Phys. J. A* **48** 59
[18] A. Muoio et al. 2016 *EPJ Web of Conferences* **117** 10006

The nuSTORM experiment

Kenneth Long
Imperial College London and STFC, Exhibition Road, London SW7 2AZ, UK
E-mail: k.long@imperial.ac.uk

Abstract. The nuSTORM facility will provide ν_e and ν_μ beams from the decay of low energy muons confined within a storage ring. The instrumentation of the ring, combined with the excellent knowledge of muon decay, will make it possible to determine the neutrino flux at the %-level or better. The neutrino and anti-neutrino event rates are such that the nuSTORM facility serving a suite of near detectors will be able to measure $\nu_e N$ and $\nu_\mu N$ cross sections with the %-level precision required to allow the next generation of long-baseline neutrino-oscillation experiments to fulfil their potential. By delivering precise cross section measurements with a pure weak probe nuSTORM may have the potential to make measurements important to understanding the physics of nuclei. The precise knowledge of the initial neutrino flux also makes it possible to deliver uniquely sensitive sterile-neutrino searches. The concept for the nuSTORM facility will be presented together with an evaluation of its performance. The status of the planned consideration of nuSTORM at CERN in the context of the Physics Beyond Colliders workshop will be summarised.

1. Introduction

nuSTORM ("Neutrinos from Stored Muons") is a facility based on a low-energy muon decay ring (see figure 1) [1, 2]. Pions, produced in the bombardment of a target, are captured in a magnetic channel. The magnetic channel is designed to deliver a pion beam with central energy E_π and energy spread $\sim \pm 20\% E_\pi$ to the muon decay ring. The pion beam is injected into the production straight of the decay ring. Roughly half of the pions decay as the beam passes through the production straight. At the end of the straight, the return arc selects a muon beam of central energy $E_\mu < E_\pi$ and energy spread $\sim \pm 10\% E_\mu$ that then circulates. Undecayed pions are directed to a beam dump. A detector placed on the axis of the production straight will receive a bright flash of muon neutrinos from pion decay followed by a series of pulses of muon and electron neutrinos from subsequent turns of the muon beam. Appropriate instrumentation in the decay ring and production straight will be capable of determining the integrated neutrino flux with a precision of $\lesssim 1\%$ [2]. The flavour composition of the neutrino beam from muon decay is known and the neutrino-energy spectrum can be calculated precisely using the Michel parameters and the optics of the muon decay ring. The pion and muon energies (E_π and E_μ) can be optimised to:

- Measure $\nu_e A$ ($\bar{\nu}_e A$) and $\nu_\mu A$ ($\bar{\nu}_\mu A$) interactions with per-cent-level precision; and
- Search for sterile neutrinos with exquisite sensitivity.

IOP Conf. Series: Journal of Physics: Conf. Series **1056** (2018) 012033 doi:10.1088/1742-6596/1056/1/012033

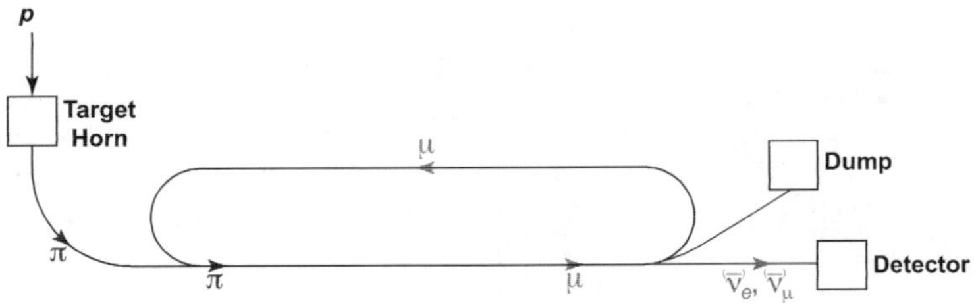

Figure 1. Schematic of the nuSTORM neutrino-beam facility.

2. Neutrino-nucleus scattering

2.1. Impact on searches for leptonic CP-invariance violation

The search for CP-invariance violation (CPiV) in present and planned long-baseline neutrino-oscillation experiments relies on the measurement of the rate of ν_e ($\bar{\nu}_e$) appearance in ν_μ ($\bar{\nu}_\mu$) beams. The phenomenological description of the effect relies on the assumption of three neutrino-mass eigenstates that mix to produce the three neutrino flavours [3, 4, 5, 6]. CPiV arises in this framework if the value of a phase parameter, δ, is such that $\sin\delta \neq 0$.

The oscillation probability is a function of the source-detector distance (the baseline) and the neutrino energy. Typical baselines range from 295 km for T2K [7] and the proposed Hyper-K experiment [8, 9, 10, 11], 800 km for NOνA [12] and 1300 km for the DUNE experiment [13, 14, 15, 16]. Neutrino interactions that occur as the neutrino beam passes through the earth introduce a "matter effect" that causes the oscillation probability of neutrinos to differ from that of anti-neutrinos. This introduces an "apparent" CPiV effect that depends on the neutrino mass hierarchy. The discovery of CPiV in neutrino oscillations requires that the "true" CPiV that depends on δ be distinguished from the apparent CPiV that arises from neutrino interactions with the earth.

The projected sensitivity to CPiV of the DUNE experiment is plotted as a function of exposure in figure 2 [14]. An exposure of 288 kt MW years will be achieved after seven years of running, with the planned staging to reach a total detector mass of 40 kt and a proton beam-power of 1.2 MW [17]. Equal exposures in neutrino and antineutrino mode have been assumed. The DUNE collaboration presents the sensitivity as a function of the assumed normalisation uncertainties on the ν_e and $\bar{\nu}_e$ appearance signals. Reducing the $\bar{\nu}_e$ normalisation uncertainty from 3% to 1% brings the exposure required to exclude CP invariance at the 3σ confidence level over 75% of all possible values of δ down from $\sim$ 1200 kt MW years to $\sim$ 600 kt MW years.

The projected sensitivity of the Hyper-K experiment, updated from [9], is also shown in figure 2. An exposure of 13 MW$\times 10^7$ s will be achieved after ten years assuming a 1:3 ratio between neutrino and anti-neutrino running. The planned staged implementation of two 187 kt detectors is indicated, a proton beam-power of 1.3 MW at 30 GeV has been assumed. The systematic uncertainties assumed by the Hyper-K collaboration in their estimation of the CPiV sensitivity of their experiment dominated by the combined "flux and near-detector" and the "cross-section model" uncertainties [9].

In addition to systematic uncertainties, a lack of knowledge of $\nu_{e,\mu}A$ ($\bar{\nu}_{e,\mu}A$) cross sections or inaccuracies in the simulation of the hadronic final state can lead to biases in the parameters extracted from the data. Such biases may arise, for example, from mis-classification of events [18] or mis-reconstruction of the energy of the incident neutrino [19, 20, 21, 22]. A discussion of possible sources of bias is presented in [23]. When searching for CPiV, any effect that differs in $\nu_e A$ and $\bar{\nu}_e A$ scattering and is not quantitatively understood is particularly pernicious since such a difference may be mis-interpreted as a signal for CPiV.

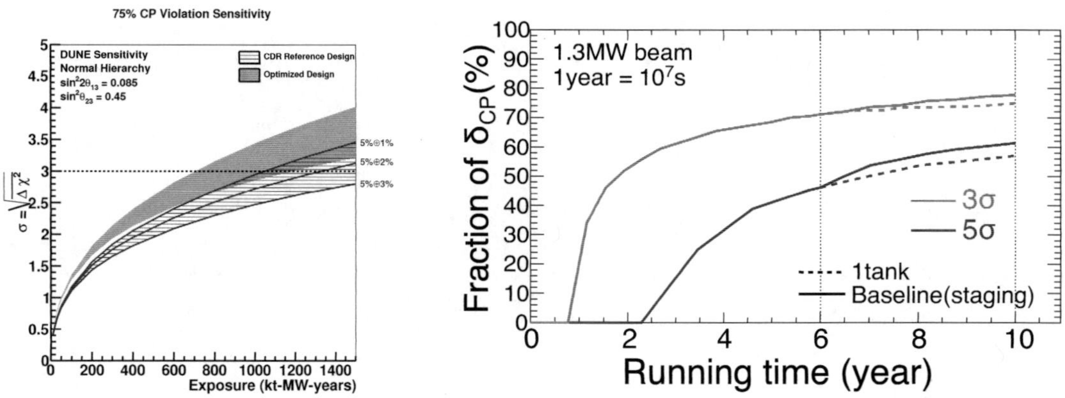

Figure 2. Left panel: Expected sensitivity of the DUNE experiment to CP-invariance violation plotted as a function of exposure in kt·MW·years assuming equal running in neutrino and antineutrino mode, for a range of values for the ν_e and $\bar{\nu}_e$ signal-normalisation uncertainties (from 5% to 1%) added in quadrature to an uncertainty of 5% on the normalisation of the background. The sensitivities shown are for the exclusion of CP-invariance conservation over 75% of the available range of values of δ assuming the normal hierarchy. The two bands represent a range of potential beam designs: the blue hashed band is for the CDR Reference Design and the solid green band is for the optimised design. The figure is taken from [14]. Right panel: fraction of all values of the CPiV phase, δ $(= \delta_{\rm CP})$, for which $\delta = \delta_{\rm CP} = 0, \pi$ can be excluded at 3σ (red) or 5σ (blue) plotted as a function of running time. An exposure of $13\,{\rm MW} \times 10^7\,{\rm s}$ is expected to be achieved after 10 years of operation. Figure updated from [9].

The next generation of long-baseline neutrino-oscillation experiments, DUNE and Hyper-K have the potential to observe CPiV violation. To maximise the scientific impact of the large, exquisitely-precise data sets that they will collect requires that the ν_e $(\bar{\nu}_e)$ and ν_μ $(\bar{\nu}_\mu)$ cross sections be known with percent-level precision and that uncertainties associated with cross-section models are also under control at the percent level.

2.2. Potential for impact on understanding of the structure of the nucleus

Theoretical understanding of the structure of the nucleon is detailed and precise and can be used to predict cross sections for a wide variety of processes over a wide kinematic range. However, a number of measurements, such as the spin structure of the nucleon, challenge the present understanding [24]. The theoretical description of the structure of the nucleus is not quite as accurate and requires development, for example to describe correlations among the nucleons that make up the nucleus [25]. Phenomenological models of lepton-nucleus scattering are based on the present understanding of nuclear physics and exploit a wealth of data to determine a number of phenomenological parameters. Such models have been shown to give a reasonable description of some of the present neutrino-nucleus scattering data but may fail when used to extrapolate beyond the range of energies, nuclei, or types of process on which they have been "tuned" [26]. A review of the challenges that must be overcome to deliver a good description of the hadronic final states is presented in [27].

A vibrant experimental programme is underway to extend and improve the scattering database on which the theoretical and phenomenological description of the nucleus relies. In the case of electron-nucleus scattering, the 12 GeV upgrade to the CEBAF accelerator at Jefferson Laboratory combined with the new experimental facilities and detector upgrades will provide a wealth of new data that will, most likely, lead to new insights (see for example [28]). Neutrino-nucleus scattering has made both seminal and historic contributions to the development of nuclear structure. The neutrino offers a probe that is polarised and is sensitive to flavour and isospin. It is conceivable that neutrino-nucleus scattering has a role to play in unravelling issues such as the orbital contribution to the spin of the nucleon and the nature of nucleon-nucleon correlations. A facility that is able to deliver a precisely calibrated flux is required if neutrino-nucleus scattering measurements are to contribute to the understanding of nuclear structure.

3. nuSTORM and the CERN Physics Beyond Colliders study

A detailed study of nuSTORM that includes consideration of the implementation of the facility at FNAL may be found in [2, 29, 30]. A preliminary investigation of possible options for siting the facility at CERN may be found in [31, 32]. In these studies the facility was optimised for the search for a light sterile neutrino. A short description of the principal considerations that will inform the re-optimisation of the facility for the study of neutrino-nucleus scattering will be given here.

In September 2016 CERN established the Physics Beyond Colliders (PBC) study group to consider ways in which the accelerators at CERN could be used or extended to support a diverse physics programme to complement the energy-frontier physics being pursued using the LHC [33]. The feasibility of implementing nuSTORM at CERN was included as a work package in the PBC study. Within the PBC context, the scientific objectives of nuSTORM are:

- To make detailed and precise measurements of neutrino-nucleus interactions not only as a service to the long- and short-baseline neutrino oscillation programmes but also as a means of studying the nucleus using a weak probe and seeking evidence for non-standard interactions; and
- To take forward the search for light sterile neutrinos should the results of the Short Baseline Neutrino (SBN) programme at FNAL [34] indicate that such a programme is required.

The potential for nuSTORM to establish a new technique for the study of fundamental particles and their interactions is recognised. The PBC study does not include consideration of a possible six-dimensional-cooling experiment (see for example [35]) to follow the demonstration of the reduction of normalised transverse emittance that will be provided by the international Muon Ionization Cooling Experiment (MICE) [36].

nuSTORM will provide a beam of precisely known flavour for which the flux will be known to 1% or better. Instrumented with a detector capable of making measurements of exclusive final states and of reconstructing Q^2 and W, nuSTORM can deliver the data required for precise models of the nucleus and of neutrino-nucleus scattering to be developed if the neutrino-beam energy spans the range $1 \lesssim E_\nu \lesssim 6\,\mathrm{GeV}$. These considerations lead to the following specification for the energy of the circulating muon beam:

- Maximum stored muon energy, $E_\mu = 6\,\mathrm{GeV}$; and
- It must be possible to vary the muon-beam energy in the range $1 \lesssim E_\mu \lesssim 6\,\mathrm{GeV}$.

Since the neutrino-energy spectrum is precisely known once the muon-beam energy is specified, the falling edge of the neutrino-energy spectrum can be used to calibrate the energy response of the neutrino detectors. Further, by combining data taken with different stored-muon energies, as described for NuPRISM in [37], cross sections may be determined in narrow neutrino-energy bands.

A sketch of the proposed siting of nuSTORM at CERN is shown in figure 3. The existing fast-extraction from the SPS that serves beam to the West Area will be exploited. Protons will be extracted into a new transfer line and delivered to a target/capture system that will select $\sim 8\,\mathrm{GeV}$ pions and transport them to the decay ring. The design of the decay ring will be

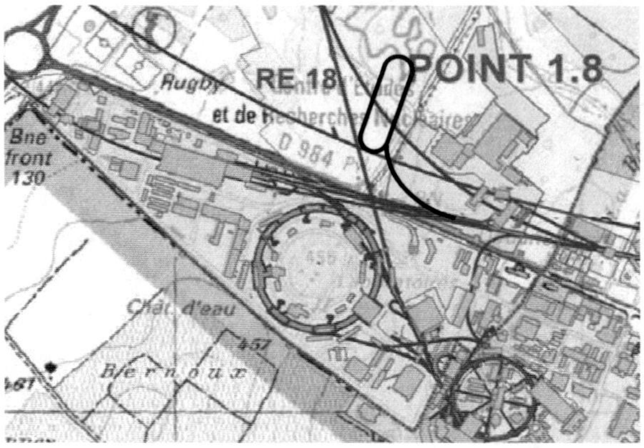

Figure 3. Preliminary sketch of the proposed implementation of nuSTORM at CERN. The blue solid lines show the existing tunnels and transfer lines serving the West Area at CERN. The proposed transfer line to serve nuSTORM and the nuSTORM storage ring location are indicated by the black solid lines. (I. Efthimiopoulos, private communication.)

revised to accommodate the maximum muon-beam energy of 6 GeV and to provide the ability to store muon beams with energies in the range $1 \lesssim E_\mu \lesssim 6$ GeV. The feasibility study will take into account the number of protons on target required and the extracted-beam parameters and will consider the fast extraction from the SPS and the proton-beam transport to the pion-production target. Preliminary investigations of the engineering of the target, horn, target complex and proton-beam absorber will be carried out. An important aspect of the feasibility study will be consideration of the radiation-protection issues raised by the target and the pion and muon fluxes.

The potential of the cross-section-measurement programme was evaluated in [2] assuming a neutrino detector with the performance of the HiResMν detector [38]. This analysis showed that nuSTORM has the potential to improve the systematic uncertainty on muon-neutrino (muon-anti-neutrino) CCQE cross section measurements by a factor of $\sim 5-6$. The $\nu_e A$ ($\bar{\nu}_e A$) cross-section measurements that can be made with nuSTORM will be unique.

4. Conclusions

The Neutrinos from Stored Muons (nuSTORM) facility is capable of delivering measurements of $\nu_{e,\mu}N$ ($\bar{\nu}_{e,\mu}N$) scattering for which the flux uncertainty can be reduced to 1% or better. Such measurements will reduce the systematic uncertainties and biases in future long-baseline neutrino-oscillation experiments thereby enhancing their sensitivity to leptonic CP-invariance violation and improving the precision of their measurements of the oscillation parameters. The cross-section measurement programme at nuSTORM has the potential to contribute to the understanding of nuclear physics through the use of a pure weak probe that is 100% polarised. Should the results of the FNAL Short Baseline Neutrino Program indicate that future searches for light sterile neutrinos are required, nuSTORM can support an exquisitely sensitive sterile-neutrino search programme.

The implementation of nuSTORM at CERN is being studied within the Physics Beyond Colliders Study Group. The goals of the study are to provide a preliminary proposal for siting the facility at CERN, to re-optimise the facility for the neutrino-scattering programme and to demonstrate through simulation that the normalisation of the neutrino flux can be constrained to $\lesssim 1\%$.

Acknowledgements
I thank the organisers for giving me the opportunity presenting nuSTORM. I gratefully acknowledge the help, advice, and support of my many colleagues within the neutrino and muon-beam communities who have freely discussed their results with me and allowed me to use their material.

References

[1] Kyberd P *et al.* (nuSTORM Collaboration) 2012 (*Preprint* 1206.0294)
[2] Adey D *et al.* (nuSTORM) 2013 nuSTORM - Neutrinos from STORed Muons: Proposal to the Fermilab PAC Tech. rep. Fermi National Accelerator Laboratory (*Preprint* 1308.6822)
[3] Pontecorvo B 1958 *Sov. Phys. JETP* **7** 172–173
[4] Pontecorvo B 1968 *Sov. Phys. JETP* **26** 984–988
[5] Maki Z, Nakagawa M and Sakata S 1962 *Prog. Theor. Phys.* **28** 870–880
[6] Patrignani C *et al.* (Particle Data Group) 2016 *Chin. Phys.* **C40** 100001
[7] The T2K collaboration 2013 T2K http://t2k-experiment.org
[8] Abe K *et al.* (Hyper-Kamiokande Working Group) 2014 (*Preprint* 1412.4673) URL http://inspirehep.net/record/1334360/files/arXiv:1412.4673.pdf
[9] Abe K *et al.* (Hyper-Kamiokande Proto-Collaboration) 2015 *PTEP* **2015** 053C02 (*Preprint* 1502.05199)
[10] Abe K *et al.* (Hyper-Kamiokande) 2016 URL http://www.hyperk.org/?p=215
[11] Abe K *et al.* (Hyper-Kamiokande proto-) 2016 (*Preprint* 1611.06118)
[12] The NOvA collaboration 2016 NOvA Neutrino Experiment https://www-nova.fnal.gov
[13] Acciarri R *et al.* (DUNE) 2016 (*Preprint* 1601.05471)
[14] Acciarri R *et al.* (DUNE) 2015 (*Preprint* 1512.06148)
[15] Strait J *et al.* (DUNE) 2016 (*Preprint* 1601.05823)
[16] Acciarri R *et al.* (DUNE) 2016 (*Preprint* 1601.02984)
[17] Cao J *et al.* 2017 (*Preprint* 1704.08181)
[18] Coloma P and Huber P 2013 *Phys. Rev. Lett.* **111** 221802 (*Preprint* 1307.1243)
[19] Coloma P, Huber P, Jen C M and Mariani C 2014 *Phys. Rev.* **D89** 073015 (*Preprint* 1311.4506)
[20] Ankowski A M, Coloma P, Huber P, Mariani C and Vagnoni E 2015 *Phys. Rev.* **D92** 091301 (*Preprint* 1507.08561)
[21] Ankowski A M, Benhar O, Coloma P, Huber P, Jen C M, Mariani C, Meloni D and Vagnoni E 2016 *Nuovo Cim.* **C39** 233
[22] Ankowski A M 2017 (*Preprint* 1704.07835)
[23] Ankowski A M and Mariani C 2017 *J. Phys.* **G44** 054001 (*Preprint* 1609.00258)
[24] Yang Y B, Sufian R S, Alexandru A, Draper T, Glatzmaier M J, Liu K F and Zhao Y 2017 *Phys. Rev. Lett.* **118** 102001 (*Preprint* 1609.05937)
[25] Hen O, Miller G A, Piasetzky E and Weinstein L B 2016 (*Preprint* 1611.09748)
[26] Benhar O 2016 *JPS Conf. Proc.* **12** 010001
[27] Alvarez-Ruso L *et al.* 2017 (*Preprint* 1706.03621)
[28] McKeown R 2017 Jefferson lab science opportunities at 12 gev Presented at the 28th International Symposium on Lepton Photon Interactions at High Energies, Guangzhou, 7–12 August 2017.
[29] Lackowski T, Dixon S, Jedziniak R, Blewitt M and Fink L 2013 nuSTORM Project Definition Report Tech. rep. Fermi National Accelerator Laboratory (*Preprint* 1309.1389)
[30] Kyberd P *et al.* 2013 nustorm costing document URL https://indico.fnal.gov/getFile.py/-access?resId=0&materialId=8&confId=6847
[31] Adey D *et al.* 2013 Neutrinos from stored muons (storm): Expression of interest Tech. Rep. CERN-SPSC-2013-015. SPSC-EOI-009 CERN Geneva
[32] Adey D, Agarwalla S, Ankenbrandt C, Asfandiyarov R, Back J *et al.* 2013 (*Preprint* 1305.1419)
[33] Jaeckel, J and Lamont M and Vallee, C 2016 Physics Beyond Colliders http://pbc.web.cern.ch
[34] The Short Baseline Neutrino Program 2014 The Short Baseline Neutrino Program https://web.fnal.gov/collaboration/sbn/_layouts/15/start.aspx#/SitePages/Home.aspx
[35] Stratakis D 2014 *ICFA Beam Dyn. Newslett.* **65** 54–62
[36] The MICE collaboration INTERNATIONAL MUON IONIZATION COOLING EXPERIMENT http://mice.iit.edu
[37] Wilking M 2017 E61: The j-parc intermediate water cherenkov detector Presented at the EPS Conference on High Energy Physics, Venice, Italy 5–12 July 2017.
[38] Mishra S 2010 *Prog.Part.Nucl.Phys.* **64** 202–204 URL http://www.fnal.gov/directorate/Longrange/-Steering_Public/files/LOI-08-02-18_HiResMnu.pdf

Conference on Neutrino and Nuclear Physics (CNNP2017) IOP Publishing
IOP Conf. Series: Journal of Physics: Conf. Series **1056** (2018) 012034 doi:10.1088/1742-6596/1056/1/012034

Challenges for high rate signal processing for the NUMEN experiment

D Lo Presti[1,2], C Agodi[3], D Bonanno[2], D Bongiovanni[2], D Carbone[3], F Cappuzzello[1,3], M Cavallaro[3], P Finocchiaro[3], G De Geronimo[4], G Gallo[1,2], F Longhitano[2], J Oliveira[5], L Pandola[3], D Torresi[3], S Tudisco[3] and S Reito[2] for the NUMEN collaboration

[1] Dipartimento di Fisica e Astronomia, Università di Catania, Catania, Italy
[2] Istituto Nazionale di Fisica Nucleare, Sezione di Catania, Catania, Italy
[3] Istituto Nazionale di Fisica Nucleare, Laboratori Nazionali del Sud, Catania, Italy
[4] Stony Brook University, Stony Brook, NY, USA
[5] Instituto de Fisica, Universidade de Sao Paulo, Sao Paulo, Brazil

E-mail: domenico.lopresti@ct.infn.it

Abstract. The objective of the research and development activities, regarding the upgrade of the MAGNEX focal plane detector (FPD) and the development of the γ detector array for the NUMEN project, is the construction of new detectors capable to fulfil the requirements of high event rate, radiation tolerance and data acquisition and transmission bandwidth deriving from the upgrade of the Superconducting Cyclotron at INFN Laboratori Nazionali del Sud. The design of the front-end (FE) and read-out (RO) electronics has been performed in parallel with that of the new tracker. The design of the new segmented anode and the architecture of the front-end and read-out electronics are presented.

1. Introduction

The NUMEN project [1, 2, 3] aims at accessing experimentally driven information on Nuclear Matrix Elements (NME) involved in the half-life of the neutrinoless double beta decay ($0\nu\beta\beta$), by high-accuracy measurements of the cross sections of Heavy Ion (HI) induced Double Charge Exchange (DCE) reactions. Particular attention is given to the (^{18}O,^{18}Ne) and (^{20}Ne,^{20}O) reactions as tools to emulate the $\beta+\beta+$ and β-β- decays, respectively. First evidence about the possibility to get information about NME from experiments is found for both kind of reactions [4]. In the experiments, performed at INFN - Laboratory Nazionali del Sud (LNS) in Catania, the beams are accelerated by the Superconducting Cyclotron (CS) and the reaction products are detected the MAGNEX magnetic spectrometer [5, 6, 7]. The measured cross sections are challengingly low, limiting the present exploration to few selected isotopes of interest in the context of typically low-yield experimental runs.

2. Upgrade of MAGNEX detectors

A major upgrade of the LNS facility is foreseen in order to increase the experimental yield of at least two orders of magnitude, thus making feasible a systematic study of all the cases of interest. Frontiers technologies are going to be developed, to this purpose, for the accelerator [8], the magnets [9, 10], the detection systems and the targets[11].

2.1. The new FPD

The electron multiplication region and the segmented readout board of the FPD tracker [12] have to be radically redesigned taking advantage of the possibility to design a brand new full-custom FE and RO electronics, described in the following sections. A new concept for the tracker was developed. The starting point was to extract, for each track, the position of the ions at different depth in the tracker in a fully digital way. The ionization negative charge is driven by the electric field through the gas towards a multiplying stage, a triple GEM. The resulting electron jets are, then, directed towards a first layer of 750 μm pitch strips. Each strip of this layer is capacitive coupled to a twin strip in a second layer. The charge pulse inducted in the twin strip is, then, integrated by the FE and shaped. The shaped signal is compared to a suitable threshold and the logic high output of the comparator identifies the hit strip. In this scenario the position is extracted by only one strip without the need for the calculation of the center of mass. The value of the capacitance for each channel of the tracker is optimized in reference to the selected FE ASIC performances. The drift time is also measured by the FE at sub-ns resolution. An innovative scheme for the connection of FE electronics to the anode board was developed. The main objective was to place the FE in air to simplify the heat dissipation, the maintenance and, above all, the interconnections to RO. An advantage of this strategy is also the possibility to adopt countermeasures respect to high level of radiation during the experiment. The working principle of the new segmented readout board of the FPD tracker is represented in figure 1.

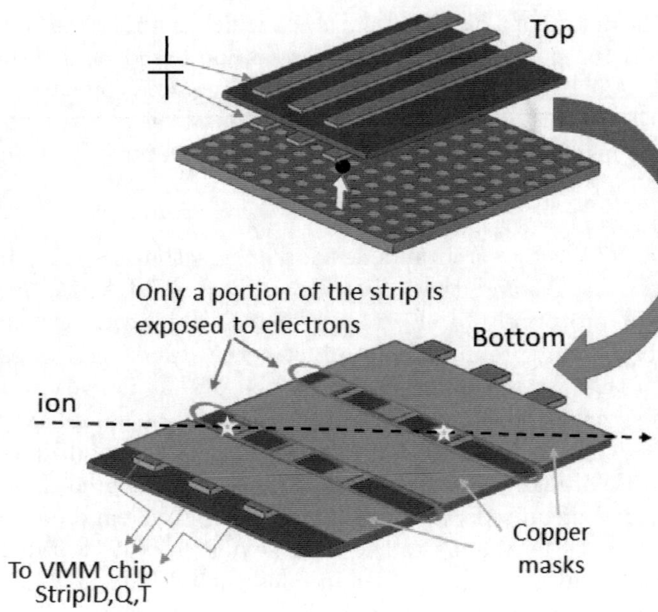

Figure 1. Schematic of the new FPD segmented anode board.

2.2. Stopping wall

The NUMEN project will also investigate promising technologies for stopping detectors, which need also to be upgraded in view of the high detection rate. Standard technologies, based on silicon pad detectors or plastic scintillators, require a high degree of segmentation (and thus high costs) in order to avoid double-hit events. At the beam currents expected for NUMEN the

IOP Conf. Series: Journal of Physics: Conf. Series **1056** (2018) 012034 doi:10.1088/1742-6596/1056/1/012034

probability of a double hit at the focal plane is considerable starting for 5 cm^2 area detectors for (^{18}O,^{18}Ne) reaction at 0°. In addition, the radiation hardness of such devices is not enough to avoid a short lifetime of these detectors. For example, in the same reaction (^{18}O,^{18}Ne) the rate limit of about 10^8 ions/cm^2, above which a silicon detector starts to deteriorate will be reached in a few days.
Interesting opportunities arise from the new technology of Silicon Carbide (SiC) crystals, which preserves many of the good properties of silicon detectors, but are much harder to radiation. Improvements in epitaxial SiC growth means that semi-insulating epitaxial SiC layers have recently become available, with thicknesses up to 100 μm. However R&D is still necessary to explore the possibility to build a reliable number of detectors for heavy ions by these epitaxial SiC. INFN is investing resources for R&D, Call SICILIA CSN V [13], in order to explore, characterize and build a wall of telescopes based on thin epitaxial SiC (100 μm thickness) for energy loss followed by thick (about 1 mm) SiC detectors for residual energy. This solution looks like to be promising because it decouples the GEM tracker from particle identification (PID) function and it is based on existing SiC technology, even if not yet implemented in commercial large area detectors. Test of characterization of epitaxial SiC under heavy ion beams are scheduled next months at the LNS in collaboration with colleagues from CNR. An ad-hoc prototype of an epitaxial SiC detector will be also built and characterized in 2018 [14]. Alternative solutions, based on phoswitch detectors, are under considerations [15, 16].

2.3. γ detector array
An important upgrade is the design of a γ calorimeter [17], which is fundamental to separate DCE cascade γ rays. A study on the performances and a prototype based on scintillator and SiPM read-out has been conducted at IFUSP in San Paolo. Particle gamma coincidence measurements are an effective tool for the investigation of nuclear reactions in the cases where the particle spectra energy resolution is insufficient to separate nearby excited nuclear states.

3. FE and RO Electronics
One of the main objectives of the research and development activities, regarding the upgrade of the focal plane detector (FPD) and the γ detector array of NUMEN, is the design and the construction of different prototypes of electronics boards which will act as a test platform for the prototype detectors. The preliminary study for the design and construction of the final electronics consists in the test of the FE board for the pre-amplification, the shaping and subsequent digitization of signals from FPD trackers, designed to be compliant with the for the high event rate and level of exposure to radiations foreseen in the final conditions; With this aim, a front-end ASIC, the VMM chip [18], identified among those documented in the literature and available in the next years, was acquired. The VMM chip is a strong candidate as the front-end of all the types of detectors in FPD, thanks to the high flexibility of its working parameters. A hosting board for the VMM2 chip was designed and constructed [19]. This board was usefull to characterize the response of the chip stand alone, for its evaluation. The board is also a useful test platform in connection with the first prototypes of trackers available. The trigger and calibration strategies, the interface to the temperature and power supply control systems and the mechanical support, will be tested. A new version of the chip, the VMM3, is now under test. We will follow the development of this device in order to evaluate if it matchs the specific requirements of the FPD front-end. The architecture, both hardware and software, is designed modular and scalable and the intelligence on board will allow for great flexibility in the overall data acquisition strategy, i.e. the trigger, the zero suppression, the data communication protocols and the slow control. In the first phase, it will be sufficient the construction, the test and the validation of each modular element ensuring, at the same time, that the built systems already exhibit the characteristics needed for the interconnections and the integration of the

final system. In Fig. 2, the architecture of the FE-RO electronics is shown.

3.1. FE Electronics

The design of the front-end electronics for the new tracker of the NUMEN Focal Plane Detector was performed in parallel with the design of the new FPD. In particular, one of the main objectives was to design a modular, scalable, radiation hard architecture, which, in addition, fulfilled the strong requirements in terms of high event rate, easy maintenance and precise synchronization.The front-end is based on the VMM chip, developed for ATLAS experiment at CERN. The architecture of the front-end electronics was designed as modular and scalable to the final dimensions of the detector. The segmented anode board was designed in order to take advantage of the unique performances of the VMM chip, allowing a digital reconstruction of the track at high event rate. In the VMM chip, each of the 64 channels provides the peak amplitude and time with respect to the bunch crossing clock or other trigger signal in a data driven mode. This is accomplished as follows. Each channel is equipped with a fast comparator with an individually adjustable threshold. When a signal crosses a set threshold a peak detection circuit is enabled. Neighbour-enable logic allows setting the threshold relatively high and yet recording very small amplitudes. At the peak a time-to-amplitude converter is started and stopped by the trigger signal. The two amplitudes are digitized and stored in a de-randomizing buffer and readout serially with a smart token passing scheme that only reads out the amplitude, timing, and addresses of the channels with information, thus dramatically reducing the data bandwidth required and resulting in a very simple readout architecture. In the selected data transfer mode, the continuous (digital) mode, a total of 38 bits are generated for each event in the VMM. The first bit is used as a readout flag, the second is the threshold crossing indicator (allows discrimination between abovethreshold and neighbour events). Next is a 6 bits word for the channel address, followed by 10 bits associated with the peak amplitude, and 20 bits associated with the timing. The 38bit word is stored in a 4events deep de-randomizing FIFO (there is one such FIFO per channel) and it is read out using a tokenpassing scheme where the token is passed firstcome firstserve only among those FIFOs that contain valid events. The first token is internally generated as needed and advanced with the token clock. The data in the FIFOs is thus sequentially multiplexed to the two digital outputs data0 and data1. The first output data0 is also used as a flag, indicating that events need to be read out from the chip. The external electronics releases a sync signal using the token clock as well (i.e. the token clock provides both advancement and data output synchronization), after which the 38bit data is shifted out in parallel to the data0 and data1 outputs using 19 clock edges of the external data clock.

3.2. RO Electronics

The main tasks of the read-out electronics for the new NUMEN FPD is the real-time data collection from the front-end boards and the high bandwidth data transmission towards data acquisition, the remote configuration and the slow control of the front-end electronics and the synchronization of the whole detector. The read-out electronics architecture, thought as modular and expandable to the final size of the detectors, is based on the System On Module (SOM) manufactured by National Instruments [20]. The SOM is a board-level circuit that integrates a system function in a single module. These very versatile devices couple high performance FPGA to powerful processor architecture and allow a graphical approach to the programming and interfacing. The tasks of SOM are the fast serial read-out of the VMM chips, the slow control of the FE and the precise synchronization of all the FE and RO boards [21].

3.3. Data Manager

The NUMEN data rate to be written on disk is estimated to be at maximum 10 Gb/s, depending on the beam configuration and on the trigger settings. Such a rate can be handled (i.e. written

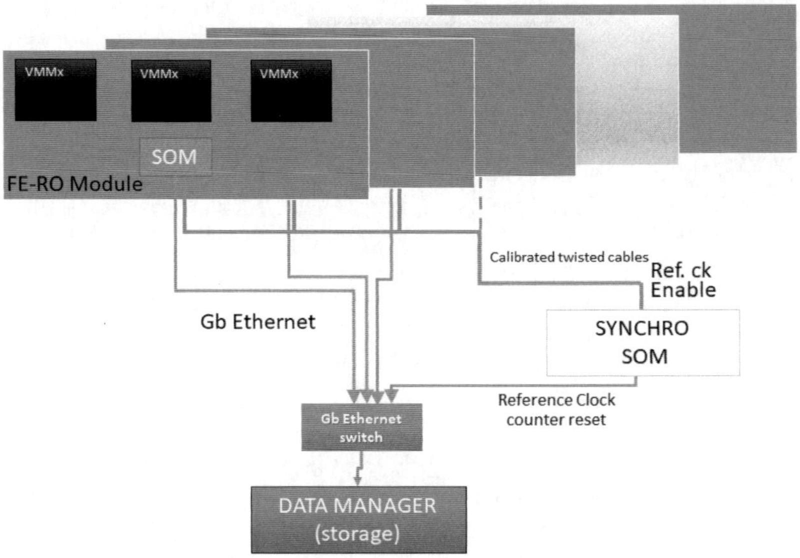

Figure 2. Architecture of FE-RO electronics and DATA Manager.

on disk) by commercial solutions, readily available on the market. The SOM modules of the electronics provide a data stream already formatted according to the TCP/IP standard protocol and transmitted over a standard Ethernet cable. One network switch (10 Gbit/s uplink) is used to collect and route the Ethernet cables coming from the SOMs. Depending on the number of cables coming from the SOMs and on the individual data rate, an additional switch could be necessary. The key component of the system, which suits all requirement and which is easily found on the market, is a one- or two-CPU 32-core server, equipped with two 10 Gbit/s Ethernet cards. Only a small fraction of the cores will be busy with the disk writing. Therefore, the free cores can be used for the event building (i.e. match the information of the same events coming from the different detector systems through different data streams) and/or for other online processing. The online processing could potentially reduce the amount of data written on disk (e.g. by compression), thus saving on the storage costs.

3.4. Radiation Tolerance and SEU

The devices to be accurately characterized from the point of view of radiation tolerance are the VMM chip and the SOM. Atlas collaboration provided the results of radiation tolerance test on VMM chip. The VMM ASIC is expected to be exposed to a total ionization dose of 100 krad according to the simulations done [22]. Deep sub-micron technologies are known to be immune to much higher Total Integrated Dose (TID) because of increasingly thinner oxide layers which can trap smaller amounts of charge. Although not expected to be a problem, the VMM3 will be tested for TID tolerance. However single event upsets (SEU) become increasingly more serious as the technology feature size decreases because of the smaller capacitance in the storage elements that need smaller energy depositions in order to flip their state. In the VMM there are two types of storage elements that require SEU protection, the configuration register, and the state machine control logic. In the data domain perhaps the 12-bit BCID register (under discussion) whereas the FIFOs need not be protected as an occasional data corruption is not an issue. To mitigate the SEU effects in the VMM storage elements two different techniques are used: the Dual Interlocked Cells (DICE) for the protection of the configuration register, and the Triple Modular Redundancy (TMR) for the state machines. The DICE uses redundancy to significantly reduce susceptibility to an upset. D flip flop based on the dual interlocked cell latches have redundant storage nodes and restore the cell's original state when an SEU error is introduced in a single node [23]. The scheme fails if multiple nodes are upset but this is far less likely. The TMR technique is used to protect the small number (less than 20) storage elements of the state machines. No data exists regarding the radiation tolerance of the SOM, despite different tests with neutrons and γ rays have been conducted on analogous devices[24]. The main criticalities are the same described for VMM, related to SEU and integrity of registers. The NUMEN collaboration plans, once accurate simulations results are available, to perform a dedicated test campaign of the overall electronics chain, FE and RO, in order to precisely determine the radiation tolerance performances and, in parallel, a possible strategy for the shielding.

4. Conclusions

Some aspects of the R&D for the upgrade of the MAGNEX detectors, looking at the high rate foreseen for the NUMEN project, has been presented. In particular, the research and development work aiming at the design of the FE-RO electronics for the NUMEN experiment was presented. Preliminary test confirms the design specifications. A complete test of a reduced scale FPD adopting all the new strategies will be performed within the end of 2018. The VMM3A chip will be available starting from spring 2018. The work is still in progress for what regards the particle identification and the γ wall. The plan is to adopt the same solution for the FE-RO electronics. A dedicated measurement campaign, aimed at the characterization of the radiation tolerance of the presented FE-RO architecture, is planned in the same period.

References

[1] Cappuzzello F *et al.* 2015 *Journal of Physics: Conference Series* **630** 012018
[2] Agodi C *et al.* 2015 *Nuclear and Particle Physics Proceedings* **265-266** 28 – 30 ISSN 2405-6014 proceedings of the Neutrino Oscillation Workshop
[3] Cappuzzello F *et al.* 2018 *Eur. Phys. J. A* **54** 72
[4] Cappuzzello F *et al.* 2015 *Eur. Phys. J. A* **51** 145
[5] Cappuzzello F *et al.* 2016 *Eur. Phys. J. A* **52** 167
[6] Cappuzzello F *et al.* 2014 *Nuclear Instruments and Methods in Physics Research A* **763** 314–319
[7] Cunsolo A *et al.* 2007 *The European Physical Journal Special Topics* **150** 343–346
[8] Calabretta L *et al.* 2017 *Modern Physics Letters A* **32**
[9] Lazzaro A *et al.* 2008 *Nucl. Inst. Meth. Phys. Res. Sect. A* **585**
[10] Lazzaro A *et al.* 2009 *Nucl. Inst. Meth. Phys. Res. Sect. A* **602**
[11] Iazzi F *et al.* 2017 *WIT Transactions on Engineering Sciences* **116** 61
[12] Cavallaro M *et al.* 2012 *The European Physical Journal A* **48** 59 ISSN 1434-601X
[13] Muoio A *et al.* 2016 *EPJ Web of Conferences* **117**
[14] Litrico G *et al. see this volume* **ibid**
[15] Carbone D *et al.* 2016 *Results in Physics* **6** 863–865
[16] Carbone D *et al.* 2018 Mini-phoswich and sipm for heavy ion detection manuscript in press at instrumentations and detectors, doi:10.1016/j.nima.2017.10.095
[17] Oliveira J R B *et al. see this volume* **ibid**
[18] De Geronimo *et al.* 2013 *IEEE Trans.Nucl.Sci.* **60** 2314–2321
[19] Bongiovanni D G *et al. see this volume* **ibid**
[20] *National Instruments website* URL `www.ni.com`
[21] Bonanno D L *et al. see this volume* **ibid**
[22] ATLAS collaboration 2013 Technical Design Report for the Phase-II Upgrade of the ATLAS TDAQ System Tech. rep. CERN-LHCC URL `https://cds.cern.ch/record/1602235/files/ATLAS-TDR-023.pdf`
[23] Faccio F *et al.* 1998 *Electronics for LHC experiments, Proceedings, 4th Workshop on Electronics for LHC Experiments, Rome, Italy* 105–113 URL `https://cds.cern.ch/record/446352`
[24] Wang J J *et al.* 1997 URL `https://ntrs.nasa.gov/archive/nasa/casi.ntrs.nasa.gov/20020060772.pdf`

IOP Conf. Series: Journal of Physics: Conf. Series **1056** (2018) 012035 doi:10.1088/1742-6596/1056/1/012035

Short-range (pairing) versus long-range (collective) correlations in two-neutron transfer reactions induced by ^{18}O

J Lubian[1], F Cappuzzello[2,3], D Carbone[2], M Cavallaro[2], M Ermamatov[1], J L Ferreira[1], R Linares[1], E Nunes[1] and B Paes[1]

[1] *Instituto de Física, Universidade Federal Fluminense, 24210-340, Niterói, Rio de Janeiro, Brazil.*

[2] *INFN, Laboratori Nazionali del Sud, I-95125, Catania, Italy.*

[3] *Dipartimento di Fisica e Astronomia, Università di Catania. I95125, Catania, Italy.*

Abstract. It is shown that the pairing correlation is very important for the elastic two-neutron transfer reactions, for reaction induced by 84 MeV ^{18}O on several targets with low collectivity in its ground state (spherical), proceeding through the one-step process. For the transition to lower excited states, the one-step process also dominated, for the final nuclei having low collectivity. On the contrary, if the collectivity of these states is considerable, the two-neutron transfer reaction is dominated by a two-step process through an intermediate partition.

1. Introduction

Transfer reactions have been used during several decades to obtain spectroscopic information of the colliding nuclei [1-4]. The single nucleon transfer reactions are used to study the single particle components of the wave functions of the initial and final states. The two-particle transfer reactions are mainly devoted to the search of signatures of pairing correlations between the two transferred nucleons. Alpha particle transfer reactions are intended to determine the cluster components of the wave functions. Finally, multinucleon transfer reactions are used to study nucleon correlations in this complex phenomenon and to produce nuclei far from the stability line because of their unique structural properties. Sometimes transfer reactions are also used to produce radioactive beams [5].

In this kind of transfer reactions, there are four important ingredients to succeed in the full description of the reaction mechanism. They are the nuclear reaction theoretical model, the optical potentials for different partitions, the spectroscopic amplitudes and the transition form-factors. The use of prior or post representations of the Hamiltonian usually helps to derive some observables, and the nonorthogonality between different partitions has to be accounted for.

Nevertheless, in some cases, the transfer reactions are the "undesired channels". This is the case of the NUMEN project [6-8] that is focused on the determinations of the double charge exchange nuclear matrix element in connection to those of neutrinoless double beta decay. Having a reaction in which the mass is conserved but the two interacting nuclei exchange two charge units trough charged meson exchange (in one- or two-step, something that still has to be determined), one has to worry about the fact that in the observed cross section may be contributions of the two-step (two correlated neutrons and two correlated protons), three-step (two correlated protons (neutrons) and two uncorrelated neutrons (protons), and four step (two uncorrelated protons and two uncorrelated neutrons in any order) transfer reaction starting and ending with the same initial and final partition [9]. In this case, the undesired contribution of transfer channel has to be determined with a high precision in order to access to the desired double charge exchange nuclear matrix element.

Having this in mind, our group has been devoted in the last couple of years to improve the reaction theory in order to access to the calculation of the transfer reaction having in mind the four factors mentioned above (proper microscopic reaction model, microscopic spectroscopic amplitudes, form factors and optical potentials) to derive reliable transfer cross sections, and to have the ability to determine transfer cross sections with confidence at least its order of magnitude. To accomplish our goal we have compared our theoretical results with high-quality data measured using the large acceptance MAGNEX spectrometer at the INFN-Laboratori Nazionali del Sud, in Catania, Italy [10,11].

In the present work we summarize some of our results concerning the two-neutron transfer reactions $^{12}C(^{18}O,^{16}O)^{14}C$ [12,13], $^{13}C(^{18}O,^{16}O)^{15}C$ [13-16], $^{16}O(^{18}O,^{16}O)^{18}O$ [17,18] and $^{64}Ni(^{18}O,^{16}O)^{66}Ni$ [19]. We will emphasize the similarities of the two-neutron elastic transfer of all these system, concerning the effect of pairing correlations between the two transferred neutrons, and the difference of the transfer to the first 2+ excited states of ^{66}Ni, for which long-range correlations prevail, to the transfer to the low-lying states of the other systems, for which the short-range pairing correlations remain important.

This work is organized as follows. The next section is devoted to the results and discussions while in the last section we give some remarks and show some perspectives on our work.

2. Results

To access to the effect of the pair correlations in the two-neutron transfer reactions we have performed two kinds of reaction calculations. The first group corresponds to the one-step (paired) two-neutron transfer models. Here the extreme cluster model and independent coordinates models were used (for details see refs. [12,14, 17-19] and references therein). The first model considers that the two neutrons are coupled antiparallel to total spin zero in the 1s state and separated to the core with different angular momenta. In the independent coordinates, the quantum numbers and coordinates of both particles are followed, and all possible combinations of the sum of the spins and angular momentum are considered according to the angular momentum and parity conservation rules. All the calculations were performed using the fresco code [20]. In the case of one-step calculations coupled reaction channels calculations (CRC) were performed. In the case of two-step transfer, coupled channel Born approximation (CCBA) calculations were performed, including couplings to all orders in the entrance partition and first-order distorted wave approximation (DWBA) between partitions.

The São Paulo potential [21] was used for both real and imaginary part of the optical potential in all partitions. In the entrance partitions, a strength factor of 0.6 was used for the imaginary part [22]. This factor accounts for the missing couplings to dissipative processes, as well as, for the coupling to continuum states, which are not explicitly considered in the calculations. In the outgoing and intermediate partitions, the strength factor for the imaginary part of the optical potential was set equal to 0.78 because no couplings were included. This coefficient was proved to be suitable for describing the elastic scattering cross section for many systems in a wide mass and energy intervals [23-25].

The spectroscopic amplitudes were determined performing extensive shell-model calculations with the NUSHELLX code [26]. Table of the spectroscopic amplitudes for the projectile and target overlaps, and the details of the interaction and model spaces used are given in refs. [12, 14, 17-19]. As mentioned above, all the experimental data were measured at Laboratori Nazionali del Sud. The details can be found in refs. [12,14, 17-19].

In Figure 1 (a) and (b) the results for the one-step, using the microscopic independent coordinates model (full line), and for the two-step (dashed lines) transfer reactions, compared to the experimental data for the ground state (g.s.) (a) and for the 2^+ (8.33 MeV) of the ^{14}C (b) in the $^{12}C(^{18}O,^{16}O)^{14}C$ [12] reaction. One can see that for the transfer to both the ground and the excited state of ^{14}C the pairing correlations prevail (especially for the excited state). The same result was found for the low-lying excited states of ^{15}C in the $^{13}C(^{18}O,^{16}O)^{15}C$ [14] reaction, showing that the extra neutron (even nucleus) was not able to considerably disturb the correlation the two neutrons in the two-neutron transfer. Also, the same result was obtained in the transfer of two neutrons to low-lying states [17] and the high natural parity excited states of ^{18}O [18] in the reaction $^{16}O(^{18}O,^{16}O)^{18}O$.

Recently [19], we also studied the two-neutron transfer reaction for the $^{64}Ni(^{18}O,^{16}O)^{66}Ni$ system. The results using the shell model for the calculations of the spectroscopic amplitudes are shown in Figure 1 for the g.s. (a) and the first excited state (b) of ^{66}Ni. One can see that in this case, the elastic transfer

IOP Conf. Series: Journal of Physics: Conf. Series **1056** (2018) 012035 doi:10.1088/1742-6596/1056/1/012035

follow the same results as for all the other more light systems mentioned above. On the contrary, the transfer to the first excited state of ^{66}Ni is dominated by the two-step process. The reason for that may lay in the fact that this state of Nickel has stronger collectivity that the excited states of the other systems studied before. To prove this we show in Table 1 the results of the reduced quadrupole transition probabilities for systems (including the ones that we are studying here).

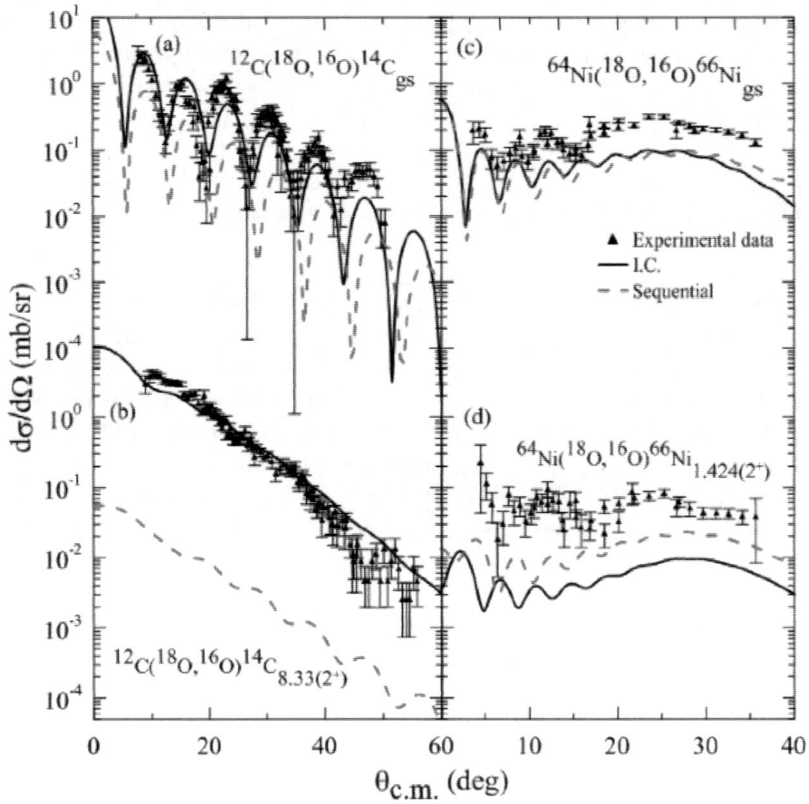

Figure1: Comparison of one-step (IC) and two-step (sequential) two-neutron transfer angular distributions with the experimental data for the ^{12}C(^{18}O,^{16}O)^{14}C [12] and ^{64}Ni(^{18}O,^{16}O)^{66}Ni [19] reactions.

Table 1. Electric quadrupole reduced transition probabilities [27].

Nucleus	**B(E2); $0^+ \rightarrow 2^+$ (e^2b^4)**
^{14}C	0.0018
^{18}O	0.0045
^{66}Ni	0.060
^{76}Ge	0.270

From Table 1 it is clearly seen that, the collectivity of the ^{66}Ni and ^{76}Ge nuclei are in fact bigger. In fact, in ref. [28] Lemaire and Low obtained similar results, as we obtained here for ^{66}Ni, for the two-neutron transfer to the g.s. and the first excited state of the ^{76}Ge nucleus.

3. Conclusions

In the present work, we have shown that the pairing correlation is very important in two-neutron transfer reactions for the final states having low collectivity, leading to the very important one-step process. Nevertheless, when the collectivity of the final states is important, the role of pairing correlations is not so relevant, and the two-step transfer reaction prevails. It would be very important to study if the pairing correlation is also important in the two-proton and deuteron transfer reactions.

Acknowledgments
The Brazilian authors acknowledge the partial financial support from CNPq, FAPERJ, and CAPES and from INCT-FNA (Instituto Nacional de Ciência e Tecnologia- Física Nuclear e Aplicações).

References:

[1] Satchler G R 1983 *Direct Nuclear Reactions* (Oxford University Press, Oxford).
[2] Broglia R A and Winther A 2004 *Heavy Ion Reactions* (Westview Press, Boulder,).
[3] Kahana S and Baltz A J 1977 in *Advances in Nuclear Physics*, edited by M. Baranger and E. Vogt (Plenum Press, New York,), Vol. 9, pp. 1–122.
[4] von Oerzen V and Vitturi A 2001 *Rep. Prog. Phys.* **64** 1247.
[5] Lépine-Szily A, Lichtenthales R, Guimarães V 2014 *Eur. Phys. J. A* **50** 128.
[6] Cappuzzello F *et al.* 2018 *Eur. Phys, J. A* **54** 72.
[7] Cappuzzello F *et al.* 2015 *J. Phys.: Conf. Ser.* **630** 012018.
[8] Agodi C *et al.* 2015 *Nucl. Part. Phys. Proc.* **265** 28.
[9] Cappuzzello F *et al.* 2015 *Eur. Phys. J. A* **51** 145.
[10] Cappuzzello F, Agodi C, Carbone C, and Cavallaro M 2016 *Eur. Phys. J. A* **52** 167.
[11] Cunsolo A *et al.* 2007 *Eur. Phys. J. Special Topics* **150** 343.
[12] Cavallaro M *et al.* 2013 *Phys. Rev. C* **88** 054601.
[13] Cappuzzello F *et al.* 2015 *Nat. Comm.* **6** 6743.
[14] Carbone D *et al.* 2017 *Phys. Rev. C* **95** 034603.
[15] Carbone D 2015 *Eur. Phys. J. Plus* **130** 143.
[16] Cavallaro M *et al.* 2016 *Phys. Rev. C* **93** 064323.
[17] Ermamatov M *et al.* 2016 *Phys. Rev. C* **94** 024610.
[18] Ermamatov M *et al.* 2017 *Phys. Rev. C* **96** 044603.
[19] Paes B *et al.* 2017 *Phys. Rev. C* **96** 044612.
[20] Thompson I J 1988 *Comput. Phys. Rep.* **7** 167.
[21] Chamon L C, Pereira D, Hussein M S, Candido Ribeiro M A, and Galetti D 1997 *Phys. Rev. Lett.* **79** 5218.
[22] Pereira D, Lubian J, Oliveira J, Sousa D and Chamon L C 2009 *Phys. Lett. B* **670** 330.
[23] Gasques L R, Chamon L C, Gomes P R S and Lubian J 2006 *Nucl. Phys. A* **764** 135.
[24] Pereira D *et al* 2012 *Phys. Lett. B* **710** 426.
[25] Oliveira J R B *et al* 2013 *J. Phys. G* **40** 105101.
[26] Rae W D M, [http://www.garsington.eclipse.co.uk].
[27] [https://www.nndc.bnl.gov].
[28] Lemaire M C and Low K S 1977 *Phys. Rev. C* **16** 183.

IOP Conf. Series: Journal of Physics: Conf. Series **1056** (2018) 012036 doi:10.1088/1742-6596/1056/1/012036

Numerical characterization of the ARAPUCA: a new approach for LAr scintillation light detection

F Marinho[1], L Paulucci[2], A. A. Machado[2] and E Segreto[3]

[1] Universidade Federal de São Carlos, Rodovia Anhanguera, km 174, 13604-900, Araras, SP, Brazil

[2] Universidade Federal do ABC, Av. dos Estados, 5001, 09210-170, Santo André, SP, Brazil

[3] Universidade Estadual de Campinas, Rua Sérgio Buarque de Holanda, 777, 13083-859, Campinas, SP, Brazil

E-mail: `fmarinho@ufscar.br`

Abstract. The ARAPUCA concept has been proposed as a simple and neat solution for increasing the effective collection area of SiPMs through the shifting and trapping of scintillation light in noble liquids, thus with great potential for improving timing and calorimetry resolution in neutrino and dark matter search experiments using time projection chambers. It is expected to achieve a single photon detection efficiency larger than 1%. The initial design consists of a box made of highly reflective internal surface material and with an acceptance window for photons composed of two shifters and a dichroic filter. The first shifter converts liquid argon scintillation VUV light to a photon of wavelength smaller than the dichroic cutoff, so the surface is highly transparent to it. When passing through the dichroic filter, it reaches the second shifter which allows the photon to be shifted to the visible region and be detected by the SiPM nested inside it. When it enters the box, the photon will likely reflect a few times, including on the dichroic filter surface, before being detected. We present a full numerical description of the device using a Monte Carlo framework, including characterization of the acceptance window, models of reflection of different materials, and sensor quantum efficiency, that can now be used to further improve the detection efficiency by comparing different geometries, positions of SiPM and materials. Estimates of simulated efficiencies, number of reflections and acquisition time are presented and compared to analytical calculations. Those are very promising results, giving a total efficiency for the detection of scintillation light in liquid argon of 1.7±0.3%. Comparison of the estimated total efficiency with a preliminary result from an experimental test with an ARAPUCA prototype made in Brazil is also presented.

1. Introduction

Investigations at the frontier of particle physics involve measurements of low frequency events, such as interacting neutrinos, possible proton decay, and detection of dark matter. New technologies are in high demand to improve the chance of detecting such events. Scintillation light detectors are important component systems for the next generation of LArTPCs experiments which can provide additional information regarding calorimetry and timing thus enhancing event reconstruction performance. To explore the full potential of these systems one must consider the best options for detector technologies under development and their impact on quality of physics measurements.

One recent suggestion for photon detection at the VUV range is the use of ARAPUCA devices [1] for trapping photons and increasing the detection effective area of silicon photomultipliers

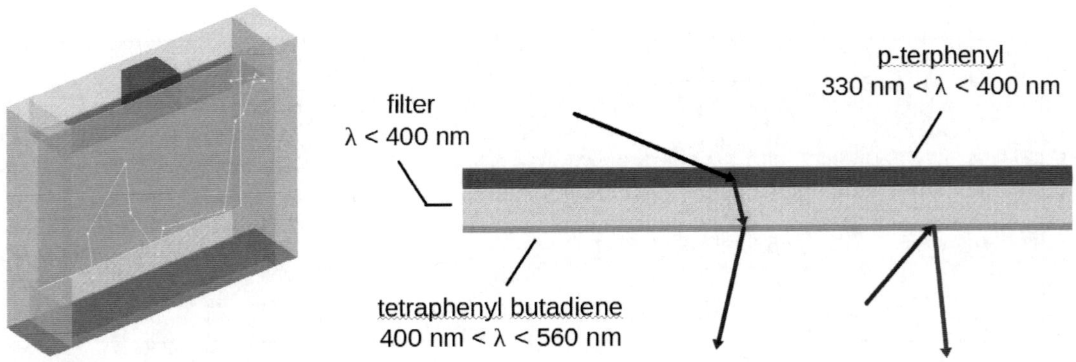

Figure 1. ARAPUCA device geometry (left): acceptance window (yellow), SiPM sensor (blue) and reflective cavity (gray). Acceptance window concept (right): PTP (dark magenta), dichroic filter (gray), TPB (blue).

(SiPMs). The concept consists of a box cavity with highly reflective internal walls inside which photons can reflect back and forth until reaching its photomultipliers. This box is equipped with an acceptance window formed by two layers of wavelength shifters and a dichroic filter which allows photons to pass efficiently inside the cavity. It also presents high internal reflectance to the accepted photons, therefore acting as a trap. For use in a liquid argon (LAr) environment, important for dark matter and neutrino searches, those layers were optimized as follows. The scintillation light emitted by argon atoms, in the vacuum ultra-violet spectral region (VUV) with a wavelength around 127 nm, are converted to the detector's range (visible). The first shifter in the acceptance window, made of p-terphenyl (PTP) [2], converts the liquid argon scintillation light to a photon of wavelength that peaks around 350 nm, smaller than the dichroic cutoff ($\sim$ 400 nm). When passing through the dichroic filter, it reaches the second shifter, made of Tetraphenyl-butadiene (TPB) which shifts to the visible region (peaking at $\sim$ 430 nm) [3], ensuring the dichroic filter surface to now become highly reflective. The photon can reflect a few times on the internal surface before being detected by the SiPM.

We present a Monte Carlo based approach to characterize the ARAPUCA light scintillation detector device as a tool for research and development of this technology.

2. Numerical simulation

The computational description of the ARAPUCA was developed in order to include its main components and their correspondent optical functions correctly. For that, the Geant4 simulation framework [4, 5] provides a suitable set of functionalities that allows the propagation of optical photons through various materials and interfaces considering different models for light reflection, refraction, absorption, emission, etc. Figure 1 shows a complete 3D model of a detector indicating its acceptance window, reflective cavity and a silicon photomultiplier (SiPM). The green line indicates a photon optical path and the yellow dots its reflection points on the cavity and filter internal reflective walls.

The acceptance window consists of a dichroic filter with one layer of $\sim$ 0.3mg/cm^2 PTP deposited on top and one layer of $\sim 7 \times 10^{-2}$mg/cm^2 TPB deposited on the bottom of the filter as illustrated in figure 1. The widths of the PTP and TPB are chosen to ensure optimal wavelength shifting [2]. The incoming VUV photons are absorbed by the PTP material which in turn emits photons isotropically in the 330-400 nm range. As a consequence, half of these emitted photons pass through the dichroic material, reach the TPB and are absorbed. The TPB emits photons in the 400-560 nm range, which therefore, are trapped within the cavity.

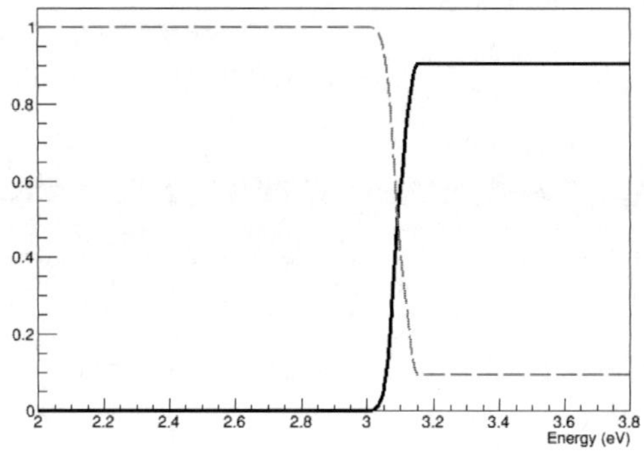

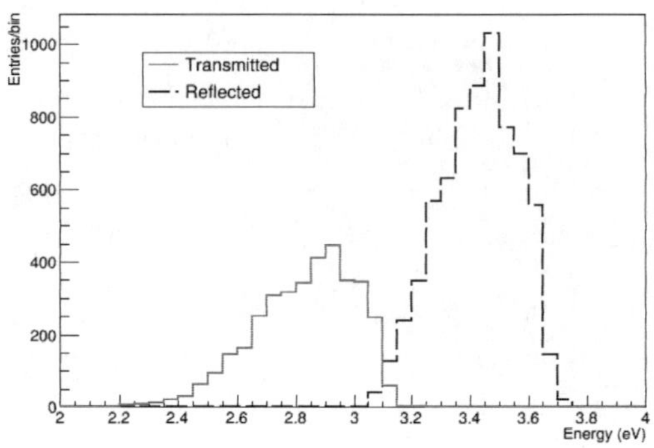

Figure 2. Dichroic filter (left): Transmission (solid black) and reflection (dashed red) [6]. Window emitted photon spectra (right).

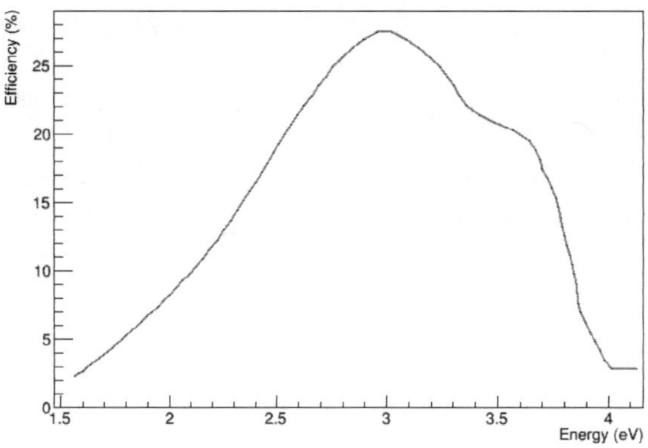

Figure 3. SiPM photon detection efficiency as function of incoming photon energy.

Figure 2 shows the average transmittance and reflectance of the dichroic filter at 45° incidence angle (left) and the spectra for the accepted and reflected light (right) both as a function of the photons' energy.

For the internal walls there are two main possibilities for modelling the reflection types occurring due to the materials envisioned for the device. The box can be made of Teflon^{TM} PTFE which offers high reflection ($R > 95\%$) with Lambertian angular distribution, however the minimum thickness of the walls is limited due to the need to avoid transmission. An alternative is to use some adequate material for mechanical support and coating the internal walls with Vikuiti^{TM} 3M which is a specular reflective film ($R \sim 98\%$) [7]. The trajectory of the photons inside the cavity is fully described in the simulation which allows to calculate the time interval between photon acceptance and detection at the SiPM surface. This is an important parameter for the detector timing evaluation.

Photon detection efficiency dependence with photon energy must be taken into account for the SiPM sensor. In particular, we used Sensl C-Series models with efficiency shown as function of energy in figure 3 in our model [8].

3. Results

A detailed description of the photons trajectories within the different materials of the device geometry allows to evaluate all the relevant parameters for its characterization and overall operation. The following sections describe results obtained with the complete simulation of the ARAPUCA or its components. Comparison with analytic calculations or measurements is provided when those are accessible.

3.1. Efficiency

The total efficiency of the detector can be defined as:

$$\epsilon_{total} = \frac{N_{detected}}{N_{total}}, \tag{1}$$

where $N_{detected}$ is the number of observed photons from the SiPM signal and N_{total} is the number of incoming photons which arrive at the acceptance window from outside.

However, it is important to characterize each component in terms of its own efficiency. Hence, one can factorize the total efficiency as:

$$\epsilon_{total} \approx \epsilon_{acceptance} \times \epsilon_{collection} \times \epsilon_{SiPM}, \tag{2}$$

where $\epsilon_{acceptance}$ is the window efficiency on converting incoming VUV photons to visible light inside the cavity and $\epsilon_{collection}$ gives the fraction of accepted photons which reach the SiPM (i.e. collected and not absorbed in the cavity walls). The SiPM efficiency is given by:

$$\epsilon_{SiPM} = \frac{1}{N_{collected}} \int_{E_0}^{E_1} \epsilon_{SiPM}(E) \frac{dN_{collected}}{dE} dE, \tag{3}$$

where $N_{collected}$ is the number of photons which arrive at the surface of the SiPM and $\epsilon_{SiPM}(E)$ is the sensor photon detection efficiency as a function of the photon energy.

Note the factorization in equation 2 is possible because $\epsilon_{acceptance}$ is constant around the VUV incoming photons energy. The $\epsilon_{collection}$ is also assumed constant as it should only depend on R, the active coverage f and geometry of the reflective cavity. In both aforementioned materials R is above 95% and approximately constant across the whole wavelength range inside the cavity. The other two parameters are independent.

Table 1 shows the value obtained for these efficiencies calculated for a device with 36 mm × 25 mm × 6 mm internal dimensions and $R = 95\%$. An experimental test realized with a prototype following the same configuration has provided a value $\epsilon_{total}^{exp} \sim 1.8\%$ which is in reasonable agreement with the Monte Carlo estimate.

Table 1. Monte Carlo estimated ARAPUCA efficiencies.

$\epsilon_{acceptance}$	$\epsilon_{collection}$	ϵ_{SiPM}	ϵ_{total}
$34.5 \pm 6.0\%$	$19.7 \pm 0.4\%$	$25.1 \pm 0.3\%$	$1.7 \pm 0.3\%$

Studies with different geometries for the cavity were also performed such that partial focusing of the light could be used for improving the efficiency. Figure 4 shows two possible alternatives which provide $\epsilon_{collection}$ estimates above analytic calculations given by [1]:

$$\epsilon_{collection}^{analytic} = \frac{f}{1 - R(1 - f)}. \tag{4}$$

Note that it is not possible to obtain efficiencies higher than the analytic estimates with the proposed non-focusing box design from figure 1. The cylindrical cavity with a disk shaped

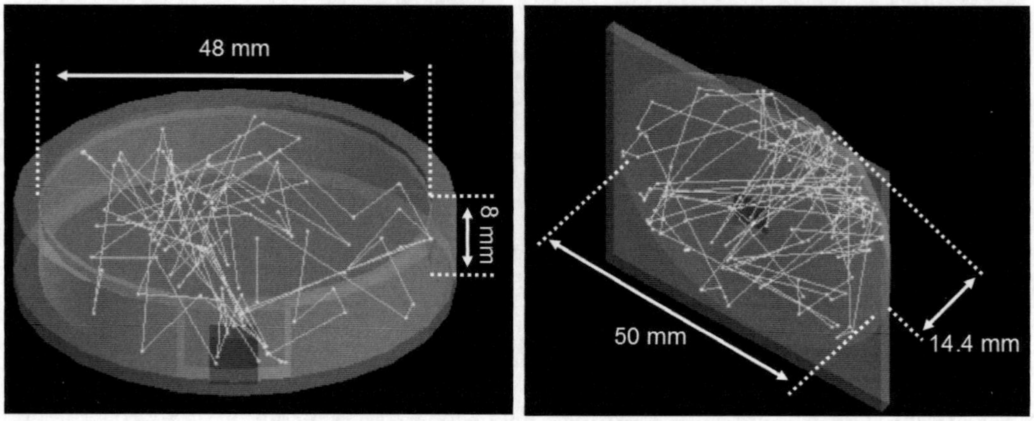

Figure 4. Focusing geometries.

acceptance window provides $\epsilon_{collection} = 14.9 \pm 0.4\%$ while the analytic estimate for an equivalent setup (same R and f) gives 13.0%. The cavity with a spherical dome shaped window gives $\epsilon_{collection} = 16.0 \pm 0.4\%$ and the analytic estimate is 13.7%. These calculations indicate that there is room for improvement of up to 15% on the overall efficiency with just geometry optimization of the original ARAPUCA proposal.

3.2. Timing

The evaluation of the time characteristics of the device is another important aspect that needs to be understood if considering it to be used for timing purposes in a LArTPC experiment. There are three major delay sources that determine the acquisition time of the device:

$$t_{acquisition} = t_{PTP} + t_{TPB} + t_{collection}, \tag{5}$$

where t_{PTP} and t_{TPB} are the emission decay time of the wavelength shifters and $t_{collection}$ is the time it takes for the photon accepted in the cavity to arrive on the SiPM after successive reflections.

All these times are random quantities in a event-by-event basis. Figure 5 on the left shows an example of the $t_{collection}$ distribution obtained from the Monte Carlo simulation. As expected the distribution has a sharp decay as the overall distance traveled is very small inside the cavity. One can evaluate a reasonable distribution for $t_{aquisition}$ assuming a mixture of exponential distributions for each of the emission times and their respective measured values [9]. Figure 5 on the right shows the obtained distribution with a most probable acquisition time of $\sim$ 4 ns. The long tail results mainly from the TPB time response function to the VUV 127 nm photons. In a LArTPC such fast acquisition may allow to determine with good resolution the time t_0 for the interacting primary particle in a physics event or to help as trigger device to distinguish beam events from other sources in a surface installed experiment.

4. Conclusion

An ARAPUCA full simulation was implemented and is currently employed as a R&D design tool for this detector technology. The total efficiency obtained from a prototype simulation and its experimental value were in good agreement. Comparison between the analytic estimates and simulation for the collection efficiency, number of internal reflections and timing as function of the f and R parameters follow the same behavior. It is a flexible software application which allows not only parameter optimization but geometrical and physical modifications with ease.

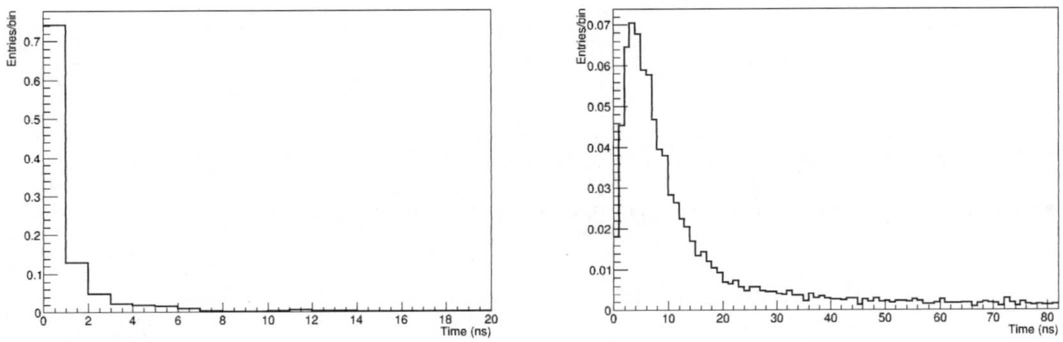

Figure 5. Expected distribution for $t_{collection}$ (left) and $t_{acquisition}$ (right).

In particular one can evaluate the performance of each component according to its features such as wavelength shifters types, width, spectra emission dependence on temperature, dichroic filter transmission and reflection as function of the angle, etc. One can also choose the main surface models according to the type of cavity walls and adjust the response of the SiPM used according to the operating parameters. In addition a set of analysis algorithms were also developed to evaluate all the relevant figures of merit of interest.

Acknowledgments

The authors would like to thank Fundação de Amparo à Pesquisa do Estado de São Paulo (FAPESP) for financial support under grants n° 2016/01106-5 and n° 2017/13942-5.

References

[1] Machado A A and Segreto E 2016 *JINST* ARAPUCA a new device for liquid argon scintillation light detection **11** C02004
[2] DeVol T A, Wehe D K, Knol G F 1993 *Nucl. Instr. Meth. Phys. Res.* A Evaluation of p-terphenyl and 2,2" dimethyl-p-terphenyl as wavelength shifters for barium fluoride **327** 354
[3] Francini R *et al* 2013 *JINST* VUV-Vis optical characterization of Tetraphenyl-butadiene films on glass and specular reflector substrates from room to liquid Argon temperature **8** P09006
[4] Agostinelli S *et al* 2003 *Nucl. Instr. Meth. Phys. Res.* A Geant4 - a simulation toolkit **506** 250
[5] Allison J *et al* 2016 *Nucl. Instr. Meth. Phys. Res.* A Recent developments in GEANT4 **835** 186
[6] Dichroic filter, edmundoptics.com.
[7] Vikuiti^{TM} Enhanced Specular Reflector Film (ESR), 3M.com/displayfilms
[8] Sensl C-Series SiPM, sensl.com/products/c-series
[9] Segreto E 2015 *Phys. Rev.* C Evidence of delayed light emission of tetraphenyl-butadiene excited by liquid-argon scintillation light **91** 035503

Is it possible to study neutrinoless $\beta\beta$ decay by measuring double Gamow-Teller transitions?

Javier Menéndez[1], **Noritaka Shimizu**[2] **and Kentaro Yako**[3]

Center for Nuclear Study, University of Tokyo, Bunkyo-ku, Hongo, 113-0033, Japan

E-mail:
[1]menendez@cns.s.u-tokyo.ac.jp
[2]shimizu@cns.s.u-tokyo.ac.jp
[3]yako@cns.s.u-tokyo.ac.jp

Abstract. Searches of neutrinoless double-beta decay require information on the value of the nuclear matrix elements that rule the process to plan and interpret experiments. At present, however, even the matrix elements obtained with the most reliable many-body approaches do not agree to each other better than a factor two or three. A usual test of the many-body calculations is the comparison to several nuclear observables, but so far no nuclear structure property has been found to show a good correlation to neutrinoless double-beta decay. Here we propose that double charge-exchange experiments can offer a very valuable tool to provide insights on neutrinoless double-beta decay. Double charge-exchange reactions are being currently performed in various laboratories worldwide and aim to find the novel nuclear collectivity given by double Gamow-Teller excitations. Our results suggest a good linear correlation between double Gamow-Teller transitions to the ground state of the final nucleus and neutrinoless double-beta decay nuclear matrix elements. The correlation seems robust across pf-shell nuclei.

1. Introduction

Neutrinoless double-beta ($0\nu\beta\beta$) decay is possibly the best probe available to test whether neutrinos are its own antiparticle. While present $0\nu\beta\beta$ decay half-life limits are impressive, exceeding 10^{25} years for ^{136}Xe, ^{76}Ge and ^{130}Te [1, 2, 3, 4], observing the decay rates predicted by the standard model of particle physics demands improving the half-life sensitivity of present experiments by two orders of magnitude, if the ordering of the neutrino masses is "inverted", or by four orders of magnitude, if it is "normal" [5]. The estimated sensitivities, however, depend critically in the nuclear matrix elements (NMEs) that control the decay, $M^{0\nu\beta\beta}$ [6]

$$[T^{0\nu}_{1/2}]^{-1} = G^{0\nu} \left| M^{0\nu\beta\beta} \right|^2 m^2_{\beta\beta} \,, \tag{1}$$

with $G_{0\nu}$ a phase-space factor, and $m_{\beta\beta} = \sum_l U_{el} m_l / m_e$ a combination of the neutrino masses and mixing matrix, normalized to the electron mass. Strategies to design future experiments exploring "inverted" and "normal" scenarios require a good knowledge of the NMEs.

Unfortunately nuclear theory is currently not in a position to provide very reliable NMEs. The best available calculations disagree between each other by a factor two or three when comparing results obtained with different nuclear many-body approaches [6]. These calculations are tested against experimental data on several nuclear structure properties: excitation spectra, electromagnetic and β decays [7], knockout and transfer reactions [8, 9], $\beta\beta$ decay with the emission of two neutrinos [10]... Various many-body approaches generally find good agreement with data when confronting these properties. By contrast, they differ significantly on their prediction of $0\nu\beta\beta$ decay NMEs. So far no nuclear observable has been found to be especially well correlated to $0\nu\beta\beta$ decay. The observation that $0\nu\beta\beta$ decay seems to be subtle on its own way has prevented to pin down the value of the NMEs.

2. Double Gamow-Teller transitions and $\beta\beta$ decay

Gamow-Teller (GT) strength distributions measured via charge-exchange experiments [11, 12, 13] are good tests of the many-body calculations. GT distributions have the advantage that they can probe transitions not accessed in β decay due to the limited Q_β value, and in addition they exhibit collective behavior best manifested in the GT giant resonance. Recently, there is a strong experimental interest in performing double charge-exchange experiments to access the novel nuclear collectivity indicated by double GT (DGT) transitions [14, 15, 16, 17]. Similarly to the GT case, this brings the opportunity to assess the validity of many-body calculations of $\beta\beta$ decays by testing predictions against measured DGT distributions.

Double charge-exchange reactions are a result of the strong interaction. On the other hand $\beta\beta$ decay is governed by the weak interaction—the reason why it is related to neutrinos. Nonetheless it is important to keep in mind that in both cases a pair of neutrons can become a pair of protons, via nucleon exchange in charge-exchange reactions, or via nucleon decay in $\beta\beta$ decay. In addition, the two processes are sensitive to the spin of the nucleons. We also note that $0\nu\beta\beta$ decay occurs at second-order in the weak interaction, and in principle it should depend explicitly on the states of the intermediate nucleus. However, it does not make much difference if one uses the so-called "closure approximation" that brings the NME only dependent on the initial (i) and final (f) nuclear states [18]. When doing so the $0\nu\beta\beta$ decay and DGT matrix elements read [6, 19]

$$M^{0\nu\beta\beta}(i\to f) = M^{0\nu}_{GT} + \frac{M^{0\nu}_F}{g_A^2} + M^{0\nu}_T = \sum_{X=GT,F,T} \langle f|\sum_{a,b} H_X(r_{ab})\, S_X\, \tau^+_a\tau^+_b\,|i\rangle\ , \quad (2)$$

$$M^{DGT}(i\to f) = \langle f|\sum_{a,b}[\boldsymbol{\sigma}_a\tau^+_a \times \boldsymbol{\sigma}_b\tau^+_b]^\lambda\,|i\rangle\ , \quad (3)$$

where the sums run over all nucleons, τ represents the isospin, and the DGT operator is coupled to $\lambda = 0, 2$. S_X denote the GT, Fermi and tensor spin structures that contribute to $0\nu\beta\beta$ decay, and $H_X(r_{ab})$ are the so-called neutrino potentials corresponding to each spin structure —divided by g_A^2 in the Fermi case. The neutrino potentials depend on the distance between the decaying nucleons, and are given by [6]

$$H_X(r_{ab}) = \frac{2R}{\pi}\int_0^\infty q^2\, dq \frac{j_X(q\,r_{ab})h_X(q)}{q\,(q+\mu)}\ , \quad (4)$$

with $\boldsymbol{q}$ the momentum transfer, $R = 1.2A^{1/3}$ fm the nuclear radius, $j_F(q\,r_{ab}) = j_{GT}(q\,r_{ab}) = j_0(q\,r_{ab})$ and $j_T(q\,r_{ab}) = j_2(q\,r_{ab})$ spherical Bessel functions, and μ the so-called closure energy [18]. The explicit form of the $h_X(q)$ functions can be found in Ref. [20].

The $M^{0\nu}_{GT}$ part with $S_{GT} = \boldsymbol{\sigma}_a \cdot \boldsymbol{\sigma}_b$ is by far dominant in $0\nu\beta\beta$ decay, accounting for almost 90% of the total NME. Therefore the operator structure that mainly drives $0\nu\beta\beta$ decay is that of DGT transitions with $\lambda = 0$ coupling. Overall, the main differences between the two processes are the presence of the neutrino potentials in $0\nu\beta\beta$ decay, and the capability of DGT transitions to connect states that differ in up to two units of angular momentum. In addition, in practice $0\nu\beta\beta$ NMEs only refer to decays to the ground state of the final nucleus. Other decays are very suppressed, or energy forbidden, due to the small—about a couple of MeV at most—$Q_{\beta\beta}$ values. In contrast, DGT transitions can populate final states excited up to several tens of MeV.

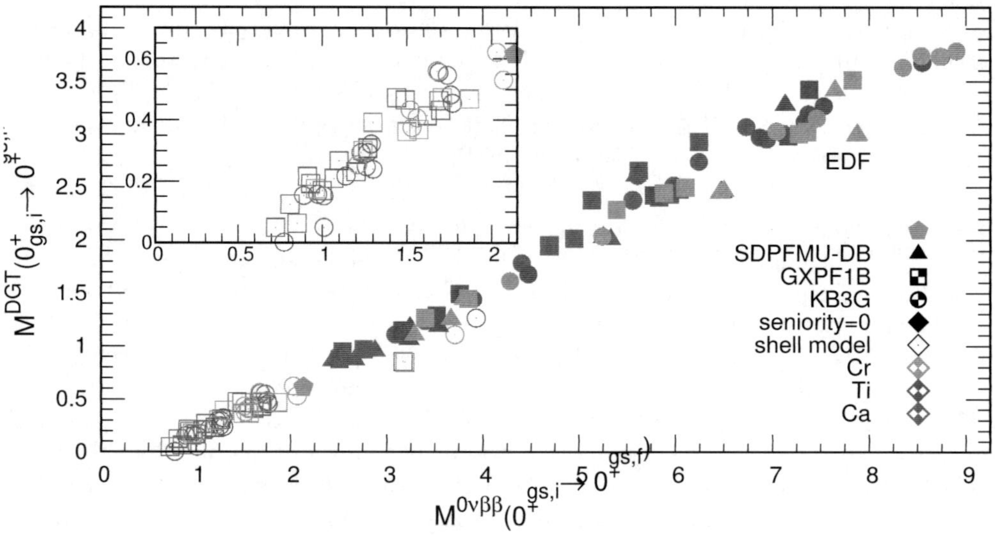

Figure 1. Correlation between the $0\nu\beta\beta$ decay NME $M^{0\nu\beta\beta}(0^+_{gs,i} \rightarrow 0^+_{gs,f})$ and the DGT transition to the ground state of the final nucleus $M^{\rm DGT}(0^+_{gs,i} \rightarrow 0^+_{gs,f})$. Shell model results for initial calcium (Ca), titanium (Ti) and chromium (Cr) isotopes are shown in red, blue and orange, respectively. Calculations with the pf-shell GXPF1B [22], KB3G [23] and $sd-pf$ SDPFMU-DB [24] interactions are denoted by squares, circles and triangles, respectively. Open symbols indicate fully correlated nuclear states, while filled ones refer to nuclei restricted to seniority-zero configurations. The shell model results are compared to the non-relativistic energy-density functional (EDF) theory ^{48}Ca result from Ref. [26] (light blue pentagon). The inset zooms in the results corresponding to fully correlated states.

3. Results for DGT and $0\nu\beta\beta$ decay matrix elements

The entire DGT distribution of the $\beta\beta$ emitter ^{48}Ca was recently calculated in Ref. [19], using the shell model many-body framework. Ref. [19] considered a typical shell model space consisting of the pf-shell—this is, including the $0f_{7/2}$, $1p_{3/2}$, $1p_{1/2}$ and $0f_{5/2}$ single-particle orbitals—and an extended one comprising the $sd-pf$ configuration space—with the $0d_{5/2}$, $1s_{1/2}$ and $1d_{3/2}$ orbitals included as well. The results predicted a DGT distribution relatively independent of the size of the configuration space and the shell model interaction used.

An analysis of the sensitivity of the total DGT strength distribution to isovector—dominated by like-particle—and isoscalar—to which only proton-neutron contribute—pairing correlations showed that isovector pairing correlations tend to push the DGT giant resonance to higher energies [19]. Combining this sensitivity with the well-known dependence of the $0\nu\beta\beta$ decay NME to isovector pairing [21], Ref. [19] showed that, for moderate increases of the isovector pairing correlations, the energy of the ^{48}Ca DGT giant resonance and the $0\nu\beta\beta$ decay NME value follow a simple linear correlation. The fact that a collective property such as the energy of the DGT giant resonance might be related to the value of an individual matrix element is very promising, and supports the similarity between DGT transitions and $0\nu\beta\beta$ decay. Further, the relation suggests that nuclear structure measurements of DGT transitions can be used to study, indirectly, $0\nu\beta\beta$ decay.

If there is a DGT transition that can be expected to be especially related to the $0\nu\beta\beta$ decay NME this is the DGT transition to the ground state of the final nucleus. In such case the final states of the $0\nu\beta\beta$ decay and DGT transitions are the same. Moreover, if the initial state is a 0^+ state—as the ground states of every even-even nucleus are— the DGT transition to the ground state can only be mediated by the $\lambda = 0$ channel, which implies that the spin structure of the DGT transition and the (dominant part of) the $0\nu\beta\beta$ decay are common as well. This implies that when only ground states are involved, and taking into account the validity of the closure approximation, the only difference between the $0\nu\beta\beta$ decay and the DGT transition matrix elements is caused by the neutrino potential $H(r_{ab})$. The potential signals the fundamental difference between the two processes: while $0\nu\beta\beta$ decay is driven by the weak interaction, DGT transitions, explored in charge-exchange experiments, are a result of the strong force.

Figure 1 shows the results of the shell model calculation of $0\nu\beta\beta$ decay NMEs compared to the DGT transitions to the ground state. The decays of calcium (shown in red), titanium (blue) and chromium (orange) isotopes are shown. Figure 1 considers a total of twenty six initial even-even nuclei, from the lightest ^{42}Ca to the heaviest ^{60}Cr. In addition, Fig. 1 explores two kind of nuclear states: the open symbols correspond to full shell model states, while the filled symbols indicate simplified states that entirely consist of pairs of neutrons and pairs of protons coupled to 0^+—known as seniority-zero states. Finally, Fig. 1 shows results obtained using different configuration spaces and shell-model interactions: the pf-shell interactions GXPF1B [22] (squares) and KB3G [23] (circles), and the extended $sd - pf$ space SDPFMU-DB interaction [24] (triangles). The results were obtained with the shell model codes KSHELL [25] and NATHAN [7].

Figure 1 highlights a very good linear correlation between $0\nu\beta\beta$ decay NMEs and DGT transitions to the ground state. The correlation seems robust. It is valid across a very wide range of nuclei in this region—mass numbers $42 \leq A \leq 60$—and it is also common to the three shell model interactions used, in the usual and extended configuration spaces. Furthermore, the linear correlation remains valid regardless of whether we consider fully correlated shell model states or fairly simple seniority-zero wave functions. Because seniority-zero nuclear states overpredict the $0\nu\beta\beta$ decay NMEs, the last property implies that the correlation is valid for a very wide range of NME values $0 \lesssim M^{0\nu\beta\beta} \lesssim 10$. Such a broad range encompasses the factor-two-or-three range of NME values that covers the results of the different many-body calculations for any $\beta\beta$ emitter.

It is interesting to note that the ^{48}Ca results obtained in the framework of energy density functional theory [26] (light blue pentagon in Fig. 1) are consistent with the linear correlation found in the shell model calculations. Reference [19] extended to heavier nuclei the comparison of $0\nu\beta\beta$ decay and DGT transitions to ground states, finding a linear correlation very similar to the one in Fig. 1 [19]. Again, for heavier nuclei the results from energy-density functional theory agree well with the correlation observed in the shell model calculations. In contrast, the results of the quasiparticle random phase approximation (known as QRPA) [27] behave very differently, with DGT transition values that tend to be persistently small [19].

How can we understand the existence of a rather universal linear correlation between $0\nu\beta\beta$ decay and DGT matrix elements? We can try to answer this question by studying the matrix element distributions as a function of the distance between decaying/transferred nucleons, and of the momentum transfer. For $0\nu\beta\beta$ decay the normalized distributions are defined as

$$C_{GT}(r) = \langle f | \sum_{ab} \delta(r - r_{ab})\, H_{GT}(r_{ab})\, \boldsymbol{\sigma}_a \cdot \boldsymbol{\sigma}_b\, \tau_a \tau_b | i \rangle / M^{0\nu}_{GT}\,, \tag{5}$$

$$C_{GT}(p) = \langle f | \sum_{ab} \delta(p - q)\, \hat{H}_{GT}(q)\, \boldsymbol{\sigma}_a \cdot \boldsymbol{\sigma}_b\, \tau_a \tau_b | i \rangle / M^{0\nu}_{GT}. \tag{6}$$

An analogous definition can be used for DGT transitions, except that in that case there is no neutrino potential and the spin structure is defined with a different normalization $[\boldsymbol{\sigma}_a \times \boldsymbol{\sigma}_b]^0 = \frac{-1}{\sqrt{3}} \boldsymbol{\sigma}_a \cdot \boldsymbol{\sigma}_b$ (the different definition does not affect the normalized distributions).

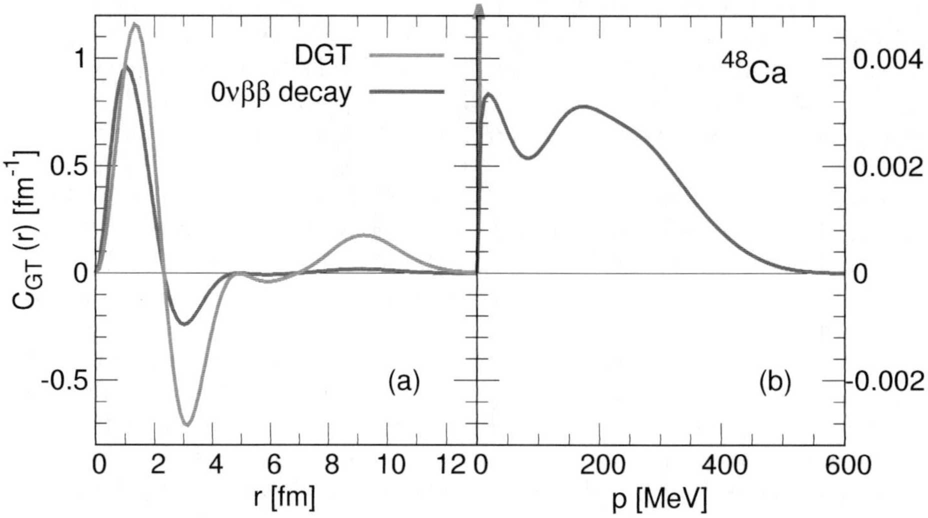

Figure 2. Normalized distributions of the Gamow-Teller part of the $0\nu\beta\beta$ decay NME $M^{0\nu}_{\rm GT}$ (red) and and the DGT matrix element $M^{\rm DGT}$ (orange), as a function of (a) the internucleon distance $C_{GT}(r)$, (b) the momentum transfer $C_{GT}(p)$. The results correspond to ^{48}Ca calculated with the pf-shell GXPF1B interaction.

Figure 2 shows the normalized radial and momentum distributions for ^{48}Ca. The results show that the momentum distribution of the two processes is very different: while the DGT transitions in Eq. (3) do not entail any momentum transfer[1], typical transferred momenta amount to $p \sim 100$ MeV in $0\nu\beta\beta$ decay. In contrast, the radial distributions of the two processes are much more similar: they are dominated by short-range internucleon distances of $r \lesssim 2$ fm. Moreover, the longer-range contributions that contribute to DGT transitions partially cancel, making the effective DGT range even more similar to that of $0\nu\beta\beta$ decay. We have observed that this partial cancellation is common to all the other DGT transitions in Fig. 1. We have to stress that the very short-range character of $0\nu\beta\beta$ decay and DGT transitions is a somewhat surprising outcome of the many-body calculations. Its validity should be carefully checked in more controlled ab initio calculations, such as those in Ref. [28].

The short-range character of $0\nu\beta\beta$ decay and DGT transitions to the ground state can explain the simple linear correlation observed in Fig. 1. References [29, 30] showed that when an operator probes only the short-range physics of low-energy states, the corresponding matrix elements factorize into a universal operator-dependent constant times a state-dependent number common to all short-range operators. Therefore, a linear relation is expected between two short-range operators, such as $0\nu\beta\beta$ decay and the corresponding DGT transition to the ground state. This explanation is also consistent with the quasiparticle random phase approximation results not being correlated: for this many-body method $0\nu\beta\beta$ decay NMEs are of short-range as in the shell model, but DGT transitions receive important contributions from much longer internucleon distances [27, 31]. To clarify this picture, further work should explore the physics of the long-range contributions in detail.

[1] The identity that replaces the neutrino potential in Eq. (6) can be seen as the Fourier transform of a delta function in momentum space.

4. Conclusions

We have used the nuclear shell model to study the similarity between $0\nu\beta\beta$ decay NMEs and DGT transitions to the ground state, for a collection of twenty six transitions in the pf-shell. We have found that the $0\nu\beta\beta$ decay and DGT matrix elements are very well correlated. The correlation is common to all the transitions studied, and does not depend on the details of the nuclear interaction used, or the nuclear structure correlations permitted in the initial and final nuclei. Moreover, a result for ^{48}Ca obtained with energy-density functional theory is consistent with the correlation found in the shell model calculations. Our findings suggest that future measurements of DGT transitions in double charge-exchange experiments can provide very valuable insights on $0\nu\beta\beta$ decay.

Acknowledgments

This work was supported by JSPS KAKENHI Grant (25870168, 17K05433) and CNS-RIKEN joint project for large-scale nuclear structure calculations. It was also supported by MEXT and JICFuS as a priority issue (Elucidation of the fundamental laws and evolution of the universe, hp170230) to be tackled by using Post K Computer.

References

[1] Gando A *et al.* (KamLAND-Zen Collaboration) 2016 *Phys. Rev. Lett.* **117** 082503
[2] Agostini M *et al.* (GERDA Collaboration) 2017 *Nature* **544** 47
[3] Albert J B *et al.* (EXO Collaboration) 2018 *Phys. Rev. Lett.* **120** 072701
[4] Alduino C *et al.* (CUORE Collaboration) 2018 *Phys. Rev. Lett.* **120** 132501
[5] Agostini M, Benato G and Detwiler J A 2017 *Phys. Rev. D* **96** 053001
[6] Engel J and Menéndez J 2017 *Rep. Prog. Phys.* **45** 014003
[7] Caurier E, Martínez-Pinedo G, Nowacki F, Poves A and Zuker A P 2005 *Rev. Mod. Phys.* **77** 427
[8] Freeman S J and Schiffer J P 2012 *J. Phys. G* **39** 124004
[9] Brown B A, Horoi M and Sen'kov R A 2014 *Phys. Rev. Lett.* **113** 262501
[10] Barabash A S 2015 *Nucl. Phys. A* **935** 52
[11] Ichimura M, Sakai H and Wakasa T 2006 *Prog. Part. Nucl. Phys.* **56** 446
[12] Frekers D, Puppe P, Thies J H and Ejiri H 2013 *Nucl. Phys. A* **916** 219
[13] Yako K *et al.* 2009 *Phys. Rev. Lett.* **103** 012503
[14] Cappuzzello F, Cavallaro M, Agodi C, Bondi M, Carbone D, Cunsolo A and Foti A 2015 *Eur. Phys. J. A* **51** 145
[15] Takaki M *et al.* 2015 *JPS Conf. Proc.* **6** 020038
[16] Uesaka T *et al.* 2015 *RIKEN RIBF NP-PAC* NP1512-RIBF141
[17] Takahisa K *et al.* *arXiv:1703.08264*
[18] Sen'kov R A and Horoi M 2013 *Phys. Rev. C* **88** 064312
[19] Shimizu N, Menéndez J and Yako K 2018 *Phys. Rev. Lett.* **120** 142502
[20] Menéndez J 2018 *J. Phys. G* **45** 014003
[21] Caurier E, Menéndez J, Nowacki F and Poves A 2008 *Phys. Rev. Lett.* **100** 052503
[22] Honma M, Otsuka T and Mizusaki T 2008 *RIKEN Accel. Prog. Rep.* **41** 32
[23] Poves A, Sanchez-Solano J, Caurier E and Nowacki F 2001 *Nucl. Phys. A* **694** 157
[24] Iwata Y, Shimizu N, Otsuka T, Utsuno Y, Menéndez J, Honma M and Abe T 2016 *Phys. Rev. Lett.* **116** 112502
[25] Shimizu N *arXiv:1310.5431*
[26] Rodríguez T R and Martínez-Pinedo G 2013 *Phys. Lett. B* **719** 174
[27] Šimkovic F, Hodák R, Faessler A and Vogel P 2011 *Phys. Rev. C* **83** 015502
[28] Pastore S, Carlson J, Cirigliano V, Dekens W, Mereghetti E and Wiringa R B 2018 *Phys. Rev. C* **97** 014606
[29] Anderson E R, Bogner S K, Furnstahl R J and Perry R J 2010 *Phys. Rev. C* **82** 054001
[30] Bogner S K and Roscher D 2012 *Phys. Rev. C* **86** 064304
[31] Šimkovic F, Faessler A, Rodin V, Vogel P and Engel J 2008 *Phys. Rev. C* **77** 045503

IOP Conf. Series: Journal of Physics: Conf. Series **1056** (2018) 012038 doi:10.1088/1742-6596/1056/1/012038

Chasing the light sterile neutrino: status of the STEREO experiment

Alessandro Minotti
LAPP, 9 Chemin de Bellevue - BP 110, 74000 Annecy-le-Vieux, France
E-mail: minotti@lapp.in2p3.fr

Abstract. This article describes the status of the STEREO experiment. STEREO is a short-baseline reactor antineutrino experiment, whose primary goal is to study oscillations involving extra sterile neutrino states with $\Delta m^2 \sim 0.1 - 1\,\mathrm{eV}$, an hypothesis suggested by existing anomalies. Such hypothesis is tested by comparing antineutrino spectra in 6 identical cells located at a increasing distance from the reactor core of the ILL facility. Calibration sources have been used to characterise the detector response, perform energy reconstruction, and study the detection efficiency. Results of these studies, along with preliminary antineutrino and background rates, are reported here. STEREO has started operating in November 2016 and will collect data for 8 reactor cycles. Preliminary results based on 1.5 reactor cycles and ongoing analysis, which are mentioned here, will be released in early 2018.

1. Introduction

Neutrino oscillation is now a well-established phenomenon: all the mixing angles of the PMNS matrix and squared-mass splittings have been measured. However, the absolute scale of neutrino masses, as well as the CP-violating phase in the PMNS matrix, are yet unknown. Moreover, some yet unexplained experimental anomalies challenge the three-family paradigm. One of these anomalies arose in 2011, when novel computations of reactor $\bar{\nu}_e$ spectra pointed out a rate excess of approximatively 3% with respect to previous measurements at close distance ($10 - 100\,\mathrm{m}$) from various reactor cores (figure 1) [1]. This discrepancy, known as reactor antineutrino anomaly, was later confirmed data collected by Double Chooz, Daya Bay and RENO near detectors, for a global $\sim 2.8\,\sigma$ statistical significance [2].

The RAA can be interpreted as an oscillation between the three currently known and an extra sterile neutrino flavour state. If we add a small component of this new flavour state to the existing mass states, as well as a fourth neutrino mass state with $\Delta m^2_{14} \sim 0.1 - 1\,\mathrm{eV}$ mainly consisting of this new flavour state, then the resulting oscillation at very short baseline

$$P_{\bar{\nu}_e \to \bar{\nu}_e}(L \lesssim 10\,\mathrm{m}) \simeq 1 - \sin^2(2\theta_{new}) \sin^2(\Delta m^2_{14} L/4E)\,, \quad (1)$$

will explain the experimental deficit of the RAA. Note that other anomalies (Gallium, LSND) can be explained by active-sterile neutrino oscillations. Nonetheless, a global simple solution is disfavoured by disappearance results [2].

2. The STEREO experiment

The STEREO detector is located in the ILL research reactor facility, at a distance of approximatively 9 m from the 58.3 MWth nominal power reactor core. The characteristics of the

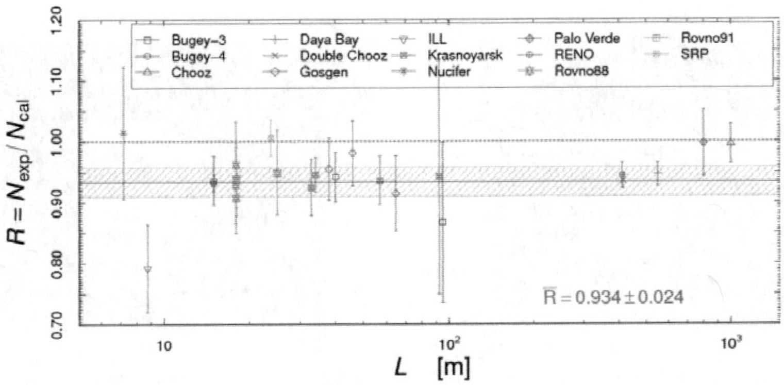

Figure 1. Ratios R between observed (N_{exp}) and predicted (N_{cal}) neutrino rates, as functions of the reactor-detector distance L, of the reactor experiments considered in [2].

ILL reactor make it a perfect candidate to study oscillations at a short distance: the compactness of the core ($\varnothing = 37\,\mathrm{cm}$) allows in fact a good resolution of the oscillation, especially at low energies (equation 1), while the high enrichment (93% ^{235}U) simplifies the issue of accounting for the fuel evolution.

2.1. The detector

The STEREO detector consists of an inner detector, an upper muon veto, calibration systems, and shielding (figure 2). In turn, the inner detector is divided in a target area surrounded sideways by an outer crown. The target is segmented longitudinally in 6 identical cells and filled with Gd-doped liquid scintillator, while the crown is segmented in its 4 sides and filled with a mixture of liquid scintillation similar to that of the target but without Gd doping. The walls that separate the target and gamma catcher, as well as the segmentation walls, consist in a foil of reflecting material sandwiches between acrylic plates. The scintillation light is thus conveyed upwards, where is collected by 48 photomultiplier detector hosted in acrylic buffers (figure 3. The optical coupling is guaranteed by mineral oil, of which these buffers are filled.

The muon veto is a water cherenkov detector, equipped with 20 photomultipliers and positioned above the inner detector. Its purpose is to tag cosmic muons and remove associated background. The inner detector individually, as well as the ensemble of inner detector and veto, are packed in multiple layers of shielding materials in order to suppress different sources of background and noise: μ-metal and soft iron to shield photomultipliers from external magnetic fields, lead for high energy γ's, B and polyethylene for neutrons (figures 4, 5). Moreover, the STEREO area at the ILL facility is bordered by walls of lead, polyethylene, and B, and surmounted by the ILL water channel, which provides a $\sim 15\,\mathrm{mwe}$ shielding against muons.

2.2. Signal and background

Antineutrinos produced in the reactor are detected via their inverse beta decay (IBD) interactions with the free protons of the STEREO detctor

$$\bar{\nu}_e \, p \rightarrow e^+ \, n \,. \tag{2}$$

The two-fold coincidence between the positron ionising the liquid scintillator and suddenly annihilating with an electron of the medium, and the delayed neutron capture, provides a clear signature of the IBD, allowing for a strong suppression of sigle-event noise. Residual background

IOP Conf. Series: Journal of Physics: Conf. Series **1056** (2018) 012038 doi:10.1088/1742-6596/1056/1/012038

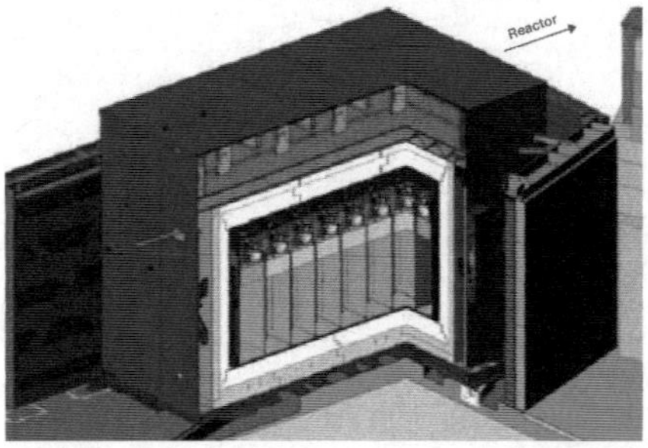

Figure 2. Outline of STEREO. The inner detector consists of six target cells and an outer crown (ochre yellow) surmounted by buffer cells (green), it is packed in multiple shielding layers, and surmounted by the muon veto (blue).

Figure 3. Top view of the STEREO inner detector, before the last two target buffers where installed. Note the 40 photomultipliers with their red bases, and the reflecting walls inside.

Figure 4. The STEREO inner detector inserted in shielding layers; from the inside out: μ-metal, borated polyethylene, lead, soft iron.

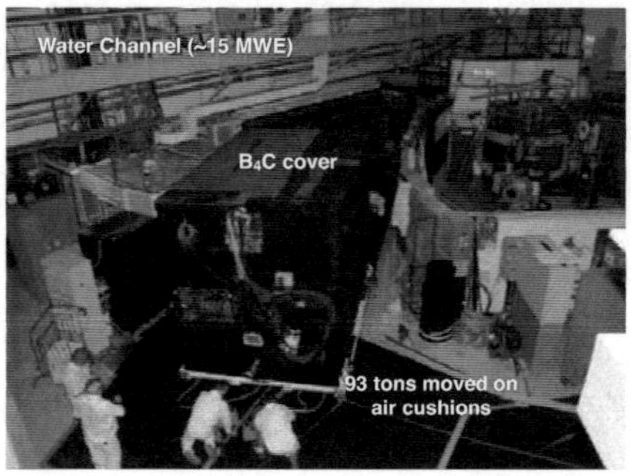

Figure 5. The B_4C-coated STEREO detector, as it moves on air cushions to its final position below the ILL water channel.

include all events that can mimic such coincidence, mostly γ's and neutrons generated in muon spallations or from the reactor, either from the same source or in accidental coincidence. The presence in the target liquid compound of Gd, a n-capturer with high cross-section and released energy ($\sim 8\,\text{MeV}$), enhances the neutron detection efficiency, while the un-doped crown allows to fully contain γ energy deposits and to additionally suppress background thanks to event topology.

2.3. Oscillation analysis

STEREO aims to test the hypothesis of oscillation between active and sterile neutrinos at short distance by comparing spectra of neutrino detected in its 6 identical target cells. Such phenomenon, in fact, would result in a distortion of the spectra at different distances from the neutrino sources (figures 6, 7). The energy of an antineutrino interacting via IBD can be inferred directly from that of the positron (equation 2).

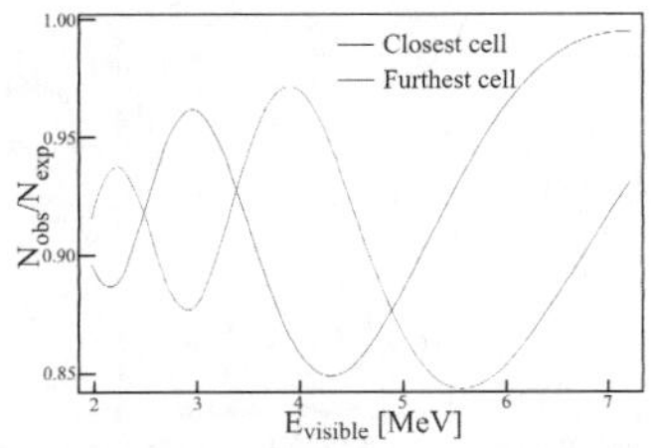

Figure 6. Ratio between neutrino observed with (N_{obs}) and without (N_{exp}) active-sterile neutrino oscillation, for the RAA best fit parameters and as a function of the positron energy (E_{vis}), for the reactor closest (blue) and furthest (red) target cell.

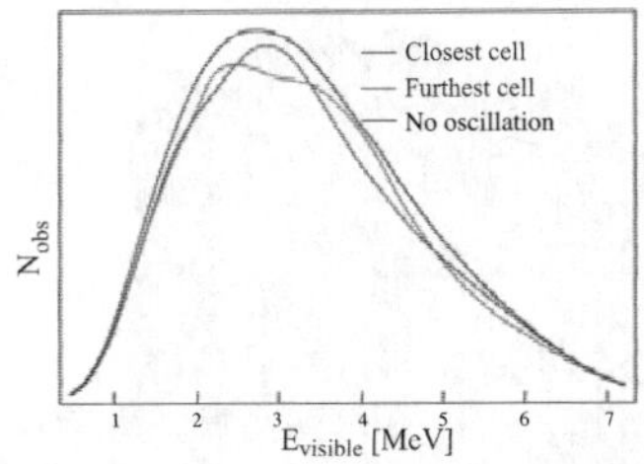

Figure 7. Expected renormalised positron spectra, for the RAA best fit parameter, in case of active-sterile oscillation for the reactor closest (blue) and furthest (red) target cell, and in case of no oscillation (black).

3. Status of the analysis of the data

The STEREO detector was fully assembled by April 2016; it was then moved to the operating position (figure 5) and started to take data in November of the same year. A first data set was collected, consisting of 70 days with the reactor operating, and 25 days of reactor off. Subsequently, a long shutdown and component replacement of the reactor compelled a disassembling of the experiment. This period was used for maintenance and improvement of the detector performance. The detector is now again operational and is collecting reactor-off data since October 2017. The ILL reactor is due to start its operation in early 2018, allowing for STEREO to collect data from additional 7 cycles in 2018-2019. What follows refers to the analysis of data collected in the first phase.

3.1. Detector calibration

The detector response of STEREO is monitored using regular calibration runs with γ and neutron sources, using three independent systems: an internal systems consisting of tubes entering the target cells where the source is deployed; an external automated system allowing for a deployment of the source around the detector, and an underneath rail that can be used to deploy sources below the detector floor (figure 8).

γ sources of various energy spanning from approximatively 0.5 to 4.5 MeV are used for the energy calibration. From each source, cell-by-cell calibration coefficients, i.e. conversion factors between collected light and deposited energy, are obtained (figure 9). The variation of collected light peaks with the source position along the vertical axis is estimated to be below 6% and is well reproduce by our Monte Carlo simulation. Calibrations with the ^{54}Mn source ($E_\gamma = 0.38$ MeV) are used to calculate the cross-talk between cells arising from optical photons that occasionally cross the reflecting walls. In order to obtain the deposited energy in each cell for a given event, we perform a matrix inversion process on the values of light collected by photomultipliers, taking into account calibration coefficients obtained with ^{54}Mn and cross-talk values. Then a correction for the non-linearity of the energy response, or quenching (figure 9), is applied.

Along with γ sources, an alpha reaction neutron source of Am-Be is used to study the detector response to neutrons. More specifically, we calculated the capture time of neutrons from the source, which is consistent with that measured for IBD candidates. Moreover, the fraction of captures on Gd and its dependency on the neutron position was measured, allowing for a correction of the Monte Carlo simulation. Such results will be used in the calculation of a global

IOP Conf. Series: Journal of Physics: Conf. Series **1056** (2018) 012038 doi:10.1088/1742-6596/1056/1/012038

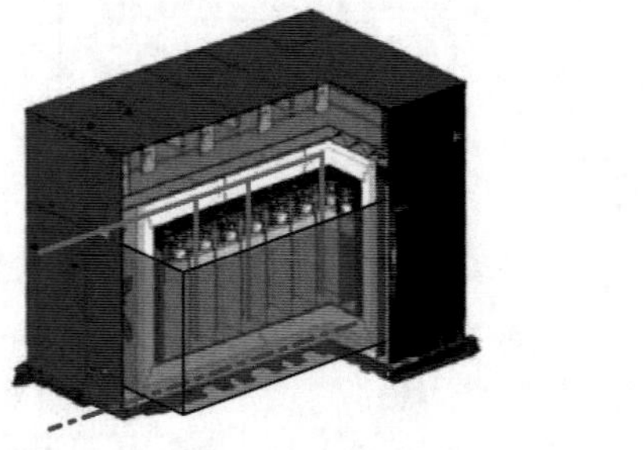

Figure 8. Scheme of STEREO calibration systems: internal (red), external (blue) and underneath (purple).

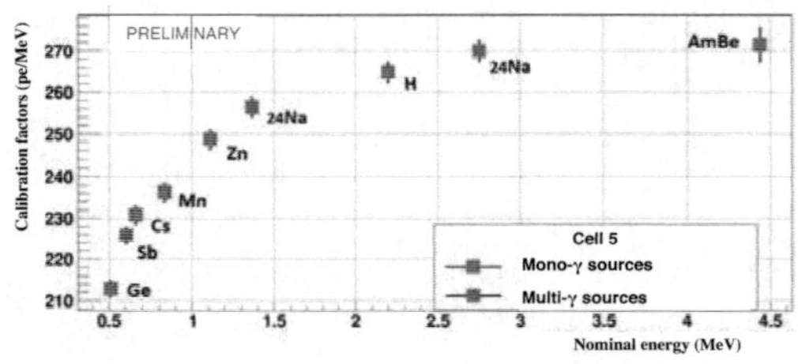

Figure 9. Calibration coefficients (pe-MeV conversion factors) for target cell 5, calculated for different calibration sources as a function of the nominal energy of the source γ('s).

Table 1. List of preliminary cuts applied on IBD candidates.

Cut	Targeted background	$\bar{\nu}_e$ inefficiency
100 μ before-prompt/after delay veto	Multiple event from spallation	negligible
100 μs after-μ veto	Fast spallation neutrons	6.5 %
Photomultipliers light homogeneity	μ's decaying in the upper target	2 %
Pulse shape discrimination	All cosmic-induced	varies
Cell multiplicity	Fast spallation neutrons	$< 1\,\%$

neutron detection efficiency for the oscillation analysis.

3.2. Neutrino selection

An IBD selection is defined basing on the following steps: first, we select pairs of events that are separated by a time interval of less than 70 μs; then we apply separate energy conditions on the prompt and delayed event

$$2\,\text{MeV} < E_{dep}^{det} < 8\,\text{MeV} \;\wedge\; E_{dep}^{GC} < 1.1\,\text{MeV} \quad \text{(prompt)} \tag{3}$$

$$5\,\text{MeV} < E_{dep}^{det} < 10\,\text{MeV} \;\wedge\; E_{dep}^{Tg} > 1\,\text{MeV} \quad \text{(delayed)}\,; \tag{4}$$

finally, additional cuts are applied to reject residual background (table 1). Accidental coincidences of neutrons/γ's interacting in the detector are estimated online using multiple off-time windows, and the amount is subtracted statistically from IBD candidates. Our analysis is based on a estimation of neutrino spectra for reactor-on and -off periods and statistical subtraction of the latter from the former to obtain a "pure" antineutrino spectrum for each target cell. In order to achieve that, IBD rates (figure 10) are corrected for the atmospheric pressure, which influences the rate of cosmic muons and therefore some background components.

3.3. Expected sensitivity

At the present state, the STEREO collaboration is working on refining the IBD selection and cuts, the systematics for the -off subtraction, and finalise the energy reconstruction for the final oscillation analysis. The signal-over-background ratio and the predicted statistics are close

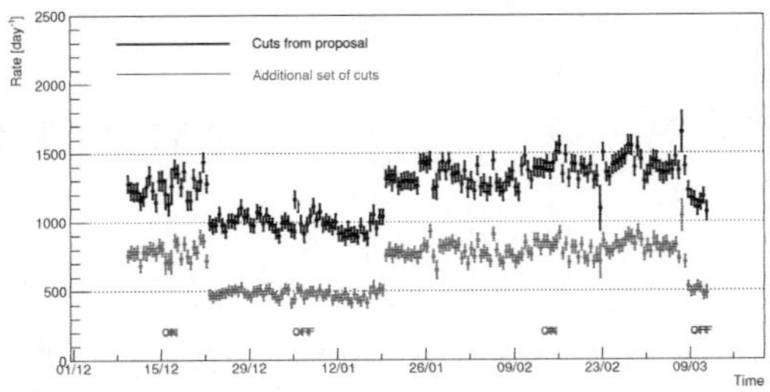

Figure 10. IBD rates for data collected during the STEREO first phase, with a correction for atmospheric pressure appleid. The refined set of cuts described in table 1 allowed to reduce the background (red points) with respect to a simple IBD selection (black points).

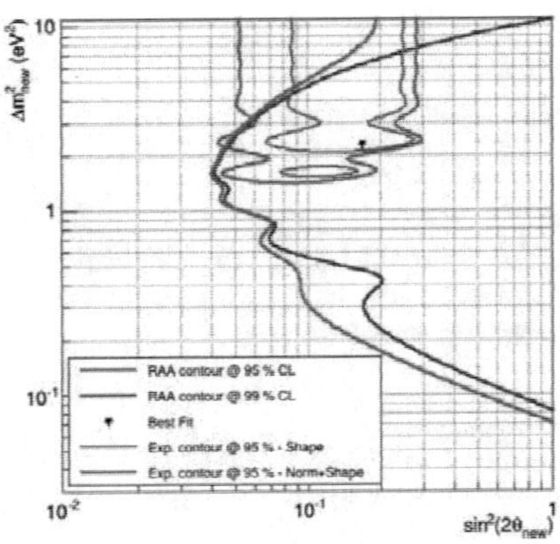

Figure 11. Predicted sensitivity of STEREO, assuming 400 $\bar{\nu}_e$/day in 300 days, a signal-over-backgound ratio of 1.5, a detection efficiency of 60 %, energy scale and flux systematics of 4 and 2 % respectively.

to expected requirements that will allow us to cover a broad spectrum of the active-sterile neutrino oscillation phase-space allowed by the RAA (figure 11). The works done during the long shutdown have seen a sizeable improvement of the detector performance, while meanwhile the IBD selection and energy reconstruction were refined.

4. Conclusions

The RAA points to physics beyond the Standard Model in the hypothesis of oscillations involving extra light sterile neutrinos. The STEREO experiment aims to test such hypothesis by searching oscillations of antineutrinos at very short distance from the ILL reactor core source. In its first phase, STEREO has collected data from 1.5 reactor cycles, plus 25 days of reactor off, that we are currently analysing. Furthermore, a second phase is currently ongoing, foreseeing the collection of data from 7 additional cycles by 2019. Although the mitigation of reactor-induced and cosmic background represents a challenge for surface-level reactor antineutrino experiments, we achieved an remarkable control of accidental coincidences, while signal-over-background ratio is close to our intended goal. Our collaboration is working on the refinement of topological cuts, while systematics uncertainties are being defined A technical paper describing the detector and analysis performances, as well as preliminary oscillation results, will be published in early 2018.

Acknowledgments

STEREO is an international collaboration of CEA-Saclay, ILL, LAPP, LPSC and MPIK, supported by ENIGMASS and ANR. The author wishes to thank P2IO and ANR for co-financing his fellowship, as well as all the STEREO collaborator, with a special mention for his former colleagues of CEA-Saclay.

References

[1] Mention G, Fechner M, Lasserre T, Mueller TA, Lhuillier D, Cribier M, Letourneau A. Reactor antineutrino anomaly. Physical Review D. 2011 Apr 29;83(7):073006.
[2] Gariazzo S, Giunti C, Laveder M, Li YF. Updated Global 3+ 1 Analysis of Short-BaseLine Neutrino Oscillations. arXiv preprint arXiv:1703.00860. 2017 Mar 2.

IOP Conf. Series: Journal of Physics: Conf. Series **1056** (2018) 012039 doi:10.1088/1742-6596/1056/1/012039

Direct neutrino mass measurement by the HOLMES experiment

A Nucciotti[1,2], B Alpert[3], M Balata[9] , D Becker[3], D Bennett[3], A Bevilacqua[4,5] , M Biasotti[4,5], V Ceriale[4,5] , G Ceruti[2], D Corsini[4,5], M De Gerone[4,5], R Dressler[6], M Faverzani[1,2], E Ferri[2], J Fowler[3], J Gard[3], F Gatti[4,5], A Giachero[1,2], J Hays-Wehle[3], S Heinitz[6], G Hilton[3], U Köster[7], M Lusignoli[8], J Mates[3], S Nisi[9], A Orlando[1] , L Parodi[4,5], G Pessina[2], A Puiu[1,2], S Ragazzi[1,2], C Reintsema[3], M Ribeiro-Gomez[10], D Schmidt[3], D Schuman[3], F Siccardi[4,5], D Swetz[3] , J Ullom[3] and L Vale[3]

[1] Dipartimento di Fisica, Università di Milano-Bicocca, Milano, Italy
[2] Istituto Nazionale di Fisica Nucleare, Sezione di Milano-Bicocca, Milano, Italy
[3] National Institute of Standards and Technology, Boulder, CO, USA
[4] Dipartimento di Fisica, Università di Genova, Genova, Italy
[5] Istituto Nazionale di Fisica Nucleare, Sezione di Genova, Genova, Italy
[6] Paul Scherrer Institut, Villigen, Switzerland
[7] Institut Laue-Langevin, Grenoble, France
[8] Istituto Nazionale di Fisica Nucleare, Sezione di Roma 1, Roma, Italy
[9] Laboratori Nazionali del Gran Sasso, INFN, Assergi (AQ), Italy
[10] Multidisciplinary Centre for Astrophysics (CENTRA-IST), University of Lisbon, Lisbon, Portugal

E-mail: `angelo.nucciotti@mib.infn.it`

Abstract. The assessment of the neutrino absolute mass scale is still a crucial challenge in today particle physics and cosmology. Beta or electron capture spectrum end-point study is currently the only experimental method which can provide a model independent measurement of the absolute scale of neutrino mass. HOLMES is an experiment to directly measure the neutrino mass by performing a calorimetric measurement of the energy released in the electron capture decay of the artificial isotope ^{163}Ho. In a calorimetric measurement the energy released in the decay process is entirely contained into the detector, except for the fraction taken away by the neutrino. This approach eliminates both the issues related to the use of an external source and the systematic uncertainties arising from decays on excited final states. HOLMES will deploy a large array of low temperature microcalorimeters implanted with ^{163}Ho nuclei. The achievable neutrino mass statistical sensitivity is expected in the eV range, thereby making HOLMES an important step forward in the direct neutrino mass measurement with a calorimetric approach as an alternative to spectrometry. HOLMES will also establish the potential of this approach to achieve a sub-eV sensitivity. HOLMES is designed to collect about 3×10^{13} decays with an instrumental energy resolution around 1 eV FWHM and a time resolution around 1 μs. To achieve this in three years of measuring time, HOLMES is going to deploy 16 sub-arrays of TES microcalorimeters. Each sub-array has 64 pixels ion implanted with ^{163}Ho nuclei to give a pixel activity of 300 Bq per pixel. The TES arrays are read out using microwave multiplexed rf-SQUIDs in combination with a Software Designed Radio data acquisition system. The commissioning of the first implanted sub-array is scheduled for 2018 and it will provide first high statistics data about the EC decay of ^{163}Ho together with a preliminary limit on the neutrino mass. In this contribution we outline the HOLMES project

IOP Conf. Series: Journal of Physics: Conf. Series **1056** (2018) 012039 doi:10.1088/1742-6596/1056/1/012039

with its physics reach and technical challenges, along with its status and perspectives. In particular we will present the status of the HOLMES activities concerning the ^{163}Ho isotope production by neutron irradiation and purification, the TES pixel design and optimization, the multiplexed array read-out characterization, the cryogenic set-up installation, and the setting up of the mass separation and ion implantation system for the isotope embedding in the TES absorbers.

1. Introduction

The HOLMES experiment aims at directly measuring the electron neutrino mass using the electron capture (EC) decay of ^{163}Ho [1]. HOLMES performs a calorimetric measurement of the energy released in the decay of ^{163}Ho to measure all the atomic de-excitation energy, except the fraction carried away by the neutrino [2]. The direct measurement exploits only energy and momentum conservation and it is therefore completely model-independent. At the same time, the calorimetric measurement eliminates systematic uncertainties arising from the use of external beta sources, as in experiments with beta spectrometers, and minimizes the effect of the atomic de-excitation process uncertainties.

The expected calorimetric spectrum is shown in figure 1: the distribution of the measured de-excitation energy, carried mostly by electrons with energies up to about 2 keV, presents lines at the binding energies of the captured electrons. These lines have a natural width of a few eV, therefore the actual spectrum is a continuum with marked peaks with Breit-Wigner shapes. The spectral end-point is shaped by the same neutrino phase space factor $(Q_{\rm EC}-E)\sqrt{(Q_{\rm EC}-E)^2-m_\nu^2}$ that appears in a beta decay spectrum, where $Q_{\rm EC}$ is the EC transition energy. A finite neutrino mass m_ν causes a deformation of the energy spectrum which is truncated at $Q_{\rm EC}$-m_ν. The sensitivity of this approach depends on the nearness of $Q_{\rm EC}$ to the binding energy of one of the captured electrons: ^{163}Ho was proposed in [3] as an ideal isotope with a $Q_{\rm EC}$ between 2.5 keV and 3.0 keV – with 2.55 keV as raccomended value [4],– close to the binding energy of about 2.0 keV for the M1 electrons. Recent measurements [5] established that $Q_{\rm EC}$ is $2833 \pm 30_{stat} \pm 15_{sys}$ eV, therefore not as close to the M1 shell binding energy as initially hoped for.

The spectrum shown in figure 1 is calculated neglecting the second order effects in the atomic de-excitation cascade – shake up and shake off – which have been recently considered in many

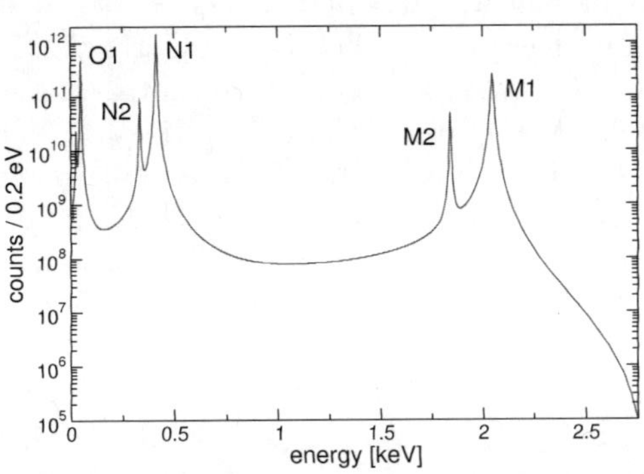

Figure 1. Calculated ^{163}Ho EC calorimetric spectrum for $Q = 2.8$ keV, $\Delta E = 2$ eV, and $N_{ev} = 10^{14}$.

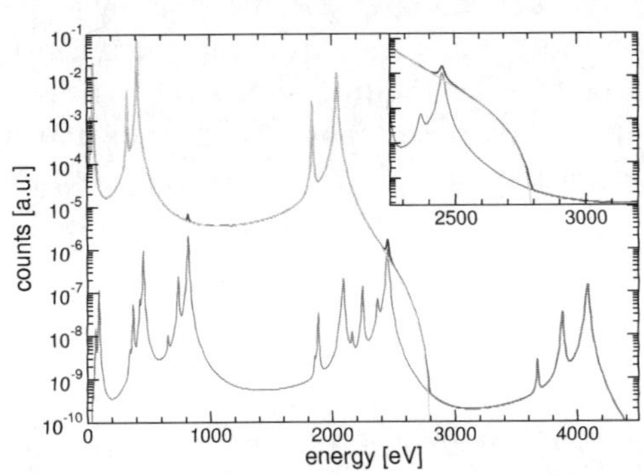

Figure 2. Calculated experimental ^{163}Ho EC calorimetric spectrum for $Q = 2.8$ keV, $\Delta E = 2$ eV, $f_{pp} = 10^{-4}$, and $N_{ev} = 10^{14}$ (blue). The pile-up spectrum is the red curve.

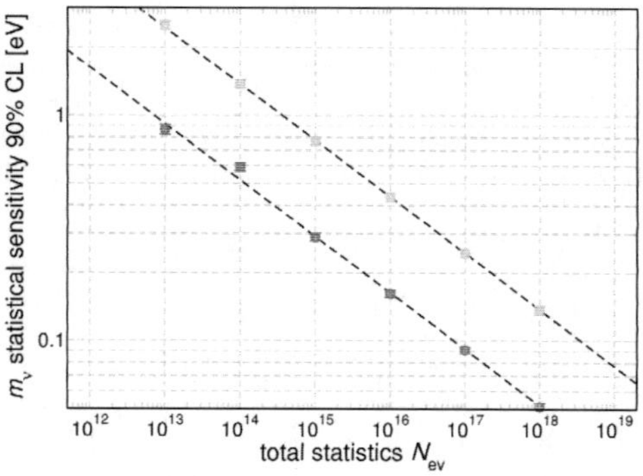

Figure 3. Monte Carlo estimated statistical sensitivity for $\Delta E = 1\,\text{eV}$, $\tau_R = 1\,\mu\text{s}$, and for both $f_{pp} = 10^{-3}$ and 10^{-6} (from top to bottom). The dashed lines are the extrapolated curves using a $N_{ev}^{-1/4}$ scaling law.

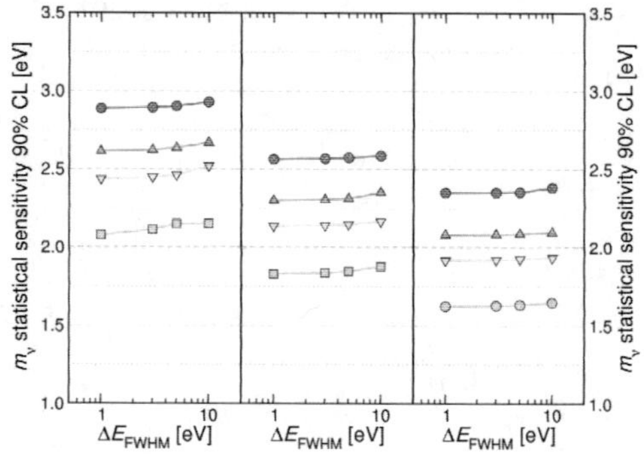

Figure 4. Monte Carlo estimated statistical sensitivity for a fixed exposure $t_M \times N_{det}$ of about 3000 detector×year and for different A_{EC} (from left to right panel: 30, 100, and 300 Bq/detector) and τ_R (from top to bottom: 10, 5, 3, and 1 μs).

publications and that are summarized in [2] and references therein.

2. The HOLMES experiment design

The sensitivity of calorimetric experiments is affected by a major drawback of this approach. In a calorimeter the whole decay spectrum is acquired and the decaying isotope activity A_{EC} must be restrained to limit the rate of accidental coincidences. Time unresolved coincidences (pile-up) show up in the spectrum as an additional background which worsens the statistical sensitivity. Figure 2 shows the pile-up spectrum whose normalization f_{pp} relative to single events spectrum (figure 1) is $f_{pp} \approx \tau_R A_{\text{EC}}$, where τ_R is the time resolution of the detector. Therefore, when planning a calorimetric experiment, a trade off must be found between the quest for collecting a large statistics in a short time and the need for limiting the pile-up spectrum amplitude. For fixed measuring time t_M and detector performance – ΔE and τ_R, i.e. energy and time resolution, respectively, – the experimental parameters to tweak are the number of detectors N_{det} and the isotope activity A_{EC} in each of them. The statistical sensitivity as a function of the above experimental parameters can be investigated exploiting the Monte Carlo methods described in [6]. Figure 3 shows that the statistical sensitivity improves with the total statistics N_{ev} as $1/\sqrt[4]{N_{ev}}$ and that reaching a sensitivity below 2 eV requires around 10^{13} decays. While matching the current 2 eV sensitivity obtained by spectrometers with ^{3}H [7, 8] is a meaningful first target for current ^{163}Ho experiments, the goal for future experiments must be at least 0.1 eV – i.e. better than the KATRIN experiment goal [9]. From figure 3 it is apparent how the 0.1 eV target calls for statistics larger than about 10^{17} events to be collected by arrays of more than 10^7 pixels.

HOLMES optimal configuration is found by means of the simulations plotted in the right side of figure 4 and by appreciating that, for fixed exposure $t_M \times N_{det}$ and time resolution τ_R, the statistical sensitivity constantly improves for increasing pixel activities A_{EC}. As a consequence, in spite of the increasing fraction of pile-up events f_{pp}, it is winning to have the largest tolerable pixel activity. With this in mind, HOLMES plans to deploy an array with 1024 pixels, each with an ^{163}Ho activity A_{EC} of about 300 Bq. The right panel of figure 4 shows that to reach a statistical sensitivity below 2 eV the time resolution τ_R plays the main role, while the energy resolution ΔE could be anywhere below about 10 eV. In particular τ_R must be better than about

$3\,\mu$s, which translates in a fraction of pile-up f_{pp} better than about 10^{-3}.

Because of the very low fraction of decays in the region of interest close to $Q_{\rm EC}$, the background is another factor impacting the statistical sensitivity of end-point neutrino mass measurements. A constant background b is negligible as long as it is much smaller than the pile-up spectrum, that is when $b \lll\approx A_{\rm EC} f_{pp}/2Q_{\rm EC}$, as it is confirmed also by Monte Carlo simulations [6]. Therefore, a large activity $A_{\rm EC}$ with a correspondingly large pile-up fraction f_{pp} makes experiments relatively insensitive to cosmic rays and to environmental radioactivity. For HOLMES this translates in the requirement that the background level at the end-point must be lower than about 0.1 count/eV/day/det. From Geant4 Monte Carlo simulations and comparison with past experiments, HOLMES' background due external sources is expected to be of the order of 10^{-4} count/eV/day/det at sea-level. Indeed HOLMES' major source of background is the internally present ^{166m}Ho, an isotope which is produced along with ^{163}Ho [10]. This β decaying isotope, with a half life of about 1200 years and a Q-value of about 1854 keV, according to Geant4 Monte Carlo simulations causes a background below 5 keV of about 0.5 count/eV/day/det/Bq(^{166m}Ho). The ^{166m}Ho activity in the pixels must be then kept lower than 0.2 Bq for a ^{163}Ho activity $A_{\rm EC}$ of 300 Bq, or a factor 1500 lower, i.e. the number of contaminating ^{166m}Ho nuclei must be about a factor 6000 lower than that of ^{163}Ho.

As discussed in more details in the following, the HOLMES project is divided in tasks which are carried out in parallel and on which the HOLMES groups are making steady progresses. The tasks are 1) the ^{163}Ho isotope production and purification, 2) the isotope embedding, 3) the TES pixel and array design and testing, 4) the microwave multiplexed read-out system, and 5) the Digital Acquisition system (DAQ).

3. ^{163}Ho production and embedding

The ^{163}Ho isotope needed to carry out the experiment is produced at the nuclear reactor operated by the Institut Laue-Langevin (ILL, Grenoble, France) by irradiating Er_2O_3 samples enriched in ^{162}Er in a thermal neutron flux of about 10^{15} n/s/cm^2. The Er_2O_3 samples are purified before irradiation and the accumulated holmium is radiochemically separated in hot-cells after the neutron irradiation. Both processes have been developed and optimized at the Paul Scherrer Institute (PSI, Zurich, CH). Two samples of enriched Er_2O_3 have been already irradiated at ILL and processed at PSI. Inductive Coupled Plasma Mass Spectroscopy (ICPMS) analysis performed at LNGS and PSI on the two samples before and after chemical purification at PSI demonstrated a production of a sufficiently radiopure sample of about 43 MBq of ^{163}Ho (along with about 50 kBq of ^{166m}Ho, measured by γ-spectroscopy). The analysis of these samples yielded also a preliminary estimate of the ^{163}Ho production yield including all these steps. We estimate that the whole HOLMES program requires a total amount of about 250 MBq of ^{163}Ho. This estimate is obtained considering the above mentioned yield and the efficiencies related to the ion implantation process. Since not all the efficiencies are known yet, for the above estimate we made conservative guesses and came up with a global efficiency of the order of 0.1%. About 540 mg of Er_2O_3 enriched in ^{162}Er at 25% along with about 100 mg of Er_2O_3 at 26.9% were irradiated at ILL for 46 days until early 2017. We estimate a production of about 150 MBq of ^{163}Ho which are expected to be enough both for testing the isotope embedding and for the production of the first 512 detectors.

^{163}Ho is introduced in the absorbers of the HOLMES' low temperature microcalorimeters by means of custom system which combines a high efficiency Penning sputter ion source, a mass analyzing magnet, an electrostatic triplet focusing stage, and an XY magnetic beam scanner. The system is designed to achieve an optimal mass separation for ^{163}Ho, thereby eliminating other trace contaminants not removed by chemical methods at PSI, such as ^{166m}Ho or stable ^{165}Ho. The magnetic selection also prevents the implantation of holmium in chemical forms other than metallic, which would likely cause chemical shifts of the end-point energy. The embedding

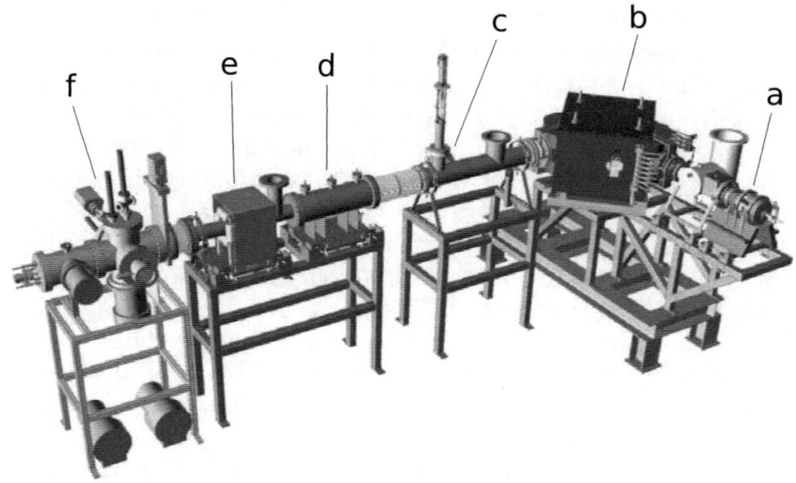

Figure 5. 3D model of the embedding system showing: (a) Penning sputter ion source, (b) analyzing magnet, (c) analyzing slit, (d) electrostatic triplet focusing stage, (e) XY magnetic beam scanner, and (f) Target Chamber.

system includes a focusing stage designed to have a beam cross-section on the detectors of about 4 mm thereby maximizing the geometrical efficiency of the ion implantation. During ion implantation, the detectors are hosted in the UHV Target Chamber which is equipped with an ion beam assisted gold sputtering system to control the ^{163}Ho concentration in the detector absorbers, to compensate the absorber atom sputtering caused by the ion implantation, and to deposit the final absorber layer to complete the ^{163}Ho embedding. The metallic cathode for the ion source is made out of an inter-metallic alloy containing ^{163}Ho in a metallic form. The metallic ^{163}Ho is produced in an evaporation chamber by thermal reduction and distillation of the Ho_2O_3 separated at PSI after neutron irradiation. The details of the ion source cathode fabrication are described in [11]. The whole embedding system is expected to be ready for detector implantation testing early in 2018.

4. TES microcalorimeters, microwave multiplexed read-out, and DAQ

The detectors used for the HOLMES experiment are microcalorimeters on Si_2N_3 membranes with Mo/Cu transition edge sensors (TES) and with 2 μm thick gold absorbers [12]. The pixel design has been optimized to match the experimental specifications in terms of energy and time resolution, pulse duration, and ^{163}Ho decay radiation full absorption [13, 14]. As explained above the most critical parameter is the time resolution. By means of Monte Carlo simulations we find that with rise and decay time constants of about 10 μs and 100 μs, respectively, and with signal sampling rate of at least 500 kHz it is possible to obtain a time resolution better than 3 μs exploiting discrimination algorithms based on Singular Value Decomposition [15] or Wiener filtering [19].

The HOLMES arrays are read-out with the microwave multiplexing (μMUX) developed by NIST, which is based on rf-SQUIDs as input devices [17]. The μMUX read-out leverages the Software Defined Radio (SDR) implemented in the firmware of a ROACH-2 board by NIST [18].

The number of pixels which can be read-out simultaneously by one ROACH-2 board is about $n_{\mathrm{TES}} \approx 0.005 f_{\mathrm{ADC}} \tau_{rise}$ for a signal sampling frequency of about 0.5 MHz, where f_{ADC} is the SDR ADC sampling frequency of 550 MHz. Considering the design τ_{rise} of about 10 μs, for the

HOLMES DAQ n_{TES} can be as high as 32 and therefore to read-out the full 1024 pixel array a total of 32 ROACH-2 systems is required.

Two 6×4 prototype arrays were fabricated at NIST (Boulder, Co, USA) with slight variations in the TES pixel designs and were characterized using both ^{55}Fe and a fluorescence multi-line source. The results of these measurements, which are described in details in [19], confirm that the baseline pixel design, while matching the desired time constants, provide an energy resolution well below 10 eV FWHM at Q_{EC}. These measurements also prove that the HOLMES DAQ is ready for the experiment.

According to these results, we have now finalized the design of the 4×16 sub-arrays which will be deployed for the HOLMES experiment: the design aims to 1) minimize the signal bandwidth limitations due to stray self-inductance of the read-out leads, 2) minimize the signal cross-talk due to mutual inductance between read-out lines, and 3) maximize the geometrical filling for an optimal implantation efficiency. The arrays are fabricated in a two step process. The arrays are provided by NIST with a 1 μm gold layer and they are further processed after ion-implantation: first in the Target Chamber the thin (few 100Å) layer of Au:^{163}Ho is covered by a second 1 μm gold layer to fully encapsulate the ^{163}Ho source, then the Si_2N_3 membranes are released by means of a Deep Reactive Ion Etching (DRIE). Preliminary tests on dummy samples from NIST are in progress to tune the DRIE process and define the details of this two-step fabrication process.

5. Future plans

With the embedding system fully operational early in 2018, HOLMES will start soon to optimize the isotope implantation process along with the two-step array fabrication process. As soon as the first implanted arrays will be available, the first high statistics calorimetric measurements of the ^{163}Ho decay will provide relevant information on the spectral shape and new competitive limit on m_ν.

Acknowledgments

The HOLMES experiment is funded by the European Research Council under the European Union Seventh Framework Programme (FP7/2007-2013)/ERC Grant Agreement no. 340321. We also acknowledge support from INFN, from the NIST Innovations in Measurement Science program, and from Fundação para a Ciência e a Tecnologia (PTDC/FIS/116719/2010).

References

[1] Alpert B *et al* 2015 *Eur. Phys. J.* C **75** 112
[2] Nucciotti A 2016 *Advances in High Energy Physics* **2016** 9153024
[3] de Rújula A and Lusignoli M 1983 *Nucl. Phys.* B **219** 277
[4] Reich C W and Singh B 2010 *Nuclear Data Sheets* **111** 1211
[5] Eliseev E *et al* 2015 *Phys. Rev. Lett.* **115** 062501
[6] Nucciotti A 2014 *Eur. Phys. J.* C **74** 3161
[7] Lobashev V M *et al* 2001 *Nucl. Phys. Proc. Suppl.* **91** 280
[8] Kraus Ch *et al* 2005 *Eur. Phys. J.* C **40** 447
[9] Weinheimer Ch 2002 KATRIN collaboration, *Prog. Part. Nucl. Phys.* **48** 141
[10] Engle J W *et al* 2013 *Nucl. Instrum. Methods* B **311** 131
[11] Pizzigoni G *et al* 2016 *Nucl. Instrum. Methods A* **824** 223
[12] Enss C 2005 *Cryogenic Particle Detection* vol 99 of *Topics in Applied Physics* (Berlin: Springer)
[13] Hays-Wehle J P *et al* 2016 *J. Low Temp. Phys.* **184** 492
[14] Faverzani M *et al* 2016 *J. Low Temp. Phys.* **184** 922
[15] Alpert B *et al* 2016 *J. Low Temp. Phys.*, **184** 263
[16] Ferri E *et al* 2016 *J. Low Temp. Phys.* **184** 405
[17] Noroozian O *et al* 2013 *Appl. Phys. Lett.* **103** 202602
[18] Mates J A B *et al* 2017 *Appl. Phys. Lett.* **111** 062601
[19] Ferri E *et al* 2018 *J. Physics: Conf. Series* this issue

IOP Conf. Series: Journal of Physics: Conf. Series **1056** (2018) 012040 doi:10.1088/1742-6596/1056/1/012040

New spectrometer projects for challenging particle-gamma measurements of nuclear reactions

J R B Oliveira[1,*], P Finocchiaro[2], C Agodi[2], I Boztosun[3], F Cappuzzello[2,4], P N de Faria[4], L Gasques[1], R Linares[5], N H Medina[1], D R Mendes[5], M Moralles[6], S O Solacki[4], V A B Zagatto[5], J A Alcántara-Núñez[1], M A Gonzalez-Alvarez[1,7], V A P Aguiar[1], R. Escudeiro[1], V Kurman[1]

[1] Instituto de Física da Universidade de São Paulo, São Paulo, SP, Brazil
[2] Istituto Nazionale di Fisica Nucleare, Laboratori Nazionali del Sud, Catania, Italy
[3] Dipartimento di Fisica e Astronomia, Università di Catania, Catania, Italy
[4] Department of Physics, Akdeniz University, Antalya, Turkey
[5] Instituto de Física da Universidade Federal Fluminense, Niterói, Brazil
[6] Instituto de Pesquisas Energeticas e Nucleares, São Paulo, SP, Brazil
[7] Universidade de Sevilla, Spain

[*] Email: zero@if.usp.br

Abstract. Two new gamma spectrometer projects are under development, both with some challenging technical characteristics in common, one for the investigation of weakly bound nuclear beam reactions at near barrier energies, and the other for the measurement of double charge exchange reactions in the 15-70 MeV per nucleon range. The first one is being developed and tested at the IFUSP, and the other is going to be built and installed at the LNS, INFN, under the NUMEN collaboration. The main characteristics of the two projects are presented and discussed.

1. Introduction

Under quite common situations charged particle detectors alone do not provide sufficient energy resolution for the measurement of reaction cross sections of specific nuclear states. In such cases the particle-gamma technique is an alternative which can adequately provide the required energy resolution to discriminate between nearby states. This technique has been implanted and tested at the *Instituto de Física da Universidade de São Paulo*, IFUSP [1,2], where the measurement of nuclear reactions with weakly bound stable beams have been performed. The beams were provided by the Pelletron Accelerator of the LAFN (*Laboratório Aberto de Física Nuclear*), DFN, IFUSP. For such measurements, the Saci-Perere spectrometer, originally developed for nuclear structure experiments

[3], was adapted with the insertion of collimator slits for the measurement of inelastic scattering and transfer angular distributions. The new project NC1 (*Nossa Caixa* phase 1) is being developed to improve the power and quality of the spectrometer for this type of measurement, and also for use with the RIBRAS [4] radioactive ion beam facility. With this system both stable and radioactive beams will be used to elucidate the complex reaction mechanisms with weakly bound nuclei at near barrier energies.

Another system in which the coincidence with gamma rays is required is the one to be used in the NUMEN project [5-8], under development at the *Laboratori Nazionali del Sud* (LNS), INFN, Italy. In this project, double charge exchange reactions are to be measured with the MAGNEX spectrometer [9-12], with beams provided by the CS (*Ciclotrone Supercondutore*) accelerator [12] in the 15-50 MeV per nucleon range. In some cases, particularly at transitional or strongly deformed mass regions, and also at the highest energies, the energy resolution is not sufficient to separate the ground state from the first excited state of the reaction products. A new gamma spectrometer array (GNUMEN) will be developed with the purpose of providing the necessary discrimination between these nearby energy states and allow for their separate cross section evaluation. With this system an extensive research program will be developed with the aim of helping to determine the nuclear matrix elements relevant to double beta decay physics. The characteristics of these new arrays will be presented in the next section.

2. General characteristics of the new spectrometers

Both spectrometers have in common that a moderate gamma energy resolution is sufficient for the discrimination of the relatively simple gamma spectra expected after coincidence with the charged particles, while high count rates and intense neutron radiation will have to be tolerated. For this reason, inorganic scintillators were chosen as the gamma-sensitive material. In principle LYSO(Ce) or LuAG(Ce) crystals will be used, but other possible materials are under consideration, such as GAGG(Ce), $LaBr_3$(Ce) or $CsBr_3$.

2.1. The NC1 project

The NC1 project comprises both the particle and gamma arrays. The particle detector system will be an array of plastic *phoswich* ΔE-E telescopes, similar to the *Saci* [3] ones, whose scintillation will be detected with 4×4 pixel SiPM (Silicon Photomultipliers) devices, for position discrimination, allowing kinematic coincidences. Only a few units will be used in phase 1, but a more complete array is under consideration, covering a large fraction of the solid angle of a sphere in advanced phases. With *phoswich* detectors, a high count rate per detector and radiation tolerance can be achieved. The energy resolution of these detectors is not very high, while sufficient for the identification of the atomic number Z of the particle, and the identification of the channel and state will be done essentially by the characteristic gamma rays which are detected in coincidence. Separate measurements with high particle energy resolution Silicon detectors are required for obtaining the elastic angular distributions, such as in [13].

The NC1 gamma array consists of 4 modules of 3×3 LYSO(Ce) scintillator crystals. Each crystal has a length of 40 mm and a cross section of 12.4×12.4 mm². The scintillation photons from each crystal are converted into electric pulses by 2×2 SiPM devices with about the same total area as the crystal cross section. Additional modules will be incorporated in advanced phases. These modules will operate in vacuum at a close packed geometry in order to maximize the geometrical efficiency. The use of SiPM allows for the operation of the array in the strong magnetic fields of the superconducting solenoid of the RIBRAS radioactive beam facility. For the stable beam experiments high resolution GeHP detectors can also be used.

2.2. The GNUMEN project

For the advanced phases of the NUMEN project [5-8], several improvements will be required with respect to the present system, among which an increase in magnetic field intensity of MAGNEX, replacement of the focal plane detectors (FPD), and a dramatic increase in beam intensity (by 2 or 3 orders of magnitude from the presently available) after the CS upgrade, which will also require cooling of the target and the construction of a bunker to suppress the gamma and neutron radiation from the beam dump. This will allow for the measurement of very low DCE cross sections and detailed angular distribution measurements of the moderately low ones. The uncorrelated gamma count rate expected from a typical experiment could be of the order of 1 GHz or more. The system should have, therefore, a large granularity in order to avoid the pile up of signals in each detector, and to keep the count rate per electronic channel manageable. The preliminary design of the array would consist of 2000 or more scintillator crystals covering 60% or more of the 4π solid angle of a sphere with a inner radius of about 25 cm. The crystals would be combined in 3×3 modules similarly to the NC1 ones, but with about two times larger volume crystals.

3. Preliminary tests

3.1. Gamma spectra with radioactive sources

Figure 1 shows typical gamma spectra from a ^{60}Co radioactive calibration source obtained with the single crystals of the NC1 project. The relative energy resolution (7.4% and 5.8% at 1.33MeV) of both types of scintillation crystals (LYSO(Ce) and GAGG(Ce), respectively) is compatible with what is expected for the respective material compositions. Further tests indicated that, while the LYSO(Ce) crystal spectra maintain the expected performance in resolution down to at least 30 keV gamma-ray energy (at least when the cross section of the SiPM is close to that of the crystal, granting a good light collection efficiency), the GAGG(Ce) measurements have shown a degradation of the timing and energy resolutions for gamma ray energies below about 200 keV, which is certainly due to the self absorption of the scintillation in this type of crystal. Further tests are planned with SiPM devices coupled at both sides of the crystal in order to increase the light collection and also obtain detection position information along the axis of the crystal. The energy and timing resolution performances at low energies are expected to be significantly improved with this approach.

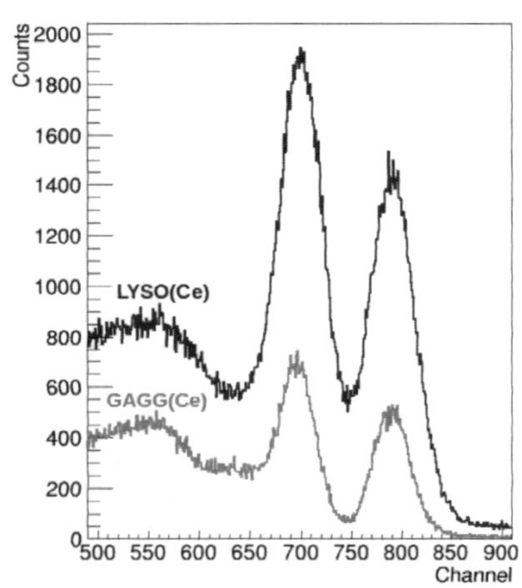

Figure 1. *Typical ^{60}Co source gamma spectra of LYSO(Ce) (upper line, in blue) and GAGG(Ce) (lower line, in red) scintillators of the NC1 project.*

3.2. In-beam timing and gamma spectrum

Figure 2 presents the random-background subtracted time spectrum (gamma pulse on start, beam bunch 20 MHz radio-frequency pulse on stop) of the ^{20}Ne +^{116}Cd reaction at 15 MeV/nucleon (from the LNS CS) taken with a 16×16×50 mm³ LYSO(Ce) crystal, outside the MAGNEX scattering chamber. The radiation from the Faraday cup (~18 cm downstream) was partially shielded with a Lead brick, and both time peaks, one from the 1.46 mg/cm² target related events, and the other from the thick Faraday cup events are clearly visible. The time-of flight difference between gamma rays generated in these separate sites (4.0 ns), is consistent with the measured time separation between the two peaks. A FWHM resolution of about 1.4 ns could be determined for these peaks. The time spectra is gated by the gamma-ray energy between 900 keV and 1060 keV. The energy and timing resolutions below about 300 keV are progressively degraded. This is due to poor light collection since the SiPM used for the scintillation signal conversion was only 6 mm wide. Indeed tests with larger SiPM devices and radioactive sources have shown a much better energy and timing resolution at low energy.

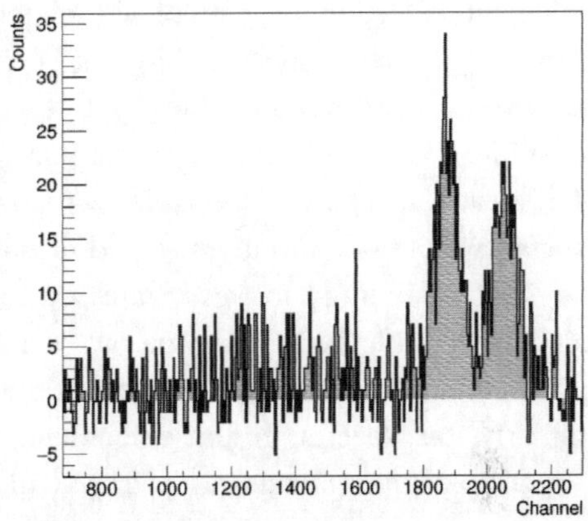

Figure 2. *In-beam time spectrum (23 ps/channel) from the ^{20}Ne experiment showing time peaks from target (right) and from the Faraday cup (left) reactions*

4. GEANT4 simulations of the detector modules

The performance of the detection modules was studied with GEANT4 [14] simulations. Figure 3 (Left) shows the perspective view of a 3×3 scintillator crystal module and (Right) simulation of 500 keV gamma rays hitting the central crystal and eventually scattering into the surrounding units.

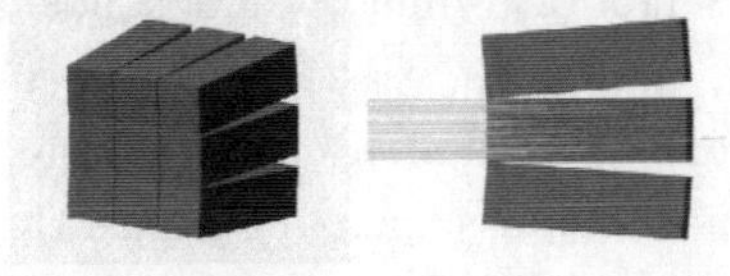

Figure 3. *Detector modules with 3 × 3 pixel LYSO crystals. Left: perspective view. Right: section view illustrating the GEANT4 simulations of the scattering of 500 keV gamma rays hitting the central pixel.*

The simulated photopeak efficiency and peak-to-total ratios after summation of surrounding first neighbours of each crystal approaches that of a monolithic one with the same total volume, at low energies. At 500 keV the intrinsic efficiency decreases from about 75% to 71%, and the peak to total ratio (P/T) from 86% to 84% for the 16 mm wide crystals, and also a few percent less for 12.4 mm crystals. With the full array of closely packed units the intrinsic efficiency could rise to 93% and P/T to 94%, but some form of gamma ray tracking will be necessary to properly recover the full energy of the gamma rays. Detailed simulations are under way towards the evaluation and optimization of this procedure.

5. Expected performance and observational limit

With the detector granularity proposed for the GNUMEN array, the pile-up probability is expected to remain well below 10% for beam intensities up to 5×10^{12} particles/s (or about 100 times the presently available one), and can be neglected in a preliminary evaluation of the performance. At such high beam intensities, however, the probability of a second nuclear reaction, besides the DCE, with production of gamma rays in a given beam bunch can be quite significant (typical total reaction cross sections are of the order of a few barns). These gamma rays will be detected within the 2 ns time window of the prompt peak and cannot be resolved. They constitute unavoidable chance coincidences. Due to that fact, the main source of background (BG) in the DCE measurement of the cross sections populating the ground state (*gs*) and first excited state (2^+) of the exit channel is expected to be the accidental detection of one of these uncorrelated gamma rays which happen to have an energy within the energy gate around the first excited state transition. The $2^+ \rightarrow 0^+$ transition gamma-gated FPD DCE reaction energy spectrum will present in general two peaks (partially resolved or unresolved), one corresponding to *gs* state in accidental coincidence with the BG gamma rays, and the other to the 2^+ state in true or accidental coincidence. Similarly, the spectra in anti-coincidence with the $2^+ \rightarrow 0^+$ transition gamma-ray energy will have two superimposed contributions, from the *gs* state peak, and from the 2^+ excited state when the corresponding transition is not detected due to the less than 100% efficiency of the array. Two main parameters control the proportion of these different contributions: the total photopeak efficiency and the probability of production of a gamma signal within the 2^+ excited state, and which is approximately proportional to the width of the energy gate and, therefore, to the energy resolution of the array. From Poisson statistics it is possible to predict analytically the *observational limit* of the array: the minimum ratio of DCE to total reaction cross section which can be measured with a specified statistical uncertainty (such as 20%). Figure 4 presents the “reduced” observational limit (B, a number proportional to the observational limit but independent of the time duration of the experiment), as a function of the average number of reactions per beam bunch $\bar{k}$ (which is proportional to the beam luminosity). The curves apply to both the *gs* and 2^+ excited state measurement since in these calculations an equal cross section was assumed for both. The probability that a spurious reaction (not the DCE) produces a gamma ray signal within the $2^+ \rightarrow 0^+$ transition energy gate was assumed to be close to 50%. This is appropriate for high gamma-ray multiplicity reactions and relatively poor energy resolutions such as for LYSO(Ce) scintillators. As it can be seen, the observational limit decreases rapidly with $\bar{k}$ up to values of the order of $\bar{k}$=1. The curves with the presence of BG go through a minimum which defines the optimum $\bar{k}$ and the absolute reduced observational limit (B = 3 at $\bar{k}$=2.8 in the 50% efficiency case, roughly corresponding to an observational limit of 6×10^{-12}, or 20pb in 3b total cross section, for a one week data taking time). Note that even in the ideal case of 100% efficiency the improvement in observational limit is not very dramatic, and the limiting factor is the growing contribution of the BG at high luminosities.

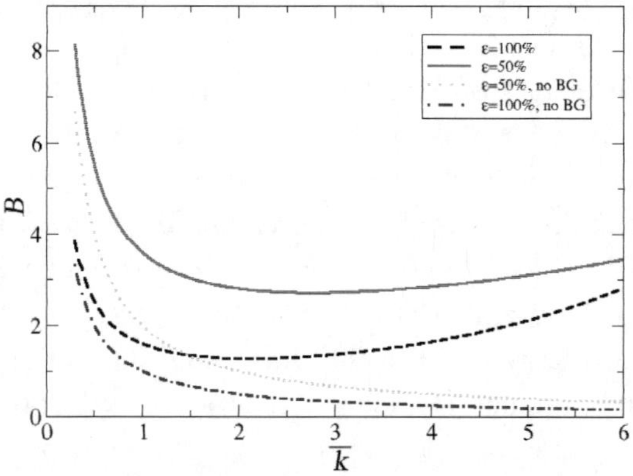

Figure 4. *Reduced observational limit as a function of* $\bar{k}$ *for 50% efficiency with BG (solid), and no BG (dot), and 100% efficiency with BG (dot) and no BG (dot-dash).*

6. *Final considerations*

According to these preliminary tests and analyses, a sufficient sensitivity can be obtained with the proposed gamma arrays to fulfil the experimental requirements. Further detailed simulations and experimental tests are planned for definition of the the details of the final design. The most difficult parameter to evaluate is the random coincidence BG which will be generated during the real experiments. For this it will be necessary to perform test experiments of the specific reactions to be measured and validate detailed simulations of the array response. The tolerance to the high neutron flux of the detectors and electronic equipment will also have to be experimentally evaluated in the near future.

Acknowledgments
We acknowledge support from the INFN, Italy, CAPES, CNPq and FAPESP (Proc. 2016/04612-9), Brazil.

References

[1] Zagatto V A B et al. 2014 *Nucl. Inst. and Meth.* A **749** 19
[2] Zagatto V A B et al. 2016 *J. Phys. G: Nucl. Part. Phys.* **43** 055103
[3] Alcántara-Núñez J A et al. 2003 *Nucl. Inst. and Meth.* A **497** 429
[4] Lichtenthäler R et al. 2005 *Eur. Phys. J.* A **25** 733
[5] Cappuzzello F et al. 2015 *Eur. Phys. J.* A **51** 145
[6] Cappuzzello F et al. 2015 *J. Phys.: Conf. Ser.* **630** 12018
[7] Agodi C et al., 2015 *Nucl. and Part. Phys. Proc.* **265** 28
[8] Cappuzzello F et al. 2018 *Eur. Phys. J.* A **54** 72
[9] Cappuzzello F et al. 2016 *Eur. Phys. J.* A **52** 167
[10] Cappuzzello F et al. 2014 *Nucl. Inst. Meth.* A **763** 314
[11] Cunsolo A et al. 2007 *Eur. Phys. J.: Spec. Top.* **150** *343*
[12] Calabretta L et al. 2017 *Mod. Phys. Lett.* A **32** 1740009
[13] Zagatto V A B et al. 2017 *Phys. Rev.* C **95** 064614
[14] Agostinelli S et al. 2003 *Nucl. Inst. and Meth.* A **506** 250

IOP Conf. Series: Journal of Physics: Conf. Series **1056** (2018) 012041 doi:10.1088/1742-6596/1056/1/012041

Structure of proton-rich nuclei via mirror β decay and charge exchange reactions

S E A Orrigo[1], B Rubio[1], Y Fujita[2,3], W Gelletly[4] and B Blank[5]
[1]Instituto de Física Corpuscular, CSIC-Universidad de Valencia, E-46071 Valencia, Spain
[2]Department of Physics, Osaka University, Toyonaka, Osaka 560-0043, Japan
[3]Research Center for Nuclear Physics, Osaka University, Ibaraki, Osaka 567-0047, Japan
[4]Department of Physics, University of Surrey, Guildford GU2 7XH, Surrey, UK
[5]Centre d'Etudes Nucléaires de Bordeaux Gradignan, CNRS/IN2P3 - Université Bordeaux 1, 33175 Gradignan Cedex, France

E-mail: sonja.orrigo@ific.uv.es

Abstract. The β decays of the proton-rich, T_z = -2 nuclei ^{48}Fe, ^{52}Ni and ^{56}Zn were investigated by decay spectroscopy at GANIL. The decay schemes were determined, together with absolute Fermi and Gamow-Teller transition strengths. New interesting results were obtained, including the first observation of a new decay mode in the *fp*-shell, the exotic β-delayed γ-proton decay seen in ^{56}Zn, and the first observation of the 2^+ isomer in ^{52}Co. These studies were complemented by Charge Exchange reactions on the stable mirror targets.

1. Introduction

One of the main topics in contemporary nuclear physics concerns the study of nuclear structure far from the valley of nuclear stability. Our knowledge of Gamow-Teller (GT) transitions when approaching the proton drip-line is still rather incomplete [1] because the production of such exotic nuclei becomes steadily more challenging. In addition to the great interest from the point of view of nuclear structure, the study of GT transitions starting from unstable proton-rich nuclei is also of crucial importance for nuclear astrophysics since many heavy proton-rich elements are produced in the *rp*-process passing through proton-rich *fp*-shell nuclei.

Decay spectroscopy with implanted Radioactive Ion Beams (RIBs) is a powerful tool to explore nuclei at the proton drip-line since the β decay has direct access to the absolute values of the Fermi B(F) and Gamow-Teller B(GT) transition strengths. Charge Exchange (CE) reactions such as (p,n) or (^{3}He,t) are the mirror strong interaction process and provide information on the relative B(GT) values [2] without energy restrictions. Hence β decay and CE studies are complementary tools. Their combined analysis allows us to investigate fundamental questions related to the role of isospin in atomic nuclei, such as isospin symmetry in mirror nuclei [3, 4]. Aiming at these goals, we have performed a series of experiments pointing to compare β decay in proton-rich nuclei and CE reactions on stable mirror targets.

The β decay of the proton-rich, T_z = -2 nuclei ^{56}Zn [5], ^{48}Fe and ^{52}Ni [6], and of the odd-odd, T_z = -1 nucleus ^{52}Co [7] was studied in an experiment performed at GANIL. New interesting results were obtained, going from the first observation of a new decay mode, the exotic β-delayed γ-proton decay seen in ^{56}Zn [5], to the first observation of the 2^+ isomer in ^{52}Co [7]. In all cases the decay schemes were established, with absolute B(F) and B(GT) strengths.

The results for the T_z = -2 nuclei [6] were compared with those obtained by complementary (^{3}He,t) CE reactions carried out on the stable mirror targets at RCNP Osaka [4, 8, 9]. In general, under the assumption of isospin symmetry, one expects that mirror Fermi and GT transitions should populate the same states in the daughter nucleus with the same probability. However, one should note that in the $T = 2$ isospin multiplet we compare transitions involving different initial and final states. For example, the mirror case of the β^+ decay of ^{56}Zn to ^{56}Cu is the ^{56}Fe(^{3}He,t)^{56}Co reaction. Thus the final nuclei (^{56}Cu and ^{56}Co) are not identical and therefore the level scheme and the strengths might be slightly different. Nevertheless, the comparison is very effective in helping shed light on the structure of such exotic nuclei and improving our understanding of their decay schemes, especially for the most exotic cases.

2. The experimental setup

The experiment was performed at GANIL. The production of several exotic nuclei was achieved by fragmenting a primary beam of ^{58}Ni^{26+} at 74.5 MeV/nucleon on a natural Ni target of 200 μm thickness. The fragments were selected by the LISE3 separator [10] and then implanted into a Double-Sided Silicon Strip Detector (DSSSD), 300 μm thick. Both the implanted heavy ions and subsequent charged-particle (β particles and protons) decays were detected by the DSSSD. Four EXOGAM Germanium clovers [11] surrounded the DSSSD in order to detect the β-delayed γ-rays. Particle identification was achieved by using the energy loss signal (ΔE) from a silicon detector placed 28 cm upstream from the DSSSD together with the Time-of-Flight (ToF) signal obtained as the time difference between the cyclotron radio-frequency and the ΔE signal. In addition, an implantation event was defined by simultaneous signals from both the ΔE and DSSSD detectors, while a decay event was defined by a signal above threshold in the DSSSD with no coincident signal in the ΔE detector. More details on the experimental setup and the procedures employed in the data analysis are available in Ref. [6].

3. The β decay of ^{56}Zn

^{56}Zn, lying at the proton drip-line, has been one of the most intriguing cases under study in our experimental campaign. In this nucleus we observed many interesting and unusual features [5], reflecting the peculiarities that one might expect when investigating the most exotic systems.

The ^{56}Zn decay scheme is shown in figure 1. One can see that there is competition between β-delayed proton and β-delayed γ emission in the decay. Moreover, the β-delayed γ rays populate levels in the daughter nucleus, ^{56}Cu, which lie above the proton separation energy and hence are unbound. Thus these levels then decay by proton emission to the ground state of ^{55}Ni. Therefore the sequence observed is β-delayed γ-proton decay, a new and exotic decay mode seen here for the first time in the fp-shell. In ^{56}Zn we observed three such sequences, involving the γ rays at 1835, 861 and 309 keV and the ^{56}Cu levels at 1691, 2661 and 1391 keV, respectively.

The comparison between the β decay of ^{56}Zn to ^{56}Cu [5] and the mirror CE process, the ^{56}Fe(^{3}He,t)^{56}Co reaction [8], shows a remarkable isospin symmetry where all the dominant transitions are observed in both cases and with very similar strengths. Hence we exploited the higher energy resolution ($\sim$ 30 keV) of the CE reaction to clarify other aspects of the level structure in ^{56}Cu which, without this comparison, would have remained unclear. The Fermi transition feeds the Isobaric Analogue State (IAS), which has $T = 2$ in this isobaric multiplet. It was observed in ^{56}Co [8] that the IAS is fragmented in two 0^+ levels which mix (with 28% in the $T = 1$ state) and both contribute to the Fermi strength. Based on this evidence we assumed that there is a similar situation in ^{56}Cu. In other words the Fermi strength is again divided between two states and we found that isospin mixing of 33% is required to obtain the expected value of B(F), given by the sum rule $|N - Z| = 4$. Later on these findings were also confirmed by two independent shell model calculations [12, 13].

IOP Conf. Series: Journal of Physics: Conf. Series **1056** (2018) 012041 doi:10.1088/1742-6596/1056/1/012041

Another intriguing feature emerged when looking in more detail at the decay of the IAS in ^{56}Cu. The proton decay from the $T = 2$ IAS would be isospin-forbidden, however it may proceed thanks to the 33% mixed $T = 1$ component. At this point the much faster proton decay ($t_{1/2} \sim 10^{-18}$ s) should dominate over the γ de-excitation ($t_{1/2} \sim 10^{-14}$ s in the mirror). This is not the case since the competing γ decay from the IAS is still observed. The reason for such behaviour lies in the nuclear structure. Both shell model calculations [12, 13] found the proton decay of the $T = 1$ IAS component hindered by a factor of 10^3.

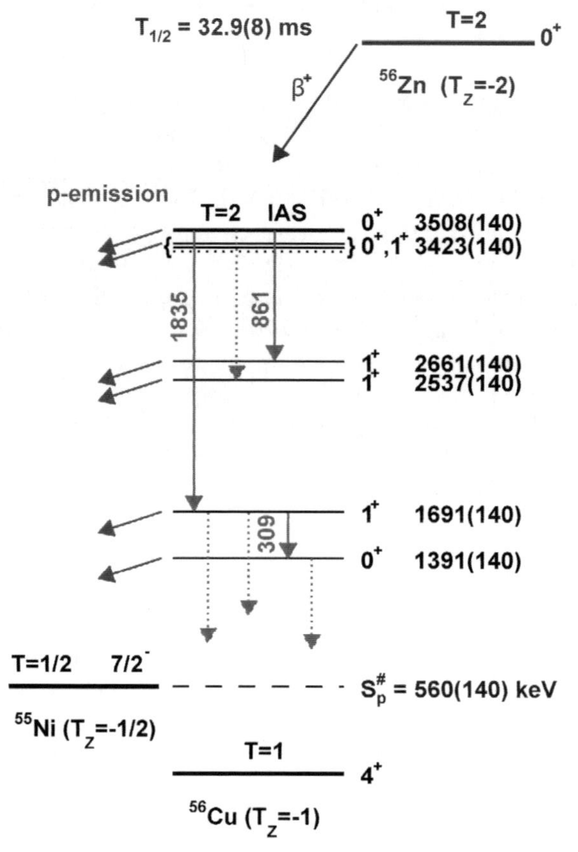

Figure 1. The decay scheme of ^{56}Zn. Transitions corresponding to those observed in the mirror ^{56}Co nucleus are represented by dotted lines.

4. The β decay of ^{48}Fe and ^{52}Ni

In the same experiment we studied the β decay of the T_z = -2 nuclei ^{48}Fe and ^{52}Ni [6]. Individual β-delayed protons and γ rays were measured also for these decays. The higher energy resolution achieved for protons, in comparison to a previous study [14], allowed us to identify many new states populated in the decays of ^{48}Fe and ^{52}Ni and to determine the related branching ratios for the first time. The DSSSD spectra were compared with the results from the mirror CE data from the (^{3}He,t) reaction on ^{52}Cr [4] and ^{48}Ti [9], respectively. We observed a good isospin symmetry in the mass 52 case, while the comparison does not work as well in the mass 48 case. The decay schemes obtained for ^{48}Fe and ^{52}Ni are shown in figure 2 and 3, respectively.

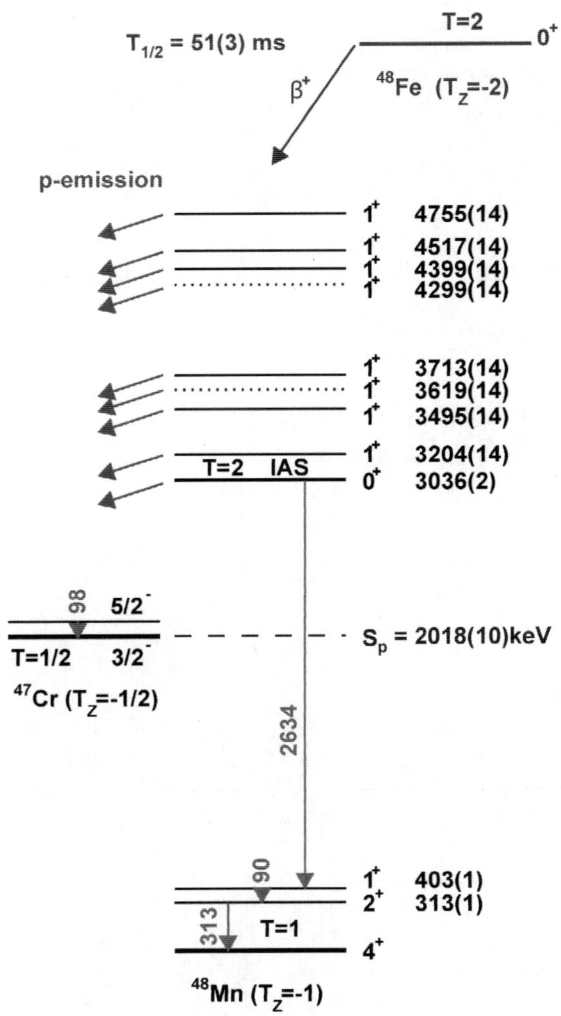

Figure 2. The decay scheme of ^{48}Fe. For details on the dotted levels see Ref. [6].

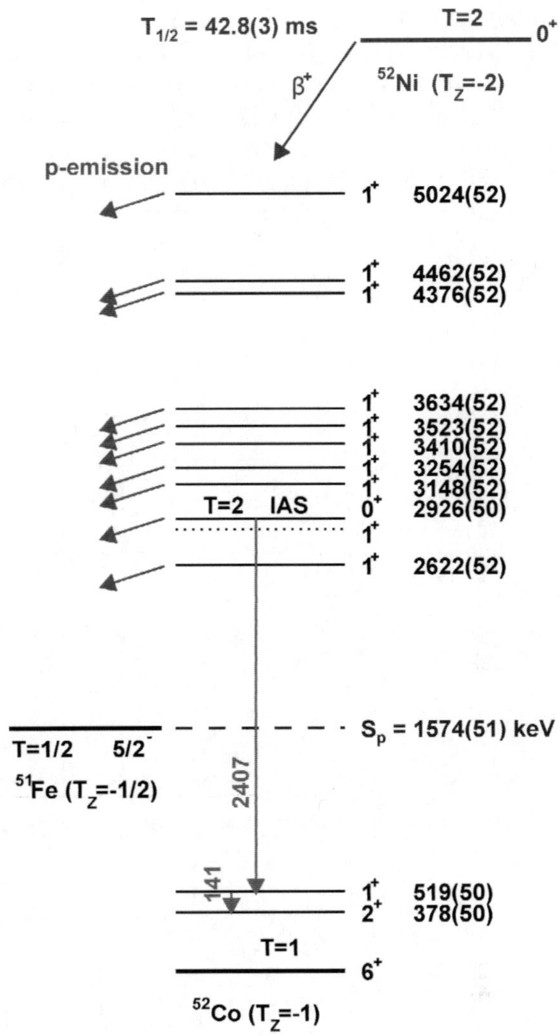

Figure 3. The decay scheme of ^{52}Ni. The dotted 1^+ state was seen in the mirror ^{52}Mn.

In the decay of ^{48}Fe and ^{52}Ni we also observed competition between β-delayed protons and β-delayed γ rays in the de-excitation of the $T = 2$ IAS. The β-delayed proton emission would again be isospin-forbidden. It was observed, however, even although it is weaker than in ^{56}Zn and is attributed to a $T = 1$ isospin impurity mixed with the IAS. For ^{48}Fe and ^{52}Ni the β-delayed proton emission constitutes only 14% and 25% of the respective totals compared with 44% in ^{56}Zn. There are two reasons for this difference. Firstly, for the ^{56}Zn daughter there is another 0^+ state less than 100 keV from the IAS and the mixing depends on how close the two states are [15], thus the strong mixing favours proton decay. Secondly it is partly due to the relatively low proton energies involved for the ^{48}Fe and ^{52}Ni daughters. As a result the calculated proton half-lives are of the same order-of-magnitude as the γ-decay Weisskopf transition probabilities.

At the time of our study the masses of the ^{52}Co ground state and its 2^+ isomer were unknown. Hence we estimated the energy of the 2^+ level in figure 3 starting from the mirror ^{52}Mn, as

IOP Conf. Series: Journal of Physics: Conf. Series **1056** (2018) 012041 doi:10.1088/1742-6596/1056/1/012041

explained in Ref. [6]. Now that these masses have been measured [16] we find an energy of 387(13) keV for the 2^+ state which agrees well with our estimated value of 378(50) keV. In Ref. [16], based on the new mass measurement, it is proposed that the energy of the IAS in ^{52}Co is 2935(13) keV. It is further suggested that the γ and proton emission do not both arise from the IAS. Instead the proton emission is from a 1^+ state lying below the IAS at 2800(16) keV. However without the proton emission from the IAS 25% of the Fermi strength would be missing. Further experiments are required to address this issue.

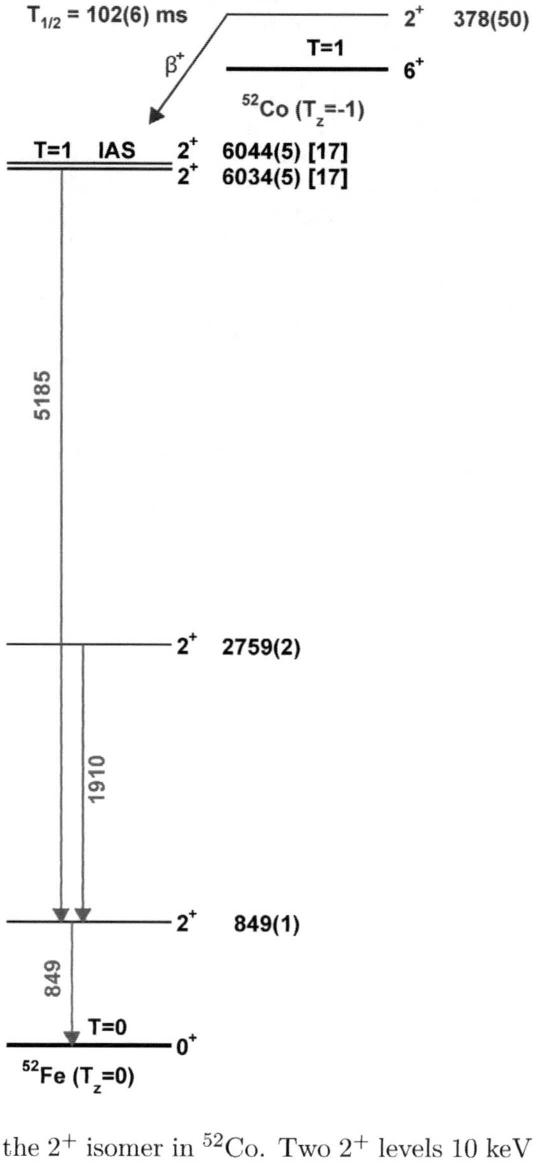

Figure 4. The decay scheme of the 2^+ isomer in ^{52}Co. Two 2^+ levels 10 keV apart are reported as IAS candidates in ^{52}Fe [17].

5. First observation of the 2^+ isomer in ^{52}Co

Odd-odd nuclei are particularly difficult to study because there are often two long-lived states, one of which is the ground state, with similar half-lives. One such case is the T_z = -1, ^{52}Co nucleus where it was speculated that an isomeric 2^+ state exists that had not been observed prior to the present work [7]. The difficulty was that when one attempted to populate ^{52}Co directly both the ground state and the isomer are produced and implanted together. We succeeded in observing the isomer for the first time by creating it as a product of the decay of ^{52}Ni (see figure 3). This decay process directly populates the $0^+(T=2)$ IAS in ^{52}Co, which then de-excites by emitting two γ rays of 2407 and 141 keV, populating in a clean way the 2^+ isomeric state. In this way we were able to disentangle the β decays of the 2^+ isomer and the 6^+ ground state in ^{52}Co. We then observed the β de-excitation of the isomer, shown in figure 4, which is followed by the emission of three γ rays of energy 849, 1910 and 5185 keV. We measured for the first time the half-life of the isomeric state, 102(6) ms, and improved the half-life measurement for the ground state, 112(3) ms [7].

6. Conclusions

We studied the β decay of a variety of proton rich-nuclei, namely ^{48}Fe, ^{52}Ni, ^{52}Co and ^{56}Zn, establishing the decay schemes and absolute values for the transitions strengths. For the T_z = -2 cases we have seen that the comparison with the CE data was very useful to better understand some aspects of the decays. Mirror symmetry works well in general, but some differences remain. The decay of ^{56}Zn showed many exotic features, such as the IAS fragmentation due to strong isospin mixing, the unexpected competition between β-delayed protons and β-delayed γ rays from the IAS, later explained as due to the nuclear structure of the states involved, and the first observation of the β-delayed γ-proton decay in the fp-shell. Another remarkable result is the discovery of the 2^+ isomer in ^{52}Co, with the first measurement of its half-life and de-exciting γ rays.

Acknowledgments

This work was supported by the Spanish MICINN grants FPA2008-06419-C02-01, FPA2011-24553, FPA2014-52823-C2-1-P; Centro de Excelencia Severo Ochoa del IFIC SEV-2014-0398; CPAN Consolider-Ingenio 2010 Programme CSD 2007-00042; *Junta para la Ampliación de Estudios* Programme (CSIC JAE-Doc contract) co-financed by FSE; ENSAR project 262010; MEXT, Japan 18540270 and 22540310; Japan-Spain coll. program of JSPS and CSIC; UK Science and Technology Facilities Council (STFC) Grant No. ST/F012012/1; FONDECYT/INICIACION No. 11130049; Région Aquitaine. W. G. acknowledges the support of the University of Valencia. We acknowledge the EXOGAM collaboration for the use of their clover detectors.

References

[1] Langanke K and Martínez-Pinedo G 2003 *Rev. Mod. Phys.* **75** 819-62
[2] Taddeucci T N *et al* 1987 *Nucl. Phys.* A **469** 125-72
[3] Fujita Y *et al* 2005 *Phys. Rev. Lett.* **95** 212501
[4] Fujita Y, Rubio B and Gelletly W 2011 *Prog. Part. Nucl. Phys.* **66** 549-606
[5] Orrigo S E A *et al* 2014 *Phys. Rev. Lett.* **112** 222501
[6] Orrigo S E A *et al* 2016 *Phys. Rev.* C **93** 044336
[7] Orrigo S E A *et al* 2016 *Phys. Rev.* C **94** 044315
[8] Fujita H *et al* 2013 *Phys. Rev.* C **88** 054329
[9] Ganioğlu E *et al* 2016 *Phys. Rev.* C **93** 064326
[10] Anne R and Mueller A C 1992 *Nucl. Instrum. Methods Phys. Res.* B **70** 276-85
[11] Simpson J *et al* 2000 *Acta Phys. Hung.* A **11** 159-88
[12] Rubio B *et al* 2016 *Nucl. Phys. Review* **33** 225-9
[13] Smirnova N A, Blank B, Brown B A, Richter W A, Benouaret N and Lam Y H 2016 *Phys. Rev.* C **93** 044305
[14] Dossat C *et al* 2007 *Nucl. Phys.* A **792** 18-86
[15] Dzubay T D, Sherr R, Becchetti F D and Dehnhard D 1970 *Nucl. Phys.* A **142** 488-98
[16] Xu X *et al* 2016 *Phys. Rev. Lett.* **117** 182503
[17] Decowski P, Benenson W, Brown B A and Nann H 1978 *Nucl. Phys.* A **302** 186-204

IOP Conf. Series: Journal of Physics: Conf. Series **1056** (2018) 012042 doi:10.1088/1742-6596/1056/1/012042

Improved method for the detection of low-energy astrophysical neutrinos bursts

G Pagliaroli[1,2],C Casentini[3,4],C Vigorito[5] and V Fafone[3,4,1]

[1]Gran Sasso Science Institute, Viale Francesco Crispi 7, I-67100 L'Aquila, Italy
[2]INFN-LNGS, I-67100 L'Aquila, Italy
[3]Universitá di Roma Tor Vergata, via della Ricerca Scientifica, I-00133 Roma, Italy
[4]INFN Sezione di Roma Tor Vergata, via della Ricerca Scientifica, I-00133 Roma, Italy
[5]Universitá di Torino & INFN, via Pietro Giuria 1, I-10125 Torino

Abstract. In this paper we describe an improved method for the search of astrophysical bursts of low-energy neutrinos. We use the different temporal behaviour describing impulsive signal with respect to the flat temporal expectation for background events. We show that this approach strongly reduces the misidentification probability without decreasing the detection probability. In other words, this method allows to neutrino detectors to work at lower threshold with the same statistical confidence.

1. Introduction

Core-Collapse Supernovae (CCSNe) [1] represent the final explosive fase of massive stars and the detection of a galactic event could be the unique opportunity for us to grasp the physical mechanism driving the final explosion of the structure. On the other hand "Failed" Supernovae [2] are collapsing stars that fail to explode forming an inner black hole. The lack of the final explosion make these sources optically silent and, at the present, have never been directly observed. Finally, Quark Novae [3], are expected when a neutron star suddenly converts into a quark star. Their existence is strongly related to the fundamental state of the matter and their detection could provide the first clear evidence of the presence of strange matter in the universe.

A common signature for all these catastrophic astrophysical phenomena is expected to be an impulsive ($\sim$ 10 seconds) emission of low-energy, ($\sim$ 10 MeV), neutrinos [4, 5, 6]. Despite the large amount of total energy ($\sim 10^{53}$ergs) released in neutrinos, when the source distance increases and/or the average energy of emitted neutrinos decreases the signal statistics drops and the identification of these astrophysical bursts embedded into the detector noise could be challenging.

In this paper, we improve the detectors capability to disentangle real astrophysical bursts of low-energy neutrinos from background signals. This powerful method exploits the different temporal structure expected for an astrophysical burst with respect to background fluctuations that are near uniformly distributed in a time window. This characteristic, described with a new parameter, can be used as an additional degree of freedom that added to the statistical requirement improves our capability to identify real signals, allowing the detection of weaker/far away astrophysical sources.

2. Emission Model

For all the astrophysical sources we are interested in, we assume that the total energy radiated in neutrinos is $\mathcal{E} = 3 \times 10^{53}$ erg. We consider a very general description of an astrophysical burst of low-energy neutrinos characterised by the following temporal evolution:

$$f(t) = (1 - \exp(-t/\tau_1)) \exp(-t/\tau_2), \tag{1}$$

where $\tau_1 = (10 - 100)$ms is the rising time and $\tau_2 \geq 1$ s represents the decaying time of the signal.

As spectral shape for the signal we assume quasi-thermal spectra and the neutrino fluence, differential in the neutrino energy E, is described by

$$\Phi_i^0 = \frac{\mathcal{E}_i}{4\pi D^2} \times \frac{E^\alpha e^{-E/T_i}}{T_i^{\alpha+2}\Gamma(\alpha+2)} \quad i = \nu_e, \nu_\mu, \nu_\tau, \bar{\nu}_e, \bar{\nu}_\mu, \bar{\nu}_\tau, \tag{2}$$

where the energy radiated in each specie is $\mathcal{E}_i = \mathcal{E}/6$ due to the equipartition hypothesis, Γ is the Gamma function and the 'temperature' is $T_i = \langle E_i \rangle/(\alpha + 1)$. The average energy per flavour is $\langle E_i \rangle$ and the parameter $\alpha = 3$ represents a mild deviation from a thermal distribution. As average energies we set $\langle E_{\nu_e} \rangle = 9$ MeV, $\langle E_{\bar{\nu}_e} \rangle = 12$ MeV and non-electronic temperature 30% higher than the one of $\bar{\nu}_e$. All these values are compatible with SN1987A[7]. As interaction channel we consider the main interaction channel, i.e., the inverse beta decay (IBD) $\bar{\nu}_e + p \to n + e^+$. Due to neutrino oscillations the $\bar{\nu}_e$ fluence at the detector is an admixture of the unoscillated flavour fluences at the source: $\Phi_{\bar{\nu}_e} = P\Phi^0_{\bar{\nu}_e} + (1-P)\Phi^0_x$, where x indicates the non-electronic flavours and P is the survival probability for the $\bar{\nu}_e$. Depending on the neutrinos mass hierarchy, this probability can be $P = 0$ for Inverted Hierarchy (IH) or $P \simeq 0.7$ for Normal Hierarchy (NH).

The expected number of IBD interactions is $S(E_\nu, D) = N_p \sigma_{\bar{\nu}_e p}(E_\nu) \Phi_{\bar{\nu}_e}(E_\nu, D) \epsilon(E_{vis})$, where D is the source distance, N_p is the number of target protons within the detector, $\sigma_{\bar{\nu}_e p}$[8] is the process cross section and ϵ is the detector efficiency as a function of the visible energy E_{vis}. The visible energy is the portion of interaction energy released inside the detector that can be observed; for the detectors considered in this paper, it is the energy converted in light inside the sensitive volume. We show our results for the NH case with $\langle E_{\bar{\nu}_e} \rangle$=15 MeV.

3. Method

The aim of this paper is to show an improved way to discriminate astrophysical burst of low-energy neutrinos from accidental burst of events due to background. For this reason background knowledge and characterization is fundamental to demonstrate the potential of this method. Low-energy neutrinos detectors at the present on data-taking like Super Kamiokande[9], LVD[11], Borexino[12] and KamLAND[13] provide all the information needed to perform this study and results are reported for these detectors both considering that they are working alone and by exploiting combined coincident search.

In all the considered detectors the search of astrophysical bursts of low-energy ν is based on the definition of clusters of events and then by a statistical selection of the most significant clusters. Following [15] we define a cluster as the group of the events contained in consecutive time windows of $w = 20$ seconds. Each cluster is characterised by its multiplicity m_i and its time duration Δt_i, defined as the time difference among the first and the last event detected inside the window. In order to increase the detection probability this search is performed one more time by shifting the consecutive time windows of 10 seconds with respect to the first search, for more details see [15].

A Monte Carlo simulation of 10 years of data-taking has been performed in order to reproduce the background fluctuations of each detector according to the parameters (frequencies and energy thresholds) reported in Tab.1.

For the observed background clusters of events we calculate its imitation frequency $f_i^{im}(\text{day}^{-1})$, i.e. how many times in a day background events produce a cluster with the same multiplicity. In the following we set as working threshold for this statistical cut the value $f^{im} \leq 1/\text{day}$ in order to test the new method to discriminate signal from background. Then we simulate the signals expected in each neutrino detector by assuming the temporal behaviour in Eq. 1 and the energy spectra in Eq.2. The Monte Carlo extraction for the signals are performed considering different source distances D, in the range $8.5 - 500$ kpc. As further step, simulated signals are randomly injected inside the background. Once that clusters are obtained following the previous procedure only clusters with $f^{im} < 1/\text{day}$ are selected. For each cluster we define the parameter, ξ_i as the ratio between the cluster multiplicity and the cluster duration:

$$\xi_i = \frac{m_i}{\Delta t_i}. \qquad (3)$$

As a function of ξ we study the distributions of pure background clusters and background plus signal clusters in term of normalised Probability Density Functions (PDF). In Panel (a) of Fig 1 we show the result obtained for SuperK detector. The distribution of clusters due to pure

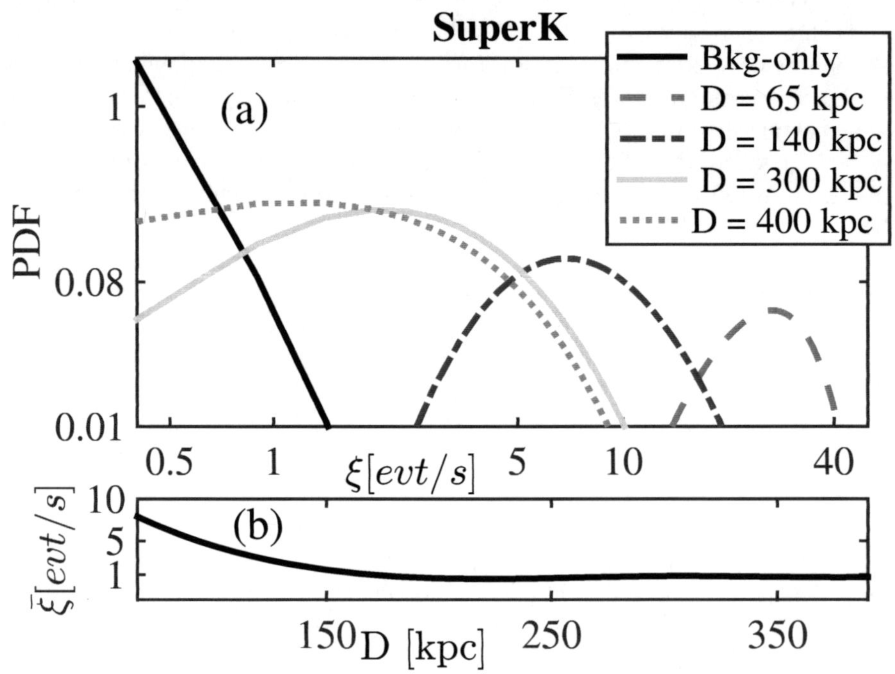

Figure 1. *Panel (a): Probability density functions for background plus signal clusters as functions of the ξ parameter and for different distances in the case of SuperK detector. The black solid line shows the PDF for pure background clusters; Panel (b): The optimal cut value for the ξ parameter, $\bar{\xi}(D)$, as a function of the source distance D for SuperK detector.*

background events is reported with a black solid line and is characterised by very small values of the ξ parameter as expected for events with a temporal uniform distribution, i.e., $\Delta t_i \to w$ seconds. On the other hand, clusters where also an astrophysical signal is present show a PDF shifted at higher values of the ξ axis, as expected for clusters with events that cumulate faster in time $\Delta t_i < 20$ seconds. For any fixed source distance the PDF is different and the Panel (a) of FIg.1 we show with different color the case of D=65, 140, 300 and 400 kpc as expected in

SuperK. For any fixed source distance the PDF is different and a different optimal cut value for the ξ parameter can be defined in order to separate background and signal distributions. By performing several simulations we define the function $\overline{\xi}(D)$, reported in Panel (b) of Fig.1 for the case of SuperK. Finally, we define as optimal value of the cut parameter $\overline{\xi}_X$ the smaller one allowing a clear separation between background and signal PDFs. Following this prescription, in the last column of Tab.1, we report the optimal cut parameters found for all the detectors considered.

So that, we add as new cut on the previous selected clusters, with $f_i^{im} < 1/\text{day}$, the condition $\xi_i \geq \overline{\xi}_X$ and we investigate the impact of this on the detection probability, η, and on the misidentification probability ζ. The detection probability is defined as the ratio between the number of signal clusters recovered by the analysis that survive after the cuts and the number of signal clusters that was initially injected into the background. In a similar manner the misidentification probability, ζ, is obtained as the fraction of background clusters that survive all the cuts over the total number of clusters observed.

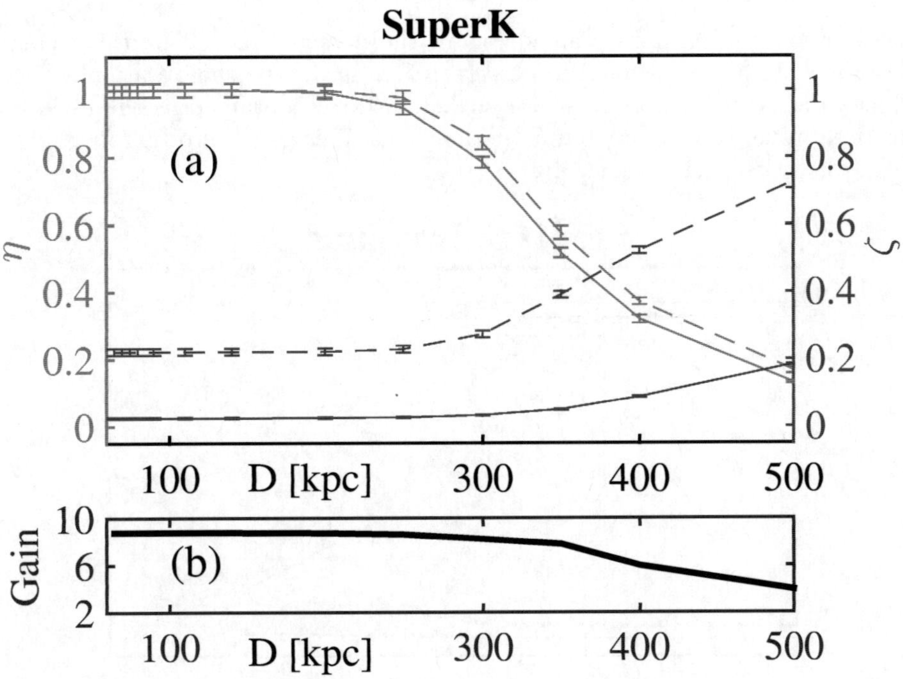

Figure 2. *Panel (a): Red lines show the detection probability curves, η, whereas blue lines show the misidentification probability curves, ζ, as a function of the source distance D for SuperK detector working at $f^{im} \leq 1/day$. Solid (Dashed) lines are obtained by following the new proposed (standard) method for background reduction. Panel (b): The gain factor for SuperK as defined in the text versus the source distance D.*

As a leading example we apply our search procedure to SuperK detector and we compare the detection probability obtained with the new procedure with respect to the standard procedure only based on statistical cut. To show the improvement provided by our method we plot in Panel (a) of Fig.2 the detection efficiency η with red lines and the misidentification probability ζ with blue lines. In particular dashed lines are obtained by using the standard procedure only based on statistical cut as described above while the solid lines are obtained by applying our additional selection criterium. As evident by the picture the efficiency is unchanged till a distance of ~ 200

kpc while the misidentification that was around a 23% by using the standard procedure drops to very small value around 3% by applying the $\bar{\xi}$ cut.

Detector	M(kton)	E_{thr}(MeV)	f_{bkg} (Hz)	$\bar{\xi}$(Hz)	$\bar{D}$(kpc)	G
Borexino	0.3	1	0.048	0.65	20	6.9
SuperK	22.5	7	0.012	0.72	200	8.9
KamLAND	1	1	0.015	0.77	50	13.4
LVD	1	10	0.028	0.72	40	14.0

Table 1. Columns in order show: sensitive detector mass in kton; energy threshold considered for the analysis in MeV; average background frequency in Hz[9, 11, 12, 13]; value for the $\bar{\xi}$ parameter that maximise the signal to noise ratio, as described in the text; maximal distance $\bar{D}$ without efficiency loss after the new cut; gain factor obtained by using the new proposed method.

In Tab.1 we show the results obtained for all the detectors considered. In particular, the maximal distance $\bar{D}$(kpc) for which no efficiency loss is due to the additional $\bar{\xi}$ cut and the corresponding gain, G $= \zeta/\zeta'$, obtained for such a distance, obtained as the ratio between the misidentification probability before and after the $\bar{\xi}$ cut. This gain factor is reported in the lower panel of Fig.2 for the SuperK case study.

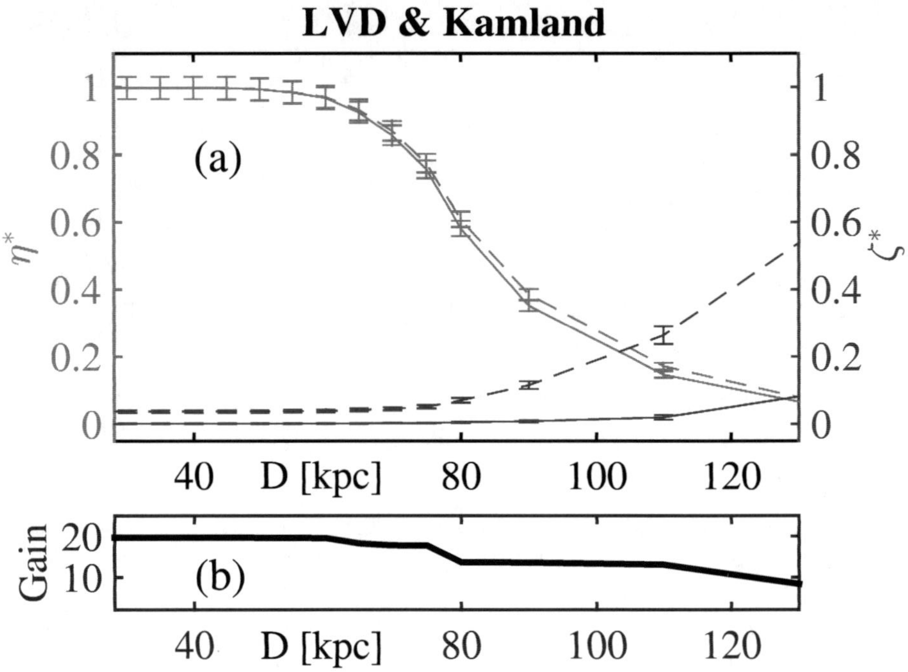

Figure 3. *Panel (a): Red lines show the detection probability η^* whereas blue lines show the misidentification probability ζ^* for the network LVD & Kamland. Solid (Dashed) lines are obtained by following the new proposed (standard) method for background reduction. Panel (b): The gain factor for the network LVD & Kamland as defined in the text as a function of the source distance D.*

When two or more detectors work together, the new method proposed can be extended. In this case, in order to construct a list of candidate clusters a further step is required, i.e. the temporal coincidence among clusters in different detectors within a time window that we assume to be $w_c = 10$ seconds [14]. In this case the procedure for the clusters definition is the same as described in the previous section and astrophysical signals are injected taking into account the time of flight between the detectors in order to simulate a real astrophysical event. The next step is the definition of the coincidences among the detectors considered. Once that coincidences were found, we require that the product of the ξ_X of coincident clusters is greater then the global cut value:

$$\bar{\xi}^* = \sqrt[N]{\prod_{X=1}^{N} \bar{\xi}_X}, \quad (4)$$

where N is the number of detectors in the network. The sensitivity of the neutrino network can be obtained by using extended definition of the detection efficiency η^*, i.e. the number of astrophysical clusters passing the statistical cut on f_{im}^X that are found in coincidence and are characterised by a global $\bar{\xi}^*$ greater then the cut value defined in Eq.4 over the total injected signals. In a similar manner the network definition of the misidentification probability ζ^* become the ratio among the background coincidences and the total number of coincidences found. As a leading example we show the case of LVD and Kamland working together. The detection efficiency and the misidentification probability of this network are showed in the upper panel of Fig.3. As in the previous plot dashed lines represent the old method based on statistical cut plus temporal coincidence search, while solid lines show the same quantities obtained by adding the $\bar{\xi}^*$ cut described above.

In particular the misidentification probability nearly constant until 75 kpc, drops from a value around 4% to a value around 0.2%, in other words the gain factor obtained in this distance range is around ~ 20 and is reported in the lower panel of Fig.3 for different distances. This reduction of the misidentification can be also converted in term of a reduction of the FAR for the network. In other word, a network like LVD plus Kamland working at a FAR of 0.001/day(0.365/year) with the inclusion of our method based on the ξ^* cut can reach the same background level of the same network working at the SNEWS threshold of 1/1000 years where only the statistical selection is applied.

We apply this extended procedure to all the possible sub-networks of detectors possible including LVD, Borexino, LVD, Kamland, and SK. In any case the improvement obtained is of the same order, indeed also for the cases of combined search between LVD and Borexino or Kamland and Borexino the gain obtained is of a factor ~ 19, however with a reduced distance $\bar{D} \sim 50$ kpc due to the lower sensitivity of Borexino.

References

[1] Janka H T *et al.*, 2007 *Phys. Rept.* **442** 38
[2] Adams S M *et al.*, 2017 *Mon. Not. Roy. Astron. Soc.* **469** 1445
[3] Ouyed R, Dey J and Dey M, 2002 *Astron. Astrophys.* **390** L39
[4] Totani T *et al.*, 1998 *Astrophys. J.* **496** 216
[5] Sumiyoshi K, Yamada S and Suzuki H, 2007 *Astrophys. J.* **667** 382
[6] Keranen P, Ouyed R and Jaikumar P, 2004 *Astrophys. J.* **618** 485
[7] Pagliaroli G *et al.*, 2009 *Astropart. Phys.* **31** 163.
[8] Strumia A, Vissani F, 2003 *Phys. Lett. B* **564** 42.
[9] Abe K *et al.* [Super-Kamiokande Collaboration], 2016 *Astropart. Phys.* **81** 39
[10] Abe K *et al.* [Super-Kamiokande Collaboration], 2016 *Phys. Rev. D* **94** no.5, 052010
[11] Agafonova N Y *et al.* [LVD Collaboration], 2015 *Astrophys. J.* **802** no.1, 47
[12] Alimonti G *et al.* [Borexino Collaboration], 2009 *Nucl. Instrum. Meth.* **A 600** 568
[13] Eguchi K *et al.* [KamLAND Collaboration], 2003 *Phys. Rev. Lett.* **90** 021802
[14] Antonioli P *et al.*, 2004 *New Journal of Physics* **6** 114
[15] Agafonova N Y *et al.*, 2008 *Astropart. Phys.* **28** 516

IOP Conf. Series: Journal of Physics: Conf. Series **1056** (2018) 012043 doi:10.1088/1742-6596/1056/1/012043

Neutrino Energy Reconstruction Methods Using Electron Scattering Data

Afroditi Papadopoulou
Massachusetts Institute of Technology
E-mail: apapadop@mit.edu

Abstract

The precise measurement of neutrino properties is among the highest priorities in fundamental particle physics, involving many experiments worldwide. Since the experiments rely on the interactions of neutrinos with bound nucleons inside atomic nuclei, the planned advances in the scope and precision of these experiments requires a commensurate effort in the understanding and modeling of the hadronic and nuclear physics of these interactions, which is incorporated as a nuclear model in neutrino event generators. This model is essential to every phase of experimental analyses and its theoretical uncertainties play an important role in interpreting every result.

Any nuclear model used to describe neutrino-nucleus scattering should first be validated against these data. Since the vector part of the weak response is related to the electro-magnetic response through CVC, such a test is necessary, but not sufficient, to ensure the validity of a model for given kinematics, namely given values of the transferred energy $\omega(= \nu$ for neutrinos) and momentum q.

The main challenges in connecting electron and neutrino reactions:

- matching models used to predict neutrino-nucleus observables to electron scattering data.
- expanding theory to include more semi-inclusive predictions.
- provide semi-inclusive neutron, proton and pion data sets with broad angular range.

The cross section for neutrino scattering from nuclei is sensitive to the same underlying structure determined by QCD, and as probed with pure electromagnetic processes, such as charged lepton scattering from nucleons and nuclei. As such, there are a number of ways that electron scattering data inform ν - A cross section modeling, as well as providing a test-bed for model validation. In contrast to past and current neutrino beams, charged lepton scattering has the distinct advantage of nearly monochromatic beams with well determined energies, allowing for a significantly cleaner kinematic separation of the various production mechanisms in inclusive scattering, such as reso- nance production and nucleon elastic scattering. In addition to providing important experimental input such as nucleon isovector elastic form factors and resonance transition form factors, electron scattering data provide critical information on the distributions of initial state momentum and energy for nucleons in nuclei, the importance of 2-body currents and final state interaction effects. In this analysis, we will give a brief overview of the experimental input provided by electron scattering data.

1. Introduction

According to the Standard Model (SM) ([1],[2]), the elementary particles can be divided into two main groups: fermions, which are the constituents of matter, and bosons, which are the carriers of interactions. The neutrinos are the lightest fermions and, to the best of our knowledge, exist in three types. They interact only via the weak interaction making the observation of neutrinos an experimental challenge. While many properties of the neutrino are still unknown, the known ones (e.g. limits on their mass, mass difference and mixing angles) are very hard to explain within the SM. This makes measuring various neutrino properties fruitful ground when searching for new physics beyond the SM. Specifically, measurements of the neutrino charge-parity (CP) symmetry-violation phase may help explain the matter anti-matter asymmetry of our universe.

Each successive generation of neutrino oscillation measurements has improved in sensitivity in order to look for more and more subtle effects. It is becoming clear that in order to realize the next generation of experiments, improvements must be made in understanding and modeling of neutrino-nucleus interactions, both in order to correctly quantify the systematic uncertainties they induce and to avoid possible biases.

Various neutrino experiments are currently taking places around the world, using different neutrino sources and detector materials. While accelerator-based experiments traditionally use water/liquid scintillator-based detectors, a new generation of state-of-the-art Liquid Argon Time Projection Chamber (LArTPC) detectors is coming online at FNAL, with the recent commencement of data-taking of the MicroBooNE experiment [5], the forthcoming SBN program, and the future DUNE experiment [6]. LArTPC detectors are showing great promise by producing bubble-chamber-like images of neutrino-scattering events that allow accurate measurements of neutrino-interactions, even in cases of very complex final states. Such measurements give more information on the underlying nuclear interactions, which could allow a reduction of both the systematical effects and the possible bias that neutrino-nucleus interaction models might induce.

2. Benchmarking Neutrino Event Generators Using Electron Scattering Data

In accelerator-based experiments, neutrinos are traditionally detected via their nuclear interaction with the detector material. These interactions are simulated using so-called "event generators" that provide a framework that aims to model a complete set of interaction processes on a wide range of target materials for an arbitrary neutrino beam energy. All charged-current and neutral-current processes can be simulated, with the full kinematics of all particles exiting the nucleus provided on an event-by-event basis. However, the generators typically employ models that are semiclassical, i.e. they work with cross sections, rather than amplitudes. This is a significant limitation since the nucleus is a quantum mechanical system and many of the interesting reaction mechanisms are known to interfere strongly. Furthermore, they typically treat the primary interaction and final state interactions separately and incoherently.

At the same time, there are crucial assumptions inherent in generators and theoretical models which have implications for neutrino experiments. For instance, many of the models implemented in event generators do not include the most up-to-date theory understanding. Furthermore, event generators are a combination of many different (possibly inconsistent) models. Such "Franken-models" may not produce correct total or differential cross sections. In addition, many event generators approximate final state interactions using semi-classical cascade models. As mis-modelling of nuclear interactions can bias the extracted neutrino cross sections and oscillation parameters, this issue must be studied in detail.

Experiments like MINERvA and other short-baseline detectors all have extensive cross-section measurement programs which are used to constrain event generators and, at the same time, study their biases. However, obtaining high-precision constraints using measurements alone is non-

IOP Conf. Series: Journal of Physics: Conf. Series **1056** (2018) 012043 doi:10.1088/1742-6596/1056/1/012043

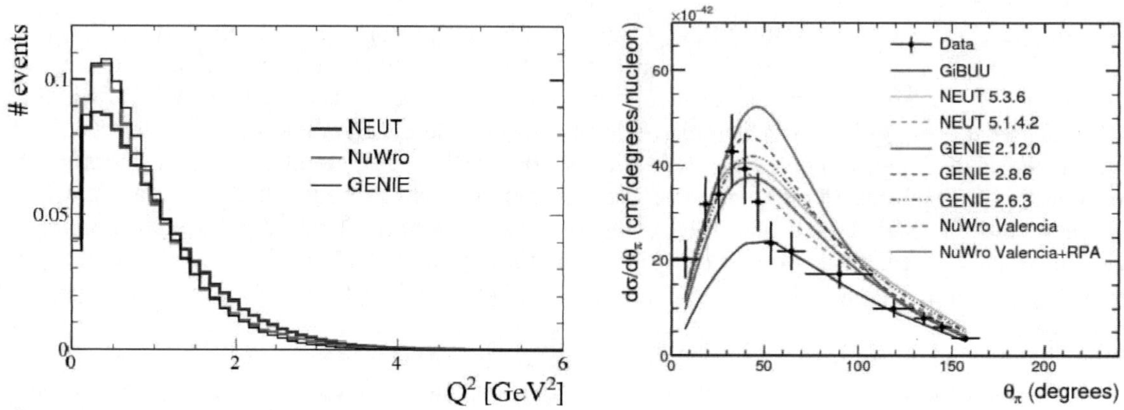

Figure 1. Different neutrino generators result into different predictions.

trivial since the measurements contain an unavoidable convolution of the (usually wide) energy spread of the incoming neutrino beam, the different vector and axial response functions of the nucleus, multinucleon effects, nuclear-final state interactions in addition to any measurement resolution or other detector-specific effects. De-convoluting all of these effects that contribute to the measured cross-section is a hard task that limits the ability of data to constrain the individual components of event generators.

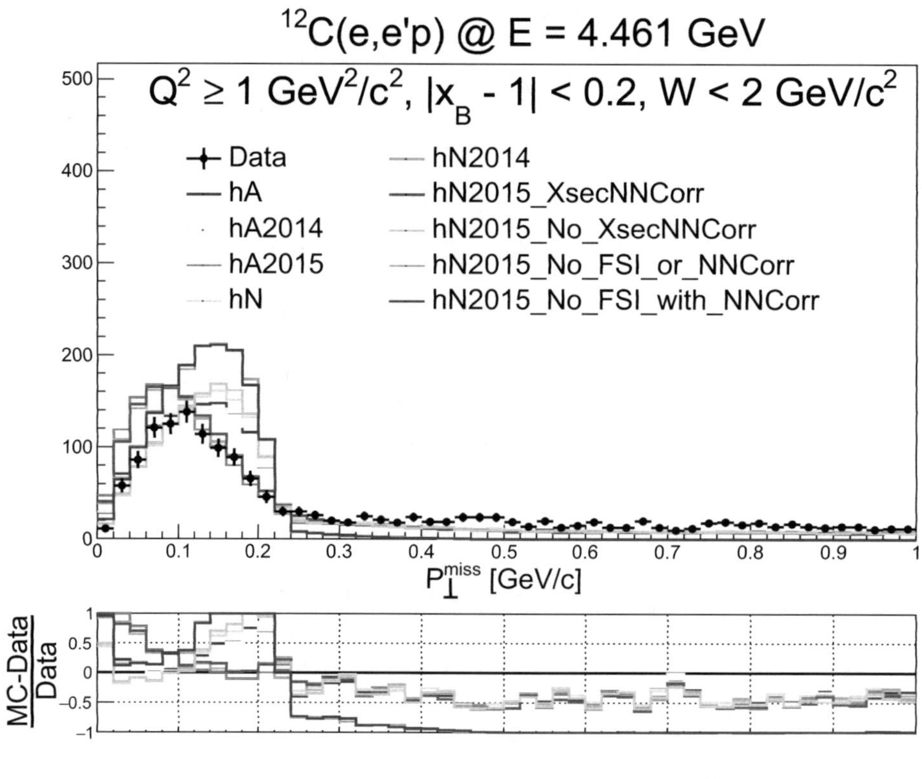

Figure 2. Data - GENIE comparison for the Q^2 distribution.

IOP Conf. Series: Journal of Physics: Conf. Series **1056** (2018) 012043 doi:10.1088/1742-6596/1056/1/012043

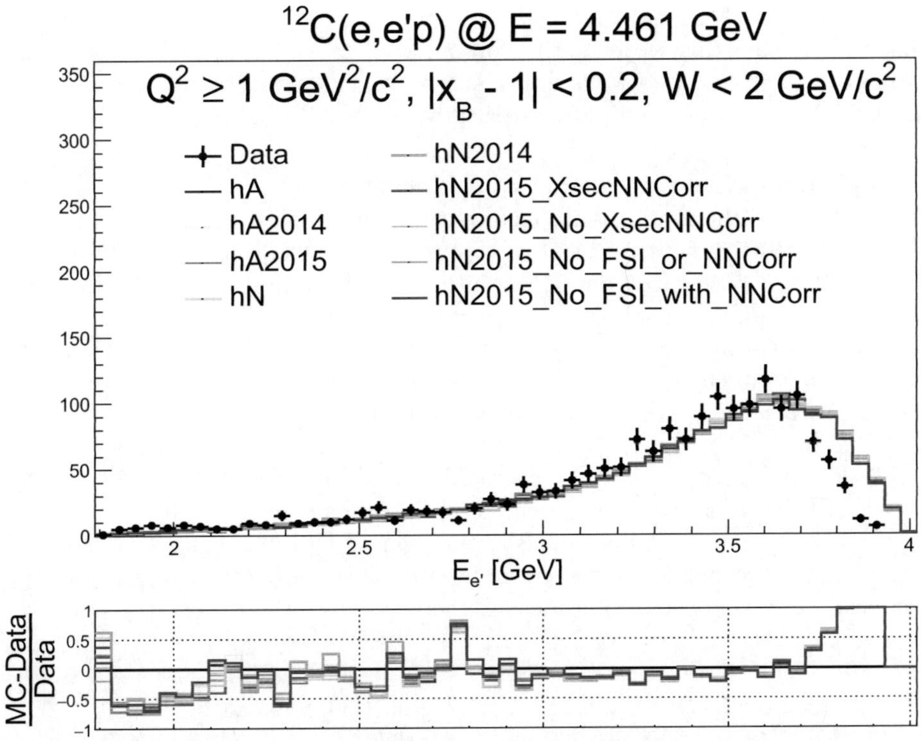

Figure 3. Data - GENIE comparison for the Q^2 distribution.

Testing neutrino event generators against mono-energetic, wide phase-space, electron scattering data can complement these activity. While electrons and neutrinos do not have identical nuclear interactions, there are many similarities that allow electron data to constrain parts of the generator model space. These constraints can effectively complement those coming from short base-line neutrino cross-section measurements, such as the axial nuclear response.

A concrete example can be seen in the benchmarking of models of nuclear final state interactions. The propagation of a nucleon through the nuclear medium is a strongly depends on the momentum transfer of the reaction and is relatively independent of whether the interaction was initiated by an electron or by a neutrino. As the majority of generators used for neutrino scattering can also be run under an electron scattering configuration, we can mitigate the residual difference between electron and neutrino interactions by comparing electron to neutrino mode. This connection can thus be exploited to test and tune the models in event generators. Furthermore, the same parameters used to quantify the agreement with electron scattering data can be provided to neutrino oscillation experiments, so that the impact of mis-modelling (see e.g. Fig. 1) can be quantified and reduced. This makes electron scattering data a critical input to those physics programs.

Currently, the Jefferson-Lab data mining collaboration is working on comparisons of their electron scattering data measured using the CLAS detector with the predictions of the GENIE event generator running in electron scattering mode. To properly compare the generator output to data, I passed the generated events through acceptance and efficiency maps before applying the same event selection cuts that were applied to the data. Figures 2 and 3 shows an example from a recent comparison of QE $(e, e'p)$ events measured in ^{12}C using a 4.461 GeV electron beam. The data is compared to a simulation where we used the GENIE neutrino event generator and

IOP Conf. Series: Journal of Physics: Conf. Series **1056** (2018) 012043 doi:10.1088/1742-6596/1056/1/012043

a Correlated Fermi Gas (CFG) model that has been implemented for the nuclear ground state. We are planning to continue this work in the coming year to ensure detailed comparisons of electron scattering data will be done with the relevant event generators. At the moment only GENIE is being used but we are planning to expand to using GiBUU in the near future.

3. Projected Impact on Neutrino Oscillation Uncertainties

A concrete example of how this analysis impacts current and future experiments can be seen when considering semi-inclusive neutron measurements to study the difference between neutrino and antineutrino oscillations to search for CP violation. CP violation is not the only cause for asymmetry between neutrinos and anti-neutrinos and their interactions with nuclei. Protons and neutrons already have different weak couplings, different final state interactions and also occur in different numbers in the case of asymmetric nuclei (of which Ar in LArTPC detectors is an example). Achieving the systematic uncertainty requirements of the DUNE experiment will require modeling of these nuclear physics effects in great detail and doing so to very high accuracy. In addition, the development of observables like the ratios mentioned above—with reduced sensitivities to the details of the nuclear models—are also crucial.

The work that has been presented here takes a significant step in this direction, since the use of electron scattering data to benchmark event generatorsis an important step in our quest to perform high-precision neutrino oscillation measurements.

References

[1] K. A. Olive et al. "Review of Particle Physics," *Chin. Phys.*, **vol. C38**, **p. 090001**, 2014.

[2] M. E. Peskin and, D.V.Schroeder, "An Introduction to Quantum Field Theory", Advanced book program, Boulder (Colo.): Westview Press Reading (Mass.), 1995. Autre tirage : 1997.

[3] A. Goldschmidt for the IceCube Collaboration, *Proceedings of the 27th International Cosmic Ray Conference, Hamburg, Germany, 7-15 August 2001*, **1237-1240**.

[4] C.W. Walter, "The Super-Kamiokande Experiment" (*arXiv:0802.1041*).

[5] H. Chen et al. for the MicroBooNE Collaboration, "A Proposal for a New Experiment Using the Booster and NuMI Neutrino Beamlines: MicroBooNE", *FERMILAB-PROPOSAL-0974*.

[6] Long-Baseline Neutrino Facility (LBNF) and Deep Underground Neutrino Experiment (DUNE) (*arXiv:1512.06148*).

$0\nu\beta\beta$ decay: the CUPID-0 experiment

O Azzolini[1], M T Barrera[1], J W Beeman[2], F Bellini[3,4], M Beretta[5,6], M Biassoni[6], C Brofferio[5,6], C Bucci[7], L Canonica[7,8] , S Capelli[5,6] , L Cardani[4], P Carniti[5,6], N Casali[4], L Cassina[5,6], M Clemenza[5,6], O Cremonesi[6], A Cruciani[4], A D'Addabbo[7], I Dafinei[4], S Di Domizio[9,10], M L di Vacri[7], F Ferroni[3,4], L Gironi[5,6], A Giuliani[11,12], P Gorla[7], C Gotti[5,6], G Keppel[1], M Maino[5,6], M Martinez[3,4], S Morganti[4], S Nagorny[7,13], M Nastasi[5,6], S Nisi[7], C Nones[14], E Olivieri[11], D Orlandi[7], L Pagnanini[13], M Pallavicini[9,10], V Palmieri[1], L Pattavina[7,13], M Pavan[5,6], G Pessina[6], V Pettinacci[4], S Pirro[7], S Pozzi[5,6], E Previtali[6], A Puiu[5,6], F Reindl[4], C Rusconi[7,15], K Schäffner[7,13], C Tomei[4], M Vignati[4], A Zolotarova[14]

[1]INFN - Laboratori Nazionali di Legnaro, Legnaro (Padova) I-35020 - Italy
[2]Materials Science Div., Lawrence Berkeley National Laboratory, Berkeley, CA 94720 - USA
[3]Dipartimento di Fisica, Sapienza Università di Roma, Roma I-00185 - Italy
[4]INFN - Sezione di Roma, Roma I-00185 - Italy
[5]Dipartimento di Fisica, Università di Milano-Bicocca, Milano I-20126 - Italy
[6]INFN - Sezione di Milano Bicocca, Milano I-20126 - Italy
[7]INFN - Laboratori Nazionali del Gran Sasso, Assergi (L'Aquila) I-67010 - Italy
[8]Massachusetts Institute of Technology, Cambridge, MA 02139 - USA
[9]Dipartimento di Fisica, Università di Genova, Genova I-16146 - Italy
[10]INFN - Sezione di Genova, Genova I-16146 - Italy
[11]CSNSM, Univ. Paris-Sud, CNRS/IN2P3, Université Paris-Saclay, 91405 Orsay, France
[12]DiSAT, Università dell'Insubria, 22100 Como, Italy
[13]Gran Sasso Science Institute, 67100, L'Aquila - Italy
[14]IRFU, CEA, Universit Paris-Saclay, F-91191 Gif-sur-Yvette, France
[15]Dept. of Physics and Astronomy, University of South Carolina, Columbia, SC 29208 - USA

E-mail: `maura.pavan@mib.infn.it`

Abstract. CUPID-0 is an array of 24 Zn^{82}Se scintillating bolometers used to search for the $0\nu\beta\beta$ decay of ^{82}Se. It is the first large mass $0\nu\beta\beta$ experiment exploiting a double read-out technique: the phonon signal to accurately measure particle energies and the light signal to reject the α-induced background. Its success might open the road to a next generation project of ton mass scale, CUPID. The array is in operation at Laboratori Nazionali del Gran Sasso since the beginning of 2017, in this paper we present the preliminary results obtained with an exposure of 10.45 kg· y.

1. Introduction

The search for $0\nu\beta\beta$ decay is challenging experimentalists since about half a century [1]. It is today commonly accepted that the key features of a ton-scale next generation experiment are isotopic enrichment and active background rejection. In the case of phonon detectors these two requirements can be simultaneously matched by the use of "scintillating bolometers" [2, 3]. This technique dates back to the 90es but its application to $0\nu\beta\beta$ decay is recent, CUPID-0 is the first kg-scale $0\nu\beta\beta$ experiment ever realized. It is today in operation at Laboratori Nazionali del Gran Sasso (LNGS), L'Aquila, Italy.

The detector is an array of 26 ZnSe crystals, 24 made of ^{82}Se enriched material. Its design exploits a successful experience in operating TeO_2 bolometers arrays [4, 5, 6] and several years of R&D activities dedicated to the development of ZnSe enriched detectors [7, 8, 9].

The main physics goal is the investigation of $0\nu\beta\beta$ decay in ^{82}Se, signature of the decay is a monochromatic line, equivalent to an energy deposition of 2998 keV in the crystal (the ^{82}Se Q-value [10]). Exploiting the particle discrimination capability of the detector, events with a similar energy deposition but produced by α particles can be rejected with high efficiency. If successful, CUPID-0 would open the way for a CUORE Upgrade with Particle Identification (CUPID) [11].

2. The array

CUPID-0 array includes 26 ZnSe scintillating-bolometers and 31 Ge light detectors. 24 out of the 26 ZnSe crystals are made with 96% ^{82}Se enriched material. The total mass of Zn^{82}Se is 9.65 kg (corresponding to 5.28 kg of ^{82}Se) while the two natural crystals sum to 0.85 kg (corresponding to 40 g of ^{82}Se).

The ZnSe cylindrical single-crystals are instrumented to read-out the phonon signal produced by particle traversing their volume (with a design almost identical to that of TeO_2 bolometers). Disk shaped Ge wafers are used to read out the scintillation light emitted by the ZnSe crystals. These light detectors (LD) are operated as bolometers, they use a phonon sensor and a read-out chain that is very similar to the one serving the ZnSe bolometers but – thanks to the very small volume and hence the low heat capacity – are sensitive to energy depositions as low as few tens of eV.

The double read-out allows a precise measurement of particle energy deposition (using the phonon signal) and the discrimination of particles having different light yields (using the light signal). Indeed, while the the phonon signal is dependent only on the amount of energy deposited in the ZnSe bolometer, the scintillation photon production depends on the particle ionization properties. The result is a light signal with amplitudes and shapes that depend on the kind of interacting particle. Either the ratio between the light and the phonon signals or the shape of the light signal itself can be effectively used to discriminate highly ionizing particles like β/γ's (including $0\nu\beta\beta$ generated βs) from α's and nuclear recoils.

In CUPID-0, crystals are organized in 5 columns that hung from a copper disk (the detector support plate) mechanically secured to the cold finger of the dilution refrigerator through an anti-vibration suspension system. In each column, the ZnSe crystals are stacked one over the other being separated by a LD. In this way each ZnSe bolometer is viewed by two LD's. A reflective foil surrounds the lateral part of the ZnSe cylinders to improve light collection.

All the materials used in the construction of the array (mainly made of PTFE[1] and NOSV[2] copper) were selected for their low radioactive contamination in Th and U. Assembly procedures followed very strict protocols in order to avoid recontaminations. The array stands, at about 10 mK, in the inner vacuum chamber of the Oxford Instrument dilution refrigerator formerly used by Cuoricino [4] and CUORE-0 [5], the same shielding system used in these experiment serves now CUPID-0 (details on cryostat and shield design and contamination can be found [12]).

[1] PolyTetraFluoroEthylene.
[2] A special copper alloy suitable for cryogenic use produced by Aurubis company.

3. Commissioning and operation

Commissioning started in October 2016 with an optimization of the vibrational and cryogenic configurations of the refrigerator. Following, the electronic read-out and DAQ systems were optimized. The first physics run started at the beginning of 2017, it resulted characterized by large instabilities of the cryogenic system therefore requiring an "ad hoc" tuning of the signal processing algorithms which is still not perfectly satisfactory. The following run started only after a further optimisation of the system that fixed most of the instability problems.

Here we report the results collected in these two runs, totalling an exposure of 1.46 kg·y on enriched ZnSe crystals. One out of the two natural crystals and two enriched ones had poor performances and are excluded from the analysis. These results are really preliminary, in particular for the above mentioned reasons the first physics run will be likely excluded from the final data-release.

4. Preliminary results of CUPID-0

The phonon signal, used for energy measurement, shows a noise figure corresponding to a 0.4 keV FHWM, mainly dominated by vibrational noise. We observe a worsening of the energy resolution with increasing energy, likely due to crystal quality. In the ^{232}Th calibration, we measure an average FWHM of ~25 keV for the 2615 keV peak due to ^{208}Tl. The FWHM extrapolated at the ^{82}Se Q-value is ~28 keV. A study for a better description of the peak shape (here simply represented with a gaussian function) is under-way and may lead to slightly different FWHM values.

As already done for a small test array operated in 2015 [9], we verified that the most efficient and reliable way to discriminate between β/γ and α particles relies on the shape of the LD signal (ascribed to the different characteristic time associated to the scintillation light emitted by ZnSe crystals when excited by β/γ or α particles).

The processing-chain applies to triggered events a set of cuts consisting in:

(i) a single-hit cut that removes events where more than one ZnSe crystal triggered. These are either spurious pulses or radioactivity induced signals.

(ii) pulse shape cuts that reject pulses deformed by noise and pile-up (since the algorithm used to evaluate pulse amplitude could fail).

(iii) a cut based on the shape of the light signal and used to reject α induced events.

The energy spectrum of events surviving the first two cuts is shown in figure 1 in gray, those surviving also the third cut are shown in blue.

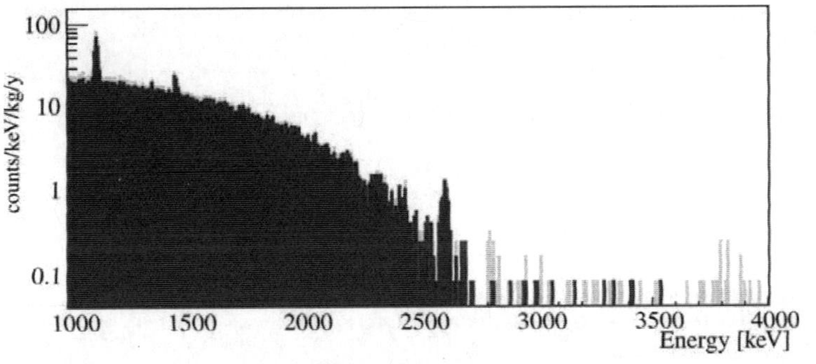

Figure 1. Energy spectrum of single-hit events surviving pulse shape cuts (gray) and superimposed that of events surviving also the α rejection cut (blue).

The spectrum is clearly dominated by the $2\nu\beta\beta$ decay of ^{82}Se. Visible peaks are the lines of ^{40}K at 1460 keV and that of ^{208}Tl at 2615 keV, both due to natural radioactivity. The line at 1115 keV is ascribed to ^{65}Zn, a short lived isotope produced in ZnSe by cosmogenic activation. At the right of the 2615 keV line, the counting rate is dramatically reduced.

Between 2.8 and 3.2 MeV the counting rate is $(52 \pm 9) \cdot 10^{-3}$ cts/(keV·kg·y) on the gray spectrum and is reduced to $(15 \pm 5) \cdot 10^{-3}$ cts/(keV·kg·y) in the blue one (i.e after α particle rejection). In this region we observe an α peak (at about 2.8 MeV) and an α continuum (gray events) plus few β/γ events (blue events) that appear to form a flat continuum. Most of the residual β/γ events are removed applying a delayed coincidence cut. This is optimized to remove what we think is the major source of these counts: β or $\beta + \gamma$ events from ^{208}Tl decay. We use the α particle emitted by ^{208}Tl precursor, ^{212}Bi, to tag and remove events that are likely due to this decay chain. Between 2.8 and 3.2 MeV only three events survive this cut (see figure 2) therefore the counting rate is in the range $[1.8 - 8.6] \cdot 10^{-3}$ cts/(keV·kg·y)(Poisson 68% C.L. interval).

The probability that a $0\nu\beta\beta$ event is accidentally removed from the final spectrum by one of the above mentioned cuts is ~85% (however we are working on an improvement of the efficiency of the cuts and very likely we can approach a better than 90% efficiency). Finally the probability of a full containment of both the two electrons emitted in a $0\nu\beta\beta$ decay is ~80% (in the residual 20% cases at least one of the two electrons escape the source crystal that consequently records an event below the Q-value).

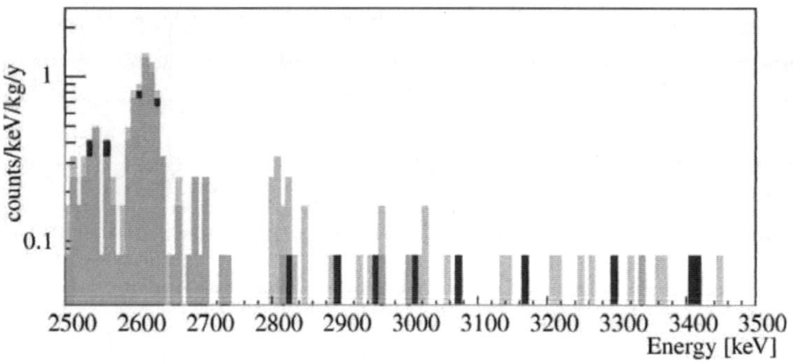

Figure 2. Detailed view near the $0\nu\beta\beta$ region of the energy spectrum of figure 1. Events surviving the delayed coincidence cut are shown in yellow. The $0\nu\beta\beta$ decay peak should appear at 2998 keV.

5. Conclusions

CUPID-0 started the physics run in early 2017, it is expected to run at least for 1 y of live-time to prove the potentialities of the scintillating bolometers technology. The high number of emitters will allow to reach a remarkable sensitivity on the ^{82}Se $0\nu\beta\beta$ decay. Preliminary result on the background counting rate near the region of interest indicate the achievement of an extremely low background.

References

[1] O. Cremonesi M. Pavan 2014 *Adv. High En. Phys.* Article ID 951432
[2] S. Pirro et al. 2006 *Phys. Atom. Nucl.* **69** 2109
[3] D. R. Artusa et al. (CUORE Collaboration) 2014 *Eur. Phys. J. C* **74** 3096
[4] C. Arnaboldi et al. (Cuoricino Collaboration) 2005 *Phys. Rev. Lett.* **95** 14501
E. Andreotti et al. (Cuoricino Collaboration) 2011 *Astropart. Phys.* **34** 822
[5] K. Alfonso et al. (CUORE Collaboration) 2015 *Phys. Rev. Lett.* **115** 102502
[6] C. Arnaboldi et al. (CUORE Collaboration) 2003 *Astropart. Phys.* **20** 91
C. Arnaboldi et al. (CUORE Collaboration) 2004 *Nucl. Instr. Meth.* A **518** 775
[7] C. Arnaboldi et al. 2011 *Astropart. Phys.* **34** 344
[8] J.W. Beeman et al. 2013 *J. Inst.* **8** P05021
[9] D. R. Artusa et al. 2016 *Eur. Phys. J.* C **76** 364
[10] D. L. Lincoln et al. 2013 *Phys. Rev. Lett.* **110** 012501
[11] G. Wang et al. (CUPID Interest Group) 2015 arXiv:1504.03599
[12] C. Alduino et al. (CUORE Collaboration) 2016 *Eur. Phys. J.* C **77**, 13

IOP Conf. Series: Journal of Physics: Conf. Series **1056** (2018) 012045 doi:10.1088/1742-6596/1056/1/012045

Experiments on the Competitive Double-Gamma ($\gamma\gamma/\gamma$) Decay

N Pietralla and H Scheit

Institut für Kernphysik, Technische Universität Darmstadt, 64289 Darmstadt, Germany

E-mail: pietralla@ikp.tu-darmstadt.de

Abstract. The recent observation of the competitive double-γ decay process is presented. It is a second-order electromagnetic decay mode. The 662-keV decay transition from the $11/2^-$ isomer of ^{137}Ba to its ground state has been studied. The emission of a single 662-keV γ quantum is a factor of 5×10^5 more likely than the simultaneous emission of two γ quanta instead of one. The observed angular correlation and energy distribution of these rare coincident double-γ quanta are well described by a dominant $M2 - E2$ and a sub-dominant $E3 - M1$ contribution to the double-γ decay branch.

1. Introduction

The Conference on Neutrino and Nuclear Physics addresses progress and challenges at the common frontier of these scientific sub-disciplines. Several contributions address the rare nuclear double-β decay process because of the paramount importance of the pending observation of its neutrinoless variant for our understanding of lepto-genesis and of the particle character and the mass of the neutrino.

Double-β decay processes (with the emission of two neutrinos) have first been considered theoretically by Maria Göppert some 80 years ago [1]. They have later on been found to be responsible for the finite lifetimes of some even-even nuclei at the edges of the valley of stability, such as e.g., ^{76}Ge, ^{100}Mo, ^{116}Cd, ^{124}Sn, ^{128,130}Te, ^{136}Xe, or ^{150}Nd, whose single β-decays are forbidden by energy conservation. The first direct detection of $2\nu\beta\beta$-decay reactions was finally achieved by Elliott, Hahn and Moe [2] not earlier than 1987.

While $\beta\beta$-decay processes are nuclear reactions of second order in the electroweak interaction it is surprising to find that even less data exist for nuclear decays that proceed in second order in the electromagnetic interaction where two γ-quanta are simultaneously emitted in a single quantum transition from one quantum state to another. In a $\gamma\gamma$-decay (speak: '*double-gamma decay*') both γ quanta with energies E_1 and E_2 share the total transition energy $E_0 = E_1 + E_2$, while the energy spectrum of the individual quanta is continuous, $0 < E_i < E_0$. These $\gamma\gamma$-decays are formally analogous to $0\nu\beta\beta$-decay processes where in the latter two β-particles and in the former two γ-quanta appear in the final state and share the total transition energy. Indeed, $\gamma\gamma$-decay processes have first theoretically been postulated and studied by Maria Göppert in her PhD thesis [3] with Max Born in Göttingen, even before discussing $\beta\beta$-decay processes.

Up to recently, $\gamma\gamma$-decay reactions in nuclei were known, only, in three particular cases, ^{16}O [4, 5, 6] and ^{40}Ca, ^{90}Zr [5, 7], where the first excited states of these even-even nuclei have spin and parity quantum numbers 0^+ and a single-γ decay is strictly forbidden by helicity conservation.

Searches for the more general situation, in which $\gamma\gamma$-decays occur in a nuclear transition which could proceed by a single-γ decay in competition, the so-called *competitive double-γ decay* ($\gamma\gamma/\gamma$-decay), was not reported in peer-reviewed literature before our recent observation [11]. The main obstacle in previous unsuccessful attempts [5, 8, 9, 10] for experimental discovery of the $\gamma\gamma/\gamma$-decay has been the low statistics and the background from much more abundant single-γ decays.

It is the purpose of this contribution to bring our observation to the attention of this cross-disciplinary audience. It is based on our previous publications [11, 12, 13] and on the material presented in the doctoral thesis of Dr. Christopher Walz [14].

2. Experiment

The γ radiation from a standard sealed ^{137}Cs radiation source with an activity of 16.3(5) μCi has been studied. The radioactive ^{137}Cs nuclei undergo β-decay reactions with a half life of 30.07(3) years and populate to 5.3(2)% directly the stable $J^\pi = 3/2^+$ ground state of the daughter nucleus ^{137}Ba and to 94.7(2)% its $J^\pi = 11/2^-$ second excited state at an excitation energy of $E_0 = 661.659(3)$ keV [15]. This $11/2^-$ isomer has a halflife of 2.552(1) minutes. It decays to 89.86% by the emission of a 661.657(3) keV γ-ray line with multipolarity $M4$ and a reduced transition strength of $B(M4; 11/2^- \to 3/2^+) = 2.725(12)$ W.u. to the ground state and with a γ-decay branching ratio of $1.12(9)\times10^{-7}$ [16] by an $E5$ transition to the $1/2^+$ first excited state of ^{137}Ba at 283.54 keV. The remaining decay intensity of the 662-keV state is converted with a conversion coefficient of $\alpha = 0.1124$ for the 662-keV $M4$ transition.

This γ-ray source was surrounded by a ring of five large-volume $LaBr_3$:Ce scintillators at a distance of approximately 22 cm to the detector face. Every two LaBr-detectors observed the radiation source at a relative mean angle of either 72° or 144°. Compton-scattering cross-talk between the detectors was suppressed by lead walls with a typical thickness of 12 cm. The entire set-up was covered by scintillator bars operated in anti-coincidence mode for the data acquisition system in order to suppress prompt coincidences between the LaBr detectors originating from shower events induced from energetic cosmic rays. The absolute photo-peak efficiency of the LaBr-detector array was measured to $\epsilon_{\rm abs} = 1.50(5)\%$ at 662 keV. An energy calibration point was continuously provided by the 662-keV γ-ray line from the ^{137}Cs source. Additional data points for the energy calibration and for the definition of coincidence events were obtained from measurement runs with ^{60}Co γ-radiation standards. Single γ-ray spectra and $\gamma\gamma$-coincidence events were recorded for 1,273 hours corresponding to a total of 53 days of continuous data taking.

The detected γ events that have not been vetoed by anti-coincidence with signals from the plastic detectors were analyzed with respect to their time difference. Clear signals of prompt $\gamma\gamma$ coincidences were observed on a considerable background of random coincidences. The latter were subtracted from the prompt coincidences by appropriate time gates. The resulting background-subtracted sum-energy spectrum obtained with the detectors at a relative observation angle of 72° is shown in figure 1. A significant sum-energy peak was obtained above a background which otherwise smoothly decreases as a function of energy. A Gaussian function consistent with the calibrated detector response was fitted to the peak above a smooth phenomenological background, as indicated by the solid and dotted curves in figure 1. A peak area of 693(95) counts with a centroid at 661.6 ± 1.6 keV was obtained from the fit. The position is in very good agreement with the $11/2^- \to 3/2^+_{\rm gs}$ transition energy of $E_0 = 661.66$ keV, albeit obtained here from the sum of two γ-ray energies of less than 481 keV, each. The statistical significance of this signal amounts to 7.3 standard deviations, substantially exceeding the required significance for a three-star discovery in particle physics.

A detailed analysis of the time difference spectrum obtained by gating on the sum-energy events rather than on the time differences, has firmly excluded [11] that the sum-energy peak at

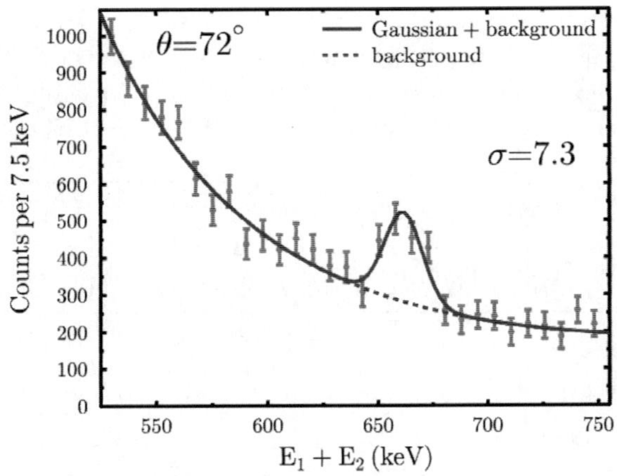

Figure 1. Sum-energy histogram for events recorded with LaBr-detectors located at a mean relative angle of 72° after subtraction of random background. For further reduction of background it has been required that neither single-γ energy was smaller than 181 keV (or larger than 481 keV, correspondingly). The error bars indicate the statistical uncertainties for each channel.

662 keV resulted from Compton scattering of a single 662-keV γ quantum from one LaBr-detector to another because of the additional time delay of $\Delta t > 0.8$ ns such a process would induce. This information fully proves that the sum-energy peak had originated from the simultaneous emission of two γ-quanta that share the total transition energy, thereby firmly establishing the discovery of the $\gamma\gamma/\gamma$-decay process.

For the group of detector pairs with relative mean observation angle of 144° a corresponding peak area of 307(78) counts has been observed at a centroid of 664.2±2.8 keV. The strong count rate asymmetry for both detector groups points at a pronounced anisotropy of the $\gamma\gamma/\gamma$-angular correlation function.

3. Formulation of the $\gamma\gamma/\gamma$-decay process

The total count rate $n_{\gamma\gamma/\gamma}$ of $\gamma\gamma/\gamma$-coincidence events is given by

$$n_{\gamma\gamma/\gamma} = A \cdot \frac{1}{\Gamma} \int_{4\pi} d\Omega \int_{4\pi} d\Omega' \int_0^{E_0} d\omega \frac{d^5\Gamma_{\gamma\gamma}}{d\omega d\Omega d\Omega'} \epsilon(\omega, \Omega)\, \epsilon(E_0 - \omega, \Omega') \tag{1}$$

where A is the activity of the source with respect to the $11/2^-$ state of ^{137}Ba at 662 keV, Γ is its total decay width, ω denotes the energy of one of the two simultaneously emitted γ quanta, $\Omega = (\theta, \phi)$ and Ω' are their angular directions, ϵ is the energy-dependent and angle-dependent absolute detection efficiency, and $\Gamma_{\gamma\gamma}$ denotes the partial decay width for double-γ decay. For an initially unoriented ensemble the quintuple-differential partial decay width in equation (1) can depend - besides on ω - only on the relative angle θ_{12} between the two γ quanta.

Following the work of Göppert and starting from equations (A.15, A.34a) from Ref. [5] one can derive [11]

$$\frac{d^5\Gamma_{\gamma\gamma}}{d\omega d\Omega d\Omega'} = \frac{\omega\omega'}{96\pi^3} \sum_{\substack{JS_1'L_1'S_1L_1 \\ S_2'L_2'S_2L_2}} P_J'(S_1'L_1'S_1L_1) P_J'(S_2'L_2'S_2L_2) \sum_l a_l^{J\xi} P_l(\cos\theta_{12}) \tag{2}$$

where ξ stands for a full set of parameters $\{S_1'L_1'S_1L_1S_2'L_2'S_2L_2\}$ specifying the parities and angular momenta of the two emitted photons with $S=0$ for electric and $S=1$ for magnetic transition characters. P_l are Legendre polynomials and $a_l^{J\xi}$ are further coefficients from angular momentum coupling and tabulated in the supplements of Ref. [11]. The functions

$$\begin{aligned} P_J'(S'L', SL, \omega'\omega) &= (-1)^{S+S'} 2\pi (-1)^{I_i+I_f} \omega^L \omega'^{L'} \cdot \frac{\sqrt{(2L+1)(2L'+1)}}{(2L+1)!!(2L'+1)!!} \sqrt{\frac{L+1}{L}} \sqrt{\frac{L'+1}{L'}} \cdot \\ \sum_n \Bigg[&\begin{Bmatrix} L & L' & J \\ I_f & I_i & I_n \end{Bmatrix} \frac{\langle I_f || i^{L'-S'} M(S'L') || I_n \rangle \langle I_n || i^{L-S} M(SL) || I_i \rangle}{E_n - E_0 + \omega} \\ &+ (-1)^{L+L'+J} \cdot \begin{Bmatrix} L' & L & J \\ I_f & I_i & I_n \end{Bmatrix} \frac{\langle I_f || i^{L-S} M(SL) || I_n \rangle \langle I_n || i^{L'-S'} M(S'L') || I_i \rangle}{E_n - E_0 + \omega'} \Bigg] . \end{aligned} \tag{3}$$

are generalised polarizabilities as given in equation (A.19) in Ref. [5]. They involve coherent sums of electromagnetic transition operators connecting the initial state I_i through virtual intermediate states I_n to the final state I_f by all combinations of multipoles L and L' that are possible by angular momentum coupling. It is convenient to define the unique-multipolar generalized polarizabilities

$$\alpha_{SLS'L'}(\omega) = \sum_n \frac{\langle I_f || i^{L'-S'} M(S'L') || I_n \rangle \langle I_n || i^{L-S} M(SL) || I_i \rangle}{E_n - E_i + \omega} . \tag{4}$$

that can be approximated for $E_n \gg E_0$ to a good accuracy by

$$\alpha_{SLS'L'} = \sum_n \frac{\langle I_f || i^{L'-S'} M(S'L') || I_n \rangle \langle I_n || i^{L-S} M(SL) || I_i \rangle}{E_n - E_0/2} . \tag{5}$$

and can be calculated in standard nuclear structure models. One obtains a full account of the partial double-γ decay width, the relative energy distribution, and the angular correlation of the $\gamma\gamma/\gamma$-coincidences as a function of the sizes of the unique-multipolar generalized polarizabilities α. In the case of a unique multipolar dominance, the energy distribution of the individual γ-quanta depends on the involved multipolarities L and L' according to

$$P(\omega) \propto \left[\omega^{(2L+1)} (E_0 - \omega)^{(2L'+1)} + \omega^{(2L'+1)} (E_0 - \omega)^{(2L+1)} \right] . \tag{6}$$

This energy distribution is symmetric around $\omega = E_0/2$ and vanishes for the extreme values of $\omega = 0$ or E_0. If more than one unique-multipolar generalized polarizability contribute significantly, then the resulting energy distribution becomes a superposition including interferences according to equations (2) and (3). From a combined analysis of the relative energy distribution of the simultaneously emitted γ quanta, their angular correlation, and the absolute value of the partial double-γ decay width $\Gamma_{\gamma\gamma}$ one can determine the unique-multipolar generalized polarizabilities α absolutely, apart from an irrelevant global sign.

4. Quantitative data analysis

A quantitative analysis of the $\gamma\gamma$-angular distribution and of the distribution of energy amongst the simultaneously emitted γ quanta was restricted to the expected two dominant multipole channels: $M2-E2$ and $E3-M1$. Under this restriction equation (2) simplifies to

$$\frac{d^5\Gamma_{\gamma\gamma}}{d\omega d\Omega d\Omega'} = A_{qq}(\alpha_{M2E2}^2) + A_{od}(\alpha_{E3M1}^2) + A_x(\alpha_{M2E2} \cdot \alpha_{E3M1}) \tag{7}$$

which only depends on the absolute values of the two unique-multipolar generalized polarizabilities α_{M2E2} and α_{E3M1} and on their relative sign. Apart from that, the quadrupole-quadrupole A_{qq}, the octupole-dipole A_{od} and their interference A_x terms are known [11] functions of the correlation angle θ_{12} according to the formulae from section 3.

From a fit of equation (7) to the data on the absolute angular distribution (2 absolute data points when having taken into account the known total decay rate of the $11/2^-$ isomer) and on the energy distribution (7 data points) the following unique-multipolar generalized polarizabilities were obtained [11]

$$\alpha_{M2E2} = +33.9 \pm 2.8 \; \frac{e^2\text{fm}^4}{\text{MeV}} \tag{8}$$

and

$$\alpha_{E3M1} = +10.1 \pm 4.2 \; \frac{e^2\text{fm}^4}{\text{MeV}} \; . \tag{9}$$

Apparently, the $M2 - E2$ double-γ decay mode dominates over an alternative $E3 - M1$ decay which however contributes constructively to the $\gamma\gamma/\gamma$-decay of the $11/2^-$ isomer of ^{137}Ba. The total branching ratio into the $\gamma\gamma/\gamma$-decay branch amounts to [11]

$$\Gamma_{\gamma\gamma}/\Gamma_{\gamma} = 2.05(37) \times 10^{-6} \; . \tag{10}$$

The Quasiparticle Phonon Model [17] provides a rather satisfactory description of the structure of ^{137}Ba and was used in Ref. [11] for the first time for calculating unique-multipolar generalized polarizabilities. The calculated dominance of the $M2-E2$ contribution to the $\gamma\gamma/\gamma$-branching ratio can be understood microscopically from the dominant contribution of the first excited $7/2^+$ state of ^{137}Ba at an excitation energy of 1252 keV [15]. Its structure is calculated as a superposition of a neutron hole in the $\nu(1g_{7/2})$ orbital outside the semi-magic core ^{138}Ba and the 2^+ core excitation at 1436 keV coupled to the $\nu(2d^{-1}_{3/2})$ ground state configuration of ^{137}Ba. Consequently, there exists a sizeable $M2$ matrix element between the $11/2^-$ isomer and this $7/2^+_1$ state of ^{137}Ba with a strength of the order of a single-particle transition. At the same time there exists a sizeable $E2$ matrix element from the $7/2^+_1$ state to the ground state of ^{137}Ba corresponding to the annihilation of the quadrupole core excitation. Since, in addition, the energy denominator is small the virtual contribution of the $7/2^+_1$ state dominates the α_{M2E2} coefficient and enhances it sufficiently such that it dominates the $\gamma\gamma/\gamma$-decay intensity to about 80%.

5. Summary

The first observation for the competitive double-γ decay process has been presented. This discovery was made on the 662-keV decay transition from the $11/2^-$ isomer of ^{137}Ba to its ground state. The $\gamma\gamma/\gamma$-branching ratio was found to be on the order of 10^{-6}. The observation became possible due to the employment of large-volume LaBr-detectors. Thanks to their good time resolution it was possible to demonstrate that the peaks in the coincident background-subtracted sum-energy spectra could not originate from sequential Compton scattering between the detectors which has been the main experimental obstacle in previous attempts for observing the $\gamma\gamma/\gamma$-branching ratio. The observed angular distribution and energy distribution of coincident γ quanta provide evidence for dominant $M2-E2$ and a minor $E3-M1$ contribution to the studied double-γ decay branch in competition to the 662-keV $M4$ single-γ transition in ^{137}Ba. The data were well accounted for by a calculation using the Quasiparticle Phonon Model.

IOP Conf. Series: Journal of Physics: Conf. Series **1056** (2018) 012045 doi:10.1088/1742-6596/1056/1/012045

6. Outlook

As discussed above, the phenomenon of double-γ decay gives experimental access to generalized nuclear polarizabilities, nuclear observables that have been studied very little up to now, and only in the dipole-dipole sector. Our observation of the $\gamma\gamma/\gamma$-decay demonstrates the feasability to study generalized nuclear polarizabilities in a broader scope and potentially not limited to the particular decays of even-even nuclei with a first excited 0^+ state. We are looking into developing more efficient experimental approaches for the detection of $\gamma\gamma/\gamma$-decay reactions. Still, due to the long measurement time and the significant difficulties in reducing the background it remains to be seen how these measurements can develop.

Furthermore, theoretical work is needed for clarifying the significance of the generalized nuclear polarizabilities for the advance of physics. One obvious route of studies may address the relation of the generalized nuclear polarizabilities with the dipole polarizability of the nuclear ground state which itself has been shown to be related to the symmetry energy parameters of the nuclear equation of state. It would also be desirable to clarify whether the nuclear models that are used to predict the $M^{(0\nu)}$ NMEs of $0\nu\beta\beta$-decay are capable of quantitatively describing the formally analogous generalized nuclear polarizabilities that in contrast to the $M^{(0\nu)}$ NMEs are experimentally accessible nuclear observables. The recent progress nurtures hope that a new sub-field of γ-ray spectroscopy may open up.

7. Acknowledgements

We thank C. Walz for his dedication to this research project. Long-standing collaborations with T. Aumann and V. Yu. Ponomarev are gratefully acknowledged. We thank A. Arsenyev and A. Severyukhin for calculations of generalized polarizabilities and D. Frekers, F. Iachello, J. Menendez, T. Otsuka, and A. Schwenk for discussions on the similarities and the differences between $\gamma\gamma/\gamma$-decay and $0\nu\beta\beta$-decay reactions and J. J. Valiente-Dobon, P. John, and P. Napiralla for discussions on future experiments on the $\gamma\gamma/\gamma$-decay. The presented research would not have been possible without the contribution of the German Research Council (DFG). The presentation was supported by the Collaborative Research Center SFB 1245 at TU Darmstadt.

References

[1] Göppert M 1935 *Phys. Rev.* **48** 512
[2] Elliott S R, Hahn A A and Moe M K 1987 *Phys. Rev. Lett.* **59** 2020
[3] Göppert M, *Über Elementarakte mit zwei Quantensprüngen* (Doctoral thesis, Universität zu Göttingen, 1930)
[4] Watson B A, Bardin T T, Becker J A, Fisher T R 1975 *Phys. Rev. Lett.* **35** 1333
[5] Kramp J *et al.* 1987 *Nucl. Phys.* A **474** 412
[6] Hayes A C *et al.* 1990 *Phys. Rev.* C **41** 1727
[7] Schirmer J *et al.* 1984 *Phys. Rev. Lett.* **53** 1897
[8] Beusch W *et al.* 1960 *Helv. Phys. Acta* **33** 363
[9] Basenko V K *et al.* 1992 *Bull. Russ. Acad.* **56** 94
[10] Lister C J *et al.* 2013 *Bull. Am. Phys. Soc.* **58**(13), DNP.CE.3
[11] Walz C, Scheit H, Pietralla N, Aumann T, Lefol R and Ponomarev V Yu 2015 *Nature* **526** 406
[12] Pietralla N, Walz C, Scheit H, Aumann T, Lefol R and Ponomarev V Yu 2016 *EPJ Web of Conferences* **107** 03002
[13] Pietralla N, Walz C, Scheit H, Aumann T, Lefol R and Ponomarev V Yu 2016 *J. Phys.: Conf. Series* **724** 012039
[14] Walz C 2014 *The two-photon decay of the* $11/2^-$ *isomer of* ^{137}*Ba and mixed-symmetry states of* ^{92,94}*Zr and* ^{94}*Mo* (Doctoral thesis, Technische Universität Darmstadt, 2014, and Springer International Publishing, Switzerland, 2016; DOI 10.1007/978-3-319-27183-5_1)
[15] Browne E, Tuli J K 2007 *Nucl. Data Sheets* **108** 2173
[16] Moran K, McCutchan E A, Lister C J, Zhu S, Carpenter M P, Chowdhury P, Greene J P, Lauritsen T, Merchan E and Shearman R 2014 *Phys. Rev.* C **90** 041303(R)
[17] Soloviev V G 1992 *Theory of Atomic Nuclei: Quasiparticle and Phonons* (Bristol, IOP)

IOP Conf. Series: Journal of Physics: Conf. Series **1056** (2018) 012046 doi:10.1088/1742-6596/1056/1/012046

Project of Thin Targets for the NUMEN Experiment

F Pinna[1,2], F Iazzi[1,2], R Introzzi[1,2], D Calvo[2] and M Fisichella[2]
for the NUMEN Collaboration

1) DISAT - Politecnico di Torino, Torino, Italy
2) INFN - Sezione di Torino, Torino, Italy

fpinna@to.infn.it

Abstract. The goal of the NUMEN collaboration is the measurement of the cross sections of Double Charge Exchange reactions for several couple of ion projectile-target, in order to provide helpful data to study the nuclear matrix elements of the neutrino-less double β-decay. The need of big statistics and high precision in the measurements require the use of high intensity beams and very thin targets. This creates some problems to the design of the target frame and to the dissipation of the heat generated by the beam. The present paper reports a possible solution for the cooling system and the production technique of a tin target, together with the results of the preliminary tests of heat dissipation.

1. Introduction

The NUMEN project [1][2][3] aims to get information on the nuclear matrix elements of neutrino-less double beta decay from experimental measurements of heavy-ion induced Double Charge Exchange reactions (DCE). A statistically significant collection of DCE data will be obtained taking advantage of the high ion beam current (up to 50 μA), which will be available in the next future at INFN-LNS (Catania, Italy) [4], and of the MAGNEX spectrometer [5][6], which allows to measure high resolution energy spectra and absolute cross sections at very forward angles even for very rare processes [7][8][9]. This requires that the targets be thin and uniform, to keep a good energy resolution. On the other hand, the large beam current together with the small thickness require that the generated heat inside the target be efficiently dissipated to avoid target damages.

The experiment will investigate the following target nuclei: ^{116}Sn, ^{116}Cd, ^{76}Ge, ^{76}Se and ^{130}Te using ^{18}O and ^{20}Ne beams. Their thickness will vary from 250 nm to 500 nm. Preliminary calculations addressed the solution of the heat dissipation problem toward the deposition of the thin target on a highly conductive substrate, as explained in Section 2. The design of the tin target has been completed and some prototypes have been realized with different deposition techniques, as illustrated in Section 3. The results of a preliminary test of the heat dissipation in the deposited tin using a LASER beam as heater are reported in Section 4. The conclusions are drawn in Section 5, together with the plan of the future work.

2. The Target Cooling System

An ion beam produces, by ionization, a rate of heat inside a target, which is usually dissipated through the surrounding cold frame. In the case of NUMEN, the beam intensities are so high that this dissipation technique is not sufficient to take away the total power. Simple calculations show that the thermal stationary state of a self-sustained target surrounded by a cold frame reaches temperatures widely exceeding the melting point in the central region under the beam spot [10]. A more complex system, made by a deposition of the target material on a thin layer of pyrolytic graphite, ensures a large heat

IOP Conf. Series: Journal of Physics: Conf. Series **1056** (2018) 012046 doi:10.1088/1742-6596/1056/1/012046

transfer from the central region to the surrounding cold frame, thanks to the high thermal conductivity of the graphite. A sketch of the system is shown in Figure 1.

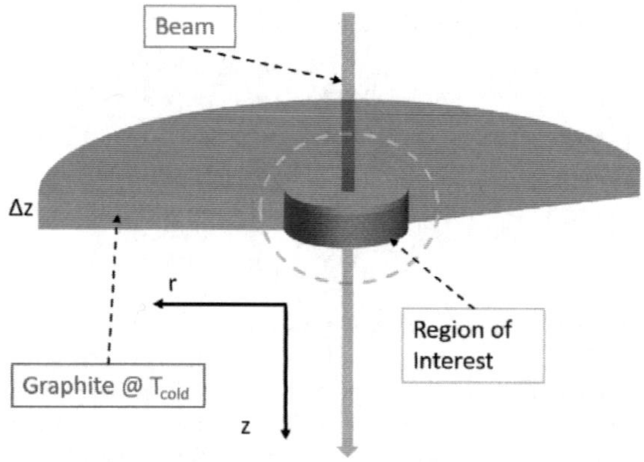

Figure 1: scheme of the target/graphite system: a disk-shaped film of target material (thickness in the range 250-500 nm, diameter about 1 cm, green region) is deposited on a pyrolytic graphite disk of 5 cm diameter and 10 µm thickness. The graphite not covered by the film (in blue) is kept at fixed temperature T_{cold}. The ion beam (in red), with Gaussian profile of 2 mm FWHM, passes through both target and not-cooled graphite (in grey), generating a source of heat, which flows toward the cold region.

The solution of the heat equation gives the time evolution of the temperature profile along the radius and has been calculated using a MatLab code for several target materials, beam currents and intensities. The result of these calculations is that, after few milliseconds, the temperature is nearly uniform along the beam axis (z-axis), in both target and graphite, while the spatial distribution of the temperatures reaches a stationary state after few seconds. Moreover, the temperature of the target looks only slightly warmer than the substrate at the same radius, showing that the heat flows preferably from the target to the substrate. Finally, the maximum temperature is always along the beam axis, as expected. Therefore, it is sufficient to maintain the temperature along the z-axis below the melting point in order to avoid the target damage. Figure 2 shows these results for a ^{116}Sn target, 500 nm thick, illuminated by a 50 µA beam current of $^{18}O^{8+}$ ions at 15 MeV/u and cooling temperature T_{cold} = 100 K.

3. Tin Deposition on Graphite Substrate

The first prototype of the target-substrate system has been obtained by using Electron Beam Deposition of Sn on graphite. Since in the above calculations the target film was supposed uniform and perfectly adherent to the graphite, several attempts (substrate at various temperatures, annealing after deposition) were made to obtain a high-quality film of tin on pyrolytic graphite.

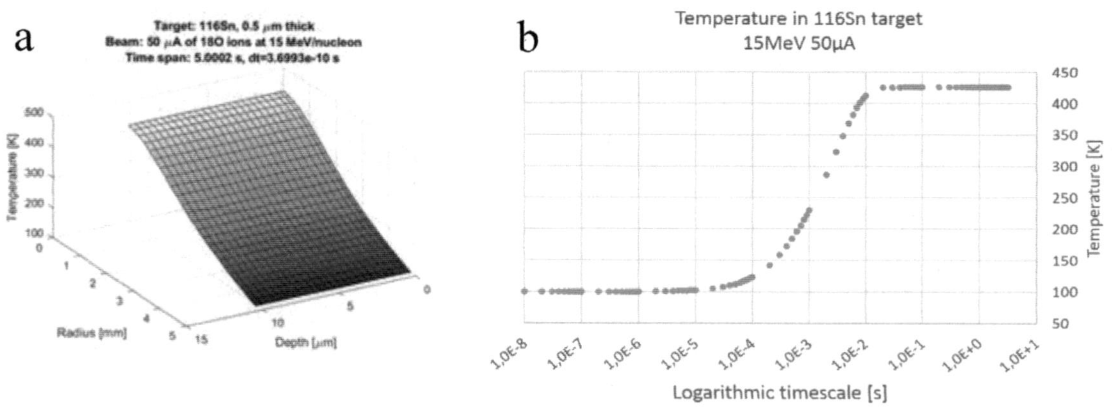

Figure 2: a) the distribution of the temperature after 5 s is shown as a function of the radius and of the depth; b) the maximum temperature (at r = 0, z = 0) is reported as a function of the time.

IOP Conf. Series: Journal of Physics: Conf. Series **1056** (2018) 012046 doi:10.1088/1742-6596/1056/1/012046

Each result was examined by Field Emission Scanning Electron Microscope technique (FESEM), to check the homogeneity of the deposited samples.

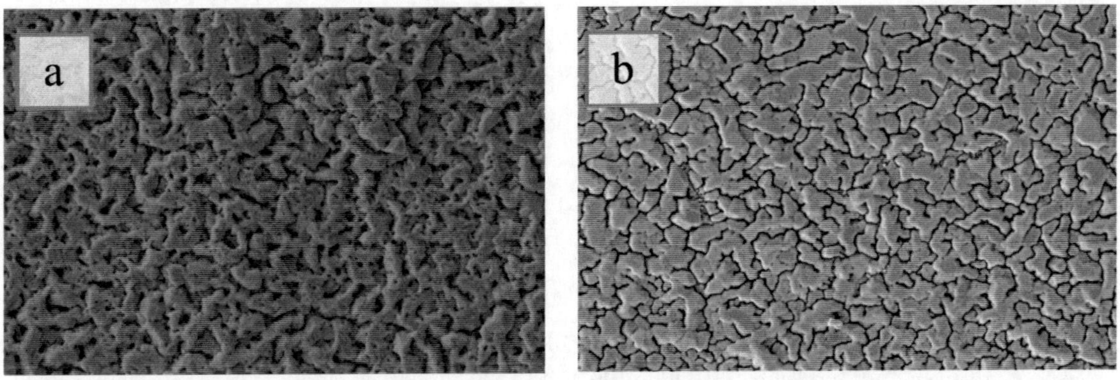

Figure 3: FESEM images: a) top view of a 500 nm thick Sn sample, deposited at room temperature; b) top view of a 500 nm thick Sn sample, deposited at 175°.

In Figure 3a the image of a deposition at room temperature shows an unsatisfactory result, because the black areas indicate that quite large empty spaces separate the grains. After several trials with different substrate temperatures, with and without annealing, the best deposition was obtained at 175°, without annealing. In Figure 3b, the surface appears much more compact and uniform. In order to check the adhesion of the tin on the graphite, some FESEM images were taken on the side of the deposition: even if they looked better, nevertheless the effects of the deposition on the border could mask the true situation. Therefore, the adhesion was tested in the experiment described in the next Section.

4. Preliminary Test of Heat Transfer Inside the Target

A preliminary test of the heat dissipation inside the Sn-graphite system has been performed using a LASER as heat source. The LASER spot had a radius of ≈1 mm and penetrated only for a few nm in the Sn. Therefore, the heat source was not uniform along the beam axis, while it is in the case of the heat generated by the ion beam. The setup of the experiment consisted of a metal frame in contact with the Sn-graphite system, maintaining the blue region of the graphite at room temperature. An IR LASER (808 nm wavelength) irradiated the center of the system on the side of the tin, with power ranging from 1.5 to 12 W. A thermal camera, looking at the Sn target, monitored the temperature of the tin every 3 ms, while a pyrometer, located beyond the target, monitored the temperature of the graphite in the center of the disk. The stationary temperature along the beam axis was calculated for this configuration using the same Matlab code of Section 2.

The results of the measurements are the following: a) the temperatures measured by the thermal camera and pyrometer reach the stationary state within 1 s, in agreement with the calculations, b) the stationary temperature is the same (within the instrumental errors) on the tin and graphite surfaces. These results show that the calculation technique is correct and that the heat, generated in tin, quickly transfers to the graphite along the beam axis. This suggests that the contact between the tin and graphite surfaces is uniform, as supposed in the calculations.

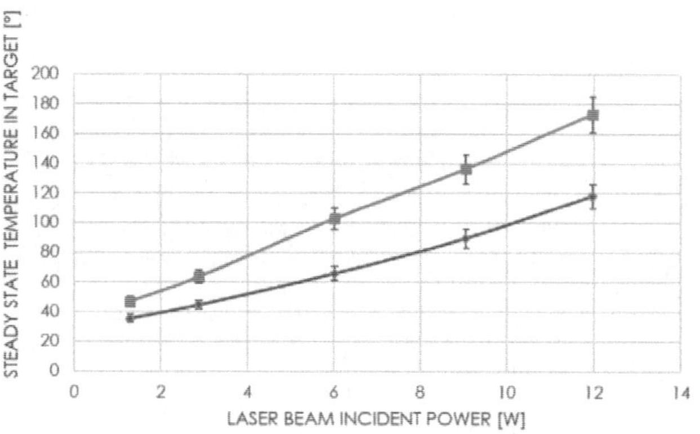

Figure 4: calculated (orange) and experimental (blue) values of the stationary temperatures as a function of the LASER power.

Figure 4 shows the trend of the measured (blue) and calculated (orange) stationary temperature as a function of the LASER power. Both look linear, but the slope of the data is lower. It is easy matter to demonstrate that the stationary temperature is proportional to the power entering in the system, therefore the disagreement is probably due to the reflection of the LASER light on tin, which decreases the true entering power with respect to the nominal one. It must be remarked that the temperatures reached in this experiment are much lower than in the future NUMEN measurements, therefore more severe tests are needed, with suitable ion beams.

5. Conclusions

A method for cooling thin targets under an intense ion beam has been studied. Such method involves the usage of a large pyrolytic graphite substrate. The target material is deposited on such substrate by means of the Electron Beam Deposition in order to enhance the heat dissipation. To stay within the smoothness requirements of the Sn target, several depositions were carried out with different process parameters, eventually obtaining satisfactory results. Concerning the heat dissipation, the effectiveness of the solution has been studied with a MatLab code and tested with a laser beam, giving encouraging results. More severe tests with ion beams of energy and intensity close to the future NUMEN beams are planned in the next future.

Acknowledgments

Authors would like to warmly thank Mr. Mauro Raimondo for his skillful work in characterizing the samples with FESEM. A particular acknowledgement is due to Mr. Carmelo Marchetta for his precious suggestions. The most part of the depositions where performed by the staff of Trustech Company (Chivasso, Italy), whose help is highly appreciated.

References

[1] Cappuzzello F et al 2015 *EPJ Web of Conferences* **630** 12018
[2] Agodi C et al 2015 *Nucl. Part. Phys. Proc* **265-266** 28
[3] Cappuzzello F et al 2018 *Eur. Phys. J* A **54** 72
[4] Calabretta L et al 2017 *Modern Phys. Lett. A* **32** 1740009
[5] Cappuzzello F, Agodi C, Carbone D and Cavallaro M 2016 *Eur. Phys J.* A **52** 167
[6] Cappuzzello F et al. 2014 *Nucl. Instr. Methods* A **763** 314
[7] Cappuzzello F et al. 2015 *Eur. Phys. J.* A **51** 145
[8] Oliveira J R B et al 2013 *J. Phys. G: Nucl. Part. Phys.* **40** 105101
[9] Cavallaro M et al. 2016 Phys. Rev. C **93** 064323
[10] Iazzi F *et al* 2017 *WIT Transactions on Engineering Sciences* **116**, pp 61-70

IOP Conf. Series: Journal of Physics: Conf. Series **1056** (2018) 012047 doi:10.1088/1742-6596/1056/1/012047

Status of the ENUBET project

G Ballerini [1,2], A Berra [1,2], R Boanta [2,8], M Bonesini [2], C Brizzolari [1,2], G Brunetti [10], M Calviani [13], S Carturan [20], M G Catanesi [12], S Cecchini [3], F Cindolo [3], A Coffani [2,8], G Collazuol [10,11], E Conti [10], F Dal Corso [10], G De Rosa [16,17], A Gola [15], R A Intonti [12], C Jollet [4], Y Kudenko [18], M Laveder [10,11], A Longhin [10], P F Loverre [6,14], L Ludovici [6], L Magaletti [12], G Mandrioli [3], A Margotti [3], V Mascagna [1,2], N Mauri [3], A Meregaglia [19], M Mezzetto [10], M Nessi [13], A Paoloni [5], M Pari [10,11], G Paternoster [15], L Patrizii [3], C Piemonte [15], M Pozzato [3], M Prest [1,2], F Pupilli [10], E Radicioni [12], C Riccio [16,17], A C Ruggeri [16], M Soldani [1,2] G Sirri [3], M Tenti [3,7], F Terranova [1,2], E Vallazza [9], M Vesco [20], L Votano [5] and E Wildner [13]

[1] Phys. Dep. Università degli Studi dell'Insubria, via Valeggio 11, Como, Italy
[2] INFN, Sezione di Milano-Bicocca, piazza della Scienza 3, Milano, Italy
[3] INFN, Sezione di Bologna, viale Berti-Pichat 6/2, Bologna, Italy
[4] IPHC, Université de Strasbourg, CNRS/IN2P3, Strasbourg, France
[5] INFN, Laboratori Nazionali di Frascati, via Fermi 40, Frascati (Rome), Italy
[6] INFN, Sezione di Roma 1, piazzale A. Moro 2, Rome, Italy
[7] Phys. Dep. Università di Bologna, viale Berti-Pichat 6/2, Bologna, Italy
[8] Phys. Dep. Università di Milano-Bicocca, piazza della Scienza 3, Milano, Italy
[9] INFN Sezione di Trieste, via Valerio, 2 - Trieste, Italy
[10] INFN Sezione di Padova, via Marzolo, 8 - Padova, Italy
[11] Phys. Dep. Università di Padova, via Marzolo, 8 - Padova, Italy
[12] INFN Sezione di Bari, via Amendola, 173 - Bari, Italy
[13] CERN, Geneva, Switzerland
[14] Phys. Dep. Università La Sapienza, piazzale A. Moro 2, Rome, Italy
[15] Fondazione Bruno Kessler (FBK) and INFN TIFPA, Trento, Italy
[16] INFN, Sezione di Napoli, via Cinthia, 80126, Napoli, Italy
[17] Phys. Dep. Università degli Studi di Napoli Federico II, via Cinthia, Napoli, Italy
[18] Institute of Nuclear Research of the Russian Academy of Science, Moscow, Russia
[19] CENBG, Université de Bordeaux, CNRS/IN2P3, 33175 Gradignan, France
[20] INFN Laboratori Nazionali di Legnaro, Viale dell'Università, 2, Legnaro (PD), Italy
Speaker, E-mail: michele.pozzato@bo.infn.it

Abstract. The next generation of neutrino experiments requires measurements of absolute neutrino cross sections at the GeV scale with high precision (~1%) presently limited by the uncertainties on neutrino flux. Monitoring the lepton production in the decay tunnel of neutrino beams is the most straightforward way to measure the neutrino flux at source. The ENUBET Collaboration develops novel technologies to monitor positrons from $K^+ \rightarrow \nu_e e^+ \pi^0$ decays on an event by event basis. This technique can achieve a precision in the ν_e flux below 1% and enable a new generation of cross section and short baseline experiments. In this paper, we present the achievements of the first year of the Project on beamline simulation, rate and dose assessment, detector prototyping and evaluation of the physics reach.

IOP Conf. Series: Journal of Physics: Conf. Series **1056** (2018) 012047 doi:10.1088/1742-6596/1056/1/012047

1. The ENUBET project

Neutrinos played a fundamental role in the study of weak interactions and, after the observation of the oscillation phenomenon, they opened a window on physics beyond the standard model. The next generation of oscillation physics experiments aiming at the study of CP violation and mass hierarchy problems will rely on precise measurements of neutrino interaction cross sections.

The prediction of the neutrino flux in conventional beam is a challenging task which relies on the detailed simulation of the beamline (constrained by data from beam monitoring devices and measurements such as proton intensity, horn currents,...) and on the hadro-production data coming from experiments which provide the particle yields from the target. Neutrino experiments are affected by an intrinsic limitation given by a large uncertainty of the overall neutrino flux (~10%) directly reflecting on the cross section measurements. Moreover the ν_e and $\overline{\nu_e}$ cross sections are poorly known despite of their relevance in the next generation long-baseline experiments.

A change in the method used to determine the neutrino flux is needed: the direct monitoring of leptons produced in the decay tunnel together with neutrinos can overcome these limitations reducing the flux-related systematics: a technique specifically focused on ν_e cross sections has been proposed in [1]. This new method has the advantage of providing a source of ν_e from $K^+ \rightarrow e^+ \nu_e \pi^0$ allowing to study directly their interactions (without relying on extrapolations from ν_μ) by using an observable (the positron rates) directly linked to the ν_e rate at far detector. The direct monitoring of the neutrino rate at the source is expected to be capable to provide a flux measurement error of ~ 1%. The ENUBET project (*"Enhanced NeUtrino BEams from kaon Tagging"*) [2] aims at demonstrating the technical feasibility of monitored neutrino beams. The project is funded by the European Research Council (ERC Consolidator Grant, PI A. Longhin, host institution INFN) from June 2016 to May 2021. ENUBET is also framed in the CERN Neutrino Platform under the NP3-Plafond R&D programme since December 2016.

2. Design of the hadronic beamline

In the proposed setup primary protons are impinging on a target to produce secondary hadrons which are charge and momentum selected and transported to a secondary instrumented beam line. Two possible solutions are under study by the ENUBET collaboration in designing the focusing system for incoming proton momentum of 30, 120, 400 GeV/c: a conventional pulsed horn and a static system. The optimization of the focusing system and the design of the transfer line confirm (or even improve) the collection efficiency assumed in [1]: the optimization of the line (dipoles and quadrupoles) is performed with TRANSPORT [3]. G4Beamline is used for the implementation and full simulation of particle transport, collimation and re-interaction. The dose assessment is addressed by using FLUKA [4,5]. The positron tagger can operate safely in terms of pile-up if the local particle rate is below 500 kHz/cm^2. This can be achieved with two possible solutions under study:

- multi-Hz slow extraction lasting O(ms) with ~10 Hz repetition and horn-based focusing;
- slow extraction of the order of 1s with DC operated focusing magnets.

3. Instrumented decay tunnel

A cost-effective solution capable of working in the harsh environment of a decay tunnel, with rates up to 500 kHz/cm^2 comes from employing compact shashlik calorimeters read out with fast wavelenght shifting fibers each coupled with its own silicon photomultiplier (SiPM). This detector, effective in e/π separation, will be used together with a plastic scintillator tracker acting as a photon veto (t_0-layer). The use of SiPM embedded inside the module allows to achieve the desired longitudinal segmentation without dead zones in the longitudinal sampling thus achieving a remarkable compactness.

A GEANT4 simulation of the decay tunnel and instrumentation has been performed in parallel with the simulation of the transfer line: a reconstruction algorithm for particle identification has been developed to cope with pile-up in realistic conditions and it is based on the energy deposit pattern in the calorimeter modules to discriminate electromagnetic from hadronic showers. Timing

IOP Conf. Series: Journal of Physics: Conf. Series **1056** (2018) 012047 doi:10.1088/1742-6596/1056/1/012047

information coming both from shashlik calorimeter and the photon veto is exploited to clusters modules belonging to the same event ("event building") and to reject electromagnetic shower generated from π^0 decays. The ENUBET instrumentation provides positron identification with 25% of efficiency while keeping signal-to-noise ratio near to 1: these performances are estimated by taking into account beam related background profiting of the input from the transfer line simulation. The assessment of systematics in the neutrino flux due to the detector response is ongoing.
Ionizing and non-ionizing doses (from neutrons) have been evaluated with FLUKA 2011 at the inner surface of the target for different choices of its radius (40, 80, 100 cm) as shown in figure 1 (left). For 1 m radius the integrated doses (correspond to $10^4\ \nu_e^{CC}$ events at the detector) amounts to < 0.05 kGy and 1.9×10^{11} n/cm^2 (1 MeV equivalent): an additional mitigation can be achieved by placing the photosensors outside the calorimeter with a non-shashlik solution.

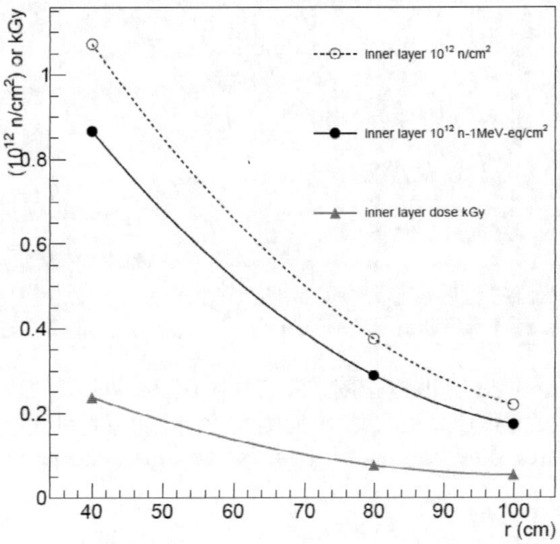

Figure 1. (Left): Ionizing dose (red line, kGy), neutron fluence (black dashed line, in n/cm^2) and 1 MeV equivalent neutron fluence (black continuous line) in the innermost modules of the calorimeter as a function of the tagger radius.

Different scenarios corresponding to 2-3 months of data taking at the present running performance of major accelerating facilities have been considered: a 500 t neutrino detector (transverse size of 6×6 m^2) placed 50 m downstream of the tagger is able to record $10^4\ \nu_e^{CC}$ events with 1.0×10^{20} proton on target (pot) at J-PARC synchrotron (30 GeV). Same results can be obtained at Fermilab Main Ring (120 GeV) with 2.4×10^{19} pot and at CERN SPS (400 GeV) with 1.1×10^{19} pot.

4. Prototypes tests at test beams

The basic calorimetric unit (Ultra-Compact Module - UCM) is made of five, 15-mm thick, iron layers interleaved by 5-mm thick plastic scintillator tiles. The transverse dimension is 3×3 cm^2 for a 10 cm length (4.3 X_0). Nine wavelength shifting (WLS) fibers crossing the UCM are connected directly to 1 mm^2 SiPM produced by the Fondazione Bruno Kesserl (FBK) avoiding the occurrence of large passive regions usually needed to bundle the fibers to a common photo-sensor. UCM prototypes and photon-veto prototypes have been characterized with cosmic rays and charged particles (1-5 MeV range) at CENR-PS East Area facility (T9) in several exposure 2016 (July-November) and 2017 (August-October)[6, 7, 8].

4.1. UCMs

IOP Conf. Series: Journal of Physics: Conf. Series **1056** (2018) 012047 doi:10.1088/1742-6596/1056/1/012047

Test beams aim to characterize UCMs in terms of energy resolution, response linearity for electromagnetic showers, modules uniformity and the effectiveness of minimum ionizing particles for self-calibration for different setup. Moreover a comparison between data a GEANT4 simulation has been deeply investigated to check the capability to reproduce data distributions. Different technological solutions have been studied: the choice of plastic scintillators (EJ200, EJ204, Uniplast) and thickness (5 - 10 mm) as well as different WLS fibers (Kuraray Y11-MC, St. Gobain BCF-92-MC). Injection molded scintillators (suitable for large scale production) have been successfully produced by Uniplast (Russia) and tested at CERN in July 2017.

During the November 2016 CERN test beam a module composed of 56 UCM in 7 longitudinal layers (~30 X_0, figure 2) and of an outer module acting as an energy catcher, were exposed to e^-/π^- with grazing incidence at various incidence angles from 0 to 200 mrad.

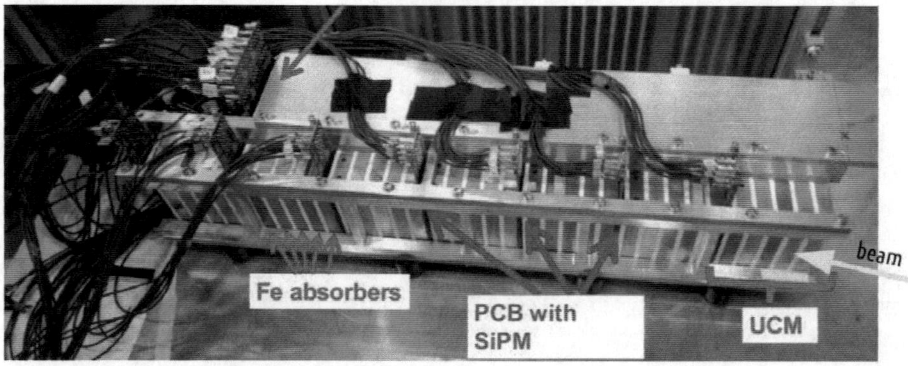

Figure 2. The calorimeter exposed at the CERN-PS T9 particle beams. The prototype is composed of an inner part (bottom in the picture) seven calorimetric blocks and a 60 cm long "hadronic" block for the outer part (up in the picture). The compact layout of the light readout is visible through the cabled PCBs.

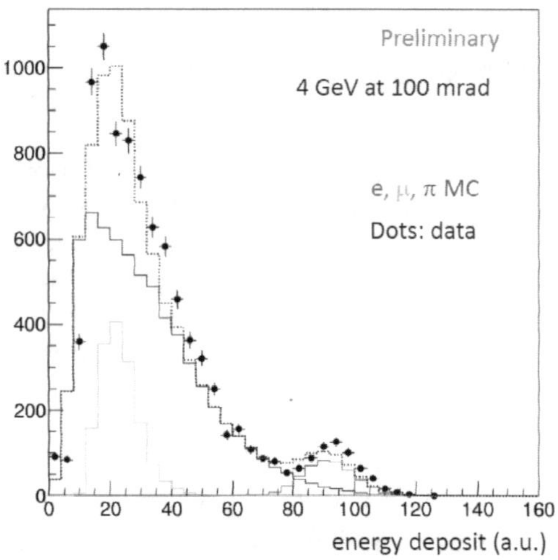

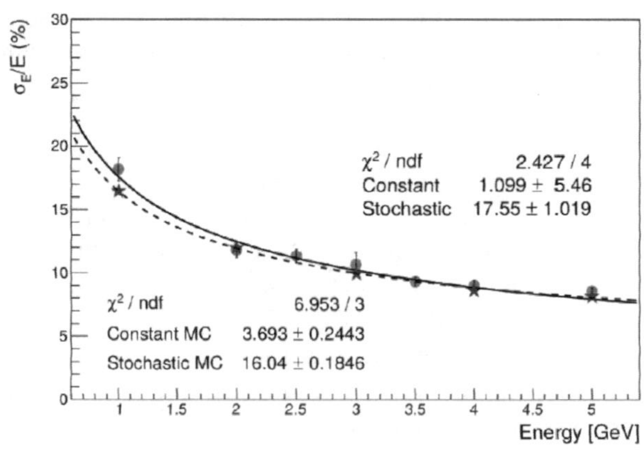

Figure 3. Data and Monte Carlo distributions of the total visible energy for a 100 mrad tilt of the calorimeter and a 4 GeV beam of negative pions, muons and electrons in the CERN-PS T9 beamline. (Right): Energy resolution for electrons in data (red point) and Monte Carlo (blue). The fitted parametrization follows the standard form: $\frac{\sigma(E)}{E} = \frac{\sigma}{\sqrt{E}} + c$.

The electron/pion separation capabilities in data and Monte Carlo have been determined at various incident angles. In figure 3 (left) the distribution of the total visible energy for a 100 mrad tilt of the calorimeter and a 4 GeV beam of negative pions, muons and electrons is shown for data (bullets) and Monte Carlo (histograms, broken down by particles). Figure 3 (right) shows the energy resolution for electrons in data and Monte Carlo. Non conventional options based on polysiloxane scintillators [9] to avoid drilling or injection molding and allowing an improved radiation hardness are also being scrutinized: a prototype polysiloxane shashlik calorimeter composed of three modules (13 X_0 in total) consisting of 12 UCM units and a 15 mm scintillator thickness was built at INFN-LNL and exposed to particle beams at the CERN East area. The absolute light yield and uniformity are very promising.

4.2. Photon Veto

The baseline option for the photon-veto (plastic scintillator doublets) has also been tested with several layouts for light collection. The requirements for this detector are the precise timing and the 1 mip/2mip separation. A custom 2 stages wideband amplifier based on HMC58-2 and Gali TB0447-05 chips has been tested to readout a SiPM SenSL-30020J (3x3 mm^2). First measurements performed on this SiPM with a PiLas laser (407 nm) showed that the time resolution is below 100 ps (matching the requirements).

During test beam at CERN in 2017 with pions, several light collection configurations have been checked as well as different scintillators/WLS fibers (EJ200 coupled with Kuraray-Y11 MC and EJ204 coupled to St. Gobain BCF92MC): all configurations have a good response in terms of efficiency (> 95%) and uniformity. In order to perform a first test on 1 particle /2 particles separation we have enhanced the π^0 production with a Delrin cylinder and let them decay through an iron slab: in figure 4 it is clearly visible the separation provided by a single t_0-layer. A first timing measurement has been performed on a single tile with EJ-204 coupled with BCF92: the results are really promising (~400 ps).

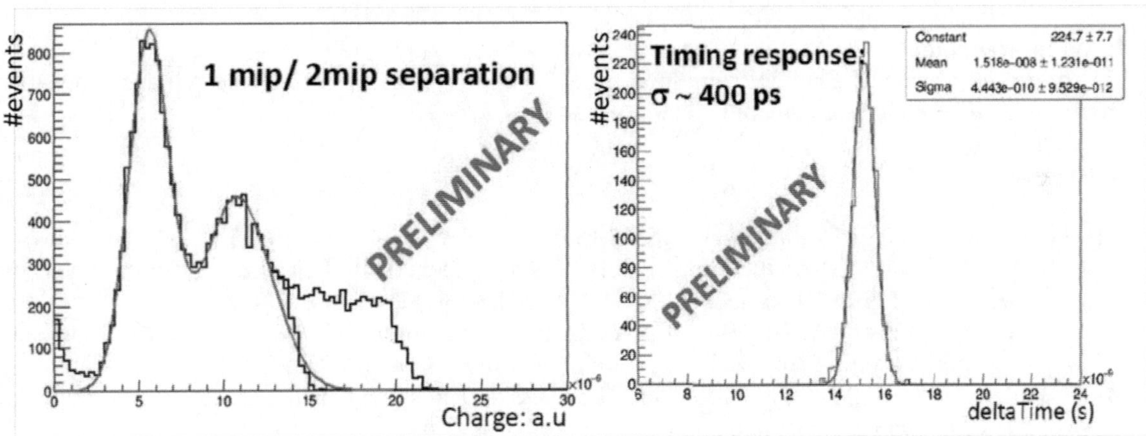

Figure 4. (left) e^+e^- couples produced from the π^0 decay: it's visible the separation between 1 particle and 2 particles. (right) timing measurements performed with a fast scintillator used as trigger, the resolution of ~400 ps.

5. SiPM neutron irradiation at INFN-LNL

In the ENUBET reference design the SiPMs will be exposed to radiation. Since such devices are sensitive in particular to neutron damages, we have performed iraadiation test in June 2007 at the CN facility at INFN-LNL (Legnaro).

Neutron fluences up to 10^{12} n/cm^2 were achieved by sending 5 MeV protons on a Beryllium target. The target area with concrete shielding is shown in figure 5, left. SiPM of different pixel sizes (12, 15, 20 μm) produced by FBK were exposed and characterized in terms of I vs V curves at

IOP Conf. Series: Journal of Physics: Conf. Series **1056** (2018) 012047 doi:10.1088/1742-6596/1056/1/012047

increasing irradiation (figure 5, right). Noise waveforms were also sampled before and at intermediate irradiation levels. Data are under analysis and will be presented in a full paper.

Figure 5. (Left) The target region at INFN-LNL used to produce large neutron fluxes for SiPM irradiation: the Be target is visible at the center of the picture. (Right) IV curves for a 20 μm cell SiPM from FBK for various values of the estimated neutron irradiation (neutrons 1 MeV-eq/cm^2).

6. Conclusions

The final goal of ENUBET (2016-2021) is to demonstrate that e^+-monitored ν_e beams can be built using existing technologies and hosted at CERN, Fermilab or J-PARC for a new generation of experiments on neutrino cross sections. The results obtained in the preparatory and initial phase of the project are very encouraging. The design and simulation of the hadronic beamline has started and the preliminary results support or improve the initial estimates of the achievable event rates. Full simulations support the viability and effectiveness of the calorimetric approach whereas prototypes tests demonstrate that shashlik calorimeters with longitudinal segmentation can fulfill the requests for e^+ monitoring in the relevant energy region of few GeV.

7. Acknowledgment

This project has received funding from the European Union's Horizon 2020 Research and Innovation programme under Grant Agreement no. 654168 and no. 681647

References

[1] Longhin A et al. 2015 *Eur. Phys. J.* C **75** 155
[2] Berra A et al. 2016 Enabling precise measurements of flux in accelerator neutrino beams: the ENUBET project CERN-SPSC-2016-036 SPSC-EOI-014
[3] Brown K L et al. TRANSPORT 1983 A computer program for designing charged particle beam transport systems http://lss.fnal.gov/archive/nal/fermilab-nal-091.pdf
[4] Böhlen T T et al. 2014 The FLUKA Code: Developments and Challenges for High Energy and Medical Applications *Nuclear Data Sheets* **120**, pp 211-214
[5] Ferrari A et al. 2005 FLUKA: a multi-particle transport code, CERN- 2005-10 INFN/TC_05/11 SLAC-R-773
[6] Berra A et al. 2016 *Nucl. Instrum. Methods* A **830** 345
[7] Berra A et al. 2017 *Nucl. Instrum. Methods* A **845** 511
[8] Berra A et al. 2017 *IEEE Trans. Nucl. Sci.* **64** 1056
[9] Carturan S et al. 2011 Novel polysiloxane-based scintillators for neutron detection *Radiation protection dosimetry* **143** pp 471-476.

Status and perspectives of the JUNO experiment

Gioacchino Ranucci on behalf of the JUNO Collaboration

Istituto Nazionale di Fisica Nucleare
Via Celoria 16
20133 Milano - Italy

Gioacchino.ranucci@mi.infn.it

Abstract. The JUNO (Jiangmen Underground Neutrino Observatory), a 20 kton multi-purpose underground liquid scintillator detector, has been proposed and approved for realization in the south of China. After an intense design phase, the overall concept of the structure of the detector has been finalized, paving the way towards the construction of the several components and subsystems, which will compose it. Meanwhile, the excavation of the site which will host the experiment has been started and is rapidly progressing. The main physics target of JUNO is the determination of the neutrino mass hierarchy, which will be accessible through the measurement of the pattern of antineutrino spectrum from two high power nuclear complexes under installation 53 km away from the experimental site. In this work I describe the broad physics capabilities of the experiment, which include in addition to the crucial measure of the neutrino hierarchy the high precision determination of three oscillation parameters, as well as a rich astroparticle program, and I illustrate the main technical characteristics of the detector.

1. Introduction

The successful saga of neutrino oscillations, culminated in the 2015 Nobel Prize, paves the way to a future, rich and diverse experimental program of precision experiments tasked to complete the determination of the elements of the PNMS oscillation matrix with unprecedented accuracy, and to unravel the yet undiscovered features of neutrino properties. Therefore, mass hierarchy (MH) determination, octant of θ_{23}, violating δ_{cp} phase and improved precision of the mass-mixing parameters (mixing angles as well as squared mass differences), are the core of the ambitious worldwide neutrino oscillation program shaped for the forthcoming two decades. At the same time the Dirac or Majorana nature of the neutrino mass term, together with its absolute value, will be probed by a suite of dedicated experiments.

In this global context, the JUNO detector [1] will play a central role on two aspects: the determination of mass hierarchy and the precise measurements of the solar oscillation parameters, i.e. Δm^2_{21}, $sin^2\theta_{12}$, as well as of the atmospheric squared mass difference Δm^2_{31}.

JUNO will be designed and realized as a huge liquid scintillator detector, therefore exploiting a mature and well proved technology, which has already provided fundamental contributions to the neutrino oscillation study through several experiments (Borexino [2], KamLAND [3], Daya Bay [4], Reno [5] and Double Chooz [6] being the most recent examples). It will base its measurements on the detection of the global antineutrino flux coming from the cores of nearby nuclear complexes.

The program will be complemented by an ensemble of astroparticle physics measurements, which will significantly enhance the physics potential of JUNO.

Overall requirements, technical features and the current status of the experiment are described in the following.

2. Summary of characteristics and of physics goals

JUNO will be a new member of the renowned, long tradition family of reactor neutrino experiments based on the scintillation technology, whose first well known example was the Savannah River experiment, with which Cowan and Reines revealed for the first time the (anti)neutrino particle.

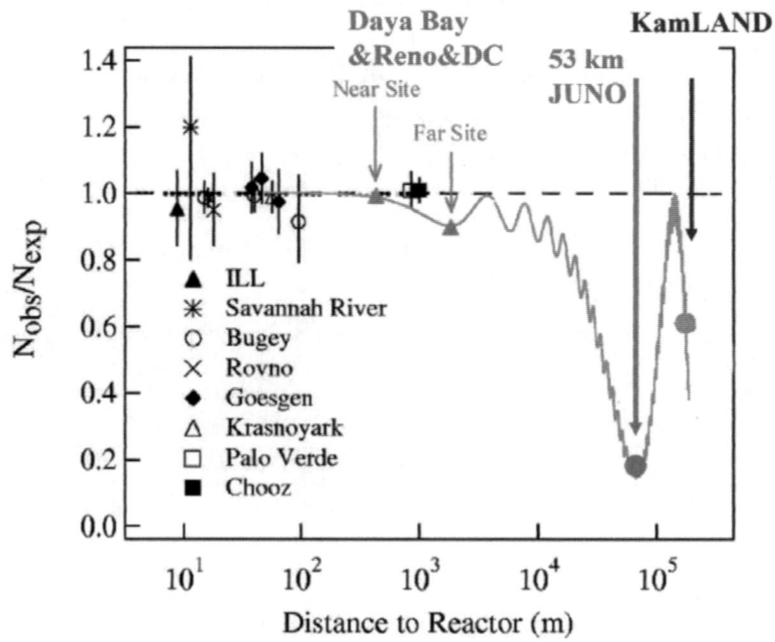

Figure 1. Summary of past reactors' results as ratio of observed to expected count rate, together with the predicted JUNO point

In Figure 1 there is the summary of reactors' results accumulated so far, expressed as ratio of observed over expected events, contrasted with the prediction from the oscillation survival probability function. On the horizontal axis the reactor-detector distance is displayed; the plot reports the well-known fact that at small distance the impact of the oscillation phenomenon on the detector count rate is not visible, while it starts to manifest from slightly less than 1 km baseline. At the distance of 53 km the count rate suppression, mainly driven by the solar oscillation parameters, is maximal, therefore creating the best condition to study the interference effect governed in turn by the atmospheric mass squared difference, which is responsible for the ripple superimposed on the count rate suppressed profile.

To fully exploit this optimal baseline to perform an effective, and successful measurement of the mass hierarchy, the detector must be endowed with two essential characteristics: large mass to perform a high statistic measurements, and stringent energy resolution to clearly distinguish the ripple induced by the atmospheric mass squared term. The two key numbers in this respects are the total mass of 20 kton of liquid scintillator, and the energy resolution of 3% at 1 MeV, which represent, therefore, the major technical features which characterize the experiment.

In term of physics reach, such a high mass detector can tackle a plurality of measurements: beyond mass hierarchy and precision determination of neutrino oscillation parameters, it can provide fundamental results concerning many hot topics in the astroparticle field, like supernova burst neutrinos, diffuse supernova neutrinos, solar neutrinos, atmospheric neutrinos, geo-neutrinos, sterile neutrinos,

IOP Conf. Series: Journal of Physics: Conf. Series **1056** (2018) 012048 doi:10.1088/1742-6596/1056/1/012048

nucleon decay, indirect dark matter search, as well as a number of additional exotic searches, as thoroughly illustrated in the physics program of the experiment (yellow book), published in [7].

3. Basic features of the program: detector structure, location and Collaboration

In terms of implementation characteristics, JUNO is a spherical unsegmented liquid scintillator detector that will push such a technology beyond the present limit, as far as the mass (20 kton) and the resolution (3%) are concerned. Briefly, the detector (see Figure 2) can be described as a large spherical acrylic vessel, which will hold the scintillator volume, contained in turn in a water pool, to ensure adequate shielding against the gamma radiation and neutrons from the rock.

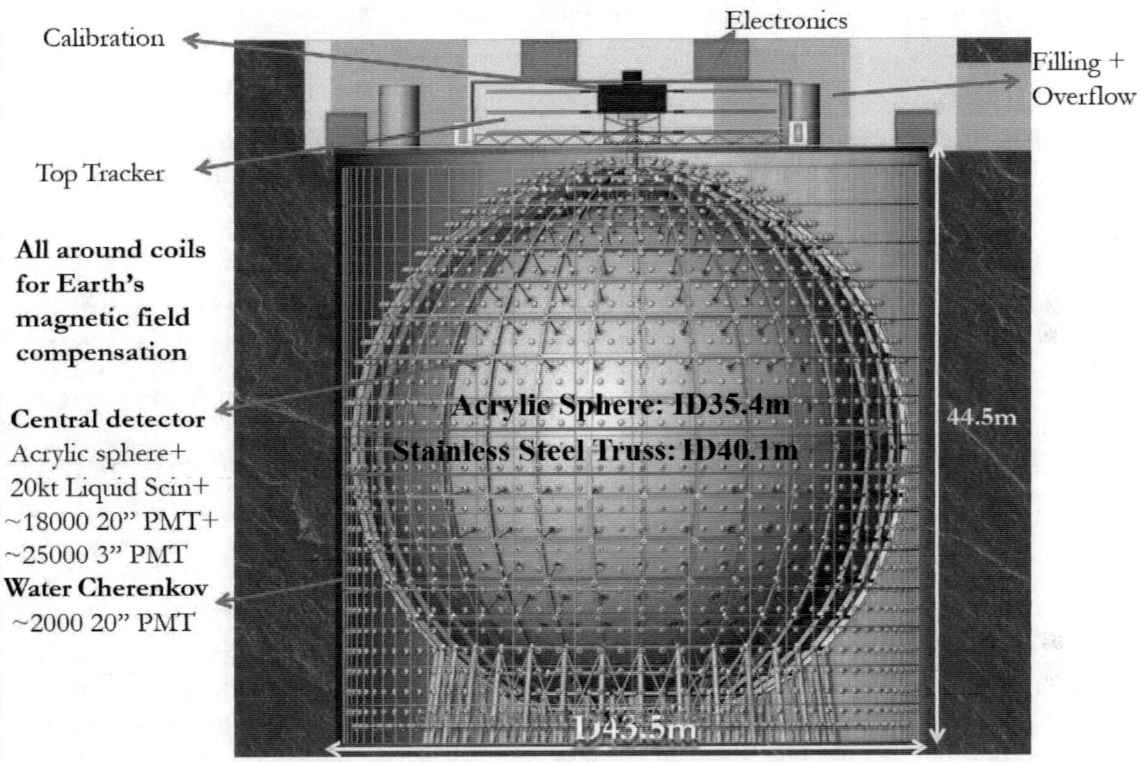

Figure 2. Schematic view of the JUNO configuration

The vessel will be surrounded by a stainless steel truss which will perform the twofold task to sustain the vessel, by relieving its internal stress, and to provide the anchor support for the 18000 20" photomultipliers observing the scintillation photons. The light detection system will comprise also an additional set of 3" PMTs, up to 25000, which will be used mainly for calibration purpose (the additional coverage is minimal) and to cross check the performances of the main PMTs, with the scope to control and reduce the systematic effects of the measurements performed by the main 20" PMT system.

Moreover, the shielding water around the acrylic vessel will be used as a Cherenkov detector, being instrumented with about 2000 phototubes, which will detect the muon induced Cherenkov light. Such an arrangement, together with the top tracker that will be deployed on the roof of the detector itself, will allow an efficient muon veto capability, an essential feature at the planned shallow depth of the experiment, i.e. about 700 m.

JUNO has been approved in China at the beginning of 2013 and has been later joined by groups from all over the world. Currently the Collaboration encompasses 71 institutions from Asia, Europe and America, with more than 500 researchers, and it is still expanding.

IOP Conf. Series: Journal of Physics: Conf. Series **1056** (2018) 012048 doi:10.1088/1742-6596/1056/1/012048

The experiment is located in the South of China, Guangdong province, Jianmeng County, Kaiping city, at 53 km from the two sites of Yangjian and Taishan, where 6 and 4 nuclear cores are planned, respectively. By 2020 according to the construction schedule of the plants 26.6 GW will be installed (2 cores will be missing at Taishan), while eventually the total power of 35.8 GW will be available.

4. How to infer the mass hierarchy

The observable quantity from which the mass hierarchy will be inferred is the positron spectrum detected in the liquid scintillator, stemming from the Inverse Beta Decay reaction through which antineutrino detection will occur. Specifically, the determination of the mass hierarchy relies on "inprinting" of the anti-ν_e survival probability P_{ee} on such spectrum.

The Inverse Beta Decay Reaction a là Cowan Reines is the following

$$\bar{\nu}_e + p \rightarrow e^+ + n$$

The energy deposited by the positron in the scintillator, i.e. its kinetic energy plus the total 1.022 keV energy of the two annihilation gammas, reflects faithfully the energy of the incoming anti-neutrinos

$$E_{vis}(e^+) = E(\nu) - 0.8\,MeV$$

$E_{vis}(e^+)$ is, thus, the specific measurement output to be analysed for the hierarchy evaluation.

The time coincidence (mean difference of the order of 250 μs) between the positron event and the γ ray from the subsequent neutron capture on protons allows to identify effectively the occurrence of neutrino detection and to pick up the positron scintillation signal, even in presence of uncorrelated background.

It can be shown that through suitable approximations the survival probability P_{ee} can be written as:

$$P_{ee} = 1 - cos^4\theta_{13} sin^2 2\theta_{12} sin^2(\Delta_{21}) - sin^2\theta_{13} sin^2(|\Delta_{31}|) \\ - sin^2\theta_{12} sin^2 2\theta_{13} sin^2(\Delta_{21}) cos(2|\Delta_{31}|) \pm \frac{sin^2\theta_{12}}{2} sin^2 2\theta_{13} sin(2\Delta_{21}) sin(2|\Delta_{31}|)$$

The sign flip in front of the last term is due to the hierarchy: positive for direct hierarchy, negative for the inverse one. The presence of the multiplicative factor $sin^2 2\theta_{13}$ in this term questioned the effectiveness of this methodology, proposed for the first time in [8], until the experimental determination of the θ_{13} (about 8°) by Daya Bay, Reno and Double Chooz. Indeed, with θ_{13} very close to 0, the last term of the P_{ee} expression would have been vanishing small, making the proposed approach unfeasible. Since instead θ_{13} is actually almost coincident to the previous Chooz limit [9] and thus sizably different from 0, the method is realizable in practice.

The effect of P_{ee} on the reactor spectrum is shown in Figure 3; the y axis is proportional to the event rate, while on the x axis the ratio L/E_ν is reported. The dashed line is the un-oscillated spectrum; the continuous black line is the spectrum distorted and suppressed as an effect of the "solar" oscillation: this large effect is the key for the very precise determination of the two "solar" mixing parameters Δm^2_{21} and $sin^2\theta_{12}$.

The blue and red lines superimposed on the smooth black line, instead, display the effect of the interference term driven by the atmospheric mass squared difference. The frequency of the ripple depends on $|\Delta m^2_{31}|$ (which therefore can also be determined with high accuracy from the precise "tracking" of the ripple itself), while its phase is linked to the MH, as shown by the reciprocal shift of the blue and red lines in the figure. Unraveling the phase of the ripple, hence, is the clue for the MH determination.

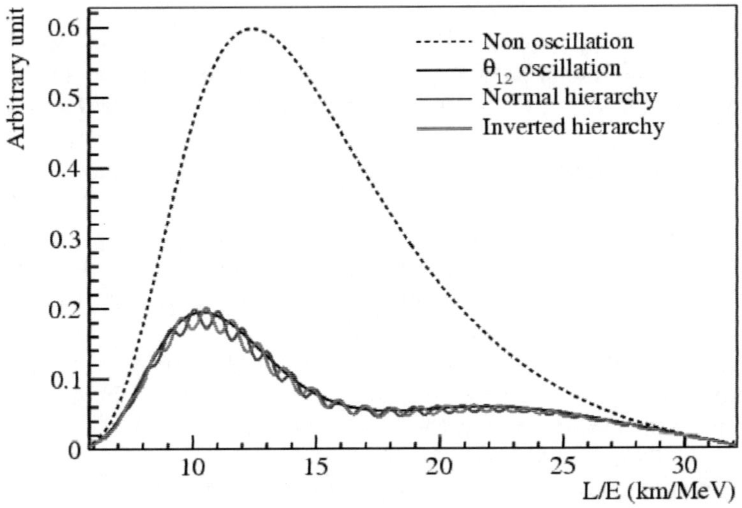

Figure 3. Effect of P_{ee} electron neutrino survival probability on the reactor spectrum

However essential prerequisite to accomplish meaningfully the hierarchy determination is the energy resolution: while in an ideal antineutrino spectrum computed for infinite resolution throats and peaks induced by the electron neutrino survival probability are perfectly visible, in a concrete measured spectrum these features are significantly smoothed as effect of the detector resolution.

For example, below 3% at 1 MeV of energy resolution, throats and peaks in the measured spectrum are still distinguishable and consequently the discrimination between the two hierarchies feasible. If, instead, the same measure would be performed with 5 or 6% resolution, which are the values of the state of the art liquid scintillator technology of large scale experiments, peaks and throats would completely disappear, making any attempt to unravel the hierarchy impossible. From several numerical evaluations it stems that the 3% energy resolution considered in this example is just the limit value above which the hierarchy discrimination can be meaningfully performed and is therefore assumed as the design goal of the experiment, representing by far its greatest challenge in term of improvement over past experiences.

Realistic χ^2 calculations performed with the input parameters related to JUNO detector and site (i.e. baseline 53 km, fiducial volume 20 kt, thermal reactor power 36 GW, exposure time 6 years, proton content 12%, energy resolution 3%) indicate that the statistical discrimination power of the experiment amounts to a $\Delta\chi^2$ equal to 16 between the true and wrong hierarchy hypothesis.

However, if systematic effects are considered, there is unavoidably a loss of discrimination power. The most important effect in this sense is the non-exactly equal baselines from the nuclear cores to the experiment, characterized by a spread of about 500 m, which is responsible for a loss of 4 of the $\Delta\chi^2$ indicator. Other adverse effects detrimental to the discovery power are the 1% shape uncertainty of the reactor spectrum, and the background uncertainty (rate 4.5%, shape 0.35%). All in all, these effects bring the discrimination power down to $\Delta\chi^2 = 10.4$.

On the other hand, a recovery of this loss can be obtained including in the χ^2 analysis the a-priori information on the value of the atmospheric mass square difference, which will be available from the LBL experiments with 1% precision at the time of the JUNO data release. Such an information can be best incorporated through the effective mass square difference, $\Delta m^2_{\mu\mu}$, as defined in [10]. The net effect of this analysis approach is to recover a value of about 8 in the $\Delta\chi^2$, bringing back the discovery power of the experiment to the $\Delta\chi^2$=16 realm.

5. Precision measurement of oscillation parameters and other physics reaches

The huge effect of the survival probability on the reactor spectrum and the large amount of data that will be accumulated (JUNO plans to record 100000 events in 6 years of data taking) make it possible to measure three of the mass-mixing parameters with unprecedented sub-percent precision: the two solar parameters Δm^2_{21} and $sin^2\theta_{12}$, and the effective parameter Δm^2_{ee} defined as (see [10]) $\Delta m^2_{ee} = cos^2\theta_{12}\Delta m^2_{31} + sin^2\theta_{12}\Delta m^2_{32}$.

From pure statistical considerations, the uncertainty of the measurements of these parameters is predicted to be very limited, i.e. 0.54%, 0.24% and 0.27% for $sin^2\theta_{12}$, Δm^2_{21} and Δm^2_{ee} respectively, taking also into account the correlation among them. Even adding background and several systematic effects, like the spectrum bin to bin uncertainty, 1%, the uncertainty on the absolute energy scale, 1%, and the energy non linearity, again 1%, the overall errors on the parameters remain below the 1% target: 0.67% for $sin^2\theta_{12}$, 0.59% for Δm^2_{21}, and 0.44% for Δm^2_{ee}.

For lack of space, I do not report here the results of many studies carried out to assess quantitatively the JUNO capabilities with respect to the several chapters of its vast astroparticle program, which are all illustrated in details in the already cited yellow book [7].

6. JUNO progress and schedule

The experiment is expected to start data taking at the beginning of the next decade. The ground breaking signaling the startup of the excavation occurred in January 2015. So far, the slope tunnel (1266 m) and the vertical shaft (564 m) have been fully excavated. The former will allow to bring the scintillator underground, the latter will enable access of personnel and construction materials. The excavation is now progressing towards the location where the experimental hall will be realized.

The civil construction is foreseen to be completed by about middle of 2019. The preparation of the detector components, e.g. phototubes, acrylic panels, etc., has started in the past year 2016 and will encompass the whole 2017 and 2018, and part of 2019, while the global onsite installation will be completed by the end of the current decade. All this is in line to ensure scintillator fill and startup of data taking at the beginning of the next decade.

7. Conclusion

The vast potential physics reach of very large liquid scintillator detectors, mass hierarchy determination and beyond, is the foundational motivation of JUNO conceived and planned to mark significant breakthroughs for the ultimate quest of the neutrino properties.

The Collaboration is rapidly progressing toward the construction of the detector with all the important design decisions already taken, the prototyping phase marking important steps forwards for all the subsystems and with the excavation of the site going ahead.

The JUNO exciting science program will start at the dawn of the next decade, when the experiment will be filled.

References

[1] Adam T et al. (JUNO Collaboration) 2016 *The JUNO Conceptual Design Report* arXiv:1508.07166
[2] Alimonti G et al. (Borexino Collaboration) 2009 *Nucl. Instr. and Meth and Meth. A* vol 600, p 568
[3] Abe S et al (KamLAND Collaboration) 2008 *Phys. Rev. Lett.* vol 100, p 221803
[4] An F P et al. (Daya Bay Collaboration) 2012 *Phys. Rev. Lett.* vol 108, p 171803
[5] Ahn J K et al., (RENO Collaboration) 2012 *Phys. Rev. Lett.* vol 108, p 191802
[6] Abe S et al., (DOUBLE CHOOZ Collaboration) 2013 *Phys. Lett. B* vol 723, p 66
[7] An F P et al. (JUNO Collaboration) 2016 *J. of Phys. G (Nucl. and Part. Phys.)* vol 43, p 030401
[8] Petcov S T and Piai M 2002 *Physics Letters B* vol 553 pp 94-106
[9] Apollonio M et al. (Chooz Collaboration) 2003 *The European Physical Journal C* vol 27, pp 331-374
[10] Nunokawa H, Parke S, Zukanovich Funchal R 2005, *Phys.Rev.D* vol 72, p 013009

Nuclear structure studies relevant to ^{136}Xe $\beta\beta$ decay

B M Rebeiro[1], S Triambak[1,2], P E Garrett[3], R Lindsay[1], P Adsley[2,4], G C Ball[5], V Bildstein[3], C Burbadge[3], A Diaz-Varela[3], T Faestermann[6], R Hertenberger[7], B Jigmeddorj[3], M Kamil[1], K G Leach[8], P Z Mabika[1,9], J C Nzobadila[1], J N Orce[1], A J Radich[3], E Rand[3] and H-F Wirth[7]

[1] Department of Physics & Astronomy, University of the Western Cape, P/B X17, Bellville ZA-7535, South Africa.
[2] iThemba Laboratory for Accelerator Based Sciences, P.O. Box 722, Somerset West 7129, South Africa.
[3] Department of Physics, University of Guelph, Guelph, Ontario N1G 2W1, Canada.
[4] Department of Physics, University of Stellenbosch, Private Bag X1, 7602 Matieland, South Africa.
[5] TRIUMF, 4004 Wesbrook Mall, Vancouver, British Columbia V6T 2A3, Canada.
[6] Physik Department, Technische Universität München, D-85748 Garching, Germany.
[7] Fakultät für Physik, Ludwig-Maximilians-Universität München, D-85748 Garching, Germany.
[8] Department of Physics, Colorado School of Mines, Golden, Colorado 80401, USA
[9] Department of Physics, University of Zululand, Private Bag X1001, Kwadlangezwa 3886, South Africa.

E-mail: b.rebeiro@gmail.com

Abstract. In these proceedings we briefly discuss preliminary results from ^{138}Ba(d,α) and ^{138}Ba(p,t) reactions performed using the Q3D magentic spectrometer at the Maier-Leibnitz-Laboratorium (MLL) tandem accelerator facility in Garching, Germany. Our results aim to provide useful spectroscopic information for the calculation of the ^{136}Xe $\rightarrow^{136}$ Ba double beta decay matrix elements.

1. Introduction

The existence of massive neutrinos open up the possibility of observing neutrinoless double beta ($0\nu\beta\beta$) decays, which are lepton number violating (LNV) processes that would affirm the Majorana nature of neutrinos. Unlike the standard model allowed two-neutrino double beta ($2\nu\beta\beta$) decay process, the (yet-to-be-confirmed) detection of the exotic neutrinoless double beta ($0\nu\beta\beta$) decay mode would require a paradigm shift in our understanding of neutrino properties. The $0\nu\beta\beta$ decay rate for an atomic nucleus can be written as

$$\Gamma^{0\nu} = G^{0\nu}(Q,Z)|\sum_i \eta_i^2 M_i^{0\nu}|^2, \tag{1}$$

where $G^{0\nu}(Q,Z)$ is a phase space factor, $M_i^{0\nu}$ are the nuclear matrix elements (NMEs) for the decay and η_i are the LNV particle physics parameters that depend on the mechanism driving the

IOP Conf. Series: Journal of Physics: Conf. Series **1056** (2018) 012049 doi:10.1088/1742-6596/1056/1/012049

process. If the dominant mechanism responsible for the decay is the exchange of light left-handed Majorana neutrinos, the amplitude for the decay reduces to just one term,

$$\eta_\nu M_\nu^{0\nu} = \frac{1}{m_e}\left(\sum_{k=1}^{3} m_k U_{ek}^2\right) M^{0\nu} = \frac{m_{\beta\beta}}{m_e} M^{0\nu}. \tag{2}$$

In the above equation, the U_{ek} are elements of the Pontecorvo-Maki-Nakagawa-Sakata (PMNS) neutrino mixing matrix, m_e is the electron mass and $m_{\beta\beta}$ represents the effective Majorana mass of the electron neutrino. In such a scenario, a measured $0\nu\beta\beta$ decay half-life would also allow a determination of the absolute neutrino mass scale. However, this requires accurate values of the NMEs, which can only be obtained using nuclear structure dependent theoretical calculations. It is now well established that the matrix elements calculated for various isotopes, using different theoretical techniques disagree with one another at a significant level [1, 2]. Since next generation $0\nu\beta\beta$ decay experiments would delve into more interesting regions of $m_{\beta\beta}$ parameter space[1], it is important at this stage to make every possible effort to reduce the uncertainties in calculated NMEs for various $0\nu\beta\beta$ decay candidates.

This work relates to NME calculations for the decay of ^{136}Xe, which is a promising candidate in several ongoing $0\nu\beta\beta$ decay experiments [3, 4, 5]. We focus on two aspects of the calculations.

(i) Going beyond the closure approximation, which is explained below.

(ii) Testing the Bardeen-Cooper-Schiffer (BCS) approximation for pairing between neutrons in the daughter (^{136}Ba) nucleus.

The NMEs for $0\nu\beta\beta$ decays are calculated as a sum of virtual transitions over states up to high excitations in the intermediate odd-odd nucleus [1, 6]. This makes the calculations computationally expensive and challenging [1], particularly when using the shell model, where one is limited by the dimensionality of the valence space used. To bypass this problem, most calculations have resorted to the closure approximation [1, 7, 8], wherein the energies of the individual states in the intermediate nucleus are replaced by an average energy $\langle E\rangle$. It is now well known that the closure approximation introduces an $\sim$ 10% uncertainty in the final results [7, 9]. Recent large-scale shell model calculations have tried to address this issue by evaluating NMEs beyond the closure approximation [7, 9, 10], using knowledge of intermediate nuclear states. Compared to the shell model and other calculations (such as the generator coordinate method, projected Hartree-Fock-Bogoliubov method, etc.), the quasiparticle random-phase approximation (QRPA) approach to calculate NMEs seems to be least sensitive to the closure approximation [11]. However, the QRPA calculations assume the ground states in the parent and daughter nuclei to be BCS condensates of proton and neutron pairs. It is known that the BCS approximation for pairing correlations breaks down in nuclei due to changes in deformation or when there is a large gap in the single particle states, such as near a shell closure [12]. Good experimental probes for studying pair-correlations in even-even nuclei are pair-transfer processes such as (p,t) or $(^3\mathrm{He},n)$ reactions [13, 14, 15]. If all the $L=0$ pair-transfer strength in these reactions will proceed to the ground states of the nuclei under study, this would validate the BCS approximation.

In light of the above, any experimental knowledge of excited states in the intermediate nuclei and pairing properties of the ground state wavefunctions would be important to place $0\nu\beta\beta$ decay matrix element calculations on a more secure footing. Below we briefly discuss preliminary results from $^{138}\mathrm{Ba}(d,\alpha)^{136}\mathrm{Cs}$ and $^{138}\mathrm{Ba}(p,t)^{136}\mathrm{Ba}$ reactions that would be relevant for future matrix element calculations of $^{136}\mathrm{Xe} \rightarrow {}^{136}\mathrm{Ba}$ $\beta\beta$ decays.

[1] For example, if neutrinos were Majorana particles and have an inverted mass hierarchy, a successful experiment would be imminent if $m_{\beta\beta} \approx 10$ meV.

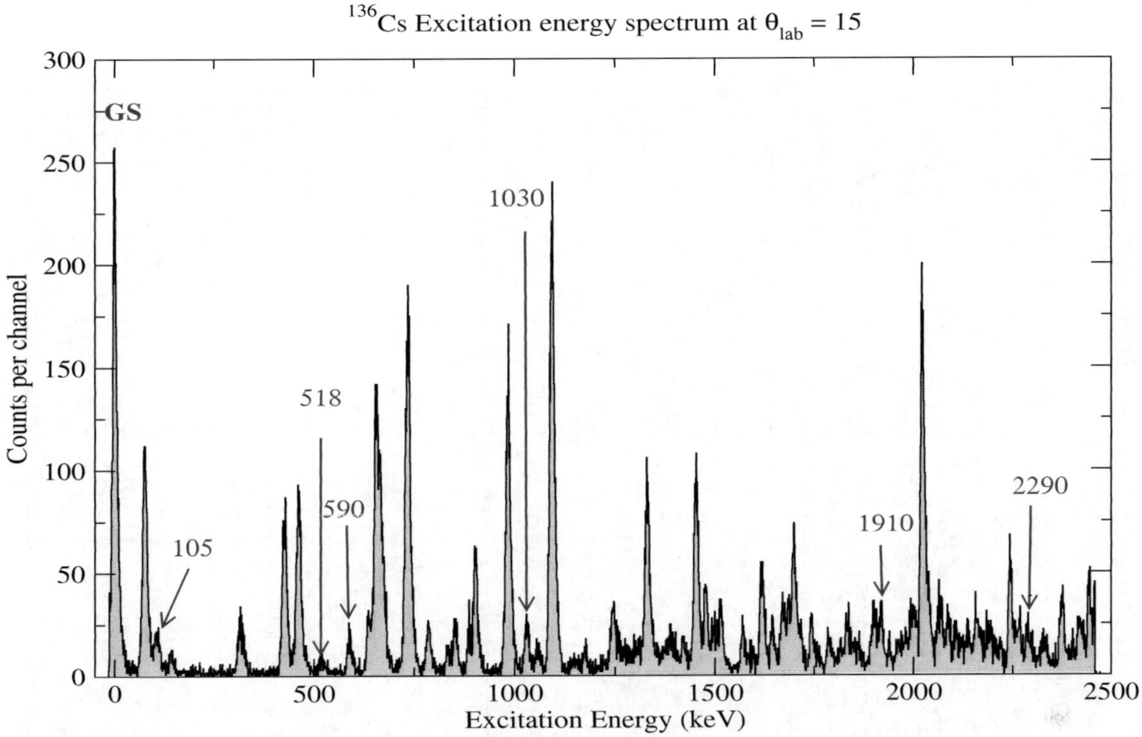

Figure 1. Preliminary excitation energy spectrum for ^{136}Cs from the ^{138}Ba(d,α) reaction. The focal plane spectrum was energy calibrated using well known states in ^{92}Nb from ^{94}Mo(d,α). In this figure we only label those states that have been observed previously. To the best of our knowledge, all the other states have been observed for the first time.

2. Experimental details

The experiments were performed at the tandem accelerator facility at Maier-Leibnitz-Laboratorium (MLL) in Garching (Germany). For the measurements we collected triton and alpha particle spectra at laboratory scattering angles of 5°-50° and 5°-45° respectively, using the high resolution Q3D magnetic spectrometer [16]. The focal plane detector [17] of the spectrometer comprised of two proportional counters filled with isobutane gas at $\sim$ 500 mbar (to measure the partial energy loss of the ejectiles) and a 7 mm thick plastic scintillator to completely stop them. The second proportional counter is coupled to a cathode-strip foil that determined the position of the light ejectiles with high resolution. The charged ejectiles were identified from the energy losses in the proportional counters $(\Delta E_1, \Delta E)$ and the total energy deposited in the plastic scintillator (E). Schematic views of the Q3D magnetic spectrometer, focal plane detector and target/beam information are listed in Ref. [8]. Well known states in ^{92}Nb from ^{94}Mo$(d,\alpha)^{92}$Nb reactions on 100 μg/cm^2 thick ^{94}MoO$_3$ target were used to calibrate the ^{138}Ba(d,α) spectrum, while the (p,t) spectra were internally calibrated using well known states in ^{136}Ba.

3. Preliminary results

Fig. 1 shows a calibrated focal plane spectrum with the α peaks corresponding to states in ^{136}Cs. We identify more than 50 new states up to approximately $E_x = 2.4$ MeV. This is because our reaction offers a different selectivity compared to previous recent work [18] that strongly favored the population of 1^+ states. More explicitly, the (d,α) reaction can be safely approximated as a single-step stripping of a 'deuteron' [19], so that the transfered proton-neutron pair *only* couple to $S=1$ and $T=0$ [19]. Furthermore, the large (positive) Q-value for the (d,α) reaction

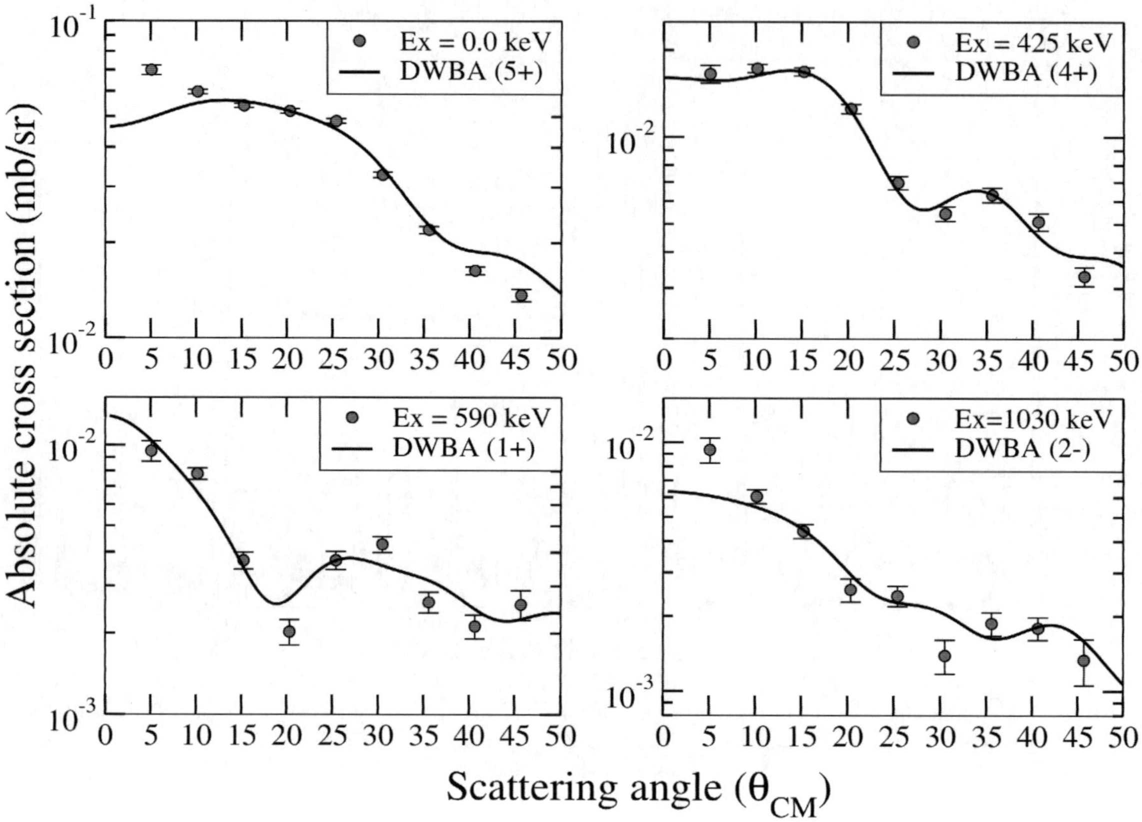

Figure 2. Comparison of DWBA calculations with measured ^{138}Ba$(d,\alpha)^{136}$Cs angular distributions.

allows for larger L values to be transfered, favoring transitions to states with (reasonably) higher angular momentum [20].

The measured angular distributions from ^{138}Ba(d,α) were compared to predictions from Distorted Wave Born Approximation (DWBA) calculations [21]. The global optical model potential (OMP) parameters best suited for the DWBA analysis were deduced from ^{138}Ba(d,d) elastic scattering data for the incoming channel [22] and from McFadden and Satchler [23] for the outgoing $(\alpha +^{136}$ Cs) channel.

Since both natural and unnatural parity states are populated by the (d,α) reaction, the total angular momentum of the final states satisfy the conditions $J = L$ and $J = L \pm 1$. Using these and the chosen OMP parameters, we obtained the DWBA predictions for the angular distributions, which were normalized to the data using a least squares minimization routine, so that

$$\left(\frac{d\sigma}{d\Omega}\right)_{\rm expt} = \alpha \left(\frac{d\sigma}{d\Omega}\right)_{\rm DWBA:J=L} \qquad (3)$$

for natural parity states and

$$\left(\frac{d\sigma}{d\Omega}\right)_{\rm expt} = \beta \left(\frac{d\sigma}{d\Omega}\right)_{\rm DWBA:J=L+1} + \gamma \left(\frac{d\sigma}{d\Omega}\right)_{\rm DWBA:J=L-1} \qquad (4)$$

for the unnatural parity states, where α, β and γ are relative normalizations.

IOP Conf. Series: Journal of Physics: Conf. Series **1056** (2018) 012049 doi:10.1088/1742-6596/1056/1/012049

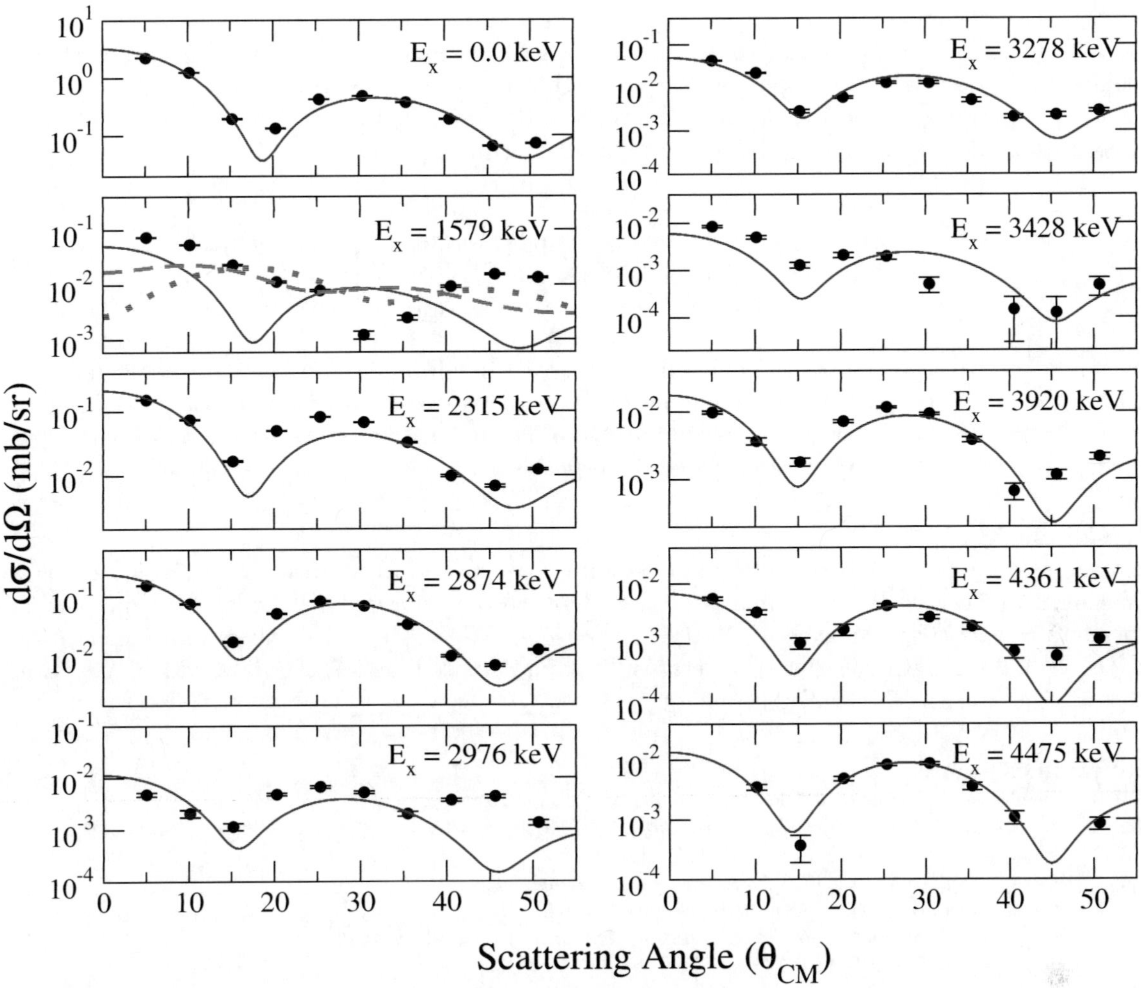

Figure 3. 0^+ states populated via the ^{138}Ba$(p,t)^{136}$Ba reaction. The measured cross-sections are compared to normalized DWBA curves. The dashed red line for the $Ex = 1579$ state is the DWBA curve for an $L = 1$ transfer, while the dotted green line represents a $L = 2$ transfer. The solid blue line in all the plots is for $L = 0$.

As examples, we show normalized DWBA angular distributions for a few states in ^{136}Cs in Fig. 2. The ground state of ^{136}Cs is known to be $J^\pi = 5^+$. Furthermore, the 590 keV state was determined to have $J^\pi = 1^+$ from a previous measurement [18], while the 1030 keV state is known to be 2^- [24]. As shown in the figure, the extraction of the spins and parities of these states using our analysis is in reasonable agreement with previous determinations. Fig. 2 also shows that we determine $J^\pi = 4^+$ for a previously unknown state at $E_x = 425$ keV. A final analysis for all other observed states shown in Fig. 1 is currently in progress.

Similar to the (d, α) analysis, we identified 0^+ states in ^{136}Ba from the (p,t) reaction by comparing the measured angular distributions with DWBA predictions. For these data we used the proton and triton global OMP sets from Ref. [25] and Ref. [26] respectively. Fig. 3 shows our measured angular distributions for the three excited 0^+ states (at $E_x = 1579$, 2315 and 2784 keV) that were previously known from the literature [27], as well as six additional 0^+ states newly identified from this experiment. However, it must be noted that our data do not confirm the

previous assignment of $J^{\pi} = 0^{+}$ for the 1579 keV state. As the figure shows, the DWBA predictions for $L = 0$, $L = 1$ and $L = 2$ transfer largely disagree with the data. We are presently looking into potential systematic effects that might be causing this discrepancy.
In order to study neutron pairing properties in ^{136}Ba, we calculate the (p,t) transfer strength to excited 0^{+} states relative to the ground state using the equation,

$$\left(\frac{d\sigma}{d\Omega}\right)_{\rm Rel} = \left(\frac{\left(\frac{d\sigma}{d\Omega}\right)^{\rm Lab}_{0^{+}_{\rm ex}}}{\left(\frac{d\sigma}{d\Omega}\right)^{\rm DWBA}_{0^{+}_{\rm ex}}}\right)\left(\frac{\left(\frac{d\sigma}{d\Omega}\right)^{\rm Lab}_{0^{+}_{\rm gs}}}{\left(\frac{d\sigma}{d\Omega}\right)^{\rm DWBA}_{0^{+}_{\rm gs}}}\right)^{-1}. \quad (5)$$

A significant cross section to the excited 0^{+} states at small angles (for $L = 0$) would indicate a deviation from the BCS-like behaviour for neutron pair-correlations in ^{136}Ba. Preliminary results from our data show that approximately 10% of the ground state strength is fragmented to the 0^{+}_{2} and 0^{+}_{3} states at 2.3 and 2.8 MeV respectively.

4. Conclusions

In conclusion, we have performed a high resolution spectroscopy of states in the odd-odd ^{136}Cs nucleus using the ^{138}Ba(d,α) reaction. We have also performed a test of neutron pairing correlations in ^{136}Ba using the ^{138}Ba(p,t) reaction. We observe a significant fragmentation of the $L = 0$ strength to excited 0^{+} states. It is anticipated that these data will be useful in constraining future NME calculations for ^{136}Xe $0\nu\beta\beta$ decays.

References

[1] Engel J and Menéndez J 2017 *Rep. Prog. Phys.* **80** 046301
[2] Vogel P 2012 *J. Phys. G: Nucl. Part. Phys.* **39** 124002
[3] Ackerman N 2011 *Phys. Rev. Lett.* **107** 212501
[4] KamLAND-Zen Collab 2013 *Phys. Rev. Lett.* **110** 062502
[5] XMASS Collab 2013 *J. Phys.: Conf. Series* **120** 042022
[6] Avignone FT, Elliott SR and Engel J 2008 *Rev. Mod. Phys.* **80** 481–516
[7] Horoi M and Stoica S 2010 *Phys. Rev.* C **81** 024321
[8] Rebeiro B *et al.* 2016 *Jour. Phys: Conf. Series* **689** 012026
[9] Sen'kov RA, Horoi M and Brown BA 2014 *Phys. Rev.* C **89** 054304
[10] Sen'kov RA and Horoi M 2013 *Phys. Rev.* C **88** 064312
[11] Šimkovic F, Hodák R, Faessler A and Vogel P 2011 *Phys. Rev.* C **83** 015502
[12] Freeman SJ and Schiffer JP 2012 *Jour. of Phys. G: Nucl. and Part. Phys.* **39** 12, 124004
[13] Yoshida S 1962 *Nucl. Phys.* **33** 685
[14] Bes DR and Broglia RA 1966 *Nucl. Phys. A* **80** 289
[15] Shihab-Eldin AA *et al.* 1995 *Int. J. Mod. Phys. E* **4** 411
[16] Löffler L *et al.* 1973 *Nucl. Instrum. Methods* **111** 1
[17] Wirth H-F *et al.* 2000. New Q3D focal plane detector with cathode-strip readout became operational, Maier-Leibnitz-Laboratorium Jahresbericht, p.71
[18] Puppe P *et al.* 2011 *Phy. Rev.* C **84** 051305(R)
[19] Glendenning NK 1965 *Phy. Rev.* B **137** 1
[20] Rivet E, Pehl RH, Cerny J and Harvey BG 1966 *Phy. Rev.* **141** 3
[21] Kunz PD 1978 DWUCK4 DWBA Program, University of Colorado, Unpublished
[22] An H and Cai C 2006 *Phy. Rev.* C **73** 054605
[23] McFadden L and Satchler GR 1966 *Nucl. Phys.* **84** 177–200
[24] Frekers D, Alanssari M, Ejiri H, Holl M, Poves A and Suhonen J 2017 *Phys. Rev. C* **95** 034619
[25] Varner RL 1991 *Phys. Rpts* **201** No.2 57–119
[26] Li X, Liang C and Cai C 2007, *Nucl. Phys.* A **789** 103–13
[27] Sonzogni AA 2002 *Nuclear Data Sheets* **95** 837

IOP Conf. Series: Journal of Physics: Conf. Series **1056** (2018) 012050 doi:10.1088/1742-6596/1056/1/012050

Solar neutrino analysis with the Borexino detector

Mariia Redchuk *

Institut für Kernphysik, Forschungszentrum Jülich, 52428 Jülich, Germany

E-mail: m.redchuk@fz-juelich.de

*** on behalf of the Borexino Collaboration:**

M. Agostini, K. Altenmüller, S. Appel, V. Atroshchenko, Z. Bagdasarian, D. Basilico, G. Bellini, J. Benziger, D. Bick, G. Bonfini, D. Bravo, B. Caccianiga, F. Calaprice, A. Caminata, S. Caprioli, M. Carlini, P. Cavalcante, A. Chepurnov, K. Choi, L. Collica, D. DAngelo, S. Davini, A. Derbin, X. F. Ding, A. Di Ludovico, L. Di Noto, I. Drachnev, K. Fomenko, A. Formozov, D. Franco, F. Froborg, F. Gabriele, C. Galbiati, C. Ghiano, M. Giammarchi, A. Goretti, M. Gromov, D. Guffanti, C. Hagner, T. Houdy, E. Hungerford, Aldo. Ianni, Andrea. Ianni, A. Jany, D. Jeschke, V. Kobychev, D. Korablev, G. Korga, D. Kryn, M. Laubenstein, E. Litvinovich, F. Lombardi, P. Lombardi, L. Ludhova, G. Lukyanchenko, L. Lukyanchenko, I. Machulin, G. Manuzio, S. Marcocci, J. Martyn, E. Meroni, M. Meyer, L. Miramonti, M. Misiaszek, V. Muratova, B. Neumair, L. Oberauer, B. Opitz, V. Orekhov, F. Ortica, M. Pallavicini, L. Papp, O. Penek, N. Pilipenko, A. Pocar, A. Porcelli, G. Ranucci, A. Razeto, A. Re, M. Redchuk, A. Romani, R. Roncin, N. Rossi, S. Schonert, D. Semenov, M. Skorokhvatov, O. Smirnov, A. Sotnikov, L. F. F. Stokes, Y. Suvorov, R. Tartaglia, J. Thurn, M. Toropova, E. Unzhakov, A. Vishneva, R. B. Vogelaar, F. von. Feilitzsch, H. Wang, S. Weinz, M. Wojcik, M. Wurm, Z. Yokley, O. Zaimidoroga, S. Zavatarelli, K. Zuber, and G. Zuzel.

Abstract. The first simultaneous measurement of the interaction rates of *pp*, ^{7}Be, and *pep* solar neutrinos has been completed using the Borexino data and will soon be published. The results were obtained by performing a global fit to the Borexino Phase-II data in an extended energy range (0.186-2.97 MeV). The scope of Phase-II extends over 1291.51 days between December 2011 and May 2016 after an extensive scintillator purification campaign. The updated results are in agreement with the previous ones and have improved accuracy. Here we will highlight some details about the data analysis on Phase-II in the solar neutrino spectroscopy with Borexino.

1. Motivation

The Sun is powered by nuclear fusion reactions that burn hydrogen into helium (Fig. 1). Neutrinos that are the products of these reactions are the only carriers of information about the processes in the *core* of the Sun. The resulting spectra of the solar neutrinos are shown in Fig. 2 in units of $cm^{-2}s^{-1}$ MeV^{-1} for continuous spectra and $cm^{-2}s^{-1}$ for monoenergetic lines.

While the spectral shapes are defined by the fusion reactions, the total flux depends on the composition of the Sun, in particular the fraction of the heavy elements. Standard Solar Models (SSMs) are divided into two classes, low and high metallicity, depending on the fraction of elements heavier than helium. Current knowledge about these models does not provide enough information to determine the metallicity of the Sun. [1]

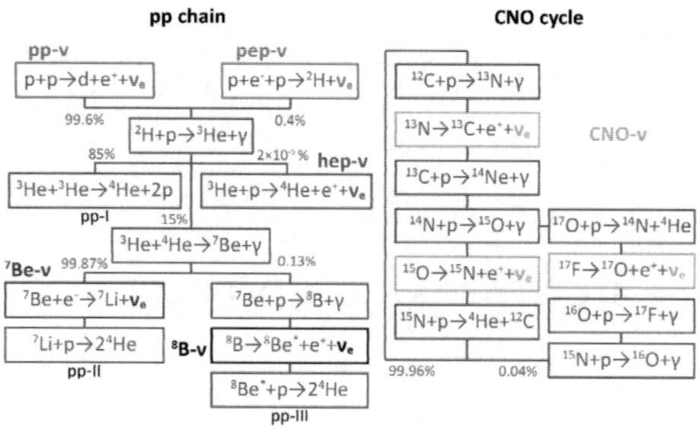

Figure 1: The *pp* chain and the CNO cycle.

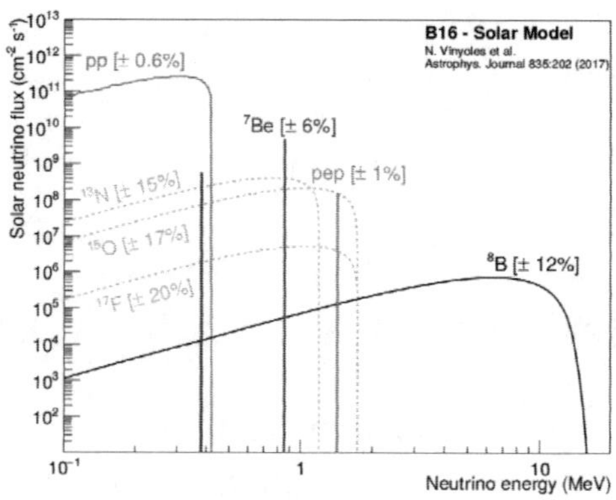

Figure 2: The spectra of the solar ν-s.

By measuring the solar neutrino fluxes we are able to study the reactions in the *pp* chain (see Fig. 1). Furthermore, using the knowledge about the fluxes and assuming a certain solar model, we test the neutrino oscillations theory; while assuming the MSW-LMA theory [2] of neutrino oscillations we test SSMs. The flux of CNO neutrinos can help us solve the metallicity puzzle described above.

2. The Borexino detector

Borexino is a liquid scintillator detector the primary goal of which is measuring the fluxes of solar ν-s [3]. It is located in the Laboratori Nazionali del Gran Sasso in the mountains of Italy at 3800 m water-equivalent depth. In 2012 Borexino started Phase-II of data taking which is characterized by its higher sensitivity. This was made possible due to extensive purification campaigns in 2010 and 2011 after which the already unprecedentedly low radioactive background of the detector was improved even more.

The schematic representation of the detector is shown in Fig. 3. It is equipped with 2200 photomultiplier tubes (PMTs) mounted on the Stainless Steel Sphere (SSS) that point inwards into the Inner Detector (ID), which contains the scintillator; and 220 PMTs pointing outwards in the Outer Detector which serves the purpose of a Cherenkov muon veto. Apart from that, the ID contains two nylon spheres which protect the scintillator from radon diffusion from the outside. [4].

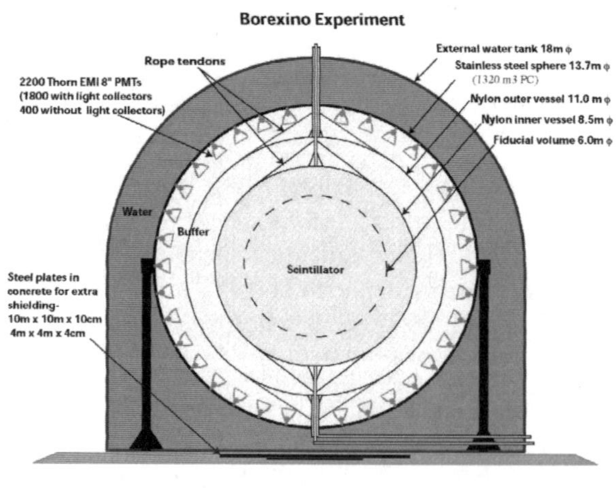

Figure 3: A schematic representation of the Borexino detector.

3. Analysis

The software selection applied to the acquired data from the Borexino detector is demonstrated on Fig. 4 and consists of:

- muon and muon daughter cut: removes the muons that pass through the ID and applies 300 ms dead time after the muon events
- fiducial volume cut: selects events farther from the edges of the SSS in order to remove background from the nylon spheres, the PMTs and the SSS.
- the three-fold coincidence (TFC) cut: selects exposure in which cosmogenic ^{11}C events (Eq. 1) are expected

The TFC cut finds coincidences of a muon, a neutron after $\sim 250\mu s$ (average time for neutron capture, Eq. 2) and a positron after 29.4 mins (^{11}C decay, Eq. 3) in a cylindrical volume following the muon path in order to determine cosmogenic ^{11}C events. Applying the TFC cut, the whole exposure is divided into two sets: ~40% of exposure containing 92% of ^{11}C, and ~60% of exposure containing 8% of ^{11}C events.

$$\mu + {}^{12}\mathrm{C} \rightarrow \mu + {}^{11}\mathrm{C} + n \tag{1}$$

$$n + p \rightarrow \mathrm{D} + \gamma \tag{2}$$

$${}^{11}\mathrm{C} \rightarrow {}^{11}\mathrm{B} + e^{+} + \nu_e \tag{3}$$

To determine the rates of the background and the *pp*, ^{7}Be, and *pep* solar species, multivariate fits have been performed on the Borexino data using Monte Carlo (MC) as well as analytical approaches [5]. The resulting multivariate likelihood is the product of the four components:

$$\mathcal{L}_{\mathrm{MV}}(\theta) = \mathcal{L}_{\mathrm{TFC\text{-}tagged}}(\theta) \cdot \mathcal{L}_{\mathrm{TFC\text{-}sub}}(\theta) \cdot \mathcal{L}_{\mathrm{PS}}(\theta) \cdot \mathcal{L}_{\mathrm{RD}}(\theta), \tag{4}$$

where TFC-tagged and TFC-sub are the two energy spectra enriched and depleted in ^{11}C respectively, obtained after the TFC cut (see above); and RD and PS are the pulse-shape and the radial distributions of events respectively. The spectral shapes used to fit both of the energy spectra are shown in Fig. 5. The pulse-shape distribution is used to discriminate between electrons and positrons in order to determine the remaining ^{11}C in the ^{11}C-depleted spectrum. The radial distribution is used to distinguish external and uniform components which plays a role in fitting external backgrounds.

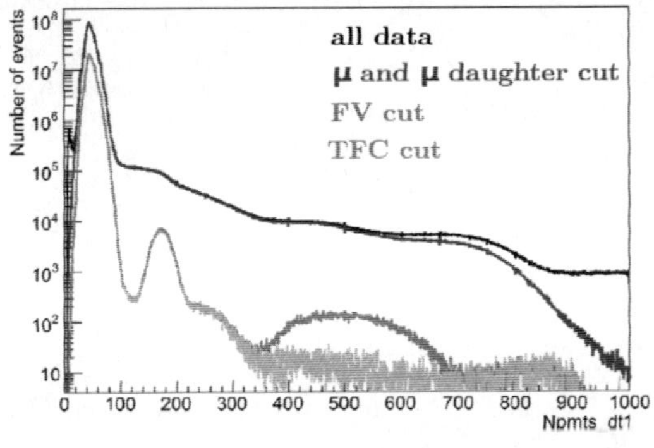

Figure 4: Data before and after cuts.

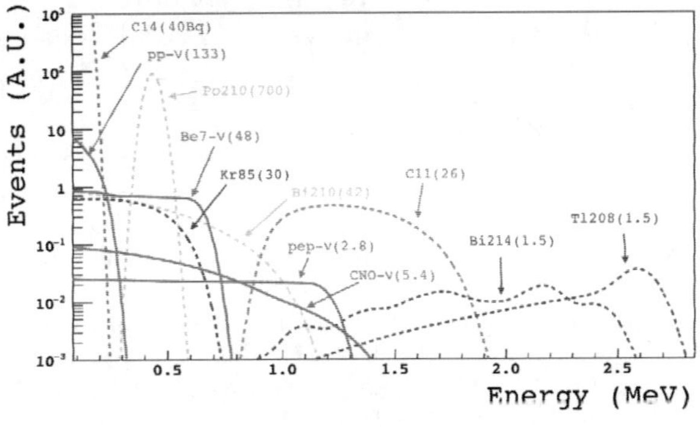

Figure 5: Spectra of recoiled electrons.

Apart from the MC method, the fit is performed using an analytical description of the detector response. In the analytical description of the energy scale a number of energy estimators is used: total number of hits (N_h), photoelectrons (N_{pe}) and triggered PMTs (N_p), as well as the number of PMTs triggered within the first 230 ns ($N_p^{dt_1}$) or 400 ns ($N_p^{dt_2}$).

The analytical model of the response describes the conversion from the deposited energy of the event to N_{pe}:

$$N_{pe}(E) = LY\left(\, Q(E)\cdot E \,+\, f_{Cher}\cdot Ch(E)\right), \tag{5}$$

where LY is the light yield, $Q(E)$ is the quenching function, f_{Cher} is a free Cherenkov tuning parameter, and $Ch(E)$ is a function describing Cherenkov radiation; and the subsequent conversion to N_p:

$$N_{\text{p}}(E) = N_{\text{live}}\left(1 - e^{-\frac{N_{\text{pe}}}{N_{\text{live}}}}\left[1 + P_t\frac{N_{\text{pe}}}{N_{\text{live}}}\right]\right)\left(1 - G_c\frac{N_{\text{pe}}}{N_{\text{live}}}\right), \tag{6}$$

where N_{live} is the average number of live PMTs, P_t is a parameter for single electron response (tuned using calibrations), and g_C is a geometrical correction coefficient (tuned with MC).

The analytical approach allows several detector response parameters to be free which gives flexibility in modelling the detector response changing in time, and enables different ways of pile-up modelling.

4. Systematic studies

In order to achieve a high precision measurement, the fit has been performed multiple times using the MC as well as analytical approach. Various sources of uncertainty have been considered such as the choice of energy estimator, pile-up modelling, energy fit range and binning, ^{85}Kr constraint, live time, scintillator density, fiducial volume, fit method (MC/analytical) and fit models. Fit models include uncertainties in the known detector response, spectra of the background and solar species based on extensive calibration campains in 2009.

5. Results and conclusions

For the first time the energy range of the fit was extended (0.186-2.97 MeV) in order to obtain information about the *pp*, ^{7}Be, and *pep* solar neutrinos simultaneously using Borexino Phase-II data (1291.51 days) [6]. The analysis yields an updated precision measurement of *pp*-ν and ^{7}Be-ν, discovery of *pep*-ν and the most stringent upper limit on CNO-ν. The measurements are consistent with Phase-I and show improved precision. The results provide valuable information which can be used to analyse SSMs and solar metallicity.

References

[1] N. Vinyoles, A. M. Serenelli, F. L. Villante, S. Basu, J. Bergström, M. C. Gonzalez-Garcia, M. Maltoni, C. Peña-Garay, and N. Song, "A new Generation of Standard Solar Models," *Astrophys. J.*, vol. 835, no. 2, p. 202, 2017.

[2] I. Esteban, M. C. Gonzalez-Garcia, M. Maltoni, I. Martinez-Soler, and T. Schwetz, "Updated fit to three neutrino mixing: exploring the accelerator-reactor complementarity," *JHEP*, vol. 01, p. 087, 2017.

[3] G. Bellini *et al.*, "Final results of Borexino Phase-I on low energy solar neutrino spectroscopy," *Phys. Rev.*, vol. D89, no. 11, p. 112007, 2014.

[4] G. Alimonti *et al.*, "The Borexino detector at the Laboratori Nazionali del Gran Sasso," *Nucl. Instrum. Meth.*, vol. A600, pp. 568–593, 2009.

[5] G. Bellini *et al.*, "Precision measurement of the 7Be solar neutrino interaction rate in Borexino," *Phys. Rev. Lett.*, vol. 107, p. 141302, 2011.

[6] M. Agostini *et al.*, "First Simultaneous Precision Spectroscopy of *pp*, ^{7}Be, and *pep* Solar Neutrinos with Borexino Phase-II," 2017.

IOP Conf. Series: Journal of Physics: Conf. Series **1056** (2018) 012051 doi:10.1088/1742-6596/1056/1/012051

New Beam lines for the NUMEN experiment at INFN-LNS

A D Russo, L Calabretta, A Calanna, G D'Agostino and D Rifuggiato for the NUMEN collaboration

INFN-LNS, Via S. Sofia 62, 95123 Catania, Italy

Corresponding autor: antonio.russo@lns.infn.it

Abstract. The NUMEN experiment at INFN-LNS demands an upgrade of the Superconducting Cyclotron (CS) to deliver beam with power up to 10 kW for ions with mass A < 40 amu. To transport these high power beams, it is mandatory to build a new extraction beam line from the CS and to upgrade the existing beam transport lines to maintain the beam losses below 100 W, both in the accelerator room and along the beam transport areas. Moreover, a new fragment in-flight separator will be built to produce radioactive ion beams, but also to perform energy selection of the beam extracted from the cyclotron, to be delivered to the MAGNEX spectrometer with an energy spread below 0.1% FWHM. The use of a 10 kW ion beam implies also to redesign of the beam transport line inside the MAGNEX experimental room to allow the installation of a new large and well shielded beam dump beyond the apparatus.

1. Introduction

The NUMEN experiment [1,2,3] needs to use mainly beams of Carbon, Oxygen and Neon with intensity up to 10^{14} pps delivered to the MAGNEX spectrometer [4,5]. The required energies for these beams are in the range 15-70 AMeV, which corresponds to a beam power in the range 1-10 kW. Due to the compactness and the relatively small extraction radius of the INFN-LNS Superconducting Cyclotron (CS) the inter-turn separation at the extraction radius stays around 1 mm. This low inter-turn separation and the features of the Electrostatic Deflectors (ED) limit the extraction efficiency to 60%. Moreover, it is convenient to maintain the beam losses inside the CS below 100 W to mitigate the activation of the machine. For these reasons the maximum beam power delivered by the CS is 140 W [6]. To overcome this limit, the extraction by stripping, just for a set of ions with A<40, was proposed [7]. Indeed, for these light ions with energies higher than 15 AMeV the stripping extraction allows to achieve extraction efficiencies higher than 99% [8], and we could accelerate and extract beams with power up to 10 kW.

Therefore, the CS will be upgraded with the installation of a new cryostat containing a new set of superconducting coils and a new extraction channel designed to allow extraction of the beam by stripping [7]. The existing extraction beam line will be also maintained to allow electrostatic extraction of all ion beams with low power.

In figure 1 the layout of the CS with the two extraction lines is shown. To connect the new extraction channel to the existing beam transport line, it is mandatory to design a new extraction line that is described in section 2.

IOP Conf. Series: Journal of Physics: Conf. Series **1056** (2018) 012051 doi:10.1088/1742-6596/1056/1/012051

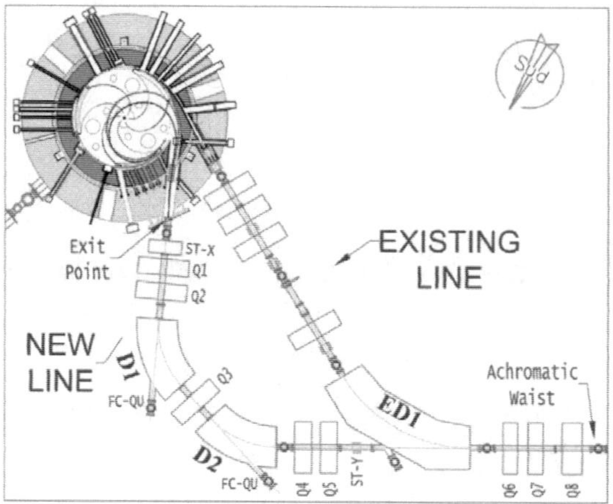

Figure 1: Layout of the Superconducting Cyclotron equipped with the existing extraction line, and the new one to deliver high power beams extracted by stripping.

The transport of a 10 kW beam along the existing beam transport lines with beam losses below 1% is challenging. To optimize the transport efficiency of the beam transport lines and at the same time to reduce the energy spread of the beam extracted by stripping to 0.1 % FWHM, we plan to build a new beam line to be used both as an energy selector and as a FRAgment In-flight SEparator (FRAISE). This new beam line is described in section 3. To install a large and safe beam dump for the 10 kW beam it is necessary to change the set-up of the beam line in the MAGNEX room as described in section 4.

2. A new extraction line for the high power beams

The difficulties to design an Extraction Line (EL) both for the electrostatic extraction and for the stripping extraction and the request of the user community to maintain the availability of all the ions, imposed to design the upgraded CS with two EL (see figure 1). The main differences between the two EL are the direction of the beam and the trajectories across the fringing field of the cyclotron. The new extraction channel crosses the fringing field in a faster way than the present one. This allows to achieve smaller transversal beam sizes along the new extraction path and moreover the values of position and angular dispersion vs. energy (element of transport matrix R_{16} and R_{26}) are lower than through the existing extraction channel [9].

Although, the tail of energy spread of the beams extracted by stripping arrives up to ±0.3%, the choice to have a new EL allows to maintain the radial beam size at values lower than 30 mm, and to mitigate the beam losses along the extraction trajectory. Moreover, the extraction channels of the new EL will be equipped with thermal shield to mitigate the effect of the beam losses due to the beam halo.

The new EL starts at the exit point shown in figure 1, and transports the beam up to the so called achromatic waist where the two extraction lines join each other. According to the layout of figure 1, the first active element of the new EL is a steering magnet, which has to adjust the beam direction in a

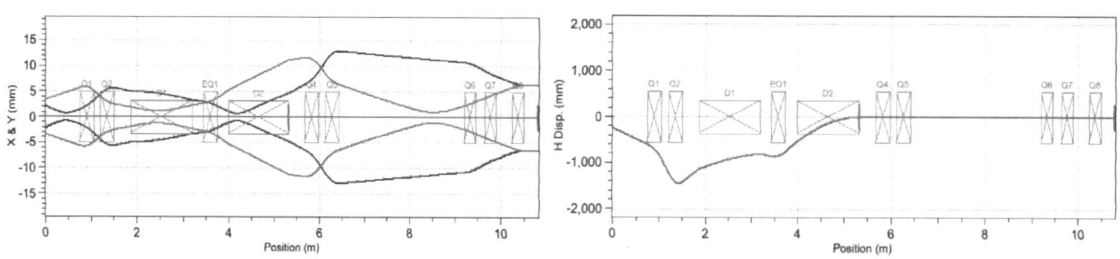

Figure 2: left side, radial (blue line) and axial (red line) beam envelope of ^{20}Ne at 71 AMeV. Right side, horizontal dispersion vs. length line.

range of ±0.5°. This steering magnet has an effective length of about 300 mm and the requested magnetic field is lower than 8 kGauss. The other magnetic elements are two quadrupoles and two symmetric bending magnets with a horizontally focusing quadrupole in between. This first part of the beam line has the role to cancel the energy-position and energy angle correlation of the beam.

After this section, two additional quadrupoles reduce the vertical beam envelope to clear the vacuum chamber of the bending magnet ED1 of the existing line, see figure 1. In particular, the vertical beam size has to be smaller than 26 mm, that is the free gap of the vacuum chamber of the bending magnet ED1.

Finally, the last three quadrupoles are the same as in the existing line and produce a small beam spot at the achromatic waist position, as shown in figure 1.

The new EL is designed to transport beams with a maximum magnetic rigidity of 2.7 Tm and with a normalized emittance of 1 π mm-mrad.

In figure 2 the radial and axial beam envelope, and the dispersion along the transport line are shown for the case of ^{20}Ne at 71 AMeV. Similar beam envelopes were achieved also for other eight simulated cases [9].

3. A new beam line as energy selector and fragment in-flight separator

To transport the 10 kW beam to the Magnex room, the transport efficiency should be increased up to 99%. This would imply to replace the existing quadrupoles and probably also some dipoles with new ones with larger diameters. This upgrading should produce a serious interference with the operations of almost all the experimental rooms of the laboratory. For these reason we choose to build a by-pass line to the existing transport line with an acceptable impact on the two experimental rooms called "20° and 40°". The new line is designed as an achromatic FRAgment In-flight SEparator (FRAISE) [10], and will be well shielded to allow to stop in this area a beam power up to 2 kW. These features allow to use this beam line also as an energy selector. Although the energy spread of the beam extracted from the cyclotron could have tails up to ±0.4%, it could be possible to reduce the energy spread of the beam delivered to the NUMEN experiment down to 0.1% FWHM.

4. Beam Lines inside the MAGNEX experimental room

In order to run experiments with 10 kW beams, a suitable well shielded beam dump is needed. In the NUMEN experiment different primary beams come out of the spectrometer on the left side of the Focal Plane Detector (FPD) or on the right side. A possible solution is shown in figure 4, where the transport beam line, the MAGNEX spectrometer and the beam dump inside the experimental room are displayed. In particular, the 0° position of the spectrometer has been rotated of 70° in respect to the present position, to allow the exit of the primary beam on the left and right side of the FPD and

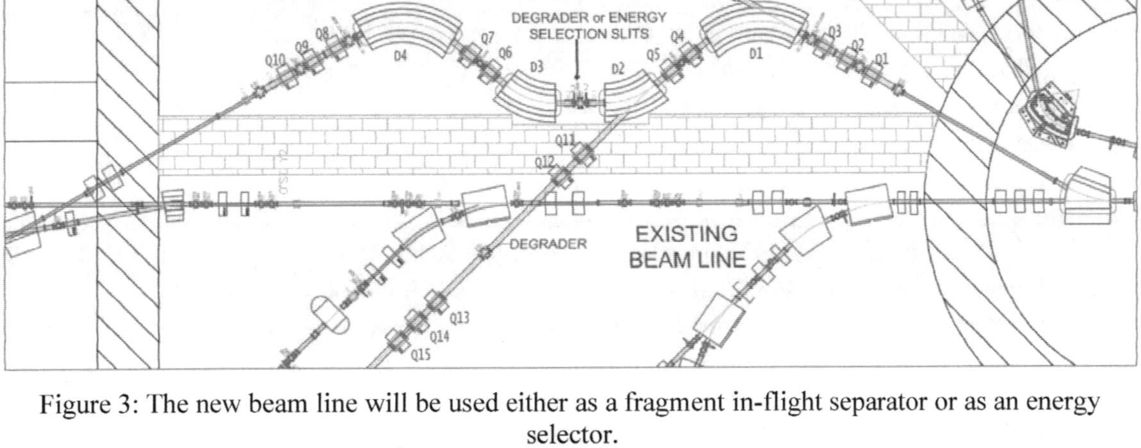

Figure 3: The new beam line will be used either as a fragment in-flight separator or as an energy selector.

maintain the beam lines from MAGNEX to the beam dump inside the north wall of the LNS. The west wall of MAGNEX room has to be shifted of about 3 m, to allow the installation of the beam dump. To mitigate the background to the FPD, the beam dump is installed underground, about 5 m below the median plane of MAGNEX and about 3.30 m below the ground. A rotating 90° bending magnet, placed over the beam dump, bends the beam in the vertical direction. The beam coming out from the right side of the FPD is transported to the beam dump by a steering magnet, a quadrupoles doublet, a 80° horizontal bending magnet and a second quadrupole doublet focuses the beam to the beam dump through the vertical 90° bending magnet. The transport of the beam coming out from the left side of the FPD is accomplished by a steering magnet and one quadrupole doublet to focus the beam on the beam dump through the vertical 90° bending magnet. To switch from the left side to the right side of the transport beam line it is mandatory to rotate the 90° vertical bending magnet of about 50°. The rotation of the MAGNEX spectrometer some time implies to dismount partially the left side or the right side beam line. The new position of MAGNEX implies to build an additional achromatic beam transport line consisting of a triplet of quadrupoles, of two symmetric 50° bending magnets with a radially focusing quadrupole in between and of a quadrupole doublet to focus the beam on the target.

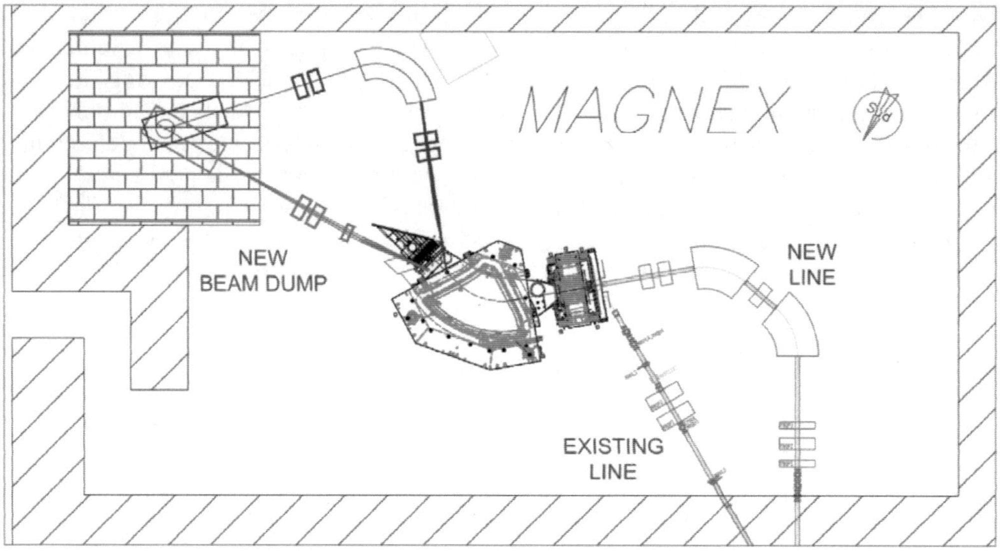

Figure 4: MAGNEX experimental hall. On the right side the new beam line is shown. On the left side, the two lines outgoing from MAGNEX and the beam dump are also shown.

References

[1] Cappuzzello F et al. 2015 *Journal of Physics: Conference Series* **630** 12018

[2] Agodi C et al. 2015 *Nuclear and Particle Physics Proceedings,* **265-266** 28

[3] Cappuzzello F et al. 2018 *European Physics Journal A* **54** 72

[4] Cappuzzello F et al. 2016 *European Physics Journal A* **52**167

[5] Cappuzzello F et al. 2014 *Nuclear Instrument and Method A* **763** 314

[6] Rifuggiato D et al. 2013 *Proceedings of Cyclotrons,* Vancouver, BC, Canada, JACOW, accelconf.web.cern.ch/AccelConf/cyclotrons2013/papers/MOPPT011.pdf

[7] Calabretta L et al. 2017 *Modern Physics Letters A* Vol. **32**, No. 17, 1740009

[8] Shima K et al. 1992 *Atomic Data and Nucl. Data Tables*, Vol. **51**, Issue 2, 173-241

[9] Calabretta L et al. 2017 *Proceedings of IPAC*, Copenaghen, Denmark, JACOW accelconf.web.cern.ch/AccelConf/ipac2017/papers/wepva053.pdf

[10] Russo A D et al. 2016 report LNS-INFN

Post-stripper study for the (^{20}Ne, ^{20}O) double charge exchange reaction at zero degrees with the MAGNEX spectrometer

G Santagati[1], F Cappuzzello[1,2], C Agodi[1], D Carbone[1], M Cavallaro[1], L Acosta[3], D Bonanno[4], D Bongiovanni[4], I Boztosun[5], S Calabrese[1,2], D Calvo[6], E R Chávez Lomelí[3], F Delaunay[6,7], N Deshmuk[1], P Finocchiaro[1], M Fisichella[6], A Foti[4], G Gallo[1,2], A Hacisalihoglu[1,8], F Iazzi[6,9], R Introzzi[6,9], G Lanzalone[1,10], R Linares[11], F Longhitano[4], D Lo Presti[2,4], N Medina[12], A Muoio[1], J R B Oliveira[12], A Pakou[13], L Pandola[1], F Pinna[6,9], S Reito[4], G Russo[2,4], O Sgouros[1,13], S O Solakcı[5], V Soukeras[1,13], G Souliotis[13], A Spatafora[1,2], D Torresi[1], S Tudisco[1], A Yildirim[5] and V A B Zagatto[11], for the NUMEN Collaboration

[1]INFN – Laboratori Nazionali del Sud, Catania, Italy
[2]Dipartimento di Fisica e Astronomia, Università di Catania, Italy
[3]Universidad Nacional Autónoma de México, Mexico
[4]INFN – Sezione di Catania, Italy
[5]Akdeniz University, Antalya, Turkey
[6]INFN –Sezione di Torino, Italy
[7]LPC Caen, Normandie Université, ENSICAEN, UNICAEN, CNRS/IN2P3, Caen , France
[8]Institute of Natural Science, Karadeniz Teknik Universitesi, Trabzon, Turkey
[9]DISAT-Politecnico di Torino, Italy
[10]Università degli Studi di Enna "Kore", Enna, Italy
[11] Universidade Federal Fluminense, Niteroi, Brazil
[12]Universidade de Sao Paulo, Brazil
[13]University of Ioannina, Ioannina, Greece.

santagati@lns.infn.it

Abstract. A study of different post-stripper materials for the (^{20}Ne, ^{20}O) double charge exchange and (^{20}Ne, ^{20}F) single charge exchange reactions at zero degrees using the MAGNEX spectrometer at 22 and 15 AMeV is presented. All these experiments belongs to the experimental campaign planned in the NUMEN project (NUclear Matrix Elements for Neutrinoless double beta decay).

1. Introduction

Recently, some experiments have been performed by the NUMEN Collaboration [1-4] at INFN - LNS to study the ^{116}Cd(^{20}Ne, ^{20}O)^{116}Sn and ^{130}Te(^{20}Ne, ^{20}O)^{130}Xe double charge exchange reactions together with the competing processes at zero degrees using the MAGNEX spectrometer [5-7].
In this kind of experiment with $^{20}Ne^{10+}$ beam, it is necessary to take into account the abundance of the beam components characterized by lower charge states ($^{20}Ne^{9+}$ and $^{20}Ne^{8+}$) produced by the interaction between the $^{20}Ne^{10+}$ incident beam and the target material. These components have in fact a magnetic

rigidity that is similar to the ejectiles of interest $^{20}F^{9+}$ and $^{20}O^{8+}$ for single charge exchange and double charge exchange reaction, respectively. Consequently, they enter in the focal plane detector acceptance, generating a large background that requires the limitation of the beam intensity.
For this reason, a system of shields upstream of the focal plane detector entrance was placed to stop such undesired background. However, due to the proximity of the $^{20}Ne^{9+}$ to $^{20}F^{9+}$ and $^{20}Ne^{8+}$ to $^{20}O^{8+}$ in the focal plane, the shields solution allows only partially the background reduction.
The horizontal angles θ_{FOC} versus the horizontal positions X_{FOC} measured at the focal plane are shown in Figure 1. Black points correspond to experimental data for the ^{20}Ne + ^{116}Cd reaction at 15 AMeV taken with the shield solutions and with MAGNEX at magnetic rigidity Bρ = 1.3764 Tm. The various colors points represent simulations [8-10] (same ^{20}Ne incident energy and MAGNEX Bρ) of different charge states of the ejectiles emerging from the following reactions:

- $^{116}Cd(^{20}Ne,^{18}O^{8+})^{118}Sn$ two-proton transfer
- $^{116}Cd(^{20}Ne,^{20}O^{8+})^{116}Sn$ double charge exchange reaction
- $^{116}Cd(^{20}Ne,^{18}F^{8+})^{118}In$ deuteron transfer
- $^{116}Cd(^{20}Ne,^{18}F^{9+})^{118}In$ deuteron transfer
- $^{116}Cd(^{20}Ne,^{19}F^{8+})^{117}In$ one-proton transfer
- $^{116}Cd(^{20}Ne,^{19}F^{9+})^{117}In$ one-proton transfer
- $^{116}Cd(^{20}Ne,^{20}F^{8+})^{116}In$ single charge exchange
- $^{116}Cd(^{20}Ne,^{20}F^{9+})^{116}In$ single charge exchange
- $^{116}Cd(^{20}Ne,^{20}Ne^{9+})^{116}Cd$ elastic
- $^{116}Cd(^{20}Ne,^{20}Ne^{8+})^{116}Cd$ elastic
- $^{116}Cd(^{20}Ne,^{21}Ne^{9+})^{115}Cd$ one-neutron transfer
- $^{116}Cd(^{20}Ne,^{22}Ne^{8+})^{114}Cd$ two-neutron transfer
- $^{116}Cd(^{20}Ne,^{22}Ne^{9+})^{114}Cd$ two-neutron transfer.

Together with the shields solution, the use of post-stripper material placed behind the reaction target was taken into account in order to minimize the amount of residual $^{20}Ne^{9+}$ and $^{20}Ne^{8+}$ beams downstream of the ^{116}Cd target.

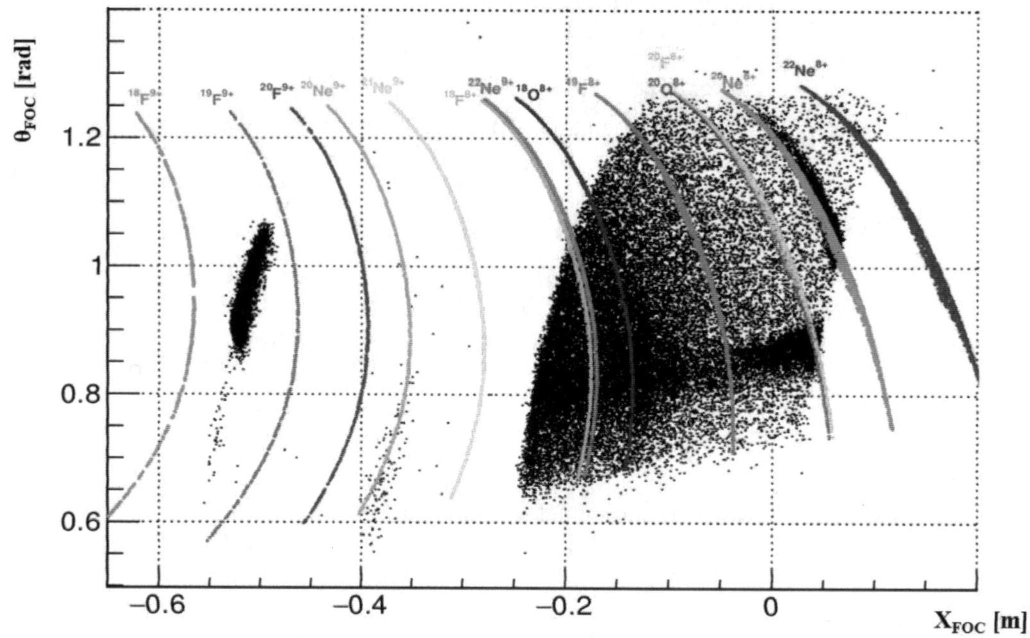

Figure 1. Comparison between data taken with the shield solution (black points) and simulations (various colors) of different ejectiles at the MAGNEX Focal Plane Detector (see text).

2. Experimental setup and results

A first test was realized concerning the study of different post-stripper materials using a beam of $^{20}Ne^{10+}$ at 22 AMeV incident energy extracted by the K800 Superconducting Cyclotron, bombarding a 77 ± 8 μg/cm² C foil positioned downstream of the 1080 ± 110 μg/cm² ^{116}Cd target.
A second test was realized with a beam of $^{20}Ne^{10+}$ at 15 AMeV incident energy extracted by the K800 Superconducting Cyclotron, bombarding a 900 ± 50 μg/cm² ^{12}C target.
The outgoing ejectiles produced in the different reactions were momentum analyzed by the MAGNEX spectrometer and detected by the Focal Plane Detector FPD [11]. The total charge collected for each run was integrated by a Faraday cup, downstream of the target.
In both these experiments, in order to estimate the amounts of $^{20}Ne^{10+}$, $^{20}Ne^{9+}$ and $^{20}Ne^{8+}$ ions produced by the interaction between the $^{20}Ne^{10+}$ incident beam and the target material, three different runs have been performed (one for each different ^{20}Ne charge state).
The three corresponding elastic peaks were momentum selected changing time by time the magnetic rigidity of the spectrometer. After a data reduction phase, consisting in the particle identification [12], the number of ^{20}Ne ions were determined counting the events belonging to the elastic peaks 10^+, 9^+ and 8^+. These quantities were normalized for the efficiency of the spectrometer and for the charge collected by the Faraday cup for the each corresponding run. In this way it was possible to determine the relative ratios $^{20}Ne^{9+}/^{20}Ne^{10+}$ and $^{20}Ne^{8+}/^{20}Ne^{10+}$ that are shown in Table 1, for the target post-stripper configurations for 15 and 22 AMeV $^{20}Ne^{10+}$ beam incident energy. In the third and fourth row of Table 1 the values of the ratios measured experimentally and published in Ref. [13] are shown.
An estimation of the errors was also done. Considering a precision of about 10% in the total charge collected, the efficiency in data collected by the spectrometer, and the systematic errors, the calculated values are affected by an uncertainty of about ± 13%.
As one can see from Table 1, the calculated values for the ^{116}Cd – C are comparable with the ones in literature, even if it is good to remind that the values in Ref. [13] were determined bombarding a C foil with a 15 and 20 AMeV $^{20}Ne^{10+}$ beam incident energy.
The 900 μg/cm² C target gives worse results respect to the literature [13]. This is likely due to the thickness which is too high and it is not suitable for the post-stripper purpose.

Table 1. Comparison between the $^{20}Ne^{9+}/^{20}Ne^{10+}$ and $^{20}Ne^{8+}/^{20}Ne^{10+}$ ratios for a $^{20}Ne^{10+}$ incident beam at 15 and 22 AMeV obtained in the tests performed at LNS and ratio values published in Ref. [13].

Target – Stripper (μg/cm²)	$^{20}Ne^{8+}/^{20}Ne^{10+}$	$^{20}Ne^{9+}/^{20}Ne^{10+}$	Energy (AMeV)
^{116}Cd (1080) – C (70)	4.09 E-06	3.84 E-03	22
C (900)	9.94 E-05	1.58 E-02	15
^{12}C Foil Ref.[5]	2.66 E-06	3.26 E-03	20
^{12}C Foil Ref.[5]	2.00 E-05	8.90 E-03	15

More recently, a systematic survey on different post-stripper materials was realized using a beam of $^{20}Ne^{10+}$ at 15 AMeV incident energy extracted by the K800 Superconducting Cyclotron, bombarding different post-stripper foils (Al, Be, C, CH_2, $C_{10}H_8O_4$, LiF, polypropylene, and others) with different thicknesses positioned downstream of some targets (Au, Cd, Te).
For each configuration three runs were performed (one for each different ^{20}Ne charge state) in order to carry out the above described procedure concerning the calculation of the charge states ratios.
The data analysis for these data are presently in progress.
Very preliminary results suggest that the best stripping ratios are obtained for light and C-like materials compatibly with information present in literature [14].

IOP Conf. Series: Journal of Physics: Conf. Series **1056** (2018) 012052 doi:10.1088/1742-6596/1056/1/012052

In particular it seems very promising the use of polypropylene. However this material has proven to be useless because of its fragility when bombarded by $^{20}Ne^{10+}$ beam.
This aspect has to be better understood and will be better investigated in the next future.
Furthermore, experiment with other C-like material (i.e. graphite [15]), will be performed in next test runs.

References

[1] Agodi C et al. 2015 *Nuclear and Particle Physics Proceedings* **265-266** 28.
[2] Cappuzzello F et al. 2015 *Journal of Physics Conference Series* **630** 012018.
[3] Cappuzzello F et al. 2018 *Eur. Phys. J. A* **54** 72.
[4] Cappuzzello F et al. 2015 *Eur. Phys. J. A* **51** 145
[5] Cappuzzello F et al. 2016 *Eur. Phys. J. A* **52** 167.
[6] Cappuzzello F et al. 2014 *Nucl. Inst. Meth. Phys. Res. Sect. A* **763** 314.
[7] Cunsolo A et al. 2007 *Eur. Phys. J. A S.T.* **150** 343.
[8] Lazzaro A et al. 2008 *Nucl. Inst. Meth. Phys. Res. Sect. A* **585** 136.
[9] Lazzaro A et al. 2009 *Nucl. Inst. Meth. Phys. Res. Sect. A* **602** 494.
[10] Carbone D 2015 *Eur. Phys. J Plus* **130** 143.
[11] Cavallaro M et al. 2012 *Eur. Phys. J. A* **48** 59.
[12] Cappuzzello F et al. 2010 *Nucl. Inst. Meth. Phys. Res. Sect. A* **621** 419.
[13] Shima K et al. 1992 *Atomic Data and Nucl. Data Tables* **51** 173-241.
[14] Baron E et al. 1979 *IEEE Trans. On Nuc. Sc.* **526** 2411.
[15] Iazzi F et al. 2017 *Transaction on engineering sciences*, **116** 61. doi:10.2495/MC170071.

IOP Conf. Series: Journal of Physics: Conf. Series **1056** (2018) 012053 doi:10.1088/1742-6596/1056/1/012053

Two-nucleon transfer and Double Charge Exchange reactions

E Santopinto, R I Magana Vsevolodovna, and H García-Tecocoatzi, within the NUMEN project

Istituto Nazionale di Fisica Nucleare, Sezione di Genova, Genova, Italy

E-mail: santopinto@ge.infn.it

Abstract. In this contribution we present a study of a direct and sequential two-nucleon transfer within the microscopic Interacting Boson Model (IBM-2) and Interacting Boson-Fermion Model (IBFM), respectively. We present our preliminary results for heavy-ion double-charge-exchange ^{40}Ca(^{18}O,^{18}Ne)^{40}Ar$_{\mathrm{G.S.}}$ differential cross section too.

1. Introduction

Neutrinoless double beta decay ($0\nu\beta\beta$) is one of the most important process that has not been observed yet, and it represents a probe of physics beyond the Standard Model. One open problem is that the $0\nu\beta\beta$ decay depends on nuclear matrix elements (NMEs) that when evaluated using different nuclear structure models can differ also by about a factor of 2-3 [1]. A better understanding of the nuclear structure will help to improve the model calculation of nuclear reactions and decay processes, such as single beta decay, double beta decay ($2\nu\beta\beta$), and $0\nu\beta\beta$. The knowledge of the internal degrees of freedom is crucial to understand nuclear structure.

In order to calculate the competing two-nucleon processes to heavy-ion double charge exchange (DCE), recently we developed the formalism using microscopic IBM-2 for the NUMEN collaboration [2, 3, 4], and this formalism has been already applied with success to ^{64}Ni (^{18}O,^{16}O)^{66}Ni two-neutron transfer as reported here in section 2 and 3. Finally, since our goal is to describe the DCE differential cross section using IBMs for the nuclear part, trying to extract information on NMEs of nuclei involved also in $0\nu\beta\beta$, we present here in the last section very preliminary results regarding DCE ^{40}Ca(^{18}O,^{18}Ne)^{40}Ar$_{\mathrm{G.S.}}$ differential cross section.

2. The ^{64}Ni (^{18}O,^{16}O)^{66}Ni two-neutron transfer

The two-neutron transfer of ^{18}O+^{64}Ni has been measured at the INFN- Laboratori Nazionali del Sud (Italy) using the MAGNEX spectrometer [5, 6], considering ^{66}Ni ground and first excited state. The two-nucleon transfer cross section has been calculated, using the spectroscopic factors obtained for the first time within microscopic Interacting Boson Model (IBM) for the direct two-neutron transfer, and the microscopic Interacting Boson Fermion Model IBFM for the two-step process. We have calculated the ^{64,66}Ni theoretical spectra using IBM2 [7] (see figure 1), and the ^{65}Ni one using the interacting boson-fermion model (IBFM) [7] (see figure 1).

IOP Conf. Series: Journal of Physics: Conf. Series **1056** (2018) 012053 doi:10.1088/1742-6596/1056/1/012053

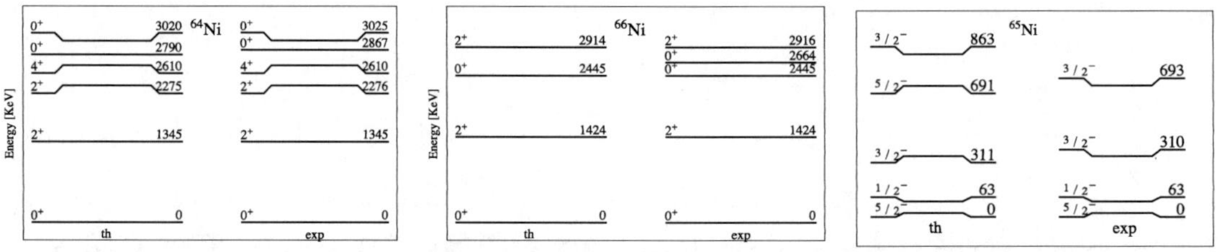

Figure 1. Theoretical spectra of the ^{64,66}Ni calculated with IBM-2 scheme, and the ^{65}Ni one with IBFM. Figure taken from Ref. [13], APS Copyright.

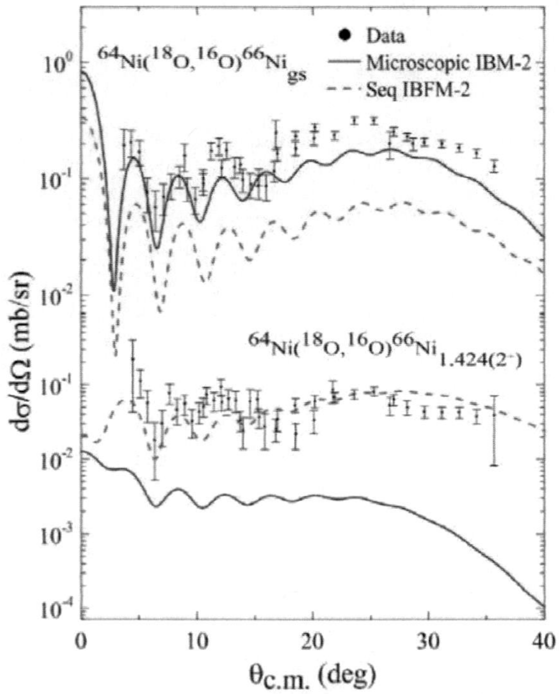

Figure 2. Differential cross section of the ^{64}Ni (^{18}O,^{16}O)^{66}Ni reaction. Figure taken from Ref. [13], APS copyright.

3. The two-nucleon transfer operator in microscopic IBM-2

The calculation of the spectroscopic amplitudes for the two-nucleon transfer reactions in the IBM-2 scheme requires the two-body matrix elements in the Generalized Seniority Scheme (GSS). The matrix elements between fermionic states in the collective subspace are identical to the matrix elements in the bosonic space. Taking into account of the Otsuka- Arima -Iachello (OAI) expansion next to leading order (NLO)[8], we can obtain the two-nucleon transfer operator [9, 10]. The advantage of this operator is that the effects of the pairing interaction are in terms of the occupation of different single-particle orbitals and the two body matrix elements take into account the non-degenerate orbits of the GS states [11]. The calculation of the spectroscopic amplitudes for single nucleon transfer reactions has been performed in the IBFM scheme[12].

The coupled channel Born approximation (CCBA) has been used for the sequential two neutron transfer, whereas the Coupled Reaction Channel (CRC) has been used for the direct two neutron transfer. In figure 2 one can see our results compared with the experimental data.

In the ^{64}Ni (^{18}O,^{16}O)^{66}Ni two neutron transfer between ground states, both two reaction mechanisms are important, while for the two-neutron transfer to the ^{66}Ni first excited state, the two-step processes are dominant (See figure 2). These results indicate that the pairing correlation effects are present in the two-neutron transfer to the ground state. It is interesting to observe that for the same nucleus different states prefer different transfer mechanisms.

This formalism for two nucleon transfer reactions [9, 10, 13] developed within the microscopic IBM-2 [14] is particularly appealing, in fact, microscopic IBM-2 allows one to calculate in a realistic way the transition matrix elements for heavy, medium nuclei and it has been exploited for the evaluation of $0\nu\beta\beta$ decay NMEs too [15]. We have just applied this formalism also to two nucleon transfer reaction that are competing processes in respect to DCE in nuclei of $0\nu\beta\beta$ type. It is interesting to observe that the two nucleon transfer is far from saturating the total detected experimental cross section for the DCE, thus pointing to a dominant role of the meson exchange double charge exchange contribution. Indeed, the calculated two nucleon transfer cross section is of the order of $10^{-4} - 10^{-3}$ nb, to be compared with the preliminary extracted experimental cross section of the order of few nb [4].

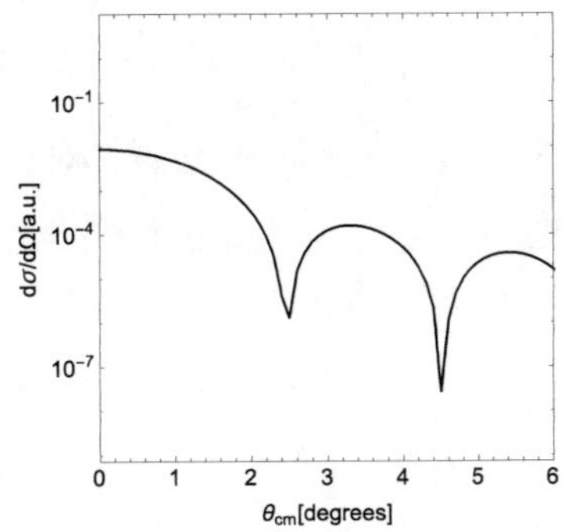

Figure 3. Preliminary ^{40}Ca(^{18}O,^{18}Ne)^{40}Ar$_{G.S.}$ DCE differential cross section in arbitrary units (a.u.).

4. Heavy-ion double charge exchange

A Double Charge Exchange (DCE) reaction is a process induced by a projectile (heavy ion) on a target, in which two protons (neutrons) of the target are converted in two neutrons (protons), being the mass number A unchanged, with opposite transition simultaneously occurring in the projectile. We study the DCE differential cross section of ^{40}Ca(^{18}O,^{18}Ne)^{40}Ar$_{G.S.}$ reaction measured by the LNS-INFN by the NUMEN collaboration [16]. The theoretical differential cross section can be obtained by an average of the initial spins and sum over final spins,

$$\frac{d\sigma}{d\Omega} = \frac{k'}{k}\frac{\mu}{2\pi^{-}\hbar^{2}}\frac{1}{(2J_P+1)(2J_T+1)}\sum_{\mathbf{m}}|M_{if}(\mathbf{m})| \qquad (1)$$

where k, k' are the initial, final relative momenta of the projectile in the center of mass frame; μ is the reduced mass of the projectile plus target system; $J_{P/T}$ are the total angular momentum of the initial projectile/target nuclei; and $\mathbf{m} = (m_T, m'_T, m_P, m'_P)$ is the set of angular-momentum

IOP Conf. Series: Journal of Physics: Conf. Series **1056** (2018) 012053 doi:10.1088/1742-6596/1056/1/012053

quantum numbers of the target and projectile wavefunction. $M_{if}(\mathbf{m})$ is the transition matrix of the DCE process. In DWBA, $M_{if}(\mathbf{m})$ is given by

$$M_{ij} = \langle \Psi^{-}_{k'} \Phi_f | V | \Psi^{+}_{k} \Phi_i \rangle \tag{2}$$

where $\Psi^{\pm}_{k}$ are the distorted wavefunctions of the nuclei in the center of mass, and $\Phi_{i,f}(r)$ are the intrinsic wavefunctions of the nuclei before and after the interaction. The interaction potential responsible for the DCE between the nuclei is given by V, that can be generated from nucleon-nucleon interaction [17]. Here we present a preliminary calculation of DCE ^{40}Ca(^{18}O,^{18}Ne)^{40}Ar$_{\text{G.S.}}$ differential cross section as shown in figure 3.

References

[1] Barea J, Kotila J, Iachello F, 2012 *Phys. Rev. Lett.* **109** 042501.
[2] Capuzzello F *et al.*, 2015 *J. Phys. Conf. Ser.* **630** 12018
[3] Agodi C *et al.*, 2015 *Nucl. Part. Phys. Proc.* **265** 28
[4] Cappuzzello F *et al.*, 2018 *Eur. Phys. J.* A **54** 72
[5] Cappuzzello F *et al.*, 2016 *Eur. Phys. J.* A **52** 167
[6] Cappuzzello F *et al.*, 2014 *Nucl. Instrum. and Meth.* A **763** 314
[7] Iachello F, P. V. Isacker, 1991 *The interacting Boson-Fermion Model* (Cambridge: Cambridge University Press)
[8] Otsuka T, Arima A, Iachello F, 1978 *Nucl. Phys.* A **309** 1
[9] Magana Vsevolodovna R, 2018 *PhD thesis* (Italy: Genoa University)
[10] Magana Vsevolodovna R, Santopinto E, Lubian J, Paes B, in preparation.
[11] Duval P D, Barrett B R, 1981 *Phys. Rev.* C **24** 1272.
[12] O. Scholten, 1980 *PhD thesis* (The Netherlands: University of Groningen)
[13] Paes B *et al.*, 2017 *Phys. Rev.* C **96** 044612
[14] Otskuka T, Arima A, Talmi I, Iachello F, 1978 *Phys. Lett.* B **76** 139
[15] Barea J, Kotila J, Iachello F, 2013 *Phys. Rev.* C **87**, 014315
[16] Cappuzzello F *et al.*, 2015 *Eur. Phys. J.* A **51** 145
[17] Santopinto E, García-Tecocoatzi H, Magaña Vsevolodovna R I, Ferretti J, *arXiv:*1806.03069

Favored neutrino mass mechanisms of the $0\nu\beta\beta$-decay unified by an interpolating formula

F Šimkovic[1,2,3], A Babič[2,4,5], S Kovalenko[6] and M I Krivoruchenko[2,7,8]

[1] Comenius University, Mlynská dolina F1, SK-842 48 Bratislava, Slovakia
[2] Boboliubov Laboratory of Theoretical Physics, JINR 141980 Dubna, Russia
[3] Czech Technical University in Prague, 128-00 Prague, Czech Republic
[4] Dept. of Dosimetry and Application of Ionizing Radiation, Czech Technical U., 115 19 Prague, Czech Republic
[5] Institute of Experimental and Applied Physics, Czech Technical University, 128 00 Prague, Czech Republic
[6] Universidad Técnica Federico Santa María, Centro-Cientifico-Tecnológico de Valparaiso, Casilla 110-V, Valparaiso, Chile
[7] Institute for Theoretical and Experimental Physics, B. Cheremushkinskaya 25, 117218 Moscow, Russia
[8] Department of Nano-, Bio-, Information, and Cognitive Technologies, Moscow Institute of Physics and Technology, 9 Institutskii per. 141700 Dolgoprudny, Moscow Region Russia

E-mail: simkovic@fmph.uniba.sk

Abstract. The neutrino mass mechanisms of the neutrinoless double beta decay ($0\nu\beta\beta$-decay) generated within the left-right symmetric models are discussed by taking advantage of an interpolating formula for nuclear matrix element depending on the neutrino mass. Its application is justified by a weak dependence of mean square neutrino momentum, which is a properly scaled ratio of nuclear matrix elements for the light and heavy neutrino mass mechanisms, on decaying isotope. Due to this fact there is a single unified lepton number violating parameter including dependendence on all neutrino masses in the decay rate and the light and heavy neutrino mass mechanisms can be hardly distinguished by observing the $0\nu\beta\beta$-decay on different isotopes. Further, a see-saw type neutrino mixing, which relates neutrino mixing in light and heavy neutrino sectors, is proposed. By considering the current limit on the $0\nu\beta\beta$-decay half-life of ^{136}Xe the regions of dominance of exchange of light and heavy neutrinos are identified.

1. Introduction

The main aim of experiments on the search for the neutrinoless double beta decay process ($0\nu\beta\beta$-decay) [1],

$$(A, Z) \rightarrow (A, Z+2) + 2e^-, \tag{1}$$

is to prove that neutrinos are Majorana particles and to measure the effective Majorana neutrino mass

$$m_{\beta\beta} = \left| \sum_{j=1}^{3} U_{ej}^2 m_j \right|, \tag{2}$$

where U_{ej} is the element of the Pontecorvo-Maki-Nakagawa-Sakata (PMNS) mixing matrix and m_j is the mass of neutrino. It is assumed that the conventional light neutrino exchange mechanism generated by left-handed V-A weak currents is the dominant mechanism of the $0\nu\beta\beta$-decay.

The value of $m_{\beta\beta}$ can be evaluated with help of neutrino oscillation parameters by making an assumption about the mass of lightest neutrino and by choosing a type of spectrum (normal or inverted) and values of CP violating Majorana phases. In future experiments, a sensitivity a few tens of meV to $m_{\beta\beta}$ is planned to be reached. This is the region of the inverted hierarchy of neutrino masses. In the case of the normal mass hierarchy $m_{\beta\beta}$ is too small, a few meV, to be probed in the $0\nu\beta\beta$-decay experiments of the next generation [1].

However, there is a possibility that the $0\nu\beta\beta$-decay is governed by another mechanism having origin in theories beyond the Standard model of particle physics. Currently, the attention is paid to the left-right symmetric theories [2, 3], which allow to understand the smallness of neutrino masses in a natural way. In this scenario, there are additional contributions to the $0\nu\beta\beta$-decay rate due right-handed currents and heavy neutrinos.

In this contribution we show that the observation of the $0\nu\beta\beta$ half-life with different isotopes will not allow to distinguish light and heavy neutrino mass mechanisms of this process as the effective neutrino momentum, which is determined through the ratio of corresponding nuclear matrix elements, depends only weakly on a given isotope. This fact allows to introduce a general lepton number violating parameter (LNV) of these mechanisms by taking the advantage of the interpolating formula for the nuclear matrix element depending on neutrino mass. Further, the dominance of light and heavy neutrino mass mechanisms is studied by considering the see-saw type neutrino mixing in which mixing of heavy neutrinos is related to the mixing of light neutrinos due to unitary condition.

2. The $0\nu\beta\beta$-decay expressed with interpolating formula

The Majorana neutrino mass mechanisms of the $0\nu\beta\beta$-decay within left-right symmetric models (LRSM) [2, 3] are considered. The left-handed ν_{eL} and right-handed ν_{eR} weak eigenstate electron neutrinos are expressed as superpositions of the light and heavy mass eigenstate Majorana neutrinos ν_j and N_j as follows:

$$\nu_{eL} = \sum_{j=1}^{3} \left(U_{ej}\nu_j + S_{ej}N_j^C \right), \qquad \nu_{eR} = \sum_{j=1}^{3} \left(T^*_{ej}\nu_j^C + V^*_{ej}N_j \right). \tag{3}$$

The 3×3 block matrices in flavor space U, S, T, V form a 6×6 unitary neutrino mixing matrix [4]

$$\mathcal{U} = \begin{pmatrix} U & S \\ T & V \end{pmatrix}. \tag{4}$$

It is a generalization of the PMNS matrix, which diagonalizes the general 6×6 Dirac-Majorana neutrino mass matrix in the Langrangian. As a result one ends up with 3 light m_i (i=1,2 and 3) and 3 heavy M_i (i=1,2 and 3) neutrino masses.

For sake of simplicity the possible $W_L - W_R$-mixing is not taken into account and the subdominant left-right interference term for the $0\nu\beta\beta$-decay half-life is neglected. Then, we have

$$[T^{0\nu}_{1/2}]^{-1} = G^{0\nu} g_{\rm A}^4 m_p^2 \left(\left| \sum_{i=1}^{3} \left(U^2_{ej}\, m_i\, M'^{0\nu}(m_i) \,+\, S^2_{ej}\, M_i\, M'^{0\nu}(M_i) \right) \right|^2 + \lambda^2 \left| \sum_{i=1}^{3} \left(T^2_{ej}\, m_i\, M'^{0\nu}(m_i) \,+\, V^2_{ej}\, M_i\, M'^{0\nu}(M_i) \right) \right|^2 \right).$$

Here, $\lambda = (M_{W_L}/M_{W_R})^2$, where M_{W_L} and M_{W_R} ($M_{W_L} < M_{W_R}$) are masses of W_L and W_R gauge bosons, respectively. $m_{\rm p}$ is the proton mass and $g_{\rm A}$ is the unquenched value of axial-vector coupling constant ($g_{\rm A} = 1.269$). The phase-space factor $G^{0\nu}$ is tabulated for various $0\nu\beta\beta$-decaying nuclei in Ref. [5]. The nuclear matrix element $M'^{0\nu}$ is given by [6]

$$M'^{0\nu}(m) = \frac{1}{m_{\rm p} m_{\rm e}} \frac{R}{2\pi^2 g_A^2} \sum_n \int d^3x\, d^3y\, d^3p e^{i\mathbf{p}\cdot(\mathbf{x}-\mathbf{y})} \frac{\left\langle 0_F^+ \right| J^{\mu\dagger}(\mathbf{x}) \left| n \right\rangle \left\langle n \right| J_\mu^\dagger(\mathbf{y}) \left| 0_I^+ \right\rangle}{\sqrt{p^2+m^2}(\sqrt{p^2+m^2} + E_n - \frac{E_I - E_F}{2})}. \tag{5}$$

Here, m_e and R are the mass of electron and the nuclear radius, respectively. Initial and final nuclear ground states with energies E_I and E_F are denoted by $\left|0_I^+\right\rangle$ and $\left|0_F^+\right\rangle$, respectively. The summation runs over intermediate nuclear states $|n\rangle$ with energies E_n. For sake of simplicity we assume that coupling constants of left- and right-handed weak one-body nuclear charged currents are the same, $J^\dagger_{LR\mu} = J^\dagger_{L\mu} \equiv J^\dagger_\mu$. $J^\dagger_\mu$ depends on the effective value of axial-vector coupling constant $g_{\rm A}^{\rm eff}$ of the nucleon in a nucleus.

Nuclear matrix elements associated with light and heavy neutrino mass mechanisms, $M'^{0\nu}_\nu$ and $M'^{0\nu}_N$, are limiting cases of $M'^{0\nu}(m)$ in Eq. (5):

$$M'^{0\nu}(m_i \to 0) = \frac{1}{m_{\rm p} m_{\rm e}} M'^{0\nu}_\nu, \qquad M'^{0\nu}(M_{\rm i} \to \infty) = \frac{1}{M_{\rm i}^2} M'^{0\nu}_{\rm N}. \tag{6}$$

To a good accuracy $M'^{0\nu}(m)$ can be approximated by the formula

$$M'^{0\nu}(m) \simeq M'^{0\nu}_{\rm N} \frac{1}{\langle p^2\rangle + m^2} \tag{7}$$

interpolating two limiting cases in (6). The parameter of this "interpolating formula",

$$\langle p^2 \rangle = m_{\rm p} m_{\rm e} \frac{M'^{0\nu}_{\rm N}}{M'^{0\nu}_\nu} \tag{8}$$

is interpreted as the mean square neutrino momentum in a nucleus. We note that the "interpolating formula" in Eq. (7) reproduces the "exact" QRPA result with rather good accuracy except for the transition region where its deviation amounts 20% - 25% [6].

The current values of the matrix elements $M'^{0\nu}_\nu$ and $M'^{0\nu}_{\rm N}$ calculated within the interacting shell model (ISM) [7], interacting boson model (IBM)[8], quasiparticle random phase approximation (QRPA) [9, 10]), projected Hartree-Fock Bogoliubov approach (PHFB)[11], and covariant density functional theory (CDFT) [12] are presented, e.g., in Tables 6 and 7 of Ref. [1]. The value of corresponding parameter $\sqrt{\langle p^2\rangle}$ is shown for various isotopes in Table 1. The unquenched value of axial-vector coupling constant and the Argonne two-nucleon short-range correlations are assumed.

For the averaged value $\sqrt{\langle p^2\rangle_a}$ and variance σ associated with considered nuclear structure approaches we find

$$\begin{aligned} \sqrt{\langle p^2\rangle_a}(\sigma) &= 155(17)\ \text{MeV}, \quad (\text{ISM}) \\ &= 120(17)\ \text{MeV}, \quad (\text{IBM}) \\ &= 175(11)\ \text{MeV}, \quad (\text{QRPA}) \\ &= 128(4)\ \text{MeV}, \quad (\text{PHFB}) \\ &= 132(5)\ \text{MeV} \quad (\text{CDFT}). \end{aligned} \tag{9}$$

IOP Conf. Series: Journal of Physics: Conf. Series **1056** (2018) 012054 doi:10.1088/1742-6596/1056/1/012054

Table 1. The values of the parameter $\sqrt{\langle p^2 \rangle}$ of the interpolating formula specified in Eq. (10) for a given isotope calculated within different nuclear structure approaches: interacting shell model (ISM) [7], interacting boson model (IBM)[8], quasiparticle random phase approximation (QRPA) [9, 10]), projected Hartree-Fock Bogoliubov approach (PHFB)[11], and covariant density functional theory (CDFT) [12]. The Argonne short-range correlations are taken into account. The non-quenched value of weak axial-vector coupling g_A is assumed.

Method	$\sqrt{\langle p^2 \rangle}$ [MeV]										
	^{48}Ca	^{76}Ge	^{82}Se	^{96}Zr	^{100}Mo	^{116}Cd	^{124}Sn	^{128}Te	^{130}Te	^{136}Xe	^{150}Nd
ISM	178	134	138				153		159	170	
IBM	113	103	103	129	136	130	109	109	109	107	155
QRPA	189	163	164	180	174	157	186	178	180	183	
PHFB				130	127			131	132		121
CDFT	122	129	131	129	131	133	138		138	137	138

We notice a large spread of values of the parameter $\sqrt{\langle p^2 \rangle_a}$ due to different nuclear structure approaches. The largest and smallest value (175 MeV and 120 MeV) is obtained within the QRPA with isospin restoration [9, 10]) and the IBM [8], respectively. However, in the case of all nuclaer structure approaches the variance σ is very small of the order of 3-10 %. It allows to conclude that $\langle p^2 \rangle$ is practically the same for all isotopes of experimental interest and can be replaced with averaged value $\langle p^2 \rangle_a$. This conclusion is also supported by the statistical treatment of $M'^{0\nu}_{\nu}$ and $M'^{0\nu}_{\rm N}$ performed in Ref. [13].

By taking the advantage of the "interpolating formula" (7) with the parameter $\langle p^2 \rangle_a$ to a good accuracy the $0\nu\beta\beta$-decay half-life in Eq. (5) can be rewritten as

$$[T^{0\nu}_{1/2}]^{-1} = \eta^2_{\nu N}\, g^4_{\rm A}\, \left| M'^{0\nu}_{\nu} \right|^2\, G^{0\nu}. \tag{10}$$

with

$$\eta^2_{\nu N} = \left| \sum_{j=1}^{3} \left(U^2_{ej}\, \frac{m_j}{m_e} + S^2_{ej}\, \frac{\langle p^2 \rangle_a}{\langle p^2 \rangle_a + M^2_j}\, \frac{M_j}{m_e} \right) \right|^2 + \lambda^2 \left| \sum_{j=1}^{3} \left(T^2_{ej}\, \frac{m_j}{m_e} + V^2_{ej}\, \frac{\langle p^2 \rangle_a}{\langle p^2 \rangle_a + M^2_j}\, \frac{M_j}{m_e} \right) \right|^2. \tag{11}$$

Here, the parameter $\eta_{\nu N}$ is a general lepton number violating parameter for light and heavy neutrino mass mechanisms in the context of the left-right symmetric models, which does not depend on the isotope under consideration.

From Eqs. (10) and (11) it follows that the dominance of light or heavy neutrino mechanisms of the $0\nu\beta\beta$-decay can not be established by an observation of this process on different isotopes. This task requires an additional information or assumption concerning neutrino masses and mixing.

3. Study of the dominance of light and heavy neutrino mass mechanisms

For sake of simplicity the flavor universal mixing between the active and sterile neutrino sectors of the seesaw mixing matrix $\mathcal{U}$ is proposed to study the dominance of light and heavy neutrino mass mechanisms in the $0\nu\beta\beta$-decay rate. We have

$$\mathcal{U} = \begin{pmatrix} U_0 & \zeta\, \mathbf{1} \\ -\zeta\, \mathbf{1} & V_0 \end{pmatrix}. \tag{12}$$

IOP Conf. Series: Journal of Physics: Conf. Series **1056** (2018) 012054 doi:10.1088/1742-6596/1056/1/012054

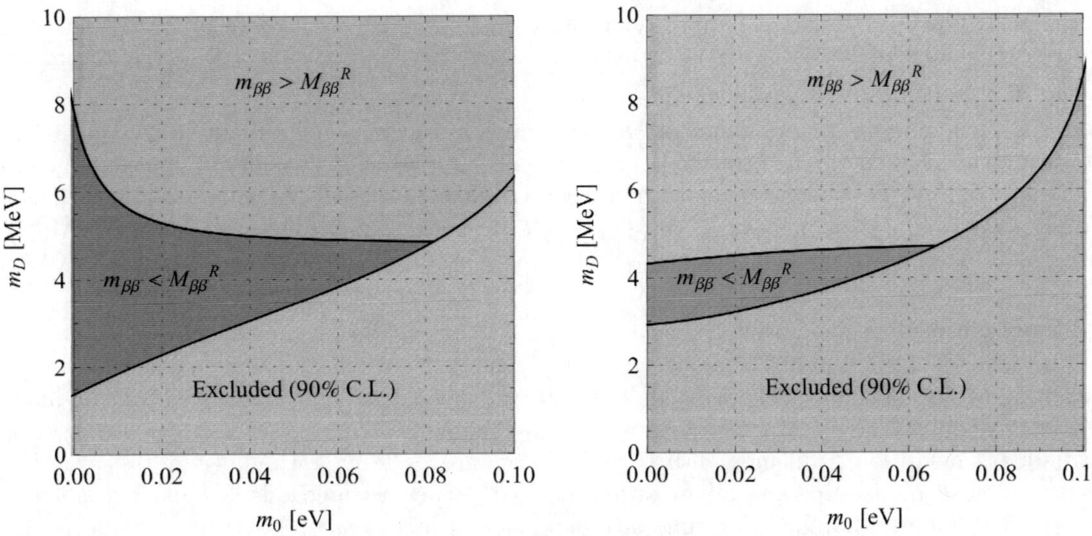

Figure 1. The region of the dominance of the $M^R_{\beta\beta}$ contribution over $m_{\beta\beta}$ contribution to the LNV parameter $\eta_{\nu N}$ for the see-saw type of neutrino mixing matrix given in Eq. (12) and by assuming $m_i \simeq m_D^2/M_i$ The cases of the normal and inverted hierarchy of neutrino masses are presented in the left and right pannels, respectively. The current constraint on the $\eta_{\nu N}$ parameter is deduced from the lower limit on the $0\nu\beta\beta$-decay half-life of ^{136}Xe [15] by using nuclear matrix element calculated within the QRPA [9]. For square ratio of masses of left and right vector bosons $\lambda = 7.7 \cdot 10^{-4}$ is considered.

Here, $\zeta = \frac{m_D}{m_{LNV}}$, where m_D is the typical scale of the charged leptons masses and m_{LNV} is the LNV scale of the order of the Majorana masses M_i of the heavy neutrinos. U_0 can be identified with the PMNS U matrix ($U = U_0$). From the unitarity conditions we find

$$V_0 = U_0^\dagger, \quad U_0 U_0^\dagger = (1-\zeta^2)\mathbf{1}, \quad V_0 V_0^\dagger = (1-\zeta^2)\mathbf{1}. \tag{13}$$

It is assumed that a small violation of the unitarity of U_0 and V_0 matrices is beyond the current accuracy of phenomenological determination of elements of the PMNS matrix.

Further, we assume that $m^2_{LNV} \gg \langle p^2 \rangle_a$ and the sees-saw relations for neutrino masses $m_i \simeq m_D^2/m_{LNV}$ and $M_i \simeq m_{LNV}$. We find

$$\eta^2_{\nu N} = \frac{1}{m_e^2}\left(m^2_{\beta\beta} + (M^{\rm R}_{\beta\beta})^2\right), \qquad M^{\rm R}_{\beta\beta} = \lambda\, \frac{\langle p^2\rangle_a}{m_D^2} \left|\sum_{j=1}^{3} (U_0^\dagger)^2_{ej}\; m_j\right|. \tag{14}$$

The matrix $U^\dagger$ expressed with three mixing angles θ_{12}, θ_{13} and θ_{23} and three CP violating phases α_1, α_2 and δ takes the form

$$V_0 = \begin{pmatrix} c_{12}\,c_{13}\,e^{-i\alpha_1} & \left(-s_{12}\,c_{23} - c_{12}\,s_{13}\,s_{23}\,e^{-i\delta}\right)e^{-i\alpha_1} & \left(s_{12}\,s_{23} - c_{12}\,s_{13}\,c_{23}\,e^{-i\delta}\right)e^{-i\alpha_1} \\ s_{12}\,c_{13}\,e^{-i\alpha_2} & \left(c_{12}\,c_{23} - s_{12}\,s_{13}\,s_{23}\,e^{-i\delta}\right)e^{-i\alpha_2} & \left(-c_{12}\,s_{23} - s_{12}\,s_{13}\,c_{23}\,e^{-i\delta}\right)e^{-i\alpha_2} \\ s_{13}\,e^{i\delta} & c_{13}\,s_{23} & c_{13}\,c_{23} \end{pmatrix}. \tag{15}$$

Here, $c_{ij} \equiv \cos(\theta_{ij})$ and $s_{ij} \equiv \sin(\theta_{ij})$. We note that each element of the first row is multiplied by the same phase $e^{-i\alpha_1}$, i.e., the Majorana phase α_1 (and α_2 as well) does not affect the heavy neutrino LNV parameter $M^R_{\beta\beta}$ within considered scenario. On the contrary, the Dirac phase δ affects $M^R_{\beta\beta}$ and has no impact on $m_{\beta\beta}$.

In Fig. 1 the region of dominance of $M^R_{\beta\beta}$ mechanism over $m_{\beta\beta}$ mechanism in the $0\nu\beta\beta$-decay rate is displayed in the plane of parameters m_D and m_0 (the lightest neutrino mass). The cases of the normal (left panel) and inverted hierarchy (right panel) of neutrino masses are discussed. We see that in the case of normal (inverted) hierarchy $m_D \leq 1.5$ MeV ($m_D \leq 2.9$ MeV) is already excluded by the present experimental lower bound on the $0\nu\beta\beta$-decay half-life of ^{136}Xe.

4. Conclusions

In summary, the light and heavy neutrino mass mechanisms of the $0\nu\beta\beta$-decay generated within the left-right symmetric models were analyzed. It was found that properly scaled ratio of nuclear matrix elements for the light and heavy neutrino mass mechanisms, which can be identified with square mean neutrino momentum $\langle p^2 \rangle$ of the interpolating formula, exhibits practically no dependence on isotope for all favoured nuclear structure methods. This fact allows to introduce new lepton number violating parameter $\eta_{\nu N}$, which is a coherent sum of square lepton number violating parameters $m_{\beta\beta}$ and $M^R_{\beta\beta}$ associated with light and heavy neutrino exchange mechanisms, respectively. It is concluded that the observation of the $0\nu\beta\beta$-decay on two and more nuclear isotopes will allow to decide about the dominance of light or heavy neutrino-mass mechanism in the decay rate.

Further, a viable see-saw type 6×6 neutrino mixing matrix (see Eq.(12)), where due to unitarity condition the 3×3 mixing matrix of heavy neutrinos is just hermitian conjugate of the 3×3 mixing matrix (PMNS matrix) of light neutrinos, was propsed. Within this neutrino mixing scenario the dominance of light and heavy neutrino mass mechanisms in the plane of parameters of Dirac mass m_D and lightest neutrino mass m_0 was established by using the experimental contraint on the $0\nu\beta\beta$-decay of ^{136}Xe and assuming $\lambda = 7.7 \cdot 10^{-4}$.

Acknowledgments

This work is supported by the VEGA Grant Agency of the Slovak Republic un- der Contract No. 1/0922/16, by Slovak Research and Development Agency under Contract No. APVV-14-0524, RFBR Grant No. 16-02-01104, Underground laboratory LSM - Czech participation to European-level research infrastructue CZ.02.1.01/0.0/0.0/16 013/0001733.

References

[1] Vergados J D, Ejiri H and Šimkovic F 2016 *Int. J. Mod. Phys. E* **25** 1630007
[2] Pati J C and Salam A 1974 *Phys. Rev. D* **10** 275; Mohapatra R and Pati J C 1975 *Phys. Rev. D* **11** 2558
[3] Senjanović G and Mohapatra R N 1975 *Phys. Rev. D* **12** 1502; Mohapatra R N and Senjanovic G 1980 *Phys. Rev. Lett.* **44** 912; 1981 *Phys. Rev. D* **23** 165
[4] Xing Zhi-zhong 2012 *Phys. Rev. D* **85** 013008
[5] Kotilla J and Iachello F 2012 *Phys. Rev. C* **85** 034316
[6] Faessler A, González M, Kovalenko S and Šimkovic F 2014 *Phys. Rev. D* **90** 096010
[7] Horoi M and Neacsu A 2016 *Phys. Rev. C* **93** 024308
[8] Barea J, Kotila J and Iachello F 2015 *Phys. Rev. C* **91** 034304
[9] Šimkovic F, Rodin V, Faessler A and Vogel P 2013 *Phys. Rev. C* **87** 045501
[10] Fang D, Faessler A and Šimkovic F 2015 *Phys. Rev. C* **92** 044301
[11] Rath P, Chandra R, Raina P, Chaturvedi K and Hirsch J 2012 *Phys. Rev. C* **85** 014308
[12] Song L S, Yao J M, Ring P and Meng J 2017 *Phys. Rev. C* **95** 024305
[13] E. Lisi, A.M. Rotunno, and F. Šimkovic, Phys. Rev. D **92**, 093004 (2015).
[14] A. Meroni, S. T. Petcov and F. Šimkovic, JHEP **1302**, 025 (2013).
[15] The KamLAND-Zen Collab., Gando A et al. 2016 *Phys. Rev. Lett.* **117** 082503

IOP Conf. Series: Journal of Physics: Conf. Series **1056** (2018) 012055 doi:10.1088/1742-6596/1056/1/012055

Geoneutrino : experimental status and perspectives

Oleg Smirnov
Dzhelepov's Laboratory for Nuclear Problems, JINR Dubna, 141980 Dubna, Moscow region, Russia

E-mail: osmirnov@jinr.ru

Abstract. The current results of the Borexino and KamLAND on geoneutrino are presented and the perspectives for measurement of geoneutrino fluxes both on existing and planned detectors are discussed. We discuss also the nuclear physics inputs relevant for precision measurement of the geoneutrino fluxes.

1. Introduction

Geoneutrino is an electron antineutrino accompanying β-decay of nuclear long-lived isotopes present in the Earth. The natural radioactivity of the Earth is a powerful source of heat, influencing the thermal history of the Earth. Radiogenic heat and primordial heat of the Earth constitutes two main contributions to the total energy loss of the Earth. Radiogenic heating helps power plate tectonics, hot-spot volcanism, earthquakes and mantle convection. Analysis of the available measurements of temperature gradients in mines gives the total thermal production of 47 ± 2 TW [1].

Geoneutrino flux measurements do provide experimental evidence for the quantity and distribution of radioactive elements internally heating the Earth, as the direct measurement of the composition of basic regions is possible for the Earth's crust only. Information on the extent and location of this heating better defines the thermal dynamics and chemical composition of Earth.

2. Models

Model predicting abundances of radioactive elements in each part of the Earth is required to calculate the geoneutrino signal. The description of the modern crust+mantle system is provided by a model of Bulk Silicate Earth (BSE), which represents a reconstruction of a primitive primordial mantle of the Earth on the base of geochemical arguments. The primitive mantle is the mantle formed immediately after the separation of the core before the crust differentiation. The composition of the primitive mantle is derived from the composition of the chondritic meteorites, which is practically identical to the chemical composition of the Sun, excepting the lightest elements, such as hydrogen and helium.

The homogeneous composition of seismically differentiated regions is usually assumed in model. The abundances of U/Th in the internal regions is based on the composition of corresponding meteorites in assumption of typical composition of the Earth within the frames of the Solar system. The BSE model provides full amount of U,Th and K in the Earth as these lithophile elements (having affinity to the silicate minerals and melts) should not

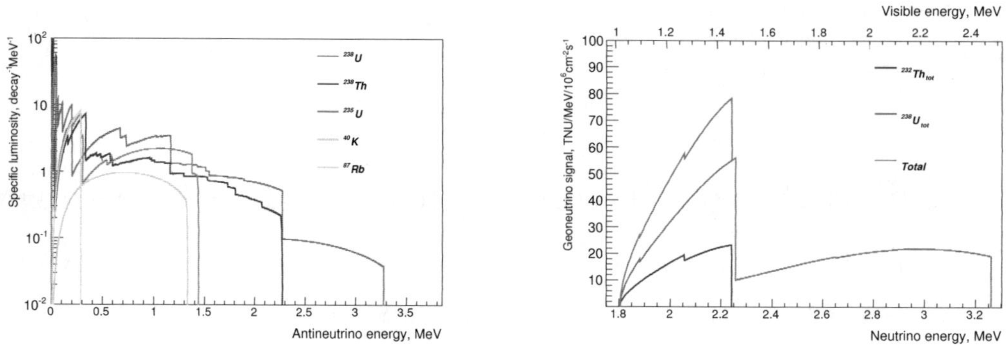

Figure 1. Left: theoretical spectra of the geoneutrino from their main sources. Right: theoretical spectra from the U and Th chains in assumption of the secular equilibrium in the chains and at the chondritic ratio of Th/U masses M(Th)/(U)=3.9.

be present in the Earth's core. The BSE model is in a good agreement with the majority of experimental observations concerning the core and upper mantle. BSE models can be classified in accordance with predicted radiogenic heat, that depends on the initial assumptions. Depending on underlying assumptions three classes of the models are considered: cosmochemical, geochemical and geodynamical ones (see [2] for more details), predicting radiogenic thermal flux of 11 ± 2, 20 ± 4 and 33 ± 3 TW correspondingly.

As the prediction for the crust contribution is bounded stronger compared to the mantle contribution, the models in fact differs even more significantly on predicted mantle contribution: low Q (3$\pm$2) for cosmochemical, medium Q (12$\pm$4) for geological and high Q (25$\pm$3) for geodynamical models.

Antineutrino spectra expected from ^{238}U and ^{232}Th chain, as well as from ^{40}K, are presented in fig.1 (left). A fraction of antineutrino spectra from ^{238}U and ^{232}Th chains exceeds the threshold of the inverse beta- decay on proton. This makes their detection by liquid organic scintillator (LOS) detectors in the inverse beta decay (IBD) reaction $\overline{\nu}_e + p \rightarrow e^+ + n$ (having a threshold of 1.8 MeV). An ideal geoneutrino signal in IBD detector is the product of the geoneutrino energy spectrum and corresponding cross section of the IBD reaction, for real-world detector signal will be smeared by resolution.

The theoretical (for ideal detector with infinite resolution) IBD spectra from U and Th chains in assumption of the secular equilibrium in the chains and at the chondritic ratio of Th/U masses M(Th)/(U)=3.9 are shown in fig. 1 (right). The visible difference in the spectra makes contributions from U and Th potentially distinguishable, though the low-energy part of the U spectrum has a noticeable contribution very similar to the Th spectrum, weakening the discrimination power of the U and Th signals.

3. Detectors and backgrounds.

At present only two collaboration presented results on the geoneutrino observation: KamLAND and Borexino. Both collaborations are using detectors on the base of the LOS. Principal backgrounds in the geoneutrino search with IBD detectors are: reactor antineutrinos, cosmic muons induced backgrounds, including cosmogenic production of (βn)- decaying isotopes and internal radioactive contamination.

Reactor antineutrinos. For KamLAND measurements they contributed 81% of the total antineutrino signal geoneutrino window of [0.9-2.6 MeV] in the initial stage of the experiment, for the Borexino the reactor contribution is 36%. After the Fukushima disaster all Japan's nuclear plants were closed, the last one went offline for maintenance on May 5, 2012. Correspondingly, the reactor background has significantly reduced at the KamLAND site.

Cosmic muons induced backgrounds, including cosmogenic production of (βn)- decaying isotopes. At LNGS the muons flux is of about factor 7 lower than at the Kamioka site. In the analysis a veto of 2 seconds is applied after each muon passing through the detector to remove events caused by short-lived cosmogenic isotopes decaying through (βn) channel. In KamLAND data analysis an additional condition on muons is applied: removed muon should generate a shower, this reduce live time losses to the moderate 4% compared to 10% of the live-time loss in Borexino analysis, despite of the lower muon flux in the Gran Sasso site. The suppression of background from the fast neutrons from buffer is achieved by applying a shorter 2 ms veto (this background is mainly caused by multiple neutron production).

Internal radioactive contamination: contributes through accidental coincidences and (αn)-reaction on ^{13}C, monoenergetic α with kinetic energy of 5.3 MeV are produced in ^{210}Po decays. The both backgrounds are much lower in the Borexino because of typical contamination of 3-4 orders of magnitude lower; later KamLAND reported factor 20 on (αn) reduction.

4. Measurements

The first indication of the non-zero geoneutrino signal was reported by KamLAND in 2005 [5]. In April 2010 the first high significance confirmation on the geoneutrino signal come from Borexino [8], the collaboration reported the observation of the geoneutrino signal at the 99.997% level.

The actual results are summarized in Table 1. The last up-to-date measurement was released by Borexino in 2015, the KamLAND analysis of the most interesting period of the data taking with reduced reactor background was presented at the "Neutrino Research and Thermal Evolution of the Earth" conference at the end of 2016.

The latest Borexino geoneutrino flux measurement was performed with 2056 days data set, the expected ratio of geoneutrinos to antineutrinos from the European reactors is about 1:2. Backgrounds from other sources for antineutrino measurement in Borexino are negligible and do not exceed one event for the measurement time. The observed geoneutrino signal is evaluated by fitting the experimental spectrum with the spectral contributions from geoneutrino (with chondritic Th/U mass ratio fixed at M(Th)/M(U)=3.9), reactor neutrino and residual backgrounds. The observed geoneutrino flux is totally consistent with the expected one for most of geophysical models. The probability of the absence of geoneutrino signal is negligible, namely, 3.6×10^{-9} (5.9σ). Moreover, for the first time in the history of geoneutrino observations, the non-zero contribution from the mantle is confirmed at 98% C.L..

The latest results of KamLAND are obtained including $\sim$3.5 years of data with the low reactor background, the evidence of the geoneutrino signal achieves 7.9σ. For the first time it is reported the positive indication of the U signal (at 3.44σ C.L.) with free U/Th ratio, and the "global" U/Th ratio measurement of M(Th)/M(U)=$4.1^{+5.3}_{-3.3}$ (with upper limit of 17.0 for the masses ratio of Th/U at 90% C.L.). All the values are consistent with chondrite data and the BSE models.

5. Near future

The Borexino detector will accumulate about 3 additional years of data in solar mode before the start of the SOX program. Further improvement of result is expected due to the tuning of the muon veto cut and applying the analysis to the whole volume of the detector.

The collaboration plans to improve the performance of cuts following KamLAND approach, this should save about 9% of live-time. Even stronger gain in statistics up to another 50% is possible if including all the data in the analysis without applying fiducial volume cut, this needs better understanding of the external backgrounds observed close to the IV walls.

Experiment/year	Live-time, days	Exposure, t·yr	number of candidates	number of geo-ν	Total geo-ν signal, TNU	Mantle geo-ν signal, TNU
KL/2013 [6]	2991	5780 ± 118	–	116^{+28}_{-27}	30.9 ± 7.3	—
KL/2016[7]	3901	7537	1130	164^{+28}_{-25}	$35.4^{+6.4}_{-5.4}$	8.8 ± 6.4
Bx/2015 [9]	2056	907±44	77	$23.7^{+6.5}_{-5.7}$	$43.5^{+12.1}_{-10.7}$	$20.9^{+15.1}_{-10.3}$

Table 1. Borexino and KamLAND results on the geoneutrino flux measurement. Results presented by KL in 2016 [7] still have preliminary status. 1 TNU, Terrestrial Neutrino Unit, corresponds to 1 event per year for 10^{32} protons in target (this corresponds to roughly 1 kton of typical LOS. Thus, a count in TNU units roughly corresponds to the expected event rate per year in 1 kton scale IBD detector.

KamLAND is continuing the data taking with Japanese nuclear reactors switched off, detecting about 14 geoneutrino events per year.

SNO+ detector should start the data taking soon. It is located at deep Sudbury mine, at 6010 m.w.e., and operates 780 tonnes LAB- based LOS detector. 29 geoneutrino events per live-year are expected compared with 26 events from reactors in the same energy range.

Multitonne detectors of the third generation reactor experiments (JUNO [14]) will be sensitive to geoneutrino. The expected geoneutrino signal is $39.7^{+6.5}_{-5.2}$ TNU [16], JUNO will acquire about 400 geoneutrino events per year. Signal from the reactors in the geoneutrino energy window will contribute 354^{+45}_{-41} TNU, making difficult the extraction of the geoneutrino signal. Nevertheless, 10% precision can be achieved with 3 years of the data taking, if the precision reactor spectrum description will be provided. The same set of data will allow to obtain the contribution of U with 20% precision and of Th with 30% precision [14].

Geoneutrino studies are planned at the China Jinping Underground Laboratory (CJPL), which is located at 2400 m depth under the Jinping Mountain in Sichuan, China. Predicted geoneutrino signal is 59.4 ± 6.8 TNU with expected background from nuclear reactors of only 6.7 ± 0.1 TNU in the geoneutrino energy range of interest [1.8, 3.3] MeV [10]. The Jinping Neutrino Experiment collaboration plans to build two detectors in the experimental hall of CJPL, using LOS or slow LS. The design mass of each of two detectors is 1.5 kt [11].

Another possible site for geoneutrino exploration is Baksan laboratory in Russia, 10 kt LOS detector is being discussed [12]. Expected geoneutrino signal is 52.6 TNU with expected background from nuclear reactors of only 14.4 TNU in the energy range of interest.

LENA was a project of 50 kt deep underground multipurpose LOS detector [13], at the moment there is no financial support of the project. About 1500 geoneutrino events per year were expected. Another project without financial support is Hanohano, which should be a portable 10 kt LOS device deployed from the barge. It would be aimed to extract mantle contribution in the total geoneutrino signal. About 100 geoneutrino events per year are expected.

In the absence of a "direct" measurement of the mantle signal with underwater detector, the only possible way to extract the mantle signal and discriminate the geophysical models would be the joint analysis of the geoneutrino data from different locations together with better understanding of the local contribution from the crust.

6. Nuclear physics uncertainties of the expected geoneutrino signal in IBD antineutrino detector

Only two isotopes in each of ^{238}U and ^{232}Th chains have enough energetic β-decays to provide a contribution to the IBD signal. In the ^{232}Th decay chain these are two branches of ^{228}Ac decay (into ground state and to the first excited one), and decay of ^{212}Bi into the ground state. In the ^{238}U chain these are the decays of ^{234m}Pa and 4 branches of the ^{214}Bi decay. The properties of β-decays in the ^{232}Th and ^{238}U chains contributing to geoneutrino signal are presented in

IOP Conf. Series: Journal of Physics: Conf. Series **1056** (2018) 012055 doi:10.1088/1742-6596/1056/1/012055

Isotope	Bq/Bq[Th]	$I \pm \delta I$	Q, MeV	Type	S,TNU	f_{Th},%	f_{tot},%
^{228}Ac→ ^{228}Th	1.000	0.07±0.05	2.0663	Allowed ($3^+ \to 2^+$)	0.3	4.8	0.8
^{228}Ac→ ^{228}Th	1.000	0.006±0.005	1.9374	Allowed ($3^+ \to 4^+$)	0.0	0.1	0.0
^{212}Bi→^{212}Po	0.6406	0.5537±0.0011	2.252026	1-st forbidden ($1^{(-)} \to 0^+$)	6.4/7.4	95.1	16.0

Table 2. Properties of β-decays in the ^{232}Th chain contributing to geoneutrino signal in IBD detectors. The second column contains activity of the corresponding reaction with respect to the activity of parent ^{232}Th. The $I \pm \delta I$ column contains the relative intensity of the corresponding decay (with respect to one parent decay) with associated experimental error. Q is the β-decay end-point. S is signal measured in TNU in assumption of 10^6 cm^{-2}s^{-1} total neutrino flux (from ^{238}U and ^{232}Th chains only) distributed between U and Th in accordance to the chondritic ratio and in assumption of the universal (allowed) shape; alternative values (separated by a slash) are calculated using shape factor corresponding to the decay type. f_{Th} and f_{tot} denote the fraction of the signal in the ^{232}Th chain and in total geoneutrino signal correspondingly in assumption of the allowed shapes. When calculating the values in the last column, the chondritic ratio of 3.9 for Th/U masses was assumed.

Isotope	Bq/Bq[U]	$I_k \pm \delta I_k$	Q, MeV	Type	S,TNU	f_U,%	f_{tot},%
^{234m}Pa→^{234}U	0.9984	0.9757±0.002	2.269	1-st forbidden ($0^- \to 0^+$)	13.0/14.6	38.4	31.9
^{214}Bi→^{214}Po	0.99979	0.1910±0.0017	3.270	1-st forbidden ($1^- \to 0^+$)	20.6/22.3	60.8	50.6
		0.0058±0.0018	2.661	1-st forbidden ($1^- \to 2^+$)	0.2/0.3	0.7	0.6
		0.0735±0.0005	1.892	1-st forbidden ($1^- \to 2^+$)	0.1/0.1	0.1	0.1
		0.0089±0.0003	1.855	1-st forbidden ($1^- \to 0^+$)	0.0	0.0	0.0

Table 3. Properties of β-decays in ^{238}U chain contributing to geoneutrino signal in IBD detectors. The notations are the same of Table 2.

Tables 2 and 3, respectively.

As one can see from Table 2, the contribution from ^{228}Ac is quite uncertain. It is contributing 0.6% in the total antineutrino signal uncertainty, because of the poorly measured relative intensity of the decay. Decay of ^{214}Bi into the ground state contributes another 0.4% into the total error of the signal. Summed quadratically, these two contributions constitute 0.7%, and other transitions in both decay chains do not contribute significantly to the total signal uncertainty. The uncertainty of the β-decay shape can, in principle, contribute much more to the signal uncertainty. The values after slashes in the S column of the tables are calculated in assumption of the β-decay shape in accordance with decay type (mostly first forbidden) and can serve as an estimation of (maximum) systematics associated with unknown precise shapes of the decay. One can see that these errors can contribute up to 10% into the signal. These errors cannot be summed quadratically because of the unknown correlations; all can contribute in the same direction. Any deviation from the allowed shape towards the forbidden one will increase the total signal. There are good qualitative reasons to assume the allowed shape for these β-decays, as discussed in [18]. At any rate, the quantitative estimation of the corresponding errors is quite complicated. It is clear that only precise experiments can provide proper shapes. The most important fraction of β-spectra to be investigated, from the point of view of geoneutrino signal calculations, lays in its low-energy region, the most difficult region to access in experimental studies (because of the detector threshold and increase of backgrounds at low energies).

An attempt to measure the shape of the ^{214}Bi $\beta + \gamma$ decay was reported in [17]. The feeding probability of the lowest state of ^{214}Bi, providing the most important contribution for geoneutrino signal, was found to be $p_0 = 0.177 \pm 0.004$ under the assumption of universal neutrino spectrum shape, consistent with the indirect estimate of the table of isotopes[1]. Larger statistics and reduction of systematics should allow for the testing of possible distortions of the neutrino spectrum from that predicted using the universal shape [17]. It was found that the effect of uncertainties on the geoneutrino signal is mostly negligible: in the fit of the ^{214}Bi decay data with the completely unconstrained shape parameter, p_0 and p_1 (feeding probabilities to the ground and the first excited states, the only one giving a contribution above 1.81 MeV) the effects of changes of shape and of p_0 on the signal are anticorrelated, if the spectrum is deformed such that there are more (less) low-energy electrons, the corresponding best-fit value for p_0 is lower (higher). While the values of p_0 range from 0.13 to 0.20, the signal changed only by about $\pm 2\%$, i.e. the resulting signal depends weakly on the spectral shape [17].

7. Conclusions

Geoneutrino existence is confirmed independently by Borexino and KamLAND. The precision of both available measurements is still too low: $\simeq$ 28% and 17% correspondingly for the U+Th signal using chondritic mass ratio. Different geological models for the moment can't be discriminated by existing measurements, more precise measurements are needed. Regional measurements in location of experiments are needed to provide more precision for the models. Independent measurements at various sites are highly desirable to check contributions from crust/mantle. For the future precision measurements one should provide corresponding precise nuclear data concerning probabilities of the transition and the shapes of the beta-decay contributing to the geoneutrino signal.

References

[1] Davies J H, Davies D R 2010 Earth's surface heat flux *Solid Earth* **115**
[2] Šrámek O *et al.* 2013 Geophysical and geochemical constraints on geoneutrino fluxes from Earths mantle *Earth and Planetary Science Lett.* **361** 356
[3] Eguchi K, *et al* (KamLAND Collaboration) 2003 First results from KamLAND: evidence for reactor anti-neutrino disappearance *Phys. Rev. Lett.* **90**, 021802
[4] Alimonti G, *et al.* (Borexino Collaboration) 2009 The Borexino detector at the Laboratori Nazionali del Gran Sasso. *NIM A* **600**, 568
[5] Araki T, *et al.* (KamLAND collaboration) 2009 Experimental investigation of geologically produced antineutrinos with KamLAND *Nature* **436** 499
[6] Gando A, *et al.* 2013 Reactor on-off antineutrino measurement with KamLAND *Phys. Rev. D* **88** 033001
[7] Watanabe H 2016 talk at the "Neutrino Research and Thermal Evolution of the Earth"
[8] Bellini G, *et al.* (Borexino collaboration) 2010 *Phys. Lett. B* **687**, 299
[9] Agostini M, *et al.* (Borexino Collaboration) 2015 Spectroscopy of geo-neutrinos from 2056 days of Borexino data *Phys. Rev. D* **92** 031101
[10] Wan L, *et al.* 2017 Geoneutrinos at Jinping: Flux prediction and oscillation analysis *Phys.Rev.D* **95** 053001
[11] Beacom J F, *et al.* 2016 Letter of intent: Jinping Neutrino Experiment arXiv:1602.01733
[12] Barabanov I R, *et al.* 2017 Large-Volume Detector at the Baksan Neutrino Observatory for Studies of Natural Neutrino Fluxes for Purposes of Geo- and Astrophysics *Physics of Atomic Nuclei* **80(3)** 446
[13] Wurm M, *et al.* (LENA Collaboration) 2011 The next-generation liquid-scintillator neutrino observatory LENA arXiv:1104.5620
[14] An F, *et al* (JUNO Collaboration) 2016 Neutrino Physics with JUNO *J. Phys. G* **43** 030401
[15] Learned J G, Dye S T and Pakvasa S 2008 Hanohano: A Deep Ocean Anti-Neutrino Detector for Unique Neutrino Physics and Geophysics Studies arXiv:0810.4975
[16] Strati V *et al.* 2015 Expected geoneutrino signal at JUNO *Progress in Earth and Planetary Science* **2**, 5
[17] Fiorentini G *et al.* 2010 Nuclear physics for geo-neutrino studies *Phys.Rev.C* **81** 034602
[18] Fiorentini G, Lissia M and Mantovani F 2007 Geoneutrinos and Earth interior *Phys.Reports* **453** 117
[19] Alkovali Y A, 1995 Nuclear Data Sheets for A=214 *Nucl. Data Sheets* **76**, 127
[20] ENSDF: Evaluated Nuclear Structure Data File Search and Retrieval *https://www-nds.iaea.org/ensdf*

[1] the consistency was checked against the value p_0 =0.182±0.006 from [19], updated data [20] gives $p_0 = 0.1910 \pm 0.0017$ that is 3.2σ away from the measured. This could indeed indicate the deviations from the allowed universal shape.

Effective value of g_A in β and $\beta\beta$ decays

Jouni Suhonen

University of Jyvaskyla, Department of Physics, P.O. Box 35, FI-40014, Jyvaskyla, Finland

E-mail: jouni.suhonen@phys.jyu.fi

Abstract. Effective value of the weak axial-vector coupling strength, g_A, in nuclear β and $\beta\beta$ (double beta) decays is discussed. Both Gamow-Teller and first-forbidden β decays are included in the analyses. Quenching of g_A in β and two-neutrino $\beta\beta$ decays is reviewed and impact of this quenching on neutrinoless $\beta\beta$ decays is addressed. New measurements of β spectra of first-forbidden non-unique β decays are encouraged, to learn about the mesonic enhancement of the weak axial charge in these decays.

1. Introduction

The impact of the quenching of g_A on the half-lives of neutrinoless double beta ($0\nu\beta\beta$) decay has recently been discussed in Ref. [1]. The related decay rates are affected by the available phase space (Q values), the nuclear matrix elements (NMEs) and the value of g_A in its fourth power [2, 3, 4, 5]. The $0\nu\beta\beta$ decay is mediated by Majorana neutrinos and the measurements of the related half-lives offer access to the absolute mass scale of the neutrinos [2]. A number of nuclear models, including configuration-interaction based models like the interacting shell model (ISM) and the proton-neutron quasiparticle random-phase approximation (pnQRPA), and various mean field models, have been adopted for the calculations [5, 6, 7].

Surprisingly little attention has been paid, at least in the theory community, to the possible (large) quenching of g_A and its possibly strong impact on the sensitivities of the present and planned $0\nu\beta\beta$-decay experiments [1]. This deviation (quenching or sometimes enhancement) from the free-nucleon value $g_A = 1.27$ can arise from the *nuclear medium effects* and the *nuclear many-body effects*. The former contain quenching related to the presence of spin-multipole giant resonances [8], non-nucleonic degrees of freedom (like the Δ isobar [9, 10]) and meson-exchange-driven two-body weak currents [11, 12, 13]. The latter relates to deficiencies of the nuclear many-body approaches used to compute the wave functions involved in the decay transitions. The effective value of g_A can also depend on the energy scale of the process in question: the effective value can be different for β decays (zero-momentum-exchange limit) and $0\nu\beta\beta$ decays (high momentum exchanges, ~ 100 MeV).

The effective value of g_A relates to the *renormalization factor* q (in case of quenching it is called *quenching factor* and in case of enhancement *enhancement factor*):

$$q = \frac{g_A}{g_A^{\rm free}}, \tag{1}$$

where $g_A^{\rm free} = 1.2723(23)$ [14] is the free-nucleon value of the axial-vector coupling as measured in neutron beta decay. Here g_A is the value of the axial-vector coupling derived from a given theoretical or experimental analysis. From (1) one can derive the *effective* value of $g_{\rm A}$ as

$$g_{\rm A}^{\rm eff} = q g_{\rm A}^{\rm free} \,. \tag{2}$$

2. Quenching of $g_{\rm A}$ in Gamow-Teller β decays

Gamow-Teller decays are mediated by the Pauli spin operator $\boldsymbol{\sigma}$ [15] and they thus change the initial nuclear spin J_i by at most one unit. The renormalization of $g_{\rm A}$ has long been studied for the Gamow-Teller β decays in the framework of the interacting shell model (ISM). In these calculations, reviewed in Table 1, it appears that the value of $g_{\rm A}$ is quenched, and the stronger the larger the nuclear mass A.

Table 1. Mass ranges and effective values of $g_{\rm A}$ extracted from the works of the last column.

Mass range	$g_{\rm A}^{\rm eff}$	Reference
Full $0p$ shell	$1.03^{+0.03}_{-0.02}$	[16]
$0p$ – low $1s0d$ shell	1.18 ± 0.05	[17]
Full $1s0d$ shell	$0.96^{+0.03}_{-0.02}$	[18] (see also [19])
	1.0	[20]
$A = 41 - 50$ ($1p0f$ shell)	$0.937^{+0.019}_{-0.018}$	[21] (see also [19])
$1p0f$ shell	0.98	[20]
^{56}Ni	0.71	[20]
$A = 52 - 67$ ($1p0f$ shell)	$0.838^{+0.021}_{-0.020}$	[22]
$A = 67 - 80$ ($0f_{5/2}1p0g_{9/2}$ shell)	0.869 ± 0.019	[22]
$A = 63 - 96$ ($1p0f0g1d2s$ shell)	0.8	[23]
$A = 76 - 82$ ($1p0f0g_{9/2}$ shell)	0.76	[24]
$A = 90 - 97$ ($1p0f0g1d2s$ shell)	0.60	[25]
^{100}Sn	0.52	[20]
$A = 128 - 130$ ($0g_{7/2}1d2s0h_{11/2}$ shell)	0.72	[24]
$A = 130 - 136$ ($0g_{7/2}1d2s0h_{11/2}$ shell)	0.94	[26]
$A = 136$ ($0g_{7/2}1d2s0h_{11/2}$ shell)	0.57	[24]

In Fig. 1 the ISM results of Caurier *et al.* [24] (red horizontal bars indicating the mass range) are contrasted against those obtained by the use of the proton-neutron quasiparticle random-phase approximation (pnQRPA) in the works [27, 28, 29] (see also [30]). The pnQRPA results constitute the light-hatched regions in the background of the ISM results. As can be seen in the figure, the ISM results and the pnQRPA results are commensurate with each other, which is non-trivial considering the large differences in their many-body philosophy.

3. Quenching of $g_{\rm A}$ in two-neutrino $\beta\beta$ decays

Recently the possibly decisive role of $g_{\rm A}$ in the half-life and discovery potential of the $0\nu\beta\beta$ experiments has surfaced [1, 31]. In Barea *et al.* [31] a comparison of the experimental and computed $2\nu\beta\beta$ half-lives of a number of nuclei yielded the rather surprising result

$$g_{\rm A}^{\rm eff}({\rm IBM\text{-}2}) = 1.269 A^{-0.18} \;;\quad g_{\rm A}^{\rm eff}({\rm ISM}) = 1.269 A^{-0.12} \,, \tag{3}$$

where A is the mass number and IBM-2 stands for the microscopic interacting boson model. The results (3), depicted in Fig. 1 as red (ISM) and blue (IBM-2) dotted curves, imply that strongly

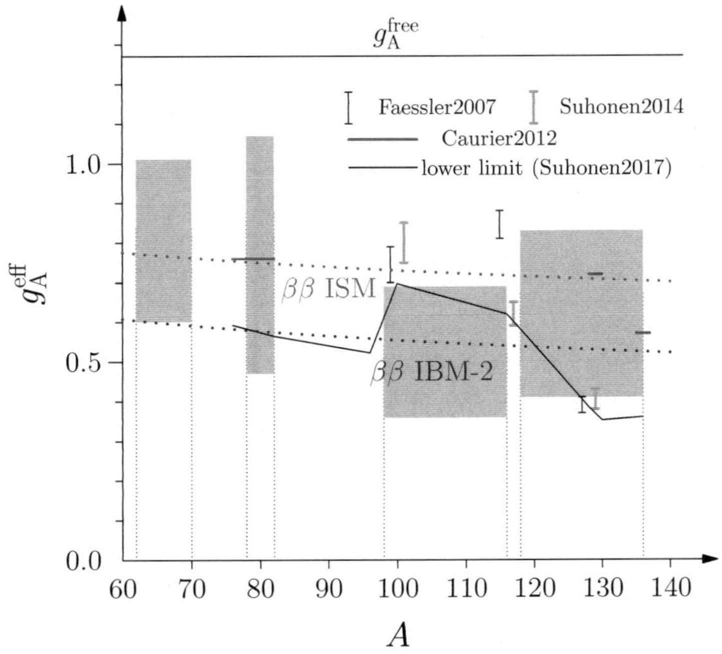

Figure 1. Effective values of g_A in different theoretical approaches for the nuclear mass range $A = 62 - 136$. The quoted references are *Suhonen2017* [1], *Caurier2012* [24], *Faessler2007* [32] and *Suhonen2014* [34]. For more information see the text.

quenched effective values of g_A are possible, thus being a threat to the discovery potential of the $0\nu\beta\beta$ experiments.

Although the study [31] was the first to draw considerable attention in the experimental $0\nu\beta\beta$ community, it was not the first one to point to a possible strong quenching of g_A. Already the pnQRPA study of Faessler *et al.* [32] gave indications of a heavily quenched effective g_A, in the range $g_A^{\rm eff} = 0.39 - 0.84$. These results, along with their 1σ errors, are shown in Fig. 1 as black vertical bars. Later a similar study was carried out in [33, 34], with results comparable with those of [32] and depicted in Fig. 1 as green vertical bars.

In Suhonen [1] a two-stage fit of the particle-particle parameter $g_{\rm pp}$ of the pnQRPA to the data on two-neutrino $\beta\beta$ decays was performed along the lines first introduced in Šimkovic *et al.* [35] and later used in Hyvärinen *et al.* [36]. The works [35, 36] were extended in [1] to include also strongly quenched values of g_A. In this analysis it turned out that there is a minimum value of g_A for which the maximum NME can fit the $2\nu\beta\beta$-decay half-life. This lower limit of the possible g_A values is presented in Fig. 1 as a solid black line and it is seen that it is consistent with the thick green vertical bars of g_A ranges obtained in [33, 34] and also commensurate with the thin black vertical bars obtained in [32]. However, the main message of Suhonen [1] is that no matter how quenched the value of g_A is, the half-lives of the present and future neutrinoless $\beta\beta$-decay measurements are only affected by factors of 6 or less.

4. Nuclear-structure effects in forbidden unique β decays

Forbidden unique β decays are the simplest beyond the allowed β decays since, in the leading order, only one nuclear matrix element is involved in the process and thus the spectrum shape is a universal one, essentially independent of nuclear structure. The relation between the corresponding half-life and the NME is

$$f_{K\mathrm{u}}t_{1/2} = \frac{\kappa}{B_{K\mathrm{u}}} \; ; \quad B_{K\mathrm{u}} = \frac{g_{\mathrm{A}}^2}{2J_i+1}|M_{K\mathrm{u}}|^2 \, , \tag{4}$$

where J_i is the angular momentum of the mother nucleus and κ is a constant with value $\kappa = 6147$ s [37]. The involved NME is given by

$$M_{K\mathrm{u}} = \sum_{ab} M^{(K\mathrm{u})}(ab)(\psi_f||[c_a^{\dagger}\tilde{c}_b]_{K+1}||\psi_i) \, , \tag{5}$$

where the factors $M^{(K\mathrm{u})}(ab)$ are the single-particle matrix elements and the quantities $(\psi_f||[c_a^{\dagger}\tilde{c}_b]_{K+1}||\psi_i)$ are the one-body transition densities, ψ_i being the initial-state and ψ_f the final-state wave function. The operator $c_a^{\dagger}$ is a creation operator for a nucleon in the orbital a and the operator $\tilde{c}_a$ is the corresponding annihilation operator. The single-particle matrix elements are given by

$$M_{K\mathrm{u}}(ab) = \sqrt{4\pi}\left(a||r^K[Y_K\boldsymbol{\sigma}]_{K+1}i^K||b\right) \, , \tag{6}$$

where Y_K is a spherical harmonic of rank K, r the radial coordinate, and a and b stand for the single-particle orbital quantum numbers. The NME (6) is given explicitly in [15].

In [38] a systematic study of the $2^-_{\mathrm{gs}} \leftrightarrow 0^+_{\mathrm{gs}}$ transitions between an odd-odd ground state 2^- and an even-even ground state 0^+ was performed in cases where experimental data on the $\log ft$ values of Eq. (4) are available. The associated NME $M_{1\mathrm{u}}$ of (5) was calculated by using a simple two-quasiparticle model ($\bar{M}_{\mathrm{qp}}$) and the pnQRPA ($\bar{M}_{\mathrm{pnQRPA}}$). Here $\bar{M} = \sqrt{M_{\mathrm{L}}M_{\mathrm{R}}}$ represents the geometric mean of the NMEs corresponding to decay transitions between the left-side (M_{L}) or right-side (M_{R}) nucleus and the central one. Use of the geometric mean makes the analysis more stable and very weakly dependent on the value of the particle-particle interaction parameter g_{pp} of the pnQRPA. Comparing the $\bar{M}_{\mathrm{qp}}$ NME, the $\bar{M}_{\mathrm{pnQRPA}}$ NME and the experimental NME $\bar{M}_{\mathrm{exp}}$ (extracted by using the free value $g_{\mathrm{A}} = 1.27$) the following overall ratios for the mass range $A = 72 - 132$

$$k_{\mathrm{NM}} = \frac{\bar{M}_{\mathrm{exp}}}{\bar{M}_{\mathrm{pnQRPA}}} \approx 0.45 \; ; \quad k = \frac{\bar{M}_{\mathrm{pnQRPA}}}{\bar{M}_{\mathrm{qp}}} \approx 0.4 \tag{7}$$

were obtained. Here "NM" denotes the nuclear-medium effects (isobaric resonances, mesonic two-body currents, etc.) combined with the nuclear-model effect related to the imperfection of the nuclear model used in the calculations.

The work of [38] was followed by the study of Ref. [39] where the comparison of the NMEs $\bar{M}_{\mathrm{qp}}$ and $\bar{M}_{\mathrm{pnQRPA}}$ was made for 148 higher-forbidden unique β decays to access the ratio k of Eq. (7). Here lack of experimental data led only to an extrapolation, based on the results of Refs. [27] and [38], for the value of k_{NM}.

5. Enhancement of the axial charge in first-forbidden $J^- \leftrightarrow J^+$ β decays

The enhancement of the axial-charge nuclear matrix element (NME) γ_5 due to meson-exchange currents was first suggested nearly four decades ago [40, 41, 42]. An enhancement of 40–70 % over the impulse-approximation value was predicted based on chiral-symmetry arguments and soft-pion theorems. This enhancement is fundamental in nature and insensitive to nuclear-structure aspects [43, 44]. Since the γ_5 NME is one of the two rank-zero matrix elements contributing to first-forbidden $\Delta J = 0$ transitions it plays an important role in many of these transitions. Therefore, a significant enhancement of this matrix element can also affect the shapes of the corresponding beta spectra.

The half-life of a first-forbidden non-unique beta decay can be written as

$$t_{1/2} = \kappa/\tilde{C}\,, \tag{8}$$

where κ was defined in connection with Eq. (4) and $\tilde{C}$ is the dimensionless integrated shape function, given by

$$\tilde{C} = \int_1^{w_0} C(w_e) p w_e (w_0 - w_e)^2 F_0(Z_f, w_e) dw_e\,, \tag{9}$$

where w_e is the total energy of the emitted electron, w_0 its maximum, p_e is the electron momentum, Z_f is the charge number of the daughter nucleus and $F_0(Z_f, w_e)$ is the Fermi function taking into account the coulombic attraction of the electron and the daughter nucleus. The shape factor $C(w_e)$ of Eq. (9) contains complicated combinations of both (universal) kinematic factors and nuclear form factors. The nuclear form factors can be related to the corresponding NMEs using the impulse approximation. For the first-forbidden non-unique decays with $J_i = J_f$, considered in this review, the relevant NMEs are those of the transition operators denoted here by $\mathcal{O}(0^-)$, $\mathcal{O}(1^-)$, and $\mathcal{O}(2^-)$ [45] and given in the leading order by

$$\mathcal{O}(0^-):\quad g_{\rm A}(\gamma_5)(\boldsymbol{\sigma}\cdot\mathbf{p}_e),\quad g_{\rm A}(\boldsymbol{\sigma}\cdot\mathbf{r})\,, \tag{10}$$

$$\mathcal{O}(1^-):\quad g_{\rm V}\mathbf{p}_e,\quad g_{\rm A}(\boldsymbol{\sigma}\times\mathbf{r}),\quad g_{\rm V}\mathbf{r}\,, \tag{11}$$

$$\mathcal{O}(2^-):\quad g_{\rm A}[\boldsymbol{\sigma}\mathbf{r}]_2\,, \tag{12}$$

where $\mathbf{r}$ is the radial coordinate and $\mathbf{p}_e$ is the electron momentum. The enhancement of the γ_5 NME ($\boldsymbol{\sigma}\cdot\mathbf{p}_e$ in non-relativistic form) is included in the coupling strength $g_{\rm A}(\gamma_5)$.

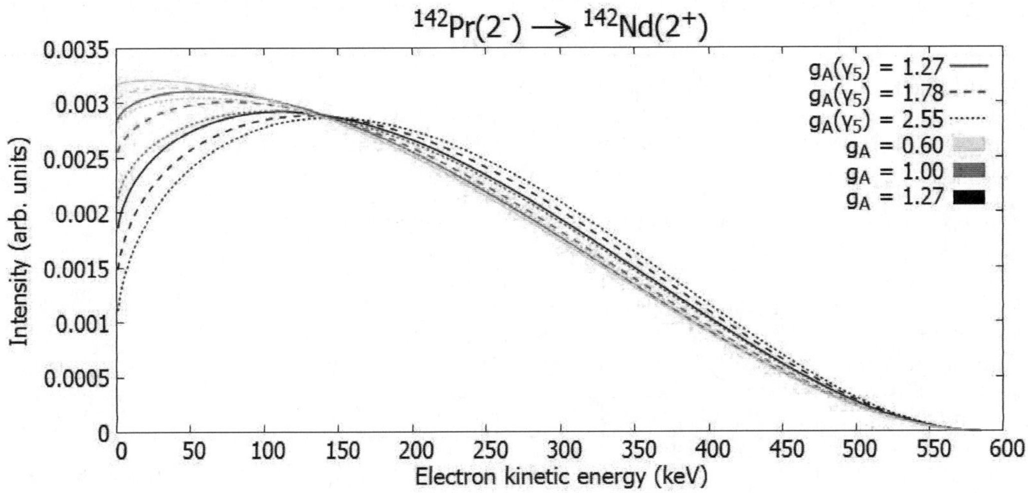

Figure 2. Electron spectrum of the ground-state-to-ground-state decay of ^{142}Pr. The value of $g_{\rm A}(\gamma_5)$ is marked by the line type. The values 1.27, 1.78, and 2.55 correspond to enhancements of 0 %, 40 %, and 100 % of the axial-charge matrix element. The color coding gives the effective value of $g_{\rm A}$ used for the matrix elements $[\boldsymbol{\sigma}\mathbf{r}]_{0,1,2}$.

The enhancement of the axial-charge NME affects not only the partial half-lives of the $J^- \leftrightarrow J^+$ first-forbidden β transitions but also the shapes of the corresponding electron spectra. An example is given in Fig. 2. A clear effect is induced by the values of both the $g_{\rm A}(\gamma_5)$ and $g_{\rm A}$, multiplying the matrix elements $[\boldsymbol{\sigma}\mathbf{r}]_{0,1,2}$ of Eqs. (10)–(12). Experimental measurements of the β spectra would thus be welcome in order to acces the γ_5 enhancement independent of the half-life considerations.

6. Final remark

The discussion above gives only scattered examples in relation to the effective values of the axial-vector coupling strength, as also to the nuclear-structure effects involved in model calculations. A more exhaustive discussion of these subjects is provided in the review [46].

Acknowledgments

This work has been partially supported by the Academy of Finland under the Finnish Centre of Excellence Programme 2012-2017 (Nuclear and Accelerator Based Programme at JYFL).

References

[1] Suhonen J 2017 *Phys. Rev.* C **96** 055501
[2] Suhonen J and Civitarese O 1998 *Phys. Rep.* **300** 123
[3] Maalampi J and Suhonen J 2013 *Adv. High Energy Phys.* **2013** 505874
[4] Vergados J D, Ejiri H and Šimkovic F 2016 *Int. J. Mod. Phys.* E **25** 1630007
[5] Engel J and Menéndez J 2017 *Rep. Prog. Phys.* **80** 046301
[6] Suhonen J and Civitarese O 2012 *J. Phys. G: Nucl. Part. Phys.* **39** 085105
[7] Suhonen J and Civitarese O 2012 *J. Phys. G: Nucl. Part. Phys.* **39** 124005
[8] Jokiniemi L and Suhonen J 2017 *Phys. Rev.* C **96** 034308
[9] Oset E and Rho M 1979 *Phys. Rev. Lett.* **42** 47
[10] Bohr A and Mottelson B R 1981 *Phys. Lett.* **100B** 10
[11] Towner L S 1997 *Phys. Rep.* **155** 263
[12] Menéndez J, Gazit D and Schwenk A 2011 *Phys. Rev. Lett.* **107** 062501
[13] Ekström A *et al.* 2014 *Phys. Rev. Lett.* **113** 262504
[14] Patrignani C *et al.* (Particle Data Group) 2016 *Chin. Phys.* C **40** 100001
[15] Suhonen J 2007 *From Nucleons to Nucleus: Concepts of Microscopic Nuclear Theory* (Berlin: Springer)
[16] Chou W T, Warburton E K and Brown B A 1993 *Phys. Rev.* C **47** 163
[17] Wilkinson D H 1974 *Nucl. Phys.* A **225** 365
[18] Wildenthal B H, Curtin M S and Brown B A 1983 *Phys. Rev.* C **28** 1343
[19] Konieczka M, Baczyk P and Satula W 2016 *Phys. Rev.* C **93** 042501(R)
[20] Siiskonen T, Hjorth-Jensen M and Suhonen J 2001 *Phys. Rev.* C **63** 055501
[21] Martínez-Pinedo G, Poves A, Caurier E and Zuker A P 1996 *Phys. Rev.* C **53** R2602
[22] Kumar V, Srivastava P C and Li H 2016 *J. Phys. G: Nucl. Part. Phys.* **43** 105104
[23] Honma M, Otsuka T, Misuzaki T and Hjorth-Jensen M 2006 *J. Phys.: Conf. Series* **49** 45
[24] Caurier E, Nowacki F and Poves A 2012 *Phys. Lett.* B **711** 62
[25] Juodagalvis A and Dean D J 2005 *Phys. Rev.* C **72** 024306
[26] Horoi M and Neacsu A 2016 *Phys. Rev.* C **93** 024308
[27] Ejiri H and Suhonen J 2015 *J. Phys. G: Nucl. Part. Phys.* **42** 055201
[28] Pirinen P and Suhonen J 2015 *Phys. Rev.* C **91** 054309
[29] Deppisch F F and Suhonen J 2016 *Phys. Rev.* C **94** 055501
[30] Delion D S and Suhonen J 2014 *Europhys. Lett.* **107** 52001
[31] Barea J, Kotila J and Iachello F 2013 *Phys. Rev.* C **87** 014315
[32] Faessler A *et al.* 2008 *J. Phys G: Nucl. Part. Phys.* **35** 075104 ; arXiv 0711.3996v1 [nucl-th] 26 Nov 2007
[33] Suhonen J and Civitarese O 2013 *Phys. Lett.* B **725** 153
[34] Suhonen J and Civitarese O 2014 *Nucl. Phys.* A **924** 1
[35] Šimkovic F, Rodin V, Faessler A and Vogel P 2013 *Phys. Rev.* C **87** 045501
[36] Hyvärinen J and Suhonen J 2015 *Phys. Rev.* C **91** 024613
[37] Hardy J C, Towner I S, Koslowsky V, Hagberg E and Schmeig H 1990 *Nucl. Phys.* A **509** 429
[38] Ejiri H, Soukouti N and Suhonen J 2014 *Phys. Lett.* B **729** 27
[39] Kostensalo J and Suhonen J 2017 *Phys. Rev.* C **95** 014322
[40] Kubodera K, Delorme J and Rho M 1978 *Phys. Rev. Lett.* **40** 755
[41] Guichon P, Giffon M, Joseph J, Laverriere R and Samour C 1978 *Z. Phys.* A **285** 183
[42] Guichon P, Giffon M and Samour C 1978 *Phys. Lett.* **74B** 15
[43] Towner I S and Khanna F C 1981 *Nucl. Phys.* A **372** 331
[44] Delorme J 1982 *Nucl. Phys.* A **374** 541c
[45] Behrens H and Bühring W 1982 *Electron Radial Wave Functions and Nuclear Beta-decay* (Oxford: Clarendon)
[46] Suhonen J 2017 *Frontiers in Physics* **5** 55

IOP Conf. Series: Journal of Physics: Conf. Series **1056** (2018) 012057 doi:10.1088/1742-6596/1056/1/012057

Upgrade of the ICARUS T600 Time Projection Chamber

F Tortorici [1,2,]. V Bellini [1,2], and C M Sutera [2] on behalf of the ICARUS Collaboration

[1] Department of Physics and Astronomy, University of Catania, Via S. Sofia 64, 95123 Catania, IT

[2] INFN, Sezione di Catania, Via S. Sofia 64, 95123 Catania, IT

E-mail: francesco.tortorici@ct.infn.it

Abstract. The ICARUS T600 detector, with about 500 tons of active mass, is the largest Liquid Argon Time Projection Chamber (LAr TPC) ever realised. In 2013 ICARUS concluded an about 4 years long experiment with the T600 detector at the LNGS underground laboratory, taking data both with the CNGS neutrino beam and cosmic rays. This very successful experiment demonstrated the high spatial and energy resolutions, electron/photon separation and particle identification capabilities (via dE/dx vs range measurements) of the LAr technology.

ICARUS Collaboration refurbished the T600 at CERN, in order to move it to FNAL in the framework of the SBN experiment, to serve as far detector in studies on the short baseline neutrino oscillations.

A fundamental part of ICARUS is the light collection system, made of 360 Hamamatsu R5912-MOD, 8 in. diameter, PMT's. This system is dedicated to three tasks: the generation of a light based trigger signal, the identification of the time of occurrence (t0) of each interaction with high time precision and the initial recognition of event topology for fast event selection purposes.

1. Introduction

ICARUS-T600 is an ideal detector for neutrino physics and search for rare events. It is a uniform, self-triggering detector, with high granularity ($\sim$ 1 mm), 3D imaging and calorimetric capabilities. It allows to accurately reconstruct a wide variety of ionizing events with complex topologies.

ICARUS is made of two identical modules, for a total of 476 t active mass. There are 2 LAr TPC's per module (internal dimensions of each chamber are 3.6 (W) x 3.9 (H) x 19.6 (L) m), with a common central cathode generating an average drift electric field E_D= 0.5 kV/cm (so the electrons have drift speed $v_D \sim$ 1.6 mm/μs and 1.5 m drift length); 3 "non-destructive" readout wire planes per TPC: $\approx$ 54000 wires at $0^o, \pm 60^o$ with respect to horizontal direction, called respectively Induction 1, Induction 2 and Collection plane, when going outward from the cathode plane; 8" PMT's arrays, used for time of occurrence t0 signal, timing and triggering of events, sensitive at 128 nm LAr scintillation light via Tetra-Phenil-Butadiene (TPB) wavelength shifter on glass window. The ionisation charge is continuously read (0.4 μs sampling time). On average, $\approx 2 \cdot 10^4$ ionisation electrons are produced per MeV.

Exposed to the Cern Neutrino to Gran Sasso (CNGS) beam, ICARUS lasted in 2013 a very successful 3-year run at Gran Sasso INFN underground laboratory, collecting $8.6 \cdot 10^{19}$ pot event statistics with a detector live time > 93% and cosmic rays to study atmospheric neutrinos (0.73 kt y exposure). Several physics/technical results have been achieved during the run at LNGS: an exceptionally low level ~20 p.p.t. [O2] eq. of electronegative impurities in LAr; the measured electron lifetime $\tau_{ele} > 15$ ms ensured few meters long drift path of the ionising electron signal without attenuation; demonstrated the detector performance, especially in ν_e identification and π^o background rejection in $\nu_\mu - \nu_e$ study to unprecedented level; performed a sensitive search for LSND-like anomaly with CNGS beam, constraining the LSND window to a narrow region at $\Delta m^2 < 1eV^2$. These results marked a milestone for LAr-TPC technology with a large impact on future neutrino and astroparticle physics projects, i.e. the current SBN short base-line neutrino program at FNAL with 3 LAr-TPC's (SBND, MicroBooNE and ICARUS) and the multi-kt DUNE LAr-TPC. T600 detector will be exposed to ~0.8 GeV Booster ν_μ beam, 600 m from target to definitely test the LSND claim searching for $\nu_\mu - \nu_e$ oscillations in the framework of SBN program.

In 2015, T600 detector was moved from LNGS to CERN to introduce some technology developments while maintaining the already achieved performance. To face the new experimental situation at FNAL, ICARUS T600 underwent an intensive overhauling at CERN in the framework of CERN Neutrino Platform (WA104/NP01 project) before being shipped to FNAL, where ICARUS will operate beneath the SBN Far Site Building. The refurbishing consisted in: new cold vessels, purely passive insulation; renovated cryogenic/LAr purification equipment; flattening of TPC cathode (the punched stainless-steel panels, 58% transparency, underwent a thermal treatment improving planarity to few mm); upgrade of light collection system, and new higher performance TPC read-out electronics.

In section 2, we will see what is the sterile neutrino puzzle, whose study is one of the main points of the SBN program and the fundamental reason for the upgrade of ICARUS. In section 3 we will review the current performance of ICARUS, achieved during the 2010-2013 Gran Sasso run. In section 4 and 5, we will talk about the light collection system and electronics, respectively. Finally, in section 6, we will analyse the essential need for a Cosmic Ray Tagger (CRT) system.

2. Sterile neutrino puzzle

Anomalies have been collected in last years in neutrino sector despite the well-established 3-flavour mixing picture within Standard Model: appearance of ν_e from ν_μ beams in accelerator experiments (LSND + MiniBooNE, combined evidence > 3σ) [1]; disappearance of anti-ν_e, hinted by nearby nuclear reactor experiments (ratio observed/predicted event rates R = 0.9384 ± 0.024) [2]; disappearance of ν_e, hinted by solar neutrino experiments during their calibration with Mega-Curie sources (SAGE, GALLEX, R = 0.84 ± 0.05). [3] These results hint to a new "sterile" flavor, described by $\Delta m^2 \sim eV^2$ and small mixing angle, driving oscillations at short distance.

ICARUS constrained $\Delta m^2_{new} \leq 1eV^2$, and small mixing; Planck data [4] and Big Bang cosmology point to at most one further flavor with $m_{new} < 0.27eV$; no evidence of ν_μ disappearance in MINOS and IceCube in 0.32-20 TeV energy range; recent reactor data (especially NEOS) are intriguing but inconclusive. New results are expected from ongoing/new experiments at reactor/radioactive source.

The experimental scenario calls for a definitive clarification.

ICARUS searched for ν_e-excess related to LSND-like anomaly with CNGS neutrino beam ($\sim$ 1% intrinsic ν_e contamination) despite the larger $L/E_\nu \sim 36.5$ m/MeV when compared to $L/E_\nu \sim 1$ m/MeV for LSND and MiniBooNE; LSND-like oscillation signal would average to $\sin^2(1.27\Delta m^2 L/E) \sim 1/2$; compared to MINOS and T2K, ICARUS operated in a L/E_ν range where contributions from standard oscillations are not yet too relevant. No excess was observed in a $7.93 \cdot 10^{19}$ pot sample: 7 ν_e charged current (CC) events compared to 8.5 ± 1.1 expected in absence of effect, providing the limits:

$$P(\nu_\mu \to \nu_e) \leq 3.85 \cdot 10^{-3}(90\%C.L.)$$
$$P(\nu_\mu \to \nu_e) \leq 7.60 \cdot 10^{-3}(99\%C.L.)$$

ICARUS restricted the allowed LSND parameters to a narrow region $\Delta m^2 < 1eV^2, \sin^2 2\theta \sim 0.005$ where all positive / negative experimental results can be coherently accomodated at 90% C.L., see figure 1.

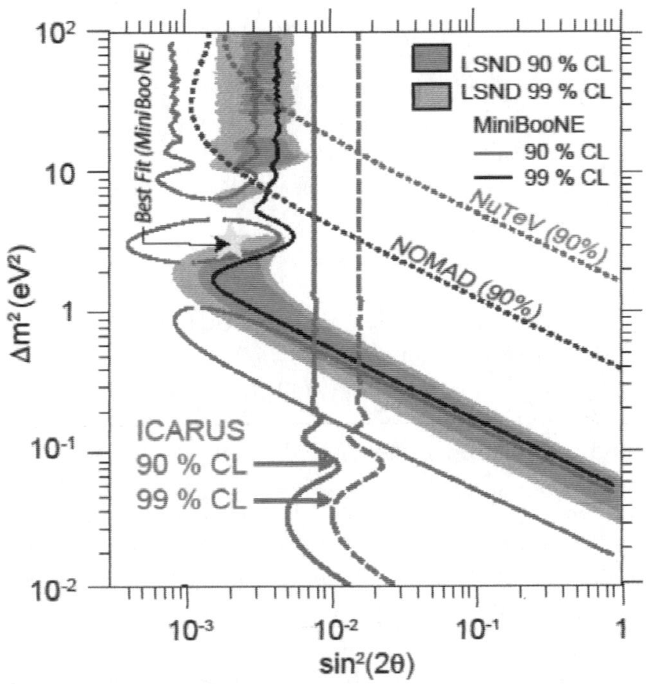

Figure 1. Parameter space experimental limits for a sterile neutrino

SBN experiment will likely clarify both LSND and reactor anomalies by precisely and independently measuring both ν_e appearance and ν_μ disappearance, mutually related through $\sin^2(2\theta_{\mu e}) \leq \frac{1}{4}\sin^2(2\theta_{\mu x})\sin^2(2\theta_{ex})$

3. Performance

In this section we will show the performance achieved by ICARUS with CNGS beam and cosmic neutrinos.

As a tracking device, T600 allows precise 3D event topology reconstruction with $\sim 1mm^3$ resolution for any ionising particle.

The detector is also a full sampling homogeneous calorimeter: total energy is reconstructed by charge integration with excellent accuracy for contained events (see eq. 1); the momentum of non contained muons is calculated by Multiple Coulomb Scattering (MCS).

$$\frac{\sigma(E)}{E} = \begin{cases} 11\%/\sqrt{E(MeV)} + 2\% & \text{low energy electrons} \\ 3\%/\sqrt{E(GeV)} & \text{electromagnetic showers} \\ 30\%/\sqrt{E(GeV)} & \text{hadron showers} \end{cases} \quad (1)$$

Muon momentum measurement by MCS is essential for escaping muons and can also used to complement the range measurement for stopping muons. RMS of muon deflection angle θ_{RMS} depends on momentum p, spatial resolution and segmentation. Such method, developed in [5] by ICARUS collaboration, has been validated comparing p_{MCS} with the corresponding calorimetric measured p_{CAL} for $\sim 10^3$ stopping muons (track length > 5 m, of which 4 m are used, corresponding to ~4-5 interaction lengths in LAr) from CNGS ν_μ interacting in upstream rock. A $\Delta p/p \sim 15\%$ resolution on average has been determined in 0.4 - 4 GeV/c range.

LAr tecnology permits a remarkable $e/\gamma/\pi^0$ separation and a powerful particle identification by dE/dx vs range (0.02 X_0 sampling, where $X_0 = 14$ cm is the interaction length in LAr). There are three "handles" to separate $e/\gamma/\pi^0$: 1) dE/dx profile corresponds to a single (for electrons) vs. double (for photons) m.i.p. (minimum ionization particle), 2) photon conversion separation from primary vertex (lower for pions), 3) invariant mass of π^0. The unique detection properties of the LAr-TPC allow to identify unambiguosly individual e-events with high efficiency combining the informations of Collection and Induction2 views.

4. Light collection system

Currently, in ICARUS, light has simply been integrated over all the PMT's. Fine, but light collection can do more than just the generation of a trigger signal. It can also be used to identify the time of occurrence t_0 of each interaction with high time precision, and to recognise the event topology for fast selection purposes.

To fulfil its goal a light collection system needs high detection coverage (to be sensitive to energy deposition in LAr down in energy to 100 MeV), high detection granularity (for space resolution purposes), fast time response ($\sim$ 1 ns) to allow accurate time tracking of each event in the T600 drift window and to take advantage of the available 2 ns/19 ns bunched beam structure of the Fermilab Booster facility. Different geometries and cathode coverage area (fraction of the wire plane surface covered by PMT windows) have been tested. The 90 Hamamatsu R5912-MOD series (8", 10 dynodes) PMT's per TPC layout (with 5% cathode coverage area) configuration shown in figure 2, has been chosen. Grey blocks represent PMT's positions. Longitudinal

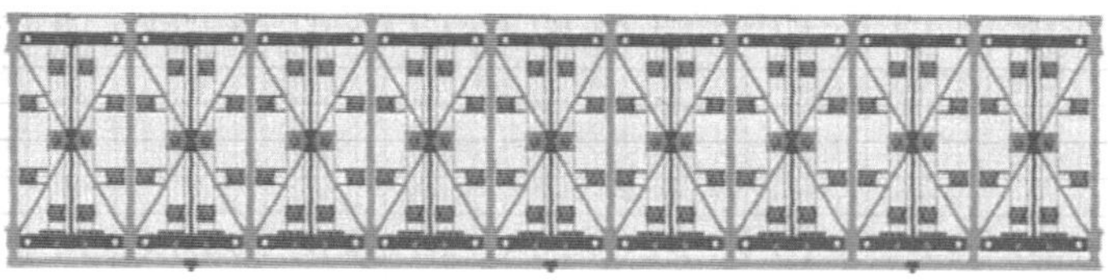

Figure 2. Chosen layout per TPC of the PMT's, represented by grey boxes. This layout includes 90 PMT's.

resolution is better than 0.5 m (effective Q.E. = 5%). A total of 4 modules (so 400 PMT's including 10% of spares) will be used for ICARUS-T600.

Hamamatsu R5912-MOD series are rated for cryogenic temperature, as they feature a cathode with platinum under-layer. PMT sand blasted glass windows is coated by $\sim 200\mu g/cm^2$ of Tetra-Phenyl-Butadiene (TPB) wavelength shifter to detect the $\lambda = 128$ nm scintillation light in LAr. Each PMT is enclosed in a wire screening cage to prevent induction of PMT pulses on the facing TPC wires. PMT timing/calibration will be provided by a LASER light system.

IOP Conf. Series: Journal of Physics: Conf. Series **1056** (2018) 012057 doi:10.1088/1742-6596/1056/1/012057

We tested all the PMT's before installation in the T600, to verify their compliance with the required functioning specifications. PMT's tests were organised in different CERN areas: tests at warm temperature were carried out in consecutive bunches of 16 samples in a dark room and a dedicated laboratory, whereas cryogenic tests were accomplished using a cryogenic facility which allowed the simultaneous measurement of 10 PMT's in LAr bath, as the producer made only a mechanical check in liquid nitrogen. Measurements included the gain as a function of the power supply, the peak-to-valley ratio, the dark count rate, the linearity of the response as a function of the light intensity and uniformity of the cathode surface. As expected the PMT's show in general an almost constant relative variation of the gain, peak to valley ratio and dark counts from room temperature down to 87K.

The achievement of the required ~1 ns timing resolution requires a PMT timing calibration system to compensate individual channel delays and transit-time drifts.

5. ICARUS Electronics

The general approach to the trigger system is the centralisation of the basic functionalities into the NI-PXI crate already used during previous ICARUS run at LNGS, with the following requirements:

- at least one FPGA will be devoted to time critical processes, as clock generation, handling of beam gates and time-stamping of signals;
- one FPGA will be dedicated to manage the PMT signals;
- one Real Time controller will handle handshake with DAQ;
- one FPGA will be dedicated to manage the signals coming from other equipment of the far detector.

Architecture of ICARUS electronics at LNGS was based on analogue low noise "warm" front-end amplifier, a multiplexed 10-bit 2.5 MHz AD converter and a digital VME module for local storage, data compression, trigger information. The signal to noise ratio $S/N \sim 9$ in Collection and ~ 0.7 mm single hit resolution, result in a precise spatial event reconstruction and muon momentum measurement by MCS. The improvements on electronics concern:

- serial 12 bits ADC, one per ch, 400 ns sampling synchronous on the whole detector, whereas previous boards were aligned within 400 ns;
- serial bus architecture with Gbit/s optical links to increase the bandwidth (10 MHz);
- new compact design to host both analogue/digital electronics (single high performance FPGA) directly on ad-hoc signal feedthrough flanges acting as electronics backplane;
- analogue front-end for a better event reconstruction
    * a faster shaping time $\sim 1.5\mu$s of analogue signals to match electron transit time in wire plane spacing;
    * a drastic reduction of undershoot in the preamp response as well as of the low frequency noise while maintaining a same or better S/N;
    * a same preamp for both Induction and Collection wires.

In addition the full 400 ns synchronous signal sampling on the whole detector will allow to slightly improve the resolution on muon momentum by MCS.

6. Cosmic ray tagger

ICARUS at FNAL will face a more challenging experimental condition than at LNGS, requiring the identification of neutrino interactions amongst 11 kHz of cosmic rays. A 3 m concrete overburden will remove contribution from charged hadrons/γ's. Moreover ~11 muon tracks will occur per triggering event in 1 ms TPC drift readout; the associated photons represent a serious background source for ν_e search since electrons produced via Compton scattering / pair production can mimic a genuine ν_e CC event.

Rejecting cosmic background, i.e. reconstructing the triggering event, would require to precisely know timing of each track in the TPC image, exploiting a much improved light detection system (described in section 4), high granularity and $\sim$ ns time resolution, and an external cosmic ray tagger to detect incoming particles and their direction of propagation by time-of-flight measurements. The latter may be done via scintillating bars surrounding T600 (aim: 98% coverage) equipped with optical fibers to convey light to SiPM arrays.

Top coverage of the CRT is under INFN/ CERN responsibility. FNAL is recovering modules by MINOS/Double Chooz for side and bottom.

7. Conclusions

LAr-TPC detection technique has been taken to full maturity with ICARUS-T600, a result of many years of R&D with continuous support of INFN. ICARUS completed in 2013 a successful continuous 3-year run at LNGS exposed to CNGS neutrinos and cosmic rays, obtaining remarkable physics and technical achievements, proving the effectiveness of the single phase LAr-TPC technology for neutrino physics. The ability in reconstructing neutrino interactions with complex topologies in a broad energy range, combined with an efficient identification of primary electrons and a unique $e/\gamma/\pi^0$ separation, allows rejecting backgrounds in the search for $\nu_\mu \rightarrow \nu_e$ transitions at an unprecedented level. ICARUS performed a sensitive search for a potential ν_e excess related to the LSND-like anomaly with CNGS beam defining a narrow region at $(\Delta m^2, \sin^2 2\theta) \sim (1eV^2, 0.005)$ which has to be investigated to definitively settle the LNSD hint of a sterile neutrino. ICARUS underwent a major overhauling at CERN and has been transported to FNAL to be exposed to Booster and NuMi neutrinos. The SBN experiment will provide a clarification of the sterile neutrino issue, both in appearance and disappearance modes.

References

[1] LSND Collaboration (A Aguilar *et al.*) 2001 Evidence for neutrino oscillations from the observation of electron anti-neutrinos in a muon anti-neutrino beam, *Phys. Rev.* D **64**112007

[2] Daya Bay Collaboration (F P An *et al.*) 2012 Observation of electron-antineutrino disappearance at Daya Bay *Phys. Rev. Lett.* **108** 171803

[3] GALLEX Collaboration (P Anselmann *et al.*) 1997 Gallium solar neutrino experiments: absorption cross sections, neutrino spectra, and predicted event rates, *Phys. Rev.* C **56** 3391
SAGE collaboration (J N Abdurashitov *et al.*) 2009 Measurement of the solar neutrino capture rate with gallium metal. III. Results for the 2002-2007 data-taking period, *Phys. Rev.* C **80** 015807

[4] Planck Collaboration (P A R Ade *et al.*) 2016 Planck 2015 results. XIII. Cosmological parameters, *A&A* **594** A13

[5] ICARUS Collaboration (M Antonello *et al.*) 2017 Muon momentum measurement in ICARUS-T600 LAr-TPC via multiple scattering in few-GeV range, *JINST* **12** no.04 P04010

The B16 Standard Solar Models

N Vinyoles[1], A Serenelli[1] and F L Villante[2,3]

[1]Institut de Ciències de l'Espai (CSIC-IEEC), Campus UAB, Carrer de Can Magrans, S/N, E-08193 Barcelona, Spain

[2]Dipartimento di Scienze Fisiche e Chimiche, Università dell'Aquila, I-67100 L'Aquila, Italy

[3]Istituto Nazionale di Fisica Nucleare (INFN), Laboratori Nazionali del Gran Sasso (LNGS), I-67100 Assergi (AQ), Italy

E-mail: `francesco.villante@lngs.infn.it`

Abstract. We describe a new generation of standard solar models (SSMs), Barcelona 2016 or B16 for short, that includes recent updates on some important nuclear reaction rates, a more consistent treatment of the equation of state and a novel and flexible treatment of opacity uncertainties. Two large sets of SSMs, each based on a different canonical set of solar abundances with high and low metallicity, are calculated and compared with different ensembles of solar observables including solar neutrinos, surface helium abundance, depth of convective envelope and sound speed profile.

1. Introduction

In the last three decades, there was an enormous progress in our understanding of the sun. The predictions of the Standard Solar Model (SSM), which is the fundamental theoretical tool to investigate the solar interior, have been tested by solar neutrino experiments and by helioseismology. The deficit of the observed solar neutrino fluxes, reported initially by Homestake [1, 2] and then confirmed by GALLEX [3], SAGE [4], GNO [5], Kamiokande [6] and Super-Kamiokande [7], generated the so-called solar neutrino problem which stimulated a deep investigation of the solar structure. The problem was solved in 2002 when the SNO experiment [8] obtained a direct evidence for flavour oscillations of solar neutrinos and, moreover, confirmed the SSM prediction of the ^{8}B neutrino flux.

Nowadays, we have a good knowledge of the solar neutrino oscillation probability and a direct experimental determination of most of the solar neutrino components. Super-Kamiokande [9] and SNO [10] have provided a high accuracy determination of ^{8}B neutrinos. Borexino has recently obtained a direct measure of the *pp*, *pep*, ^{7}Be [11, 12] and ^{8}B [13] solar neutrino fluxes and it also has the potential to provide the first direct measurements of the CNO neutrinos [14] in the next future. In addition, helioseismic observations have allowed to determine precisely several important properties of the sun, such as the depth of the convective envelope which is known at the $\sim 0.2\%$ level, the surface helium abundance which is obtained at the $\sim 1.5\%$ level and the sound speed profile which is determined with an accuracy equal to $\sim 0.1\%$ in a large part of the sun (see e.g. [15, 16] and references therein). As a results of these observations, the solar structure is now very well constrained, so that the sun can be used as a solid benchmark for stellar evolution and as a laboratory for fundamental physics.

A new solar problem has, however, emerged during the last years. Recent determinations of the photospheric heavy element abundances [17, 18, 19] indicate that the sun metallicity is lower

	S(0)	Uncert.(%)	$\Delta S(0)/S(0)$	Ref.
S_{11}	$4.03 \cdot 10^{-25}$	1	0.5%	[27, 28, 29]
S_{17}	$2.13 \cdot 10^{-5}$	4.7	+2.4%	[30]
S_{114}	$1.59 \cdot 10^{-3}$	7.5	-4.2%	[31]

Table 1. *Astrophysical S-factors (in units of MeV b) and uncertainties updated in this work. Fractional changes with respect to [32] are also included.*

than previously assumed [20, 21]. Solar models that incorporate these lower abundances are no more able to reproduce the helioseismic results. As an example, the sound speed predicted by SSMs at the bottom of the convective envelope disagrees at the 1% level with the value inferred by helioseismic data (see e.g.[22]). Detailed studies have been done to resolve this controversy (see e.g.[23]), but a definitive solution of the solar composition problem still has to be obtained.

In this brief review, which is largely based on [24], we present a new generation of SSMs, Barcelona 2016 or B16 for short, that includes recent updates on some important nuclear reaction rates, a more consistent treatment of the equation of state and a novel and flexible treatment of opacity uncertainties. Two large sets of SSMs, each based on a different canonical set of solar abundances with high and low metallicity, and with input parameters chosen randomly from their respective distributions, are computed. The predictions (and the uncertainties) for different solar observables, including solar neutrino fluxes, surface helium abundance, depth of convective envelope and sound speed profile, are presented and compared with the observational determinations.

2. The B16 standard solar models

SSMs are a snapshot in the evolution of a $1M_\odot$ star, calibrated to match present-day surface properties of the Sun. The calibration is done by adjusting the mixing length parameter ($\alpha_{\rm MLT}$) and the initial helium and metal mass fractions ($Y_{\rm ini}$ and $Z_{\rm ini}$ respectively) in order to satisfy the constraints imposed by the present-day solar luminosity $L_\odot$, radius $R_\odot$, and surface metal to hydrogen abundance ratio $(Z/X)_\odot$. The new B16 models share with previous calculations [25] much of the input physics, but include important updates. A brief account of few relevant ingredients is given in the following.

Equation of State: B16 SSMs employ, for the first time, EoS tables calculated consistently for each of the compositions used in the solar calibrations by using FreeEOS [26].

Nuclear rates: The rates of $p(p, e^+\nu_e)d$, $^7Be(p,\gamma)^8B$ and $^{14}N(p,\gamma)^{15}O$ reactions have been updated, see Tab.1[1]. For the important reaction $^3He(^4He,\gamma)^7Be$ (not included in Tab.1), two recent analyses [33, 34] have provided determinations of the astrophysical factor that differs by about 6% (to be compared with a claimed accuracy equal to 4% and 2% for [33] and [34], respectively). Considering that the results from [33] and [34] bracket the previously adopted value from [32], the latter was considered as preferred choice in B16 SSMs.

Radiative opacities: In [25] the opacity error was modelled as a 2.5% constant factor at 1σ level, comparable to the maximum difference between OP [35] and OPAL [36] opacities in the solar radiative region. It was shown, however, in [37] that this prescription underestimates the contribution of opacity uncertainty to the sound speed and convective radius error budgets

[1] For the $p(p, e^+\nu_e)d$ reaction, the quoted value for $S_{11}(0)$ underestimates the actual increase of the rate because the variation of $S_{11}(E)$ at solar energies is dominated by changes in the first and higher order derivatives of the Taylor expansion of the astrophysical factor around $E = 0$ (see [24] for details).

	GS98	AGSS09met	Obs
$\Phi(\mathrm{pp})$	$5.98(1 \pm 0.006)$	$6.03(1 \pm 0.005)$	$5.971^{+0.037}_{-0.033}$
$\Phi(\mathrm{pep})$	$1.44(1 \pm 0.01)$	$1.46(1 \pm 0.009)$	1.448 ± 0.013
$\Phi(\mathrm{hep})$	$7.98(1 \pm 0.30)$	$8.25(1 \pm 0.30)$	19^{+12}_{-9}
$\Phi(^7\mathrm{Be})$	$4.93(1 \pm 0.06)$	$4.50(1 \pm 0.06)$	$4.80^{+0.24}_{-0.22}$
$\Phi(^8\mathrm{B})$	$5.46(1 \pm 0.12)$	$4.50(1 \pm 0.12)$	$5.16^{+0.13}_{-0.09}$
$\Phi(^{13}\mathrm{N})$	$2.78(1 \pm 0.15)$	$2.04(1 \pm 0.14)$	≤ 13.7
$\Phi(^{15}\mathrm{O})$	$2.05(1 \pm 0.17)$	$1.44(1 \pm 0.16)$	≤ 2.8
$\Phi(^{17}\mathrm{F})$	$5.29(1 \pm 0.20)$	$3.26(1 \pm 0.18)$	≤ 85
Y_S	0.2426 ± 0.0059	0.2317 ± 0.0059	0.2485 ± 0.0035
R_CZ	0.7116 ± 0.0048	0.7223 ± 0.0053	0.713 ± 0.001

Table 2. *Neutrino fluxes for the two B16 SSMs and as determined by [43]. The fluxes are given in units of* 10^{10} (pp), 10^9 (^{7}Be), 10^8 (pep,^{13}N,^{15}O), 10^6 (^{8}B,^{17}F) *and* 10^3 (hep) $\mathrm{cm^{-2}s^{-1}}$. *The last two lines give the surface helium* Y_S *and the convective radius* R_CZ. *The observational values are given by [44] and [45], respectively.*

because the effects produced by opacity variations in different zones of the Sun compensate among each other and integrate to zero for a global rescaling of the opacity. Moreover this is not realistic because the accuracy of opacity calculations is expected to be better at the solar core than in the region around the base of the convective envelope. Taking this into account, the following parameterization for the opacity change $\delta\kappa(T)$ was considered:

$$\delta\kappa(T) = a + b\frac{\log\left(T_\mathrm{C}/T\right)}{\Delta} \quad (1)$$

where T is the temperature of the solar plasma, $\Delta = \log\left(T_\mathrm{C}/T_\mathrm{CZ}\right) = 0.9$, $T_\mathrm{C} = 15.6 \times 10^6\,\mathrm{K}$ and $T_\mathrm{CZ} = 2.3 \times 10^6\,\mathrm{K}$ are the temperatures at the solar center and at the bottom of the convective zone respectively. The parameters a and b are treated as independent random variables with mean equal to zero and dispersions $\sigma_a = 2\%$ and $\sigma_b = 6.7\%$, respectively. This corresponds to assuming that the opacity error at the solar center is $\sigma_\mathrm{in} = \sigma_a = 2\%$, while it is given by $\sigma_\mathrm{out} = (\sigma_a^2 + \sigma_b^2)^{1/2} = 7\%$ at the base of the convective zone, as can be motivated by the recent experimental results of [38] and the theoretical work by [39].

Surface composition: The solar surface composition is a fundamental constraint in the construction of SSMs. In this paper, we consider two different canonical sets of solar abundances which are the same employed in [25]:

- GS98 - Photospheric (volatiles) + meteoritic (refractories) abundances from [20] that correspond to metal-to-hydrogen ratio used for the calibration $(Z/X)_\odot = 0.0229$;
- AGSS09met - Photospheric (volatiles) + meteoritic (refractories) abundances from [17] that give $(Z/X)_\odot = 0.0178$.

Note that the recent results from [40, 41, 42] that have updated the abundances of [17] for all but CNO elements (which are the most abundant among the volatiles elements) do not lead to a revision of the AGSS09met composition.

3. B16-SSMs results

The main results obtained with the new generation of B16 SSMs for the two choices of solar composition, GS98 and AGSS09met, are shown in Tab.2, Fig.1 and 2 and are discussed below.

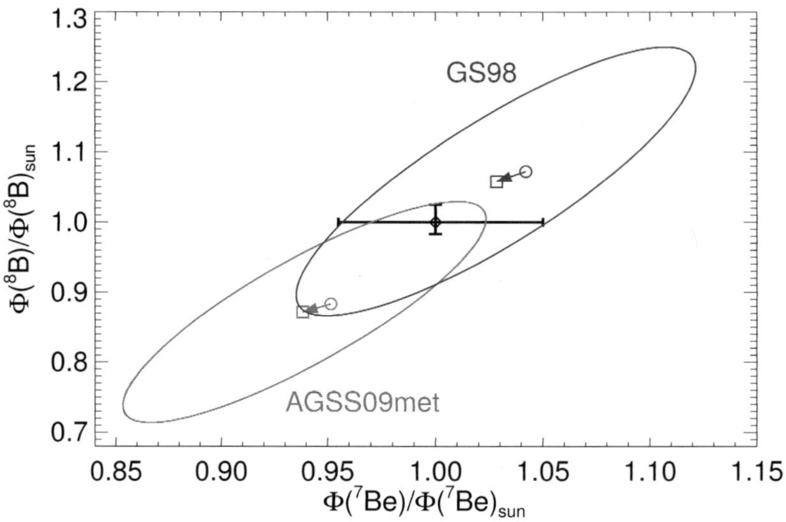

Figure 1. *$\Phi(^8$B) and $\Phi(^7$Be) fluxes normalized to solar values [43]. Black circle and error bars: solar values. Squares and circles: results for B16 (current) and (older) generation of SSMs respectively. Ellipses denote theoretical 1σ C.L. for 2 dof.*

Neutrino fluxes: The updates of nuclear reaction rates have a direct effect on neutrino production. In particular, the boron and beryllium neutrino fluxes are reduced for both GS98 and AGSS09met compositions by about 2% with respect to previous SSM calculations [25]. The overall reduction in the $\Phi(^8$B) and $\Phi(^8$Be) fluxes comes from the increase in S_{11}. In the case of $\Phi(^8$B), this is partially compensated by the 2.4% increase in S_{17}. The most important changes in the neutrino fluxes occur for $\Phi(^{13}$N) and $\Phi(^{15}$O), in the CN-cycle. The expectation values in the B16 SSMs are about 6% and 8% lower than for the previous SSMs [25]. This results from the combined changes in the p+p and ^{14}N+p reaction rates.

The predicted fluxes should be compared with the observational values in the last column of Tab.2 which have been obtained in [43] from a fit to the results of solar neutrino experiments by allowing for three-flavour neutrino oscillations. Note that observational errors for $\Phi(^8$B) and $\Phi(^8$Be) fluxes are smaller than uncertainties in theoretical predictions, as can be also appreciated in Fig.1 where we summarize the present situation for these two components of the solar neutrino spectrum. On the contrary, CN fluxes have not yet been determined experimentally and the global analysis of solar neutrino data provides only the upper limits included in Tab.2. From the comparison of predicted and observed fluxes, we see that both solar compositions lead to SSMs that are consistent with experimental results within 1σ.

Helioseismology: In the last two lines of Tab.2, we report two helioseismic quantities widely used in assessing the quality of SSMs, i.e. the surface helium abundance $Y_{\rm S}$ and the depth of the convective envelope $R_{\rm CZ}$, together with the corresponding seismically determined values. The model errors associated to these quantities are larger than previously computed because of the different treatment of uncertainties in radiative opacities. Compared to previous SSMs [25], we find a small decrease in the predicted $Y_{\rm S}$ by 0.0003 and in the predicted $R_{\rm CZ}$ by 0.0007$R_\odot$ for both compositions. These small changes together with the larger theoretical uncertainties lead B16-GS98 to a 0.9σ ($Y_{\rm S}$) and 0.3σ ($R_{\rm CZ}$) difference with respect to data while for B16-AGSS09met differences are at the 2.5σ ($Y_{\rm S}$) and 1.8σ ($R_{\rm CZ}$) level.

Finally, Fig.2 shows the fractional difference between the sound speed inferred from

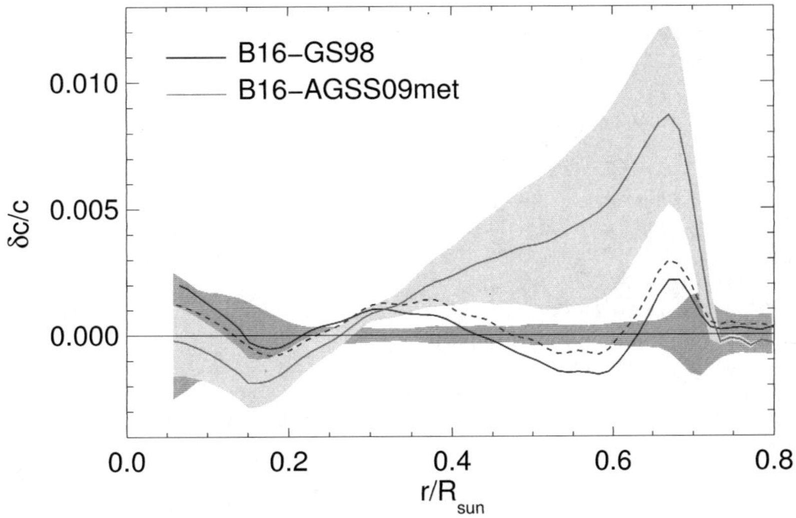

Figure 2. *Fractional sound speed difference* $\delta c/c = (c_\odot - c_{SSM})/c_{SSM}$. *The grey shaded region corresponds to errors in the helioseismic inversion procedure. The red shaded region around the AGSS09met central value (solid red line) describes uncertainties in SSM calculations. An equivalent relative error band holds around the central value of the GS98 central value (solid blue line) which we do not plot for the sake of clarity. Dashed line shows, for comparison, results for old SSM calculations [25].*

helioseismic frequencies and that predicted by B16 SSMs as a function of solar radius for the two choices of solar composition. The solar sound speed has been obtained by new inversions based on the so-called BiSON-13 dataset [46] and using consistently both B16 SSMs as reference models. Results are only slightly different with respect to previous calculations, mainly as a result of the updated $S_{11}(0)$ value. We see that B16-GS98 model yields a much better agreement, everywhere in the solar structure, with the helioseismically derived sound speed profile than B16-AGSS09met. In particular, the B16-AGSS09 model disagrees by $\sim 1\%$ with sound speed inferred from helioseismology at the bottom of the convective envelope. This has to be compared with a theoretical uncertainty of $\sim 0.3\%$ and an error in the inversion procedure smaller than 0.1%.

4. Summary and Conclusions

In this review which is largely based on the results of [24], we have presented B16-GS98 and B16-AGSS09met, a new generation of SSMs calculated for different canonical set of abundances with high and low metallicity. We summarize our most important findings here:

- Central values for $\Phi(^7\text{Be})$ and $\Phi(^8\text{B})$ in B16 SSMs are reduced by about 2% with respect to the previous generation of models that were based completely on nuclear reaction rates from [32]. The CN-cycle fluxes, $\Phi(^{13}\text{N})$ and $\Phi(^{15}\text{O})$, are reduced by 6% and 8% respectively. Solar neutrino fluxes [43] are reproduced almost equally well by both B16-GS98 and B16-AGSS09met, with only a very minor preference for B16-GS98.
- Helioseismic properties of B16 models are almost unchanged with respect to older models. However, our estimation of errors is larger due to our more pessimistic assumption of a 7% uncertainty in the radiative opacity at the base of the convective envelope. Comparison of models against Y_S and R_{CZ} and the helioseimic inferred solar sound speed profile yields a much better agreement for B16-GS98 than for B16-AGSS09met.

IOP Conf. Series: Journal of Physics: Conf. Series **1056** (2018) 012058 doi:10.1088/1742-6596/1056/1/012058

References

[1] Davis R Jr., Harmer D S and Hoffman K C 1968 *Phys. Rev. Lett.* **20** 1205
[2] Cleveland B R et al. 1968 *ApJ* **496** 505
[3] Hampel W at al. 1999 *Phys. Lett. B* **447** 127
[4] Abdurashitov J N et al. 1999 *Phys. Rev. Lett.* **83** 4686
[5] Altmann M et al. 2005 *Phys.Lett. B* **616** 174
[6] Hirata K S et al. 1989 *Phys. Rev. Lett.* **63** 16
[7] Cravens J P et al. 2008 *Phys. Rev. D* **78** 032002
[8] Ahmad Q R et al. 2002 *Phys. Rev. Lett.* **89** 011301
[9] Abe K et al. 2016 *Phys. Rev. D* **94** 052010
[10] Bellerive A et al. 2016 *Nucl. Phys. B* **908** 30
[11] Bellini G et al. 2014 *Nature*, **512** 383.
[12] Agostini M et al. 2017 *Preprint* arXiv:1707.09279 [hep-ex].
[13] Agostini M et al. 2017 *Preprint* arXiv:1709.00756 [hep-ex].
[14] Villante F L, Ianni A, Lombardi F, Pagliaroli G and Vissani F 2011 *Phys. Lett. B* **701** 336
[15] Degl'Innocenti S, Dziembowski W A, Fiorentini G, Ricci B 1997 *Astrop. Phys.* **7** 77
[16] Gough D O et al. 1996, *Science* **272** 1296
[17] Asplund M, Grevesse N, Sauval A J and Scott P 2009 *Ann. Rev. Astron. Astrophys.* **47** 481
[18] Caffau E, Ludwig H G, Steffen M, Freytag B, Bonifacio P 2011 *Solar Physics* **268** 255
[19] Asplund M, Grevesse N, Sauval A J 2005 *Astronomical Society of the Pacific Conference Series* **336** ed T G Barnes and F N Bash p 25
[20] Grevesse N, Sauval A J 1998 *Space Sci.Rev.* **85** 161
[21] Grevesse N, Noels A 1993 *Origin and Evolution of the Elements* ed N Prantzos, E Vangioni-Flam E and M Casse p 15
[22] Bergemann M and Serenelli A 2014 *Determination of Atmospheric Parameters of B-, A-, F- and G-Type Stars. Series: GeoPlanet: Earth and Planetary Sciences* (Springer, Cham) ed E Niemczura, B Smalley and W Pych p 245
[23] Basu S and Antia H M 2008 *Phys. Rept.* **457** 217
[24] Vinyoles N et al. 2017 *ApJ* **835** 202
[25] Serenelli A M, Haxton W C and Pena-Garay C 2011 *ApJ* **743** 24
[26] Cassisi S, Salaris M and Irwin A W 2003 *ApJ* **588** 862
[27] Marcucci L E, Schiavilla R and Viviani M 2013 *Phys. Rev. Lett.* **110** 192503
[28] Tognelli E, Degl'Innocenti S, Marcucci L E and Prada Moroni P G 2015 *Phys. Lett. B* **742** 189
[29] Acharya B, Carlsson B D, Ekstrm A, Forssn C and Platter L 2016 *Phys. Lett. B* **760** 584
[30] Zhang X, Nollett K M and Phillips D R 2015 *Phys. Lett. B* **751** 535
[31] Marta M, Formicola A, Bemmerer D et al. 2011 *Phys. Rev C* **83** 045804
[32] Adelberger E G et al. 2011 *Reviews of Modern Physics* **83** 195
[33] deBoer R J et al. 2014 *Phys. Rev. C* **90** 035804
[34] Iliadis C, Anderson K, Coc A, Timmes F and Starrfield S 2016 *ApJ* **831** 107
[35] Badnell N R et al. 2005 *MNRAS* **360** 458
[36] Iglesias C A and Rogers F J 1996 *ApJ* **464** 943
[37] Villante F L 2010 *ApJ* **724** 98
[38] Bailey J E et al. 2015 *Nature* **517** 56
[39] Krief M, Feigel A and Gazit D 2016 *Preprint, arXiv:1603.01153* [astro-ph.SR]
[40] Scott P et al. 2015 *Astron. and Astophys.* **573** A25
[41] Scott P, Asplund M, Grevesse N, Bergemann M and Sauval A J 2015 *Astron. and Astophys.* **573** A26
[42] Grevesse N, Scott P, Asplund M and Sauval A J 2015 *Astron. and Astophys.* **573** A27
[43] Bergström J et al. 2016 *Journal of High Energy Physics* **3** 132
[44] Basu S and Antia H M 2004 *ApJL* **606** L85
[45] Basu S and Antia H M 1997 *MNRAS* **287** 189
[46] Basu S, Chaplin W J, Elsworth Y, New R and Serenelli A M 2009 *ApJ* **699** 1403

Search for *creation of electrons* in lab

S Dell'Oro[1], S Marcocci[2] and F Vissani[3]

[1] Center for Neutrino Physics, Virginia Tech, 24061 Blacksburg, Virginia, USA
[2] Fermi National Accelerator Laboratory, 60510 Batavia, Illinois, USA
[3] INFN - Laboratori Nazionali del Gran Sasso, Assergi (L'Aquila), Italy

Abstract. We examine the physics motivations to search for lepton number violations in the laboratory, in particular by studying the important nuclear transition called neutrinoless double beta decay, which consists in the creation of a pair of electrons. We address some major shortcomings of the Standard Model of electroweak interactions, arguing that most of them can be solved simply within extended models with Majorana neutrinos. At the same time, we discuss the difficulties of obtaining precise predictions for the neutrinoless double beta decay rate and strengthen the case of probing Majorana neutrino masses of about 10 meV.

This report concerns a transition where the nucleus increases its charge by 2 units

$$(A, Z) \to (A, Z+2) + 2\mathrm{e}^- \tag{1}$$

that, by tradition, is called neutrinoless double-β decay[1], a jargonic name often abbreviated as $0\nu\beta\beta$. Notice that a more modern and fair description of this hypothetical process is simply *creation of electrons* in a nuclear transition; the traditional name is not equally transparent [1].

This work is based on a series of recent papers on the above transition, which include,

- an inquiry on the effect of nuclear physics uncertainties on the expectations, Ref. [2].
- an investigation of the impact of cosmological bounds of neutrino masses, Ref. [3].
- an extensive review on $0\nu\beta\beta$, Ref. [4].

In these papers and in the present report the leading contribution to the transition of Eq. (1) is attributed to the virtual exchange of light Majorana neutrinos, for the reasons recalled later on.

1. The components of the matter and the corresponding conservation laws

What is the nature of the constituents of Matter? Since the time of 'Rutherford experiments' conducted by Geiger and Mardsen [5], we know that any atom is made of heavy components clustered in a small nucleus (the neutrons and protons, classed as baryons), in turn surrounded by light components (the electrons, classed as leptons). The list of the known leptons – i.e., the light components of Matter – was completed later by Pauli, who hypothesized the existence of the neutrino. This particle has only weak interactions with the other particles and it is not in the nucleus, as realized by Fermi. Still, it has to be included among the Matter particles, for it

[1] Originally (1899) 'β-rays' indicated a specific type of 'uranium-rays', then 'nuclear electrons' expelled by certain nuclei. This terminology fitted a bygone theory of the nucleus, thought as an assembly of protons and electrons, and abandoned after neutron discovery (1932). Nowadays, the term 'β-ray' is an obsolete synonym of 'electron'.

allows us to introduce a quantity that is conserved in any known reaction: the net number of leptons L, strictly analogous to the net number of baryons B.

Much later, it was understood that hadrons are composed by lighter components, the quarks, which are so strongly bound that the largest fraction of the mass of the protons and neutrons is actually their binding energy. The conservation of the number of quarks holds true and implies the one of baryons. In short, to the best of current knowledge, the basic constituents of Matter are quarks and leptons. Over time, it was also understood that quarks and leptons come in three "copies" – families – among which transitions are possible. For quarks it is sufficient to recall the existence of weak decays, while for leptons, one should consider e.g. the $\nu_\mu \to \nu_e$ transitions observed by T2K [6] or the $\nu_\mu \to \nu_\tau$ one detected by OPERA [7]. The only known conserved quantities are: 1) the sum of the leptons of the electronic, muonic and tauonic families ($L = L_e + L_\mu + L_\tau$); 2) the corresponding total number of baryons B; 3) their combinations.

Is there any observational hint that B and/or L are violated at some level? Only one to date, indirect but important. This consists in the fact that the cosmos contains baryons but not any significant amount of anti-baryons. Such a circumstance is actually quite disturbing since in the earliest instants of the Universe, anti-baryons were surely present in large amount. This lead Sakharov to speculate on the existence of fundamental phenomena in which B is violated, resulting in turn into a cosmic preference for baryons [8]. Twenty years later, it was realized that the same scenario could be attributed to fundamental phenomena where L is violated [9].

The question *whether there exist phenomena observable in laboratory* where L or B are violated deserves the maximum attention. The most promising way to probe L-violation is provided by the transition of Eq. (1), both for the experimental reasons discussed in Ref. [4] and for the theoretical motivations recalled in the next section.

2. The SM, seen as the current theory of matter, and its issues

The reference theory of matter today is the Standard Model (SM) of electroweak interactions, or $SU(3)_C \times SU(2)_L \times U(1)_Y$ Glashow-Weinberg-Salam gauge theory. This stunningly successsful theory has various issues relevant for the present discussion [1], namely:

(i) neutrino masses are zero.

(ii) $B - L$, $L_e - L_\mu$, $L_e - L_\tau$, $L_\mu - L_\tau$ are exactly conserved while B, L, L_e, L_μ, L_τ only at the first order.

(iii) matter particles are absolutely distinguished – neutrinos included.

(iv) the cosmic excess of baryons cannot be explained.

Even by excluding the last point, it should be noticed that the first prediction is wrong (as recognized by the Nobel Prize in Physics 2015), the second one is mostly wrong (see the discussion in Sec. 1), while the third one is disputable: First of all, neutrinos do not have an intrinsic electric charge that distinguishes them from antineutrinos. Second, the SM has one neutrino and one antineutrino per family with opposite helicity, thus the most economical assumption is that, *in the rest frame,* these two are just the spin states of the same particle.

Incidentally, it should be remarked that: *a*) $B+L$ is not a symmetry of the SM, and this is an important reason why new phenomena where L and B are violated have very similar significance; *b*) the transition of Eq. (1) violates L and also $B - L$, so that its existence would have a major impact on the SM: it would prove that *none of its symmetries is exact.* The previous goal is distinctive of $0\nu\beta\beta$, seen as a potentially measurable transition: it would not be achieved even if we discovered $p \to \pi^0 e^+$ or $p \to K^+\bar{\nu}$ or any other $(B - L)$-conserving nucleon decay mode.

3. A powerful fix for SM issues: the hypothesis of Majorana neutrinos

The breaking of $SU(2)_L \times U(1)_Y$ due to the vacuum expectation value $\langle H^0 \rangle = 174$ GeV determines the Fermi constant $G_F = 1/(\sqrt{8}\langle H^0 \rangle^2)$ that explains weak decays at much lower energy scales.

Likewise, new physics at a scale $M \gg \langle H^0 \rangle$ can lead to neutrino masses. Indeed, by using the gauge invariant combination of one higgs and one leptonic doublet $N_\ell = (H^0, H^-)^*(\nu_\ell, \ell^-)^t$, one can construct new contributions to the Lagrangian density [10, 11]

$$\delta\mathcal{L} = \frac{1}{2}\frac{\kappa_{\ell\ell'}}{M}\, N_\ell\, C^{-1}\, N_{\ell'} \; + \; h.c. \quad \text{with} \quad \ell, \ell' = e, \mu, \tau \tag{2}$$

where C is the charge conjugation matrix, $\kappa_{\ell\ell'}$ are a-dimensional coefficients and M an explicit mass scale, needed for dimensional reasons. These terms imply a violation of the lepton number, as it is evident from the fact that they contain two lepton fields – addressing issue (ii) radically.

The previous lagrangian density gives rise to neutrino masses – solving issue (i). In fact, upon symmetry breaking, we obtain small bilinear terms for the SM neutrinos,

$$\delta\mathcal{L}_{\text{bil}} = \frac{\mathscr{M}_{\ell\ell'}}{2}\, \nu_\ell\, C^{-1}\, \nu_{\ell'} \; + \; h.c. \quad \text{where} \quad \mathscr{M}_{\ell\ell'} = 50\,\text{meV} \cdot \kappa_{\ell\ell'} \cdot \left(\frac{6 \cdot 10^{14}\,\text{GeV}}{M}\right). \tag{3}$$

Let us diagonalize the matrix $\mathscr{M}$ by using a unitary matrix U: $\mathscr{M}_{\ell\ell'} = U^*_{\ell i}\, m_i\, U^*_{\ell' i}$ with $i = 1, 2, 3$ and $m_i \geq 0$. We can introduce the new fields $\nu_i \equiv U^*_{\ell i}\nu_\ell$, which in turn allow us to define the fields $\chi_i \equiv \nu_i + C\bar{\nu}_i{}^t$. The Lagrangian density (3), rewritten with these fields, becomes simply

$$\delta\mathcal{L}_{\text{bil}} = -\frac{1}{2}\sum_{i=1}^{3} m_i\, \bar{\chi}_i\, \chi_i. \tag{4}$$

In other words, the neutrinos described by the χ_i-fields have mass m_i. We conclude that physics beyond the SM at a scale $M \gg 174\,\text{GeV}$ can actually lead to values of the neutrino masses $m_i \simeq 50\,\text{meV} = \sqrt{\Delta m^2_{\text{atm}}}$, that are needed in order to explain neutrino oscillations.

The fields with given mass are self-conjugated, $\chi_i = C\bar{\chi}_i{}^t$ – i.e., they are Majorana fields [12] – thus there is just one type of elementary oscillators with given momentum $\vec{p}$ and helicity λ

$$\chi_i = \sum_{\vec{p},\lambda} [\; u_{\vec{p},\lambda}\; e^{-ipx}\; \mathbf{a}_{\vec{p},\lambda} + C\bar{u}^t_{\vec{p},\lambda}\; e^{ipx}\; \mathbf{a}^\dagger_{\vec{p},\lambda}\;] \tag{5}$$

Simply stated, these neutrinos are matter and antimatter at the same time: – which fits issue (iii).[2]

4. Majorana neutrinos and the process of creation of electrons

We are ready to come back to the transition of Eq. (1). The key parameter that regulates its rate is:

$$m_{\beta\beta} \equiv |\mathscr{M}_{ee}| = \left|\sum_{i=1}^{3} U^2_{ei}\, m_i\right|. \tag{6}$$

In fact, an evident feature of Eq. (1) is that the electronic lepton number L_e is violated twice, and $\mathscr{M}_{ee}$ is the only element of the neutrino mass matrix which obeys this selection rule.[3] Unfortunately, it is not possible to infer its value from the available neutrino data and, even more, a theory of fermion masses is not available yet. This implies that particle physics offers only a qualitative understanding; we have the largest quantitative uncertainty. A reasonable way to proceed in the theoretical discussion is to resort to "educated guesses". Moving in this direction and following [1, 13] let us postulate that the neutrino mass matrix has the form

$$\mathscr{M} \overset{\mathcal{O}(1)}{=} 50\;\text{meV} \times \begin{pmatrix} \varepsilon & \varepsilon & \varepsilon \\ \varepsilon & 1 & 1 \\ \varepsilon & 1 & 1 \end{pmatrix}, \tag{7}$$

[2] Funnily enough, all we need to fix the major SM defects of Sec. 2 is a theoretical proposal appeared thirty years before the formulation of the SM– namely, Majorana neutrino masses.

[3] The parameter $m_{\beta\beta} = |\mathscr{M}_{ee}|$ is commonly called effective Majorana mass. For the notation, see again footnote 1.

IOP Conf. Series: Journal of Physics: Conf. Series **1056** (2018) 012059 doi:10.1088/1742-6596/1056/1/012059

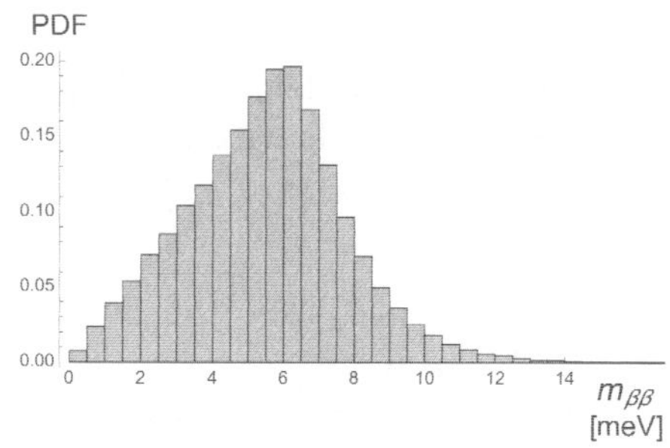

Figure 1. Monte Carlo analysis of the expectations for $m_{\beta\beta}$ in the model of Eq. (7) with $\varepsilon = \theta_C = 13°$. The $\mathcal{O}(1)$ coefficients are sampled uniformly inside the complex circle of radius 1. An expectation of $m_{\beta\beta} \sim \sqrt{\Delta m^2_{\rm atm}} \times \theta_C \sim 10\,\text{meV}$ is obtained.

assumed to be known just up to $\mathcal{O}(1)$ coefficients that multiply the individual elements. In this scheme, atmospheric neutrino oscillations are accounted for by the overall scale and by the large $\nu_\mu - \nu_\tau$ block. The small mixing $\theta_{13} \sim 8°$ indicates a similar value of the parameter ε. The values of $m_{\beta\beta}$ fall in the ballpark illustrated in Fig. 1, that points out the interest in probing values of $m_{\beta\beta}$ at or below 10 meV. This corresponds to the experimental challenge of designing, building, and operating *multi-ton* detectors for the $0\nu\beta\beta$ search [4].

5. Constraint on the neutrino masses due to cosmological measurements

When combined together, the precise observations of the large cosmic scales by Planck (via the measurement fo the Cosmic Microwave Background [14]) and those concerning smaller scales, e. g. galaxy distributions, Baryon Acoustic Oscillations or the Lyman-α forest power spectrum, allow us to probe neutrino masses with values far below the electronvolt.[4]

Several works reported that according to cosmological surveys the sum of neutrino masses, $\Sigma \equiv m_1 + m_2 + m_3$, is smaller than $(100 - 200)$ meV, that, in turn, slightly favors the NH scenario [3]. In a recent work, the authors of Ref. [15] reached analogous conclusions (see Fig. 2) and moreover their best-fit value for Σ is close to the allowed minimum, corresponding to $m_{\beta\beta} \lesssim$ a few 10 meV [3, 4]. Such an expectation is consistent with that suggested by the above theoretical argument and, again, points out that *multi-ton* detectors are necessary for the $0\nu\beta\beta$ search [3, 4]. This conclusion is so significant, that a few cautionary remarks are in order:

- In the past, several well-known works have claimed that the neutrino masses obtained from cosmology were different from zero, as summarized in fig. 8 of [4]. Today, these claims are considered premature and there is a new one [16] compatible with the new tighter limits.
- An independent indication in support of the NH case derives from the global analyses of the oscillation data [17] - note however that oscillations do not probe the absolute mass scale.
- The existence of other physics (besides Majorana neutrino masses) where L is violated in a potentially relevant manner can be probed at LHC.

In other words, it is important to be aware that the discussion is still evolving. Let us repeat that these considerations are based on the Majorana neutrino mass hypothesis, which implies the correlation between the rate of Eq. (1) and cosmological measurements.

6. The role of nuclear physics

The amplitude of the transition of Eq. (1) depends linearly upon $m_{\beta\beta}$ and the corresponding half-life time can be factorized as

$$[\, t_{1/2}\,]^{-1} = \mathscr{G} \times |\, \mathcal{M}\,|^2 \times \left(\frac{m_{\beta\beta}}{m_e}\right)^2 \tag{8}$$

[4] To get an idea, notice that at the "decoupling", the temperature of the Universe was 1,100 times larger than the present one, namely ~ 250 meV.

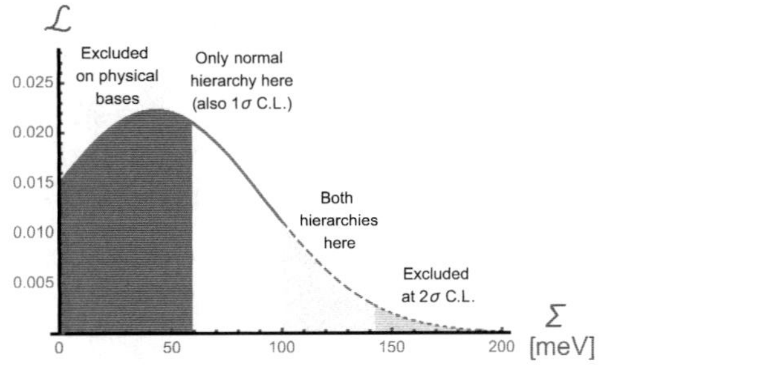

Figure 2. Likelihood $\mathcal{L}(\Sigma) \propto e^{-\Delta\chi^2/2}$ with the $\Delta\chi^2(\Sigma)$ obtained in Ref. [15]. The grey area is excluded by the Big Bang cosmology with three massive neutrinos. The 1σ region is consistent with Normal mass Hierarchy (NH) scenario and not with the Inverted Hierarchy (IH) one.

where the electron mass m_e is conventionally chosen for normalization, $\mathscr{G}$ is an overall factor that includes the phase space and the effect of the electrostatic field on the emitted electrons, while $\mathcal{M} \propto \langle A, Z+2|\text{H}_{\text{eff}}|A, Z\rangle$ is the matrix element connecting the initial final states of the transition. In particular, the latter arises at the second order of weak interactions $\text{H}_{\text{eff}} \propto G_F^2$ where G_F is the Fermi coupling and describes the residual effect of a virtual exchange of a Majorana neutrino.[5] Since the neutrons and the protons are confined inside the nucleus, the virtual momentum of the neutrino is of the order of the inverse nuclear size, $Q \simeq 100$ MeV, and it is thus much larger than typical momenta in the known nuclear weak decays.

In a conservative discussion, it is very important to quantify both the expected value of $\mathcal{M}$ and the related uncertainty. In fact, the number of signal events in a $0\nu\beta\beta$ experiment increases quadratically with this parameter whereas the corresponding increase with the detector mass (i. e. with the number of nuclei under observation) is less: linear in the optimistic situation of a background-free experiment, or as the square root of the mass in the presence of background.

7. Uncertainties in the matrix elements and the hypothesis of *quenching*

Until a few years ago, the uncertainty on $\mathcal{M}$ – estimated by comparing the published calculations – was about a factor of a few [18]. More recently, it was noted that, disregarding the Shell Model estimates, the QRPA and the IBM-2 estimates agree in a much better way, within $\sim 30\%$ for several nuclei. This suggested the possibility that the error could actually be of this size. However, when the same calculations are compared with other (measured) transitions, such as the ordinary beta decay, or the two-neutrino double beta decay, the disagreement is much larger and it is found that QRPA and IBM-2 systematically overestimate the matrix elements. These considerations have revived interest for an old hypothesis, namely that the axial coupling of charged current weak interactions, whose value is $g_{\text{axial}} \simeq 1.27$ for free nucleons, might decrease for nucleons inside the nuclear matter. It is the so called *quenching*. Scaling $g_{\text{axial}} \sim 1.27 \cdot A^{-0.18}$ with the mass number A, finds a better agreement with the measured rates for the two-neutrino double beta decay [19]. We would like to make a few comments on this important issue:

- it must be stressed that, to date, the quenching hypothesis is *not* a theory but, at most, a phenomenological model.
- if the quenching is real, it should depend upon the exchanged momentum Q and it should disappear for free nucleons. Therefore, since Q is very large in the transition of Eq. (1), the quenching could be much smaller. Indeed, model calculations indicate $g_{\text{axial}} \simeq 1$ [20], which is not as extreme as $g_{\text{axial}} = 1.27 \cdot A^{-0.18}$.
- the hypothesized quenching can be cautiously considered as a factor of uncertainty in the predictions, even if, doing so, the impact on the predictions is quite dramatic [21, 2, 4, 22]:

[5] It should be noticed that it is common to depict the Feynman diagram of the $0\nu\beta\beta$ transition in terms of free quarks or occasionally of free nucleons. However, these are poor descriptions of the actual physical situation.

e.g., for ^{130}Te, the expected $0\nu\beta\beta$ lifetime would increase as, $\sim 1/g^4_{\text{axial}}$, namely, by a factor 30.

- even if the quenching of g_{axial} is smaller than the one reported, the effect might be relevant anyway. For example, it was argued in Ref. [4] that a 20% decrease in the value implies a factor 6 increase in the detector mass to compensate for the loss of sensitivity.

In any case, the theoretical calculations cannot be considered definitive yet and, after all, no good reason is known to omit the Shell Model calculations from the general assessment concerning the matrix elements: see Ref. [23] and the discussion at the conference.

8. Summary and assessment

The reasons to search for the transition of Eq. (1) – creation of electrons [1, 24] – are stronger than ever. Particle physics theory helps for general considerations and/or for orientation, e. g. motivating the connection of Majorana neutrino masses and the extension of the SM, pointing out arguments in favor of the NH and/or suggesting ideas on flavor structure of neutrino masses. However, no model of fermion masses can be taken "too seriously", at least at this point in time.

The measurements of the neutrino oscillation parameters and, possibly, also cosmology give us a real chance to progress in the expectations for the rate of the $0\nu\beta\beta$ transition. The tightest cosmological bounds that have been discussed in the scientific literature require multi-ton detector masses, even for background-free experiments [3, 4].

The theoretical uncertainties on the expected rate are large, mostly due to particle physics, but also to nuclear physics [2, 4]. Fortunately, the importance of the discussion is understood and the situation is possibly evolving in a positive way.

Summarizing, the hypothesis of Majorana neutrino mass remains appealing and points out correlations between the transition of Eq. (1) and other phenomena, such as cosmological observations. However, if new sources of lepton number violation at low energy (TeV?) exist, these correlations could be evaded and surprises may occur: Indeed, $0\nu\beta\beta$ does not probe neutrino masses directly, but just the half-life time of the decay.

References

[1] Dell'Oro S, Marcocci S and Vissani F 2017 *PoS NEUTEL2017 (2017) 030,* `arXiv:1710.06732 [hep-ph]`
[2] Dell'Oro S, Marcocci S and Vissani F 2014 *Phys. Rev. D* **90**, *033005*
[3] Dell'Oro S, Marcocci S, Viel M and Vissani F 2015 *J. Cosm. Astropart. Phys.* **1512**, *023*
[4] Dell'Oro S, Marcocci S, Viel M and Vissani F 2016 *Adv. High Energy Phys.* **2016**, *2162659*
[5] Geiger H and Marsden E 1909 *Proc. Royal Soc.* **82**, *495*
[6] Abe K *et al.* (T2K Collaboration) 2011 *Phys. Rev. Lett.* **107**, *041801*
[7] Agafonova N *et al.* (OPERA Collaboration) 2015 *Phys. Rev. Lett.* **115**, *121802*
[8] Sakharov A D 1967 *Pisma v Z. Eksp. Teor. Fiz.* **5**, *32*
[9] Fukugita M and Yanagida T 1986 *Phys. Lett. B* **174**, *45*
[10] Minkowski P 1977 *Phys. Lett.* **B67**, *421*
[11] Weinberg S 1979 *Phys. Rev. Lett.* **43**, *1566*
[12] Majorana E 1937 *Nuovo Cim.* **14**, *171*
[13] Vissani F 2001 *Phys. Lett. B* **508**, *79*
[14] Ade P A R *et al.* (Planck Collaboration) 2016 *Astron. Astrophys.* **594**, *A13*
[15] Yèche C *et al.* 2017 *J. Cosm. Astropart. Phys.* **1706**, *047*
[16] Emami R, Broadhurst T *et al.* 2017 `arXiv:1711.05210 [astro-ph.CO]`
[17] Capozzi F, Lisi E, Marrone A and Palazzo A, `arXiv:1804.09678 [hep-ph]`
[18] Feruglio F, Strumia A and Vissani F 2002 *Nucl. Phys. B* **637**, *345* Addendum: [*Nucl. Phys. B* **659**, *359*]
[19] Barea J, Kotila J and Iachello F 2013 *Phys. Rev. C* **87**, *014315*
[20] Menéndez J, Gazit D and Schwenk A 2011 *Phys. Rev. Lett.* **107**, *062501*
[21] Robertson R G H 2013, *Mod. Phys. Lett. A* **28**, *1350021*
[22] Marcocci S, Dell'Oro S, Viel M and Vissani F 2017 *J. Phys. Conf. Ser.* **888**, *012178*
[23] Engel J and Menéndez J 2017 *Rept. Prog. Phys.* **80**, *046301*
[24] Mitra M, Senjanovic G and Vissani F 2012 *Nucl. Phys. B* **856**, *26*, see footnote 1 there

IOP Conf. Series: Journal of Physics: Conf. Series **1056** (2018) 012060 doi:10.1088/1742-6596/1056/1/012060

Neutrino astrophysics and its connections to nuclear physics

Maria Cristina Volpe
Astro-Particule et Cosmologie (APC), CNRS UMR 7164, Université Denis Diderot, 10, rue Alice Domon et Léonie Duquet, 75205 Paris Cedex 13, France
E-mail: volpe@apc.in2p3.fr

Abstract. We highlight recent developments in neutrino astrophysics. We discuss some of the connections with nuclear physics.

1. Milestones in neutrino astrophysics

Neutrino astronomy started with the pioneering measurement of solar neutrinos by R. Davis' experiment in the sixties [1] and the observation of neutrinos from the supernova 1987A located in the Large Magellanic Cloud by Kamiokande II [2], by IMB [3] and Baksan [4]. Evidence of ultra-high energy neutrinos in the PeV energy range by the ICECUBE Collaboration has initiated high energy neutrino astronomy [5]. Moreover, the measurements of gravitational waves from binary black holes [6] and neutron star mergers [7] has opened the era of multi-messenger astronomy.

Neutrinos are elementary massive particles with mixings. This is known since the discovery of vacuum oscillation by the Super-Kamiokande experiment [8], that had an impact in particle physics, astrophysics and cosmology. This phenomenon occurs because the mass or propagation basis and the flavor or interaction basis do not coincide. They are related by the Pontecorvo-Maki-Nakagawa-Sakata unitary matrix [9] which depends on three mixing angles, one Dirac and two Majorana phases, if there exists only three active neutrino flavors. If neutrinos are Dirac particles, the neutrino mass can be implemented by adding a right-handed singlet to the Glashow-Weinberg-Salam model. If neutrinos are Majorana particles, significant extensions to the standard model are necessary that produce total lepton number violation (see e.g. the book [10]).

First identified by R. Davis' experiment, the solar neutrino deficit problem has been solved with the neutrino oscillation discovery [8], the SNO measurement of the total solar neutrino flux in agreement with expectations from the Standard Solar Model [11] and with the identification of the Large-Mixing-Angle (LMA) solution by KamLAND [12] (see e.g. [13] for a review). In particular, the production of solar neutrinos and their flavor conversion are well understood. Historically, the study of solar ν has played a major role in establishing how energy is produced in low mass main sequence stars, as our Sun, and in ascertaining neutrino flavor conversion. The precise measurement of neutrinos coming from the Sun's interior have confirmed that the proton-proton (pp) reaction chain, that burns hydrogen to produce helium-4 nuclei, is responsible for 99% of energy production in the Sun [15]. The (hopefully) future measurement of neutrinos from the carbon-nitrogen-oxygen (CNO) cycle would provide us with the first direct observation of the main energy production mechanism in high mass main sequence stars [14, 13, 16].

Concerning flavor conversion in astrophysical environments, the observation of a reduced electron neutrino flux by solar experiments and of the total solar flux by the SNO experiment has demonstrated that : *i)* solar neutrinos with less than 2 MeV energy undergo averaged vacuum oscillations; *ii)* electron neutrinos than have more than 2 MeV energy convert into muon and tau neutrinos because of the Mikheev-Smirnov-Wolfenstein (MSW) effect [17, 18, 15]. This is due to neutrino-matter interactions that neutrinos encounter while propagating through the medium. They can be accounted for, by using the mean-field approximation. Due to this mean field, ν_e efficiently (adiabatically) change into ν_μ, ν_τ at the MSW resonant location that depends on the solar electron number density profile, the neutrino energy and mixing parameters. The MSW effect is analogous to the two-level problem in quantum mechanics.

Since 1998 experiments exploiting athmospheric, solar, reactor and accelerator neutrinos have allowed to determine the three neutrino mixing angles and two mass-squared differences responsible for vacuum oscillations. Still, some anomalies need to be elucidated, that do not fit into the theoretical framework of three active neutrino flavors, such as the so-called "reactor anomaly". They could be explained by the existence of sterile neutrinos that several experiments are searching for (see e.g. [19]). Key open questions on neutrino properties remain to be answered by upcoming experiments, including the absolute mass scale, the neutrino mass ordering, the neutrino nature (Dirac versus Majorana), the existence of sterile neutrinos and of CP violating phases. The combined analysis of available experimental data appear to favor normal (versus inverted)) mass ordering[1] and a CP violating Dirac phase around 1.5 π [20].

Neutrinos are copiously produced in stellar environments. Weakly interacting, they propagate from stellar cores, violent phenomena such core-collapse supernovae, or in accretion disks around compact binaries (i.e. black hole-black hole, neutron star-black hole, neutron star-neutron star). Their fluxes and spectra encode information on these environments and on flavor conversion in such media. In the last decade, theoretical studies have shown conversion phenomena that occur due to the neutrino-neutrino interaction, the presence of shock waves, of turbulence, or of new physics, like non-standard neutrino-matter or neutrino-neutrino interactions. Pinning down the underlying mechanisms and the conditions for their occurrence requires demanding simulations, much more complex than in the solar case. Such phenomena can potentially impact the explosion dynamics of iron core-collapse supernovae and the conditions for the nucleosynthesis of heavy elements (r-process) as well as induce neutrino spectral swapping(s) in future observations both of (extra)galactic supernova neutrinos and of the diffuse supernova neutrino background. The former are rare events, typically one every thirty years, while the latter should be discovered by the EGADS project [23, 22].

2. Neutrino flavor evolution as a many-body problem

A gas of self-interacting neutrinos and antineutrinos propagating in a dense astrophysical environment is a many-body system. Through many-body techniques formal connections can be established between such as neutrino gas and other many-body systems like atomic nuclei, or condensed matter. In particular the dynamics of the gas can be treated using the well known Born-Bogoliubov-Green-Kirkwood-Yvon (BBGKY) hierarchy.

BBGKY is a system of coupled first-order integro-differential equations for reduced density matrices, or equal-time correlation functions. Originally introduced for non-relativistic systems, the BBGKY hierarchy generalisation to relativistic systems involves an infinite set of equations.

[1] The mass ordering is unknown since one of the mass-squared differences has not been measured yet (referred to as the atmospheric one). As a consequence we do not yet know if the third mass eigenstate is the lightest or the heaviest.

IOP Conf. Series: Journal of Physics: Conf. Series **1056** (2018) 012060 doi:10.1088/1742-6596/1056/1/012060

The non-relativistic version of the hierarchy has been used to describe atomic nuclei, their dynamics e.g. in nucleus-nucleus collisions, and collective excitations, i.e. giant resonances [25].

The first of BBGKY equations determines the evolution of the one-body (or single-particle) density matrix leading to the time-dependent Hartree-Fock equation (TDHF), when one neglects the correlated part of the two-body density matrix [25]. Its inclusion brings to the Boltzmann equation that includes collisions in the dynamics. The linearised version of the TDHF equation is the Random-Phase-Approximation (RPA) used for the description of collective excitations such as the giant dipole resonance where protons and neutrons oscillate against each other. This collective vibration is described within RPA which is an eigenvalue equation for the particle-hole and hole-particle excitations accounted for by the forward X and backward Y amplitudes [25]. Its eigenenergies describe the frequencies of such modes. However RPA can have complex eigenvalues when the starting point (the nuclear ground state) is not the lowest energy state. Such complex values correspond to instabilities of the system.

Ref.[26] has applied the relativistic version of BBGKY to the description of a gas of neutrinos and antineutrinos evolving in an astrophysical environment. The authors have shown that the first equation gives the Liouville Von Neumann equation for single-particle densities that is commonly employed to study neutrino flavor evolution in media (see e.g. [28]). In this case, BBGKY has to be applied to reduced density matrices since off-diagonal contributions have to be introduced to account for the neutrino mixings. The mean-field associated with the particles composing the medium can be shown to give various terms. For example the well known MSW potential $V_{CC} = \sqrt{2}G_F\rho_e(r)$[2] arises because of the charged-current neutrino interaction with electrons. Similarly one obtains the non-linear neutrino self-interaction potential from neutrino interaction with the neutrinos composing the medium. Such a potential has off-diagonal contributions, as first pointed out by Pantaleone [24], because of the neutrino mixings.

Ref.[27] has provided a linearised version of the neutrino evolution equations in astrophysical media and shown that it is as RPA used for atomic nuclei, when one replaces the X and Y amplitudes with the off-diagonal terms of the neutrino and antineutrino density matrix respectively. Another difference between the nuclear case and the neutrino case is that in the latter case the starting point are the neutrino fluxes at the neutrinosphere that are quasi-stationary solutions of the evolution equation while in the former the nuclear ground state is a stationary solution. Moreover for neutrinos one seeks mostly for unstable solutions that correspond to flavor modification whereas for nuclei one looks for stable solutions corresponding to nuclear excitations.

The investigation of neutrino self-interactions has triggered a decade of studies aiming at understanding a plethora of novel flavor conversion phenomena due to its presence [29]. In fact, when a core-collapse supernova explodes, most (99%) of the gravitational energy released is taken by neutrinos of all flavors in a small burst lasting about ten seconds. As many as 10^{58} neutrinos are emitted rendering the self-interaction sizeable. Similar fluxes are found in simulations of accretion disks around compact binary systems. Several conversion mechanisms have been uncovered, such as the bipolar and the spectral split (see e.g. [29]), the matter neutrino resonance [31] or the "fast" modes [30]. The most general mean-field equations present two more kinds of contributions, i.e. wrong helicity terms due to the neutrino absolute mass [32] and pairing correlators [26]. They introduce neutrino-antineutrino coupling named helicity coherence that requires anisotropic media to be non-zero. Their impact on flavor evolution has been first studied with a one-flavor schematic model in Ref.[33]. The authors have shown that helicity coherence could produce significant flavor modification. An in-depth study based on detailed simulations in binary neutron star mergers has shown that adiabaticity is not enough to trigger efficient flavor conversion due to helicity coherence. An analytical argument is given

[2] The quantity ρ_e is the electron number density that depends on the medium.

to show that this would require a matching between the matter and the self-interaction terms that cannot take place unless specific matter density profiles are chosen. It also shows the requirements for multiple MSW resonances to take place as for the neutrino-matter resonance.

3. Observational aspects and nuclear physics

Neutrino observatories often exploit nuclei to detect astrophysical neutrinos. Scintillator detectors such as JUNO [35] use carbon, water Cherenkov like Super-Kamiokande or the future Hyper-Kamiokande use oxygen, LVD has iron nuclei [37], DUNE is based on argon [50] and the HALO observatory uses lead [39]. The SNEWS network of observatories is ready to observe a future core-collapse supernova explosion [40].

Neutrino-nucleus cross sections are still affected by significant uncertanties. The best known ones are those on deuterium, used in the SNO experiment, since they can be calculated using Effective Field Theories (see e.g. [38]). The exclusive and inclusive neutrino-carbon cross section associated with neutrinos from decay-at-rest muons have been measured by a series of experiments and are theoretically well understood (see e.g. [51]). For the other nuclei, theoretical predictions based on various approaches (mostly shell model, RPA and its variants) have been used over the years[3]. The operators involved in the nuclear matrix elements of neutrino-nucleus cross sections are of Fermi type, or of Gamow-Teller type. Constraints on these calculations can come from other weak processes such as beta-decay, muon capture, or from charge-exchange reactions. The flux-averaged neutrino-nucleus cross sections associated with typical quasi-thermal supernova neutrino fluxes[4] can vary by several tens of percent, e.g. for the case of lead [39] while differences can be even larger for the energy dependent cross sections.

Several proposals have been made over the years to measure neutrino-nucleus cross sections of interest for neutrino astrophysics. These include projects such as ORLAND at SNS based on muon decay-at-rest neutrinos, and low-energy beta-beams that would exploit the beta-decay of boosted radioactive ions to produce neutrinos in the 100 MeV energy range [41]. Experiments to precisely measure neutrino-nucleus cross sections for oxygen, argon, iron and lead are now planned by the COHERENT collaboration at SNS [36]. Such measurements will provide information on the isospin and spin-isospin response in these nuclei. In particular they will give further clues on the quenching of the axial-vector coupling constant that is essential for the search of neutrinoless double beta decay [43, 44].

The future measurement of supernova neutrinos will bring important information both for astrophysics and neutrino properties. One of the key questions is the mechanism under which iron core-collapse supernovae explode. The favored scenario is the neutrino delayed accretion shock mechanism by Wilson [45] and Bethe-Wilson [46]. This is thought to drive most of supernova explosions except probably the most energetic ones. SN1987A observations have in particular supported the delayed accretion shock model against the prompt shock one, as shown in the Bayesian analysis of Loredo and Lamb [52] and those that have followed [53].

Neutrinos and antineutrinos of all flavors are emitted during the collapse lasting about ten seconds and take away most of the gravitational binding energy of the newly formed neutron star (about 3×10^{53} erg). Current three-dimensional simulations include sophisticated neutrino transport, convection and hydrodynamical instabilities such as the Standing-Accretion-Shock-Instability (SASI). Future observation of supernova neutrinos could confirm/refute the current paradigm where the explosion is aided by the SASI by measuring its imprint on the neutrino time signal in detectors such as ICECUBE (see e.g. [42]).

[3] A measurement on ν-deuterium and one on ν-iron are also available, although the statistical and systematic errors are still large.

[4] The supernova neutrino spectra are approximated by Fermi-Dirac or power law distributions.

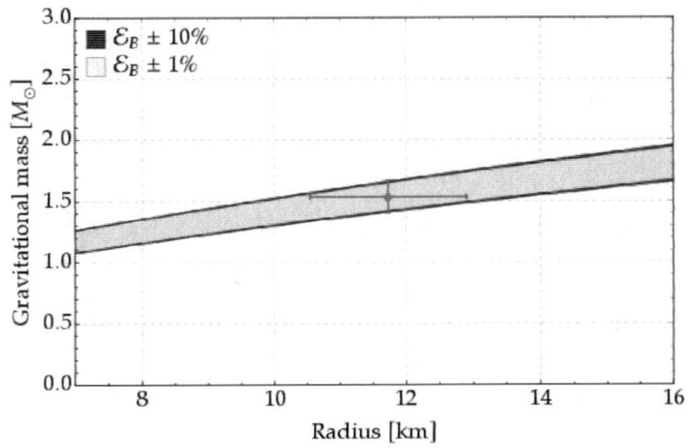

Figure 1. Gravitational neutron star mass versus radius allowed region during a core-collapse supernova explosion. The red point is a putative measurement of the neutron-star radius and the corresponding neutron star mass allowed by different EOS following the Lattimer-Prakash relation [47] (see text). The band is assuming the total gravitational binding energy of the neutron star emitted with neutrinos is measured with a precision of 10% with Super-Kamiokande or 1% in future detectors like Hyper-Kamiokande. From Ref.[48].

When neutrinos decouple from the dense supernova region and start free streaming, forward scattering that is included in the mean-field approximation produces flavor conversion and swapping(s) among the electron and the muon and tau neutrinos (see Section 2). These modify the quasi-thermal spectra that neutrinos have at decoupling. Reconstructing precisely the neutrino spectra is necessary to get insight on neutrino propagation in the dense region and on the flavor phenomena that occur in the low density layers. Several likelihood analyses have been performed to assess how precisely the neutrino spectra can be reconstructed in the case a supernova explodes. Their determination requires the neutrino luminosity, average energy and width (or pinching parameter) for the three flavors (nine free parameters overall) for quasi-thermal distributions. In particular, Ref.[54] has shown that the detection of neutrinos through inverse-beta decay in Hyper-Kamiokande allows to identify the average energies for electron antineutrinos and non-electron neutrino flavors while the presence of the pinching parameters introduces degeneracies that cannot be resolved to fully determine the widths and even the total luminosity. Other analyses have shown that, in principle, all parameters can be precisely measured by combination of several channels such as e.g. in JUNO. However many likelihood analyses make ansatz for example on an energy equipartition among the neutrino flavors, that is currently not favored by supernova simulations, or by fixing the pinching parameter. Ref.[48] has performed likelihood analyses with the 9 parameters unconstrained in order to assess how precisely the emitted gravitational energy of the newly formed neutron star can be determined. These analyses have shown that the combination of inverse-beta decay and electron scattering events allows a measurement of the total gravitational energy with 11% precision. From such a measurement, and using the relation by Lattimer and Prakash based on EOS that allow for neutron stars with more than 1.65 solar masses, the compactness of the neutron star can be obtained [48]. Similar analysis is performed for Hyper-Kamiokande showing that the electron anti-neutrino and non-electron spectra can be precisely determined [49]. Hopefully a future measurement of neutrinos from a galactic explosion will tell us more about the mass-radius relation and neutron star equation of state.

In conclusion, neutrinos keep revealing intriguing new phenomena in astrophysical environments and interesting connections with other domains, in particular nuclear physics, some of which have been highlighted in this proceedings.

IOP Conf. Series: Journal of Physics: Conf. Series **1056** (2018) 012060 doi:10.1088/1742-6596/1056/1/012060

3.1. Acknowledgments

The author acknowledges financial support from "Gravitation et physique fondamentale" (GPHYS) of the *Observatoire de Paris*.

References

[1] Davis Jr. R, Harmer D S and Hoffman K C, 1968 *Phys. Rev. Lett.* **20** 1205
[2] Hirata K et al. [KAMIOKANDE-II Collaboration], 1987 *Phys. Rev. Lett.* **58** 1490
[3] Bionta R M et al., 1987 *Phys. Rev. Lett.* **58** 1494
[4] Alekseev E N, Alekseeva L N, Krivosheina I V and Volchenko V I, 1988 *Phys. Lett. B* **205** 209
[5] Aartsen M G et al., 2014 *Phys. Rev. Lett.* **113** 101101 [arXiv:1405.5303].
[6] Abbott B P et al. [LIGO Scientific and Virgo Collaborations], 2016 *Phys. Rev. Lett.* **116** 061102 [arXiv:1602.03837]
[7] Abbott B P et al. [LIGO Scientific and Virgo Collaborations], 2017 *Phys. Rev. Lett.* **119** 161101 [arXiv:1710.05832]
[8] Fukuda Y et al., 1998 *Phys. Rev. Lett.* **81** 1562 [hep-ex/9807003]
[9] Maki Z , Nakagawa M , and Sakata S, 1962 *Prog. Theor. Phys.* **28** 870
[10] Giunti C and Kim C W , 2007 Oxford, UK: Univ. Pr. 710 p
[11] Ahmad Q R et al., 2002 *Phys. Rev. Lett.* **89** 011301
[12] Eguchi K et al., 2003 *Phys. Rev. Lett.* **90** 021802
[13] Haxton W C, Hamish Robertson R G and Serenelli A M, 2013 *Ann. Rev. Astron. Astrophys.* **51** 21 [arXiv:1208.5723]
[14] Bethe H A, 1939 *Phys. Rev.* **55** 434
[15] Bellini G et al. [BOREXINO Collaboration], 2014 *Nature* **512** 383
[16] Villante F L, 2015 *Phys. Lett. B* **742** 279 [arXiv:1410.2796]
[17] Wolfenstein L , 1978 *Phys. Rev. D* **17**
[18] Mikheev S P and Smirnov A Y , 1985 *Sov. J. Nucl. Phys.* **42** 913 [1985 *Yad. Fiz.* **42** 1441]
[19] Giunti C , 2016 *Nucl. Phys. B* **908** 336 [arXiv:1512.04758]
[20] Capozzi F, Di Valentino E, Lisi E, Marrone A, Melchiorri A and Palazzo A, 2017 *Phys. Rev. D* **95** 096014 [arXiv:1703.04471]
[21] Volpe C, 2016 *Acta Phys. Polon. Supp.* **9** 769 [arXiv:1609.06747]
[22] Beacom J F and Vagins M R, 2004 *Phys. Rev. Lett.* **93** 171101 [hep-ph/0309300]
[23] Xu C [Super-Kamiokande Collaboration], 2016 *J. Phys. Conf. Ser.* **718** 062070
[24] Pantaleone J T, 1992 *Phys. Lett. B* **287** 128
[25] Lacroix D, Ayik S, Chomaz P, 2004 *Progress in Part. and Nucl. Physics* 52 497
[26] Volpe C, Vaananen D and Espinoza C, 2013 *Phys. Rev. D* **87** 113010 [arXiv:1302.2374]
[27] Vaananen D and Volpe C, 2013 *Phys. Rev. D* **88** 065003 [arXiv:1306.6372]
[28] Serreau J and Volpe C, 2014 *Phys. Rev. D* **90** 125040 [arXiv:1409.3591]
[29] Duan H, Fuller G M and Qian Y Z, 2010 *Ann. Rev. Nucl. Part. Sci.* **60** 569 [arXiv:1001.2799 [hep-ph]]
[30] Sawyer R F, 2016 *Phys. Rev. Lett.* **116** 081101 [arXiv:1509.03323 [astro-ph.HE]]
[31] Malkus A, Kneller J P, McLaughlin G C and Surman R, 2012 *Phys. Rev. D* **86** 085015 [arXiv:1207.6648 [hep-ph]]
[32] Vlasenko A, Fuller G M and Cirigliano V, 2014 *Phys. Rev. D* **89** 105004 [arXiv:1309.2628 [hep-ph]]
[33] Vlasenko A, Fuller G M and Cirigliano V, arXiv:1406.6724 [astro-ph.HE]
[34] Chatelain A and Volpe C, 2017 *Phys. Rev. D* **95** 043005 [arXiv:1611.01862]
[35] An F et al. [JUNO Collaboration], 2016 *J. Phys. G* **43** 030401
[36] K. Scholberg, private communication.
[37] Agafonova N Y et al., 2008 *Astropart. Phys.* **28** 516 [arXiv:0710.0259 [astro-ph]]
[38] Balantekin A B and Yuksel H, 2003 *Phys. Rev. C* **68** 055801 [hep-ph/0307227]
[39] Vaananen D and Volpe C, 2011 *JCAP* **1110** 019 [arXiv:1105.6225]
[40] Antonioli P et al., 2004 *New J. Phys.* **6** 114 [astro-ph/0406214]
[41] Volpe C, 2004 *J. Phys. G* **30** L1 [hep-ph/0303222]
[42] Lund T, Marek A, Lunardini C, Janka H T and Raffelt G, 2010 *Phys. Rev. D* **82** 063007 [arXiv:1006.1889 [astro-ph.HE]]
[43] Suhonen J T, 2017 *Front. in Phys.* **5** 55 [arXiv:1712.01565]
[44] Iachello F, Kotila J and Barea J, 2015 *PoS NEUTEL* **2015** 047
[45] Wilson J R, 1971 *Astrophys. J.* **163** 209
[46] Bethe H A and Wilson J R, 1985 *Astrophys. J.* **295**, 14
[47] Lattimer J M and Prakash M P, 2001 *Astrophys. J.* **550** 426 [astro-ph/0002232]

[48] Gallo Rosso A, Vissani F and Volpe M C, 2017 *JCAP* **1711** 036 [arXiv:1708.00760]
[49] Gallo Rosso A, Vissani F and Volpe M C, 2018 *JCAP* **1804** 040 [arXiv:1712.05584 [hep-ph]]
[50] LoSecco J M, 2017 *Astron. Nachr.* **338** 984
[51] Volpe C, Auerbach N, Colo G, Suzuki T and Van Giai N, 2000 *Phys. Rev. C* **62** 015501 [nucl-th/0001050]
[52] Loredo T J and Lamb D Q, 2002 *Phys. Rev. D* **65** 063002 [astro-ph/0107260]
[53] Pagliaroli G, Vissani F, Costantini M L and Ianni A, 2009 *Astropart. Phys.* **31** 163 [arXiv:0810.0466 [astro-ph]]
[54] Minakata H, Nunokawa H, Tomas R and Valle J W F, 2008 *JCAP* **0812** 006 [arXiv:0802.1489 [hep-ph]]

IOP Conf. Series: Journal of Physics: Conf. Series **1056** (2018) 012061 doi:10.1088/1742-6596/1056/1/012061

Shape mixing in $0\nu\beta\beta$ candidates

V Werner[1], U Gayer[1], J Kleemann[1], T Beck[1], M Bhike[4], V Derya[2], J Isaak[3], Krishichayan[4] B Löher[1], L Mertes[1], H Pai[1,5], O Papst[1], N Pietralla[1], P C Ries[1], C Romig[1], D Savran[6], M Schilling[1], W Tornow[4], J Wilhelmy[2] and M Zweidinger[1]

[1] Institut für Kernphysik, Technische Universität Darmstadt, 64289 Darmstadt, Germany
[2] Institut für Kernphysik, Universität zu Köln, 50937 Köln, Germany
[3] Research Center for Nuclear Physics, Osaka University, Ibaraki, Osaka 567-0047, Japan
[4] Duke University, Durham, North Carolina 27708-0308, USA
[5] Saha Institute of Nuclear Physics, 1/AF Bidhannagar, Kolkata 700 064, India
[6] EMMI and GSI Helmholtzzentrum für Schwerionenforschung, 64291 Darmstadt, Germany

E-mail: vw@ikp.tu-darmstadt.de

Abstract. Weak processes are typically observed through nuclear effects, as they mediate between different eigenstates of either one nucleus, or a pair of nuclei. Since the derivation of important parameters of the weak interaction and weakly-interacting particles, such as their masses, spin dependencies, and alike, heavily relies on nuclear theory, it must be assured that theory properly describes the relevant wave functions. A special challenge for neutrinoless double-beta decay, for example, is the location of many candidate isotopes in regions of the nuclear chart, where nuclei may exist simultaneously in different shapes, hence, different wave function components belonging to different nuclear deformations mixing into the nuclear eigenstates. In addition, isovector parameters of nuclear models are not often well constrained, posing an additional challenge. Through the measurement of properties of the nuclear scissors mode, a magnetic isovector excitation at low energies, using photon-scattering techniques, we obtain data that is relevant to constrain the structure of the nuclei and their eigenstates in question. Furthermore, our recent research program comprises the investigation of isotopes relevant for the detection of hypothetical massive weakly-interacting particles.

1. Introduction

The broad topics of weak-interaction physics and nuclear physics strongly overlap, although not always obvious. Nuclei play a major role either as laboratories for the investigation of weak-interaction processes, or as witnesses for weakly interacting particles in sophisticated detectors. Although nuclei are bound and governed by the strong interaction, they also interact weakly, best known through the occurrence of β decay. Topics addressed include the determination of the fundamental nature of neutrinos and their masses through the search for neutrino-less double-beta ($0\nu\beta\beta$) decay. Another popular aspect is the search for new weakly-interacting particles, such as the hypothesized WIMPs, connecting nuclear and weak-interaction physics to dark matter searches. In order to be suited for such studies, though, the physics of the involved nuclei needs to be sufficiently well understood, which includes their electromagnetic response and overall structure, governed by the strong force, the classical nuclear physics. Often, information on the nuclear physics input into weak processes cannot be obtained experimentally, directly, but relies on nuclear theory. Therefore, it is of utmost importance that theory is constrained

IOP Conf. Series: Journal of Physics: Conf. Series **1056** (2018) 012061 doi:10.1088/1742-6596/1056/1/012061

by experiment. That means especially to constrain the wave functions of relevant eigenstates of the nucleus as much as possible, through the measurement of level and decay energies and the electromagnetic matrix elements among them, which carry fingerprints of their structure. In the following, especially the relation to nuclear matrix elements for the description of $2\nu\beta\beta$ and $0\nu\beta\beta$ decay, for which we obtained new experimental data, will be discussed. In the end, other implications of nuclear structure for weak processes, i.e., the scattering of weakly interacting particles off nuclei, will briefly be addressed.

2. Nuclear structure for $0\nu\beta\beta$ decay

Various experiments are underway or planned for, to search for the $\beta\beta$-decay process without the emission of neutrinos. The observation of such decay would, firstly, determine the fundamental nature of the neutrino as a Dirac- or Majorana-like particle, and if the latter, would open a possibility to determine its mass. Typically, in experiments such as GERDA [1] and MAJORANA [2] (Ge based), CUORE [3] (Te based), CUPID [4] (Te/Se based), or EXO [5] (Xe based), the decaying material is simultaneously the main component of the active detector. The signal searched for is a peak structure at the Q value of the double-beta decay, which would indicate that no energy has been transmitted to any neutrinos in the decay process. Should such a signal be separated from background, which, in some cases may stem from γ-rays from the detection isotopes themselves, the nuclear matrix elements for the $0\nu\beta\beta$ process would be needed in order to extract a neutrino mass. Those matrix elements, however, are not directly accessible experimentally, but require nuclear theory. Hence, theory needs to pinpoint the wave functions of initial and final states precisely enough to give a reliable value for the nuclear matrix element. An interesting possibility has been risen recently [6], which is the occurrence of a $0\nu\beta\beta$ transition to the excited 0^+ state in the daughter nucleus, in particular in the case of the ^{150}Nd$\to^{150}$Sm transition. We will introduce the experimental approach to address this phenomenon, based on measurements of the decay behavior of the nuclear scissors mode, in the following.

2.1. The scissors mode

The scissors mode [7, 8, 9] is an orbital M1 valence-shell excitation, which is typically observed at about 3 MeV excitation energy [10, 11]. The geometrical interpretation in well-deformed nuclei is a scissors-like counter-oscillation of the deformed proton versus the deformed neutron bodies. In other classes of nuclei the mode is also observed and can be described as a two-phonon excitation comprised of a proton-neutron symmetric quadrupole phonon and its mixed-symmetric counterpart [12]. In well-deformed nuclei, the scissors mode is the lowest-lying mixed-symmetry state. Its location in energy is sensitive to isovector model parameters, such as the Majorana parameters within the interacting boson model (IBM-2) [8]. Hence, locating the scissors mode is important to constrain the proton-neutron degree of freedom in nuclear structure models, and its M1 excitation strength depends strongly on the underlying structure and on effective g factors.

2.2. Decay behavior, shape mixing, and $0\nu\beta\beta$ rates

Like its excitation strength, also the decay behavior of the scissors mode depends on the underlying structure of the nucleus. In a deformed nucleus, the dominant decay paths are to the ground state and the first excited 2^+ states. In more spherical systems, however, the transition to the ground state should be suppressed, since it would afford the annihilation of two phonons. In this case, a strong decay to the first excited 0^+ state would be expected.

Considering the locations of nuclei which can $\beta\beta$ decay in the nuclear chart, one finds that a large number of them are located in regions which are known for fast shape changes and shape coexistence. For example, Ge and Se isotopes have long been discussed in terms of shape coexistence (e.g., recently in Ref. [13]), and in ^{76}Se significant shape mixing between the ground and first-excited 0^+ states is evident through the strong E0 strength among those states. Another example is the $\beta\beta$-decay partners ^{150}Nd and ^{150}Sm at neutron numbers $N = 90$ and $N = 88$, respectively. This is exactly in the region well-known [14] for a shape-phase transition [15] from spherical to deformed shapes, with strong mixing between both shapes at the transitional point near $N = 90$.

In a situation of configuration mixing between the lowest 0^+ states, the scissors mode may decay to both states, 0_1^+ and 0_2^+, because both share the same structure. That means the configurations that the scissors mode has been excited from, are found in both 0^+ states. Therefore, a signature of configuration mixing would be a similarly strong decay of the scissors mode into the ground and the first excited 0^+ states (compare figure 1).

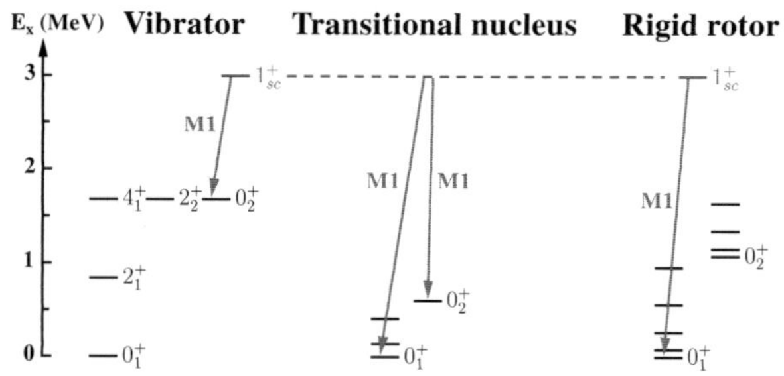

Figure 1. Decay schemes of the scissors mode in vibrator, rotor, and transitional nuclei.

For a 0ν or $2\nu\beta\beta$ decay between nuclei in such a structural region, it may occur that the ground-state wave function of the decay mother has a significant, eventually even stronger overlap with the excited-state wave function of the daughter state. Since $\beta\beta$ transition rates are enhanced if the structures of the mother and daughter states are similar, this situation can subsequently lead to an enhancement of the $\beta\beta$-decay matrix element to the excited 0^+ state in the daughter nucleus. A first hint to this mechanism, as well as to the sensitivity of the scissors mode to such cases, has firstly been found in the study of ^{154}Gd [6], where the decay of the scissors mode into both, the 0_1^+ and 0_2^+ states has been observed in a combination of photon-scattering and β-delayed γ-spectroscopy studies. The fixing of isovector model parameters and the structural location of ^{154}Gd in the IBM-2 model space led to about an order of magnitude enhancement of the $0\nu\beta\beta$-decay matrix element predicted by the model.

In the same work [6], the decay of ^{150}Nd into ^{150}Sm was identified as a case where the $0\nu\beta\beta$ matrix element to the excited state could even become larger than that to the ground state, due to a stronger wave function overlap of the ground state of ^{150}Nd with the excited state of ^{150}Sm. This fact would result from the location of these isotopes at the $N = 90$ shape phase transition, where the spherical and deformed configurations mix differently into the respective states of both isotopes. Motivated by this prediction, we performed a first photon-scattering experiment on ^{150}Sm, in order to search for a fingerprint of shape coexistence, namely, the decay of the scissors mode into both 0^+ states, as outlined below.

2.3. Experiment at HIGS

The experiment has been performed at the High-Intensity γ-Ray Source (HIγS) [16] at the Free-Electron Laser Laboratory at TUNL, on the campus of Duke University. The facility is capable of delivering γ-ray beams over a broad range of energies, with an energy spread of about 3 % (hence, near-monoenergetic), and with near 100 % polarization. The γ-rays are produced through Compton backscattering in an electron storage ring. Electron bunches in the ring produce visible laser light in the free-electron laser section, which is mirrored back and forth on the long axis of the ring, and can then interact with another electron bunch at a well-defined interaction point. The Compton-backscattered photons are then collimated and led to the experiment, located far downstream. Typical photon rates at our energy of interest around 3 MeV are 10^8 per second.

These γ-rays impinge on the probe, typically consisting of several grams of isotopically enriched material (in order to compensate for the rather small photo-absorption cross section). Target nuclei can then get resonantly excited into excited states, which due to the angular momentum carried by the incident photon mostly have angular momentum $J = 1$. Because of the high degree of polarization of the incident beam, the parities of excited states can be determined by a simple polarimeter setup with γ-ray detectors within and perpendicular to the plane spanned by the beam axis and the polarization vector. 1^+ states will radiate only within that polarization plane upon deexcitation to the ground state, whereas γ rays from the deexcitation of 1^- states to 0^+ are only emitted perpendicular to the plane.

Apart from the polarization, another advantage of γ-ray beams from Compton backscattering is the small energy spread of the beam. Compared to the traditional use of γ-ray beams from bremsstrahlung, which cover all energies from zero up to the energy from the incident electrons for beam production, a much lower background at low energies can be achieved using Compton backscattering. Therefore, the sensitivity to observe decays from resonantly-excited states into states other than the ground state is greatly enhanced as compared to bremsstrahlung experiments.

For our experiment the γ^3 setup [17] at HIγS has been employed, which consists of a combination of high-purity germanium (HPGe) and lanthanum-bromide ($LaBr_3$) detectors. Due to the very compact geometry of the setup, in principle measurements of $\gamma\gamma$ coincidences after photo-excitation are possible. In our case of investigating the scissors mode of ^{150}Sm, the photo-excitation cross section was not sufficient to allow for a coincidence measurement, however, the high-resolution detectors placed at 90° relative to the beam axis were sufficient to observe the decays of interest. It was possible to assign the observed γ-rays uniquely to the decays of specific excited states, since the level density around 3 MeV is still rather low.

The method of nuclear resonance fluorescence (NRF) can be reviewed in [18]

2.4. ^{150}Sm

^{150}Sm had previously been measured using bremsstrahlung beams [19], therefore, absolute cross sections which are necessary to extract absolute reduced transition strengths were already known. In Ref. [19] parities have not been measured directly, but rather inferred from the branching transitions of dipole-excited states to the ground and 2_1^+ states, following the Alaga rules [20]. The Alaga rules, however, require a good K quantum number, which is not ascertained in a weakly-deformed soft nucleus like ^{150}Sm. In fact, we find deviations from the previous parity assignments. Figure 2 shows the spectra taken with HPGe detectors within and perpendicular to the polarization plane. Clearly, ground-state transitions from dipole-excited states are only visible in one of the two spectra, depending on their parity. The strongest M1-excited 1^+ state in literature is the state at 3081 keV. The present data show with no doubt that this state is actually a 1^- state, hence, not belonging to the scissors mode. Nevertheless, we measure positive parity for the state at 3113 keV, identifying it as the strongest fragment of the scissors mode of ^{150}Sm.

In the γ-ray spectra of the in-plane HPGe detectors we were able to identify a peak at an energy of 2373 keV. This transition is not present in the spectra in the out-of-plane detectors, and its energy matches the transition energy of the 3113-keV state to the 0_2^+ state, which we therefore identify it with. This allows to extract the branching ratio of the scissors mode to the ground and 0_2^+ states, and, along with the known absolute excitation cross section and taking into account

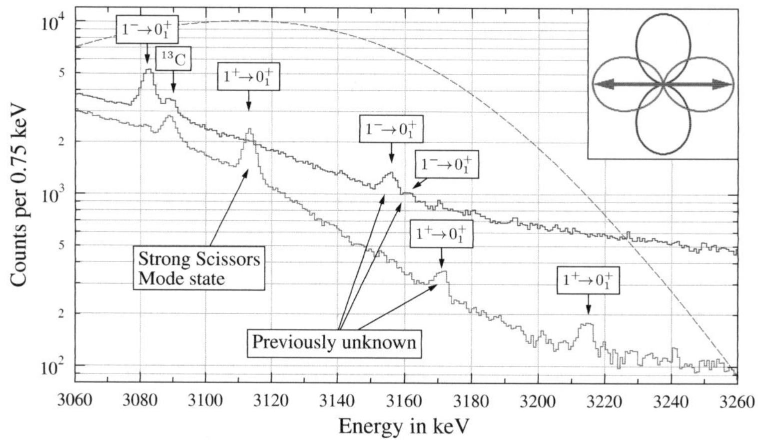

Figure 2. Spectra from the HPGe detectors within (red, bottom) and perpendicular to (blue, top) the polarization plane. Transitions from dipole-excited states are marked, including the parities of the states. The insert shows the radiation patterns from M1 (red) and E1 (blue) excited states, relative to the polarization vector indicated by the green arrow.

other observed decays of the 3113-keV state to the 2_1^+ ad 2_2^+ states, yields the absolute ground-state and 0_2^+-state B(M1) transition strengths. We obtain B(M1;$1_{sc}^+ \to 0_1^+$)= $0.071(10)\mu_N^2$ and B(M1;$1_{sc}^+ \to 0_2^+$)= $0.030(9)\mu_N^2$. Both transitions are of the same order of magnitude, and even agree within a factor of two. This is the behavior expected for the case of shape coexistence, where the scissors mode can decay with similar strengths to either 0^+ state. Also, this finding agrees with the prediction [6] that shape mixing occurs for the ^{150}Nd/^{150}Sm pair of isotopes, which would lead to a dominant $0\nu\beta\beta$ matrix element to the excited 0^+ state in ^{150}Sm. A similar study on ^{150}Nd has recently been performed, and both measurements will give input to new model calculations to properly describe the structure of both isotopes, and to give more reliable predictions for the $0\nu\beta\beta$ matrix elements between the ground state of ^{150}Nd and the ground and excited 0^+ states of ^{150}Sm.

3. Upcoming and ongoing studies

Further studies into nuclear structure relevant for weak-interaction physics, performed at TU Darmstadt, are briefly addressed in the following.

3.1. $0\nu\beta\beta$ decay

Photon-scattering experiments analog to that described above have and will be performed on other pairs of nuclei which are discussed in the context of detectors to observe $0\nu\beta\beta$ decays, e.g. ^{82}Se/^{82}Kr (SuperNEMO) or ^{76}Ge/^{76}Se (GERDA/MAJORANA). In both cases, we observe transitions from the scissors mode into the excited 0^+ states. In addition, the spectra yield information on γ-rays from the nuclei of interest, which may be close to the endpoint energy of the $\beta\beta$-decay spectrum, where the signal from $0\nu\beta\beta$ decay is expected. Such γ rays would have to be taken into account in background considerations for the respective detectors, since $J = 1$ states could be excited by background radiation penetrating their shielding, e.g., neutrinos.

IOP Conf. Series: Journal of Physics: Conf. Series **1056** (2018) 012061 doi:10.1088/1742-6596/1056/1/012061

3.2. ν scattering

From data taken at HIγS at higher energies (up to the particle thresholds of the investigated isotopes), we expect to obtain valuable information on the nuclear M1 response, in particular the spin-flip resonance. The latter is composed of states excited by exciting a single proton or neutron from a $l \pm 1/2$ to a $l \mp 1/2$ orbital. Little detail about the location and the overall shape of this resonance is known to date, with only few examples addressed by experiment (e.g., ^{136}Xe [21]). This is of interest for investigations into (ν, ν') neutrino-scattering of nuclei, which can excited quite selectively such 1^+ states from the ground state, as shown in QRPA calculations [22].

3.3. WIMP search

First measurements at the S-DALINAC facility of TU Darmstadt have just recently started, to investigate the electron-scattering form factors to the first excited states of the odd-A ^{129,131}Xe isotopes. This is of interest due to the use of natural Xe in the Xenon1T [23] WIMP detector, where the odd-A Xe isotopes make a considerable fraction of the overall detector mass. Recent works [24, 25] addressed the question of how WIMP-nucleus cross sections might be altered should WIMP-neutrino scattering be spin dependent. In this case, scattering to the first excited state could dominate the detector response. The electron-nucleus scattering experiments will test the nuclear wave functions used in the respective calculations.

Acknowledgments

We acknowledge the HIγS accelerator team for delivering excellent beam conditions. This work was supported by the German DFG within SFB 1245, by the alliance Program of the Helmholtz Association (HA216/EMMI), and by the U.S. DOE under Grant No. DE-FG02-97ER41033. We further acknowledge U.S. DOE for the supply of isotopically enriched materials through ORNL.

References

[1] Agostini M and the GERDA collaboration 2017 Nature **544** 47
[2] Abgrall N *et al.* 2014 Adv. High Energy Physics **2014** 365432
[3] Alfonso K and the CUORE collaboration 2015 Phys. Rev. Lett. **115** 102502
[4] Artusa D R *et al.* 2016 Eur. Phys. J. C **75** 591; Phys. Lett. B **767** 321
[5] The EXO-200 Collaboration 2014 Nature **510** 229
[6] Beller J *et al.* 2013 Phys. Rev. Lett. **111** 172501
[7] Lo Iudice N and Palumbo F 1978 Phys. Rev. Lett. **41** 1532
[8] Iachello F and Arima A *"The Interacting Boson Model"* (Cambridge University Press, Cambridge, 1987)
[9] Bohle D *et al.* 1987 Phys. Lett. B **195** 326
[10] Pietralla N *et al.* 1995 Phys. Rev. C **52** 2317
[11] Enders J, von Neumann-Cosel P, Rangacharyulu C and Richter 2005 A, Phys. Rev. C **71** 014306
[12] Pietralla N, von Brentano P and Lisetskiy A F 2008 Prog. Part. Nucl. Phys. **60** 225
[13] Nomura K, Rodríguez-Guzmán R and Robledo L M 2017 Phys. Rev. C **95** 064310
[14] Krücken R *et al.* 2002 Phys. Rev. Lett. **88** 232501
[15] Warner D D 2002 Nature **420** 614
[16] Weller H R *et al.* 2009 Prog. Part. Nucl. Phys. **62** 257
[17] Löher B *et al.* 2013 Nucl. Instrum. Methods Phys. Res. A **723** 136
[18] Kneissl U, Pitz H H and Zilges A 1996 Prog. Part. Nucl. Phys. **37** 349
[19] Ziegler W *et al* 1993 Nucl. Phys. A **564** 366
[20] Alaga G *et al.* 1955 Dan. Mat. Fys. Medd. **29** 1
[21] Massarczyk R *et al.* 2014 Phys. Rev. C **90** 054310
[22] Ydrefors E, Balasi K G, Kosmas T S and Suhonen J 2012 Nucl. Phys. A **896** 1
[23] Aprile E *et al.* (XENON100 Collaboration) 2012 Phys. Rev. Lett. **109** 181301
[24] Klos P, Menéndez J, Gazit D and Schwenk A 2013 Phys. Rev. D bf 88 083516
[25] Baudis L 2012 Dark Universe **1** 94

IOP Conf. Series: Journal of Physics: Conf. Series **1056** (2018) 012062 doi:10.1088/1742-6596/1056/1/012062

Borexino: new results from the high-purity phase-II data

Sandra Zavatarelli *,[1]

I.N.F.N. Sezione di Genova, via Dodecanneso 33, 16146 Genoa (Italy)

* **On behalf on the Borexino Collaboration** : M Agostini, K Altenmüller, S Appel, V Atroshchenko, Z Bagdasarian, D Basilico, G Bellini, J Benziger, D Bick, G Bonfini, D Bravo, B Caccianiga, F Calaprice, A Caminata, S Caprioli, M Carlini, P Cavalcante, A Chepurnov, K Choi, L. Collica, D DAngelo, S Davini, A Derbin, X F Ding, A Di Ludovico, L Di Noto, I Drachnev, K Fomenko, A Formozov, D Franco, F Froborg, F Gabriele, C Galbiati, C Ghiano, M Giammarchi, A Goretti, M Gromov, D Guffanti, C Hagner, T Houdy, E Hungerford, Aldo Ianni, Andrea Ianni, A Jany, D Jeschke, V Kobychev, D Korablev, G Korga, D Kryn, M Laubenstein, E Litvinovich, F Lombardi, P Lombardi, L Ludhova, G Lukyanchenko, L Lukyanchenko, I Machulin, G Manuzio, S Marcocci, J Martyn, E Meroni, M Meyer, L Miramonti, M Misiaszek, V Muratova, B Neumair, L Oberauer, B Opitz, V Orekhov, F Ortica, M Pallavicini, L Papp, O Penek, N Pilipenko, A Pocar, A Porcelli, G Ranucci, A Razeto, A Re, M Redchuk, A Romani, R Roncin, N Rossi, S Schonert, D Semenov, M Skorokhvatov, O Smirnov, A Sotnikov, L F F Stokes, Y Suvorov, R Tartaglia, J Thurn, M Toropova, E Unzhakov, A Vishneva, R B Vogelaar, F von. Feilitzsch, H Wang, S Weinz, M Wojcik, M Wurm, Z Yokley, O Zaimidoroga, S Zavatarelli, K Zuber, and G Zuzel.

Abstract. The Borexino experiment has recently celebrated the 10 years of data taking. Among the most relevant results already published, we remind the first direct measurement of ^{7}Be and pep neutrinos and the best upper limit on CNO. In 2010-11 six purification were very successful in further reducing the already low backgrounds: a new high-purity Phase II data taking was started in October 2011, aimed to perform high precision measurements and to try to aggress the weakest components of the solar neutrino flux. In this contribution the first results based on the new Phase II data are reviewed.

1. Borexino and solar neutrinos

The Sun is an intense source of neutrinos that can be usefully exploited to better access the physical properties of these elusive particles.

Since 2007 Borexino has measured the fluxes of low energy neutrinos, most notably those emitted in the pp chain nuclear fusion reactions in the Sun and the antineutrinos emitted in the beta decays along the natural radioactive chains in the Earth's rocks.

Borexino was the first experiment to make spectroscopic measurements of solar neutrinos with energy lower than 3 MeV, i.e. below the endpoint energy of natural radioactivity [1].

In 2011 thanks to extensive purification campaigns based on water extraction and nitrogen stripping, a record low scintillator contamination of less than 10^{-18} g/g was achieved for ^{238}U and ^{232}Th; the ^{85}Kr contamination decreased by a factor 5 and ^{210}Bi by a factor 2.5.

A new Phase II data taking was started and in parallel, a refined Monte Carlo code [2] was developed able to reproduce the energy response of the detector, its uniformity within the fiducial scintillator volume relevant to neutrino physics, and the time distribution of detected photons to better than 1% between 100 keV and several MeV.

In the following sections, after a brief description of the detector layout, the first results based on the new high-purity Phase II data are presented, namely the first direct measurement of pp-neutrinos, the

[1] Corresponding author (e-mail address: Sandra.Zavatarelli@ge.infn.it)

IOP Conf. Series: Journal of Physics: Conf. Series **1056** (2018) 012062 doi:10.1088/1742-6596/1056/1/012062

seasonal modulation of the ^{7}Be neutrino signal and a new upper limit on the neutrino magnetic moment.

2. The Borexino detector

Borexino, located deep underground (3,800 m water equivalent) in Hall C of the Gran Sasso Laboratory (Italy), measures solar neutrinos via their interactions with a target of 278 ton organic liquid scintillator. The ultrapure liquid scintillator (pseudocumene (1,2,4-trimethylbenzene (PC)) solvent with 1.5 g/l 2,5-diphenyloxazole (PPO) scintillating solute) is contained inside a thin transparent spherical nylon vessel of 8.5 m diameter. Solar neutrinos are detected by measuring the energy and position of electrons scattered by neutrino-electron elastic interactions. As a result of the elastic scattering, electrons gain kinetic energy up to a maximum of $E_\nu/(1 + m_e c^2/2E_\nu)$ where E_ν is the incoming neutrino energy.

The scintillator promptly converts the kinetic energy of electrons by emitting photons, which are detected and converted into electronic signals (photoelectrons (p.e.)) by 2212 photomultipliers (PMT) mounted on a concentric 13.7 m-diameter stainless steel sphere (SSS).

The volume between the nylon vessel and the SSS is filled with 889 ton of ultra pure, non scintillating fluid and acts as a radiation shield for external gamma rays and neutrons. A second, larger nylon sphere (11.5 m diameter) prevents radon and other radioactive contaminants from the PMTs and SSS from diffusing into the central sensitive volume of the detector. The SSS is immersed in a 2100 ton water Cerenkov detector meant to detect residual cosmic muons.

To better disentangle the solar neutrino signals from radioactive backgrounds over a large energy range (from 0.2 to 3 MeV) a new fitting procedure has been developed based on a multivariate approach [3]: the binned likelihood function is built by including the energy spectrum, the distance from the center and the pulse shape distributions of the selected events.

This fitting procedure is repeated by exploiting both MonteCarlo based and analytical model descriptions of the detector energy response [3,4]: the comparison of the results obtained with the two methods is a powerful check for systematics.

3. The measure of pp neutrino flux

At the temperatures typical of the Sun center (~15.7 10^6 K) hydrogen is converted into helium predominantly via the pp chain with a release of 26.73 MeV for each conversion of four protons into a helium nuclide.

The highest flux of solar neutrinos comes directly from the first reaction of the chain, the fusion of two protons into a deuteron: $p + p \rightarrow d + e^+ + \nu_e$.

The neutrinos emitted in this reaction are called pp-neutrinos and before Borexino, only the radiochemical gallium experiments [5,6] have been sensitive to this low-energy component (E < 420 keV): since the radiochemical experiments measured an integrated flux of all solar electron neutrinos above a given threshold (233 keV for ^{71}Ga) the pp neutrino signal was inferred only indirectly, by combining the radiochemical measurements with those of other experiments.

A direct study of pp neutrino is indeed essential to probe our knowledge of stellar modelling and energy production since the pp fusion is by fact the step-maker of the pp chain because of the small weak interaction cross section. By combining neutrino and optical observations it is possible to test the Sun thermodynamic equilibrium over a time scale of the order of 10^6 years, i.e. the time needed to photons emitted in the Sun's core to reach the surface after several absorption/reemission and scattering processes with the solar matter.

The pp-neutrino energy spectrum extends up to 420 keV, yielding to a maximal energy of the recoiled electrons of 264 keV: a low energy threshold and high energy resolution detector is mandatory for this measure.

In Borexino, the high scintillator light yield (~10^4 photons/MeV) resulting in ~ 500 detected photoelectrons/MeV makes possible to reach a very low energy threshold (50 keV) with a good energy resolution (~ 5% at 1 MeV). Low background is also a key point.

By fact other attempts to measure pp-neutrinos directly over the past 30 years have been hindered by the inability to sufficiently suppress radioactive backgrounds in this low-energy region: particularly relevant is the background due to ^{14}C, a β emitters intrinsic to the organic liquid scintillator whose energy spectrum extends up to 156 keV and its 'pile-up'.

In spite of its small isotopic fraction in the Borexino scintillator (^{14}C/^{12}C ~ 2.7 · 10^{-18}), ^{14}C β-decay is responsible for most of the detector trigger rate (~30 Hz at 50 keV threshold). ^{14}C and pp-neutrinos exhibit however different energy spectra and they can be therefore disentangled.

To increase the sensitivity, independent and precise data–driven methods have been developed to constrain the irreducible ^{14}C background and its pile-up [7].

The analysis was done by using the data collected from January 2012 to May 2013 for a total of 408 live days. The pp neutrino rate was extracted by fitting the measured energy spectrum of the selected events in the 165–590 keV energy window with the expected spectra of the signal and background components [7]: the main components of the fit were the solar neutrino signal (the dominant pp component and the low-energy parts of the ^{7}Be, pep and CNO components), the ^{14}C background, its pile-up, and the other relevant radioactive backgrounds (^{85}Kr, ^{210}Bi, ^{210}Po and ^{214}Pb).

The solar pp neutrino interaction rate as measured by Borexino was of 144 ± 13 (stat.) ±10 (syst.) counts/(day · 100 t) corresponding to a neutrino flux of (6.6 ± 0.7) · 10^{10} cm^{-2} s^{-1}, in accordance with the oscillation parameters reported in [8]. The survival probability was found to be P_{ee}= 0.64 ± 0.12, and it provides a constraint on the Mikheyev–Smirnov–Wolfenstein large-mixing-angle (MSW-LMA) solution in the low-energy vacuum regime.

This measurement demonstrates that about 99% of the energy of the Sun is generated by the pp fusion process.

4. The seasonal modulation of the ^{7}Be neutrino signal

In contrast with water Cherenkov detectors, Borexino cannot retain directional information of individual events: a direct solar imaging is therefore not possible. However, the eccentricity of the Earth's orbit induces a modulation of the detected signal proportional in amplitude to the solid angle subtended by the Earth with respect the Sun: this effect appears as a 6.7% peak-to-peak seasonal modulation in the measured fluxes.

A search for this signature was already done with Phase I data (collected from May 2007 to May 2010), but because of some backgrounds varying in time, the evidence for such a modulation was poorer [1].

Borexino Phase-II data, in addition to higher statistics and lower background levels, are characterized by the absence of major scintillator handling and thus display a high degree of stability of the detector, crucially important for identifying time dependent signals.

A new attempt to measure this time periodicity in the ^{7}Be solar neutrino component was made on the base of the data acquired between December 2011 and December 2015: of particular importance for this study was the reduction thanks to purification campaigns of the ^{85}Kr and ^{210}Bi concentrations, both backgrounds in ^{7}Be region.

The cuts adopted for this particular analysis are described in [9]. Recoil electrons from the elastic scattering of ^{7}Be-ν's are selected by restricting the analysis to the energy region ~ 215-715 keV. In this range, the major backgrounds are the α decays of ^{210}Po and the β decays of ^{210}Bi and ^{85}Kr. The 5.3 MeV α's appear as a peak at ~ 450 keV (after quenching) in the energy spectrum. A major change with respect to Phase I analysis was an enhanced rejection of ^{210}Po α background, based on MultiLayer Perceptron (MLP) machine learning algorithm: this method uses a neural network based on 13 α /β discriminating input variables, that are computed for each event from the pulse time distribution. The MLP provides excellent α − β discrimination. With the mlp parameter threshold set at 0.9 to retain β's, the α rejection efficiency above 99.98% was achieved for ^{214}Po candidate events (7.7 MeV).

Due to Earth's orbital eccentricity (ε = 0.0167), the total count rate is expected to vary as:

$$R(t) = R_0 + R_A \left[1 + \epsilon \cdot \cos\frac{2\pi}{T}(t - \varphi)\right] \quad (1)$$

where T is the period (one year), φ is the phase relative to the perihelion, R_A is the average neutrino interaction rate and R_0 is the time independent background rate. This formalism is consistent with the MSW solution predicting no additional time modulations at the ^{7}Be energies.
The event rate $R(t_k)$ of a series of time bins of equal length t_k was obtained as the ratio of the number of selected events and the corrected life time (subtracted of the muon veto dead time and any down-time between consecutive runs).
The time bins are too short to allow extracting a value of the ^{7}Be neutrino interaction rate via a spectral fit. The raw β-event rate was used instead, which include background contributions.

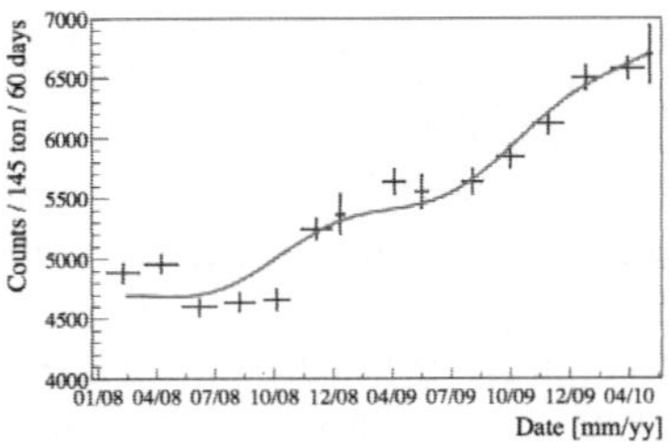

Figure 1. Phase I data : measured rate of β-like events passing selection cuts as a function of the time [1]

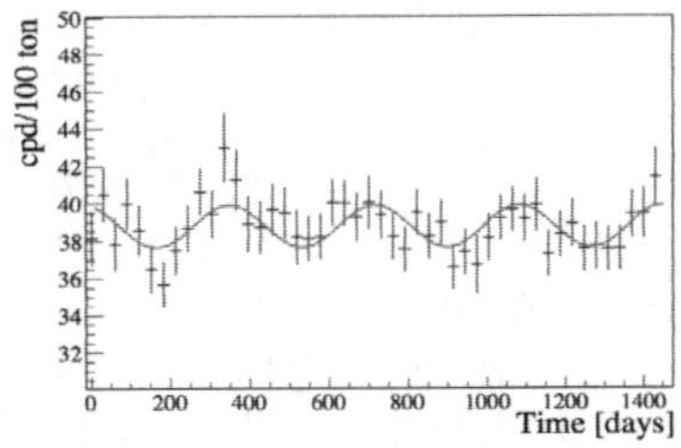

Figure 2. Phase II data : measured ra of β-like events passing selection cuts 30.4-days long bins starting from Dec 11th 2011 [9]

In a first approach the event rate as a function of the time was fit with the function defined in Eq. (2). To extract the modulation parameters, a χ^2 fit of the data with 30.43-day bins was performed without folding multiple years on top of each other. Fig. 2 shows the event rate (in cpd/100 ton) along with the best fit. The best-fit eccentricity is $\varepsilon = 0.0174 \pm 0.0045$, which corresponds to an amplitude of the modulation of (7.1 ± 1.9)%, and the best-fit period is T = 367 ± 10 days [9]. Both values are in agreement with the expected values of 6.7% and of T = 365.25 days. The fit returns a phase of $\varphi = -18 \pm 24$ days.
We note that this result does represent a measurement of the astronomic year with solar neutrinos.
The reliability of the method has been tested by following other two different analysis approaches: the Lomb-Scargle method, an extension of the Fourier Transform approach and the Empirical Mode decomposition (EMD) method, that makes no assumption about the functional form of the signal, in contrast to the Fourier analysis, and can therefore extract any time variation embedded in the data set. These alternative approaches provided perfectly consistent results. The no modulation hypothesis was excluded at 3.9 σ by comparing the χ^2 obtained with and without an annual periodicity [9].

5. A new upper limit on neutrino magnetic moment

When the standard model (SM) is extended to include neutrino masses, a neutrino magnetic moment is predicted, directly proportional to the neutrino mass: the known upper limit on neutrino mass m_ν lead to μ_ν less that 10^{-18} μ_B, which is roughly 8 orders of magnitude lower than the exiting experimental limits. The most stringent laboratory bound on μ_ν is 2.9 10^{-11} μ_B (90% C.L.) and has been obtained with reactor antineutrinos [10]. Other competitive limits come from astrophysical observations [11].
Though experimental bounds on μ_ν are far from the values by extended SM, there are more general models with right-handed bosons or with an extended sector of scalar particles predicting magnetic moments proportional to the masses of charged leptons: these values are closer to the experimental bounds, therefore they justify further experimental investigations.
In the SM the scattering of a neutrino with a not zero magnetic moment is determined by both a weak interaction and a single photon exchange term. Since the latter changes the neutrino helicity, the amplitude on the two terms does not interfere and the total cross section is at the first order is the sum of the two. The coupling of the neutrino mass eigenstates to an electromagnetic field is characterized

by a 3 x 3 matrix of the magnetic and electric dipole moments. The contribution to the ν-e⁻ scattering cross section is proportional to the square of the effective magnetic moment:

$$\frac{d\sigma_{EM}}{dT_e}(T_e, E_\nu) = \pi r_0^2 \mu_{eff}^2 \left(\frac{1}{T_e} - \frac{1}{E_\nu}\right) \quad (2)$$

$$\mu_{eff}^2 = \sum_j \left|\sum_k \mu_{kj} A_k(E_\nu, L)\right|^2 \quad (3)$$

where T_e is the electron recoil energy and r_0 is the classical electron radius.
Since neutrinos are a mixture of mass eigenstates, the effective magnetic moment for neutrino-electron scattering μ_{eff} depends on the components of the neutrino moment matrix μ_{kj} and on the amplitude of the k-mass state at the point of scattering, $A_k(E_\nu,L)$.
For $T_e << E_\nu$ the ratio between the magnetic and weak scattering cross section is proportional to $1/T_e$ and sensitivity of an experiment strongly increases at low energies.
This makes the low-energy threshold of Borexino perfectly suitable for a neutrino magnetic moment study: the effects of a non null neutrino magnetic moment can be investigated by looking for a distortion in the shape of the electron recoil energy spectrum.
A renewed search has been performed using data collected during the Borexino Phase II campaign from December 2011 to May 2016 with a live time of 1291.5 days. Events are selected within a fiducial volume corresponding to approximately ¼ of the scintillator volume in order to increase the shielding against external backgrounds.
The analytical model used to describe the energy spectrum is an improved version of the one reported in [7] with the goal of enlarging the fitting energy range. The most relevant changes concern as improved treatment of the energy scale nonlinearities and the addition of a resolution parameter to better reproduce the low energy region. In the fitting procedure the minor components of pep, ^{8}B and CNO solar neutrino were kept fixed according to the standard solar model predictions. For CNO both high and low metallicity variant was tested since the predicted rates are significantly different and the induced uncertainty added to the systematics.
Special care was taken to describe the pileup events: a "synthetic" spectrum was constructed by overlapping real events with randomly sampled data of the same time length.
Including the electromagnetic component as described in (2) in the fit leads to a decrease of the pp-neutrino component as a consequence of the larger total cross section: this correlation can be mitigated by applying the constraint coming from the result of radiochemical experiments, which are independent of the neutrino electromagnetic properties, to the sum of the neutrino fluxes detected in Borexino. If this constrain is added to the fit as a penalty term, the analysis of the likelihood profile gives a limit of $\mu_{eff} < 2.8\ 10^{-11}\ \mu_B$ at 90% C.L., already including the systematics [12].
In case of Majorana neutrinos only the transition moments of the electromagnetic matrix are non-zero, while the diagonal elements are null because of CPT. For the Dirac neutrinos all the elements may have nonzero values. The effective magnetic moment can be expanded both in terms of the mass eigen-states and of the flavour eigenstates. In the last case, the effective magnetic moment for the LMA-MSW solution is given by:

$$\mu_{eff}^2 = P^{3\nu}\mu_e^2 + (1 - P^{3\nu})\left(cos^2\vartheta_{23} \cdot \mu_\mu^2 + sin^2\vartheta_{23} \cdot \mu_\tau^2\right) \quad (4)$$

where $P^{3\nu} = sin^4\vartheta_{13} + cos^4\vartheta_{13} \cdot P^{2\nu}$ is the ν_e survival probability with $P^{2\nu}$ calculated in the standard 2-neutrino scheme; ϑ_{13} and ϑ_{23} are the corresponding mixing angles. The individual contributions in eq. (3) are positive therefore this expression can be used to compute limits on the single flavour magnetic moments.
In the approximation that the survival probability is the same for pp and ^{7}Be neutrinos and equal to $P^{2\nu} = 0.55$, from eq. (3) the following limits can be quoted at 90% C.L. [12]:

IOP Conf. Series: Journal of Physics: Conf. Series **1056** (2018) 012062 doi:10.1088/1742-6596/1056/1/012062

- $\mu_e < 3.9 \cdot 10^{-11}\, \mu_B$ (5)
- $\mu_\mu < 5.8 \cdot 10^{-11}\, \mu_B$ (6)
- $\mu_\tau < 5.8 \cdot 10^{-11}\, \mu_B$ (7)

Because the mass hierarchy is still unknown, the values above were calculated for the choice of hierarchy providing the most conservative limits, i.e. the inverted one.

In conclusion a new model independent limit of $\mu_{eff} < 2.8\ 10^{-11}\ \mu_B$ was obtained at 90% C.L. including systematics, that slightly improves the previous best limit from GEMMA [9]. This limit was obtained by constraining the sum of the solar neutrino fluxes using the results form the gallium experiment and is free from uncertainty associated with prediction from the SSM flux and systematics from the detector fiducial volume mass.

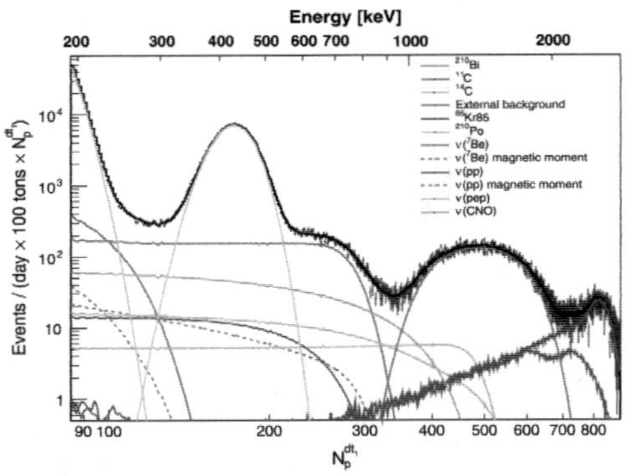

Figure 3. Spectral fit with the neutrino magnetic moment equal to $\mu_{eff} = 2.8\ 10^{-11}\ \mu_B$

6. Conclusions

The first results based on the new Borexino high-purity Phase II data have been reviewed, namely the first direct measurement of pp-neutrinos, the seasonal modulation of the ^{7}Be neutrino signal and a new and improved upper limit on the neutrino magnetic moment. New preliminary and more precise neutrino fluxes measurements obtained with a simultaneous fit of pp, pep and ^{7}Be components have been presented at this conference and they will be published soon.

7. References

[1] Bellini G et al. (Borexino Collaboration) 2014 *Physical Review D* **89** 112007
[2] Agostini M et al. (Borexino Collaboration) 2018 *Astroparticle Physics* **97** 136-159
[3] Agostini M et al. (Borexino Collaboration) 2017 arXiv: 1707.09279 [hep-ex]
[4] Redchuk M et al. (Borexino Collaboration) 2017 "Solar neutrino analysis with the Borexino detector" , Poster presented at CNNP 2017 Conference
[5] Hampel W et al., 1999 *Physics Letters B* **447** 127
[6] SAGE Collaboration (Abdurashitov J N et al.), 2002 *J. Exp. Theor. Phys.* **95** 181, arXiv: astro-ph/0204245 [astro-ph]
[7] Bellini G et al. (Borexino Collaboration) 2014 *Nature* **512** 383-386
[8] Serenelli A M, Haxton W C and Pena-Garay C, 2011 *Astrophys J.* **743** 24
[9] Bellini G et al. (Borexino Collaboration) *Astroparticle Physics* **92** 21-29
[10] Beda A G et al. (GEMMA Collaboration) 2013 *Phys. Part. Nucl. Lett.* **10** 139
[11] SNO Collaboration (Ahmad Q R et al.) 2002 *Phys. Rev. Lett.* **89** 011301 (2002)
[12] Bellini G et al. (Borexino Collaboration) 2017 *Phys. Rev. D* **96** 091103